Power System Transients

Parameter Determination

Power System Transients

Parameter Determination

Edited by
Juan A. Martinez-Velasco

CRC Press
Taylor & Francis Group
Boca Raton   London   New York

CRC Press is an imprint of the
Taylor & Francis Group, an **informa** business

MATLAB® is a trademark of The MathWorks, Inc. and is used with permission. The MathWorks does not warrant the accuracy of the text or exercises in this book. This book's use or discussion of MATLAB® software or related products does not constitute endorsement or sponsorship by The MathWorks of a particular pedagogical approach or particular use of the MATLAB® software.

Library of Congress Cataloging-in-Publication Data

Power system transients : parameter determination / editor, Juan A. Martinez-Velasco.
    p. cm.
  "A CRC title."
  Includes bibliographical references and index.
  ISBN 978-1-4200-6529-9 (hardcover : alk. paper)
  1. Electric power system stability. 2. Transients (Electricity) Mathematical models. 3. Arbitrary constants. 4. Electric power systems--Testing. I. Martinez-Velasco, Juan A.

TK1010.P687 2010
621.319'21--dc22
                                                         2009031432

**Visit the Taylor & Francis Web site at
http://www.taylorandfrancis.com**

**and the CRC Press Web site at
http://www.crcpress.com**

Contents

Preface...vii
Editor...ix
Contributors...xi

1. **Parameter Determination for Electromagnetic Transient Analysis in Power Systems**..1
   Juan A. Martinez-Velasco

2. **Overhead Lines**..17
   Juan A. Martinez-Velasco, Abner I. Ramirez, and Marisol Dávila

3. **Insulated Cables**...137
   Bjørn Gustavsen, Taku Noda, José L. Naredo, Felipe A. Uribe, and Juan A. Martinez-Velasco

4. **Transformers**...177
   Francisco de León, Pablo Gómez, Juan A. Martinez-Velasco, and Michel Rioual

5. **Synchronous Machines**...251
   Ulas Karaagac, Jean Mahseredjian, and Juan A. Martinez-Velasco

6. **Surge Arresters**...351
   Juan A. Martinez-Velasco and Ferley Castro-Aranda

7. **Circuit Breakers**..447
   Juan A. Martinez-Velasco and Marjan Popov

   Appendix A: Techniques for the Identification of a Linear System from Its Frequency Response Data...557
   Bjørn Gustavsen and Taku Noda

   Appendix B: Simulation Tools for Electromagnetic Transients in Power Systems..591
   Jean Mahseredjian, Venkata Dinavahi, Juan A. Martinez-Velasco, and Luis D. Bellomo

Index..619

v

Preface

The story of this book may be traced back to the Winter Meeting that the IEEE Power Engineering Society (now the Power and Energy Society) held in January 1999, when the Task Force (TF) on Data for Modeling Transient Analysis was created. The mandate of this TF was to produce a series of papers that would help users of transient simulation tools (e.g., EMTP-like tools) select an adequate technique or procedure for determining the parameters to be specified in transient models. The TF wrote seven papers that were published in the July 2005 issue of the *IEEE Transactions on Power Delivery*.

The determination or estimation of transient parameters is probably the most difficult and time-consuming task of many transient studies. Engineers and researchers spend only a small percentage of the time running the simulations; most of their time is dedicated to collecting the information from where the parameters of the transient models will be derived and to testing the complete system model. Even today, with the availability of powerful numerical techniques, simulation tools, and graphical user interfaces, the selection of the most adequate model and the determination of parameters are very often the weakest point of the whole task.

Significant efforts have been made in the last two decades on the development of new transient models and the proposal of modeling guidelines. Currently, users of transient tools can take advantage of several sources to select the study zone and choose the most adequate model for each component involved in the transient process. However, the following drawbacks must still be resolved: (1) the information required for the determination of some parameters is not always available; (2) some testing setups and measurements to be performed are not recognized in international standards; (3) more studies are required to validate models, mainly those that are to be used in high or very high-frequency transients; and (4) built-in models currently available in simulation tools do not cover all modeling requirements, although most of them are capable of creating custom-made models.

Although procedures and studies of parameter determination are presented for only seven power components, it is obvious that much more information is required to cover all important aspects in creating an adequate and reliable transient model. In some cases, procedures of parameter determination are presented only for low-frequency models; in other cases, all the procedures required for creating the whole model of a component are not analyzed. In addition, there is a lack of examples to illustrate how parameters can be determined for real-world applications.

The core of this book is dedicated to current procedures and techniques for the determination of transient parameters for six basic power components: overhead line, insulated cable, transformer, synchronous machine, surge arrester, and circuit breaker. Therefore, this book can be seen as a setup that has joined an expanded version of the *transaction* papers. It will help users of transient tools solve part of the main problems they face when creating a model adequate for electromagnetic transient simulations.

This book includes two appendices. The first provides updated techniques for the identification of linear systems from frequency responses; these fitting techniques can be useful both for determining parameters and for fitting frequency-dependent models of some of the components analyzed in this book (e.g., lines, cables, or transformers). The second reviews the capabilities and limitations of the most common simulation tools, taking into consideration both off-line and online (real-time) tools.

A crucial aspect that needs to be emphasized is the importance of standards. As pointed out in several chapters, standards are a very valuable source of information for the determination of parameters, for the characterization of components, and even for their selection (e.g., surge arresters). Whenever possible, recommendations presented in standards have been summarized and included in the appropriate chapter.

Although this is a book on electromagnetic transients, many of the most important topics related to this field have not been well covered or have not been covered at all. For instance, Appendix B on simulation tools contains only a short summary of computational methods for transient analysis.

Most of the topics covered require one to have some basic knowledge of electromagnetic transient analysis. This book is addressed mainly to graduate students and professionals involved in transient studies.

I want to conclude this preface by thanking the members of the IEEE TF for their pioneering work, and by expressing my gratitude to all our colleagues, friends, and relatives for their help, and, in many circumstances, for their patience.

Juan A. Martinez-Velasco
Barcelona, Spain

MATLAB® is a registered trademark of The MathWorks, Inc. For product information, please contact:

The MathWorks, Inc.
3 Apple Hill Drive
Natick, MA 01760-2098 USA
Tel: 508 647 7000
Fax: 508-647-7001
E-mail: info@mathworks.com
Web: www.mathworks.com

Editor

Juan A. Martinez-Velasco was born in Barcelona, Spain. He received his Ingeniero Industrial and Doctor Ingeniero Industrial degrees from the Universitat Politècnica de Catalunya (UPC), Spain. He is currently with the Departament d'Enginyeria Elèctrica of the UPC.

Martinez-Velasco has authored and coauthored more than 200 journal and conference papers, most of them on the transient analysis of power systems. He has been involved in several ElectroMagnetic Transients Program (EMTP) courses and has worked as a consultant for Spanish companies. His teaching and research areas cover power systems analysis, transmission and distribution, power quality, and electromagnetic transients. He is an active member of several IEEE and CIGRE working groups (WGs). Currently, he is the chair of the IEEE WG on Modeling and Analysis of System Transients Using Digital Programs.

Dr. Martinez-Velasco has been involved as an editor or a coauthor of eight books. He is also the coeditor of the IEEE publication *Modeling and Analysis of System Transients Using Digital Programs* (1999). In 1999, he received the 1999 PES Working Group Award for Technical Report, for his participation in the tasks performed by the IEEE Task Force on Modeling and Analysis of Slow Transients. In 2000, he received the 2000 PES Working Group Award for Technical Report, for his participation in the publication of the special edition of *Modeling and Analysis of System Transients Using Digital Programs*.

Contributors

Luis D. Bellomo
École Polytechnique de Montréal
Montreal, Quebec, Canada

Ferley Castro-Aranda
Universidad del Valle
Escuela de Ingeniería Eléctrica y
    Electrónica
Cali, Colombia

Marisol Dávila
Universidad de los Andes
Escuela de Ingeniería Eléctrica
Mérida, Venezuela

Francisco de León
Department of Electrical and Computer
    Engineering
Polytechnic Institute of NYU
Brooklyn, New York

Venkata Dinavahi
University of Alberta
Department of Electrical and Computer
    Engineering
Edmonton, Alberta, Canada

Pablo Gómez
Instituto Politécnico Nacional
Departmento de Ingeniería Eléctrica
México, Mexico

Bjørn Gustavsen
SINTEF Energy Research
Trondheim, Norway

Ulas Karaagac
École Polytechnique de Montréal
Montreal, Quebec, Canada

Jean Mahseredjian
École Polytechnique de Montréal
Montreal, Quebec, Canada

Juan A. Martinez-Velasco
Universitat Politècnica de Catalunya
Department d'Enginyeria Elèctrica
Barcelona, Spain

José L. Naredo
Centro de Investigación y de Estudios
    Avanzados del Instituto Politécnico
    Nacional
Guadalajara, Mexico

Taku Noda
Central Research Institute of Electric
    Power Industry
Yokosuka, Japan

Marjan Popov
Delft University of Technology
Faculty of Electrical Engineering,
    Mathematics and Computer Science
Delft, the Netherlands

Abner I. Ramirez
Centro de Investigación y de Estudios
    Avanzados del Instituto Politécnico
    Nacional
Guadalajara, Mexico

Michel Rioual
Électricité de France R & D
Clamart, France

Felipe A. Uribe
Universidad de Guadalajara
Departmento de Ingeniería Mecánica
    Eléctrica
Guadalajara, Mexico

1

Parameter Determination for Electromagnetic Transient Analysis in Power Systems

Juan A. Martinez-Velasco

CONTENTS

1.1   Introduction 1
1.2   Modeling Guidelines 2
1.3   Parameter Determination 6
1.4   Scope of the Book 11
References 15

1.1   Introduction

Power system transient analysis is usually performed using computer simulation tools like the Electromagnetic Transients Program (EMTP), although modeling using Transient Network Analyzers (TNAs) is still done, but decreasingly. There is also a family of tools based on computerized real-time simulations, which are normally used for testing real control system components or devices such as relays. Although there are several common links, this chapter targets only off-line nonreal-time simulations.

Engineers and researchers who perform transient simulations typically spend only a small amount of their total project time running the simulations. The bulk of their time is spent obtaining parameters for component models, testing the component models to confirm proper behaviors, constructing the overall system model, and benchmarking the overall system model to verify overall behavior. Only after the component models and the overall system representation have been verified, one can confidently proceed to run meaningful simulations. This is an iterative process. If there are some transient event records to compare against, more model benchmarking and adjustment may be required.

This book deals with parameter determination and is aimed at reviewing the procedures to be performed for deriving the mathematical representation data of the most important power components in electromagnetic transient simulations. This chapter presents a summary on the current status and practice in this field and emphasizes needed improvements for increasing the accuracy of modeling tasks in detailed transient analysis.

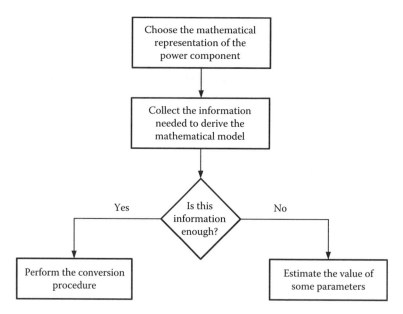

FIGURE 1.1
Procedure to obtain a complete representation of a power component. (From Martinez, J.A. et al., *IEEE Power Energy Mag.*, 3, 16, 2005. With permission.)

Figure 1.1 shows a flow chart of the procedure suggested to obtain the complete representation of a power component [1]:

- First, choose the mathematical model.
- Second, collect the information that could be useful to determine the values of parameters to be specified.
- Third, decide whether the available data are enough or not to derive all parameters.

The procedure depicted in Figure 1.1 assumes that the values of the parameters to be specified in some mathematical descriptions are not necessarily readily available and they must be deduced from other information using, in general, a data conversion procedure.

1.2 Modeling Guidelines

An accurate representation of a power component is essential for reliable transient analysis. The simulation of transient phenomena may require a representation of network components valid for a frequency range that varies from DC to several MHz. Although the ultimate objective in research is to provide wideband models, an acceptable representation of each component throughout this frequency range is very difficult, and for most components is not practically possible. In some cases, even if the wideband version is available, it may exhibit computational inefficiency or require very complex data.

Modeling of power components that take into account the frequency-dependence of parameters can be currently achieved through mathematical models which are accurate enough for a specific range of frequencies. Each range of frequencies usually corresponds to some particular transient phenomena. One of the most accepted classifications is that proposed by the International Electrotechnical Commission (IEC) and CIGRE, in which frequency ranges are classified into four groups [2,3]: low-frequency oscillations, from 0.1 Hz to 3 kHz; slow-front surges, from 50/60 Hz to 20 kHz; fast-front surges, from 10 kHz to 3 MHz; and very fast-front surges, from 100 kHz to 50 MHz. One can note that there is overlap between frequency ranges.

If a representation is already available for each frequency, the selection of the model may suppose an iterative procedure: the model must be selected based on the frequency range of the transients to be simulated; however, the frequency ranges of the test case are not usually known before performing the simulation. This task can be alleviated by looking into widely accepted classification tables. Table 1.1 shows a short list of common transient phenomena.

An important effort has been dedicated to clarify the main aspects to be considered when representing power components in transient simulations. Users of electromagnetic transient tools can nowadays obtain information on this field from several sources:

1. The document written by the CIGRE WG 33-02 covers the most important power components and proposes the representation of each component taking into account the frequency range of the transient phenomena to be simulated [2].

2. The fourth part of the IEC standard 60071 (TR 60071-4) provides modeling guidelines for insulation coordination studies when using numerical simulation, e.g., EMTP-like tools [3].

3. The documents produced by the Institute of Electrical and Electronics Engineers (IEEE) WG on modeling and analysis of system transients using digital programs and its task forces present modeling guidelines for several particular types of studies [4].

Table 1.2 provides a summary of modeling guidelines for the representation of the most important power components in transient simulations taking into account the frequency range.

The simulation of a transient phenomenon implies not only the selection of models but the selection of the system area that must be represented. Some rules to be considered in

TABLE 1.1

Origin and Frequency Ranges of Transients in Power Systems

Origin	Frequency Range
Ferroresonance	0.1 Hz to 1 kHz
Load rejection	0.1 Hz to 3 kHz
Fault clearing	50 Hz to 3 kHz
Line switching	50 Hz to 20 kHz
Transient recovery voltages	50 Hz to 100 kHz
Lightning overvoltages	10 kHz to 3 MHz
Disconnector switching in GIS	100 kHz to 50 MHz

TABLE 1.2

Modeling of Power Components for Transient Simulations

Component	Low-Frequency Transients 0.1 HZ to 3 kHz	Slow-Front Transients 50 Hz to 20 kHz	Fast-Front Transients 10 kHz to 3 MHz	Very Fast-Front Transients 100 kHz to 50 MHz
Overhead line	Multiphase model with lumped and constant parameters, including conductor asymmetry. Frequency-dependence of parameters can be important for the ground propagation mode. Corona effect can be also important if phase conductor voltages exceed the corona inception voltage.	Multiphase model with distributed parameters, including conductor asymmetry. Frequency-dependence of parameters is important for ground propagation mode.	Multiphase model with distributed parameters, including conductor asymmetry and corona effect. Frequency-dependence of parameters is important for the ground propagation mode.	Single-phase model with distributed parameters. Frequency-dependence of parameters is important for the ground propagation mode.
Insulated cable	Multiphase model with lumped and constant parameters, including conductor asymmetry. Frequency-dependence of parameters can be important for the ground propagation mode.	Multiphase model with distributed parameters, including conductor asymmetry. Frequency-dependence of parameters is important for the ground propagation mode	Multiphase model with distributed parameters. Frequency-dependence of parameters is important for the ground propagation mode.	Single-phase model with distributed parameters. Frequency-dependence of parameters is important for the ground propagation mode.
Transformer	Models must incorporate saturation effects, as well as core and winding losses. Models for single- and three-phase core can show significant differences.	Models must incorporate saturation effects, as well as core and winding losses. Models for single- and three-phase core can show significant differences.	Core losses and saturation can be neglected. Coupling between phases is mostly capacitive. The influence of the short circuit impedance can be significant.	Core losses and saturation can be neglected. Coupling between phases is mostly capacitive. The model should incorporate the surge impedance of windings.
Synchronous generator	Detailed representation of the electrical and mechanical parts, including saturation effects and control of excitation.	The machine is represented as a source in series with its subtransient impedance. Saturation effects can be neglected. The control excitation and the mechanical part are not included.	The representation is based on a linear circuit whose frequency response matches that of the machine seen from its terminals.	The representation may be based on a linear lossless capacitive circuit.
Metal oxide surge arrester	Nonlinear resistive circuit, characterized by its residual voltage to switching impulses.	Nonlinear resistive circuit, characterized by its residual voltage to switching impulses.	Nonlinear resistive circuit, characterized by its residual voltage to switching impulses, including the effect of the peak current and its front.	Nonlinear resistive circuit, characterized by its residual voltage to switching impulses, including the effect of the peak current and its front.
Circuit breaker	The model has to incorporate mechanical pole spread, and arc equations for interruption of high currents.	The model has to incorporate mechanical pole spread, the sparkover characteristic vs. time, arc instability, and interruption of high-frequency currents.	The model has to incorporate the sparkover characteristic vs. time, arc instability, and interruption of high-frequency currents.	The model has to incorporate the sparkover characteristic vs. time, and interruption of high-frequency currents.

Source: Martinez, J.A., *IEEE PES General Meeting*, Tampa, 2007. With permission.

Note: The representation of machines and transformers is not valid for calculation of voltage distribution in windings.

the simulation of electromagnetic transients when selecting models and the system area can be summarized as follows [5]:

1. Select the system zone taking into account the frequency range of the transients; the higher the frequencies, the smaller the zone modeled.
2. Minimize the part of the system to be represented. An increased number of components does not necessarily mean increased accuracy, since there could be a higher probability of insufficient or wrong modeling. In addition, a very detailed representation of a system will usually require longer simulation time.
3. Implement an adequate representation of losses. Since their effect on maximum voltages and oscillation frequencies is limited, they do not play a critical role in many cases. There are, however, some cases (e.g., ferroresonance or capacitor bank switching) for which losses are critical to defining the magnitude of overvoltages.
4. Consider an idealized representation of some components if the system to be simulated is too complex. Such representation will facilitate the edition of the data file and simplify the analysis of simulation results.
5. Perform a sensitivity study if one or several parameters cannot be accurately determined. Results derived from such a sensitivity study will show what parameters are of concern.

Figure 1.2 shows a test case used for illustrating the differences between simulation results from two different line models with distributed parameters: the lossless constant-parameter (CP) line and the lossy frequency-dependent (FD) line. Wave propagation along an overhead line is damped and the waveform is distorted when the FD-line model is used, while the wave propagates without any change if the line is assumed ideal (i.e., lossless), as expected. In addition the propagation velocity is higher with the FD-line model.

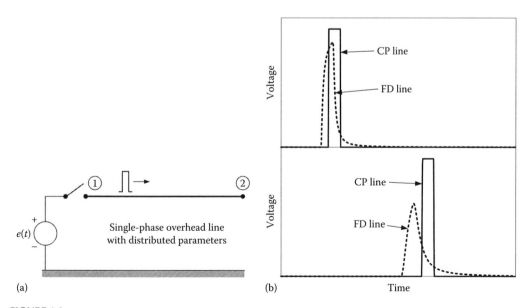

FIGURE 1.2
Comparative modeling of lines: FD and CP models. (a) Scheme of the test case; (b) wave propogation along the line.

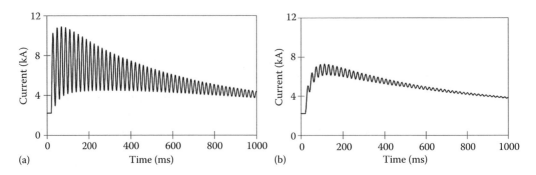

FIGURE 1.3
Field winding current in a synchronous generator during a three-phase short circuit. (a) Field current when coupling between the rotor d-axis circuits is assumed; (b) field current when coupling between the rotor d-axis circuits is neglected. (From Martinez, J.A. et al., *IEEE Power Energy Mag.*, 3, 16, 2005. With permission.)

Frequency-dependent modeling can be crucial in some overvoltage studies; e.g., those related to lightning and switching transients. In breaker performance studies, since the breaker arc voltage excites all circuit frequencies, it might be important to apply frequency-dependent models even for substation bus-bar sections where the breaker effect is analyzed.

Figure 1.3 shows the field current in a synchronous generator during a three-phase short circuit at the armature terminals obtained with two different models. The first model assumes a coupling between the two rotor circuits located on the d-axis (field and damping windings), while this coupling has been neglected in the simulation of the second case. Although the differences are important, both modeling approaches can be acceptable if the main goal is to obtain the short circuit current in the armature windings or the deviation of the rotor angular velocity with respect to the synchronous velocity.

1.3 Parameter Determination

Given the different design and operation principles of the most important power components, various techniques can be used to analyze their behavior. The following paragraphs discuss the effects that have to be represented in mathematical models and what are the approaches that can be used to determine the parameters, without covering the determination of parameters needed to represent mechanical systems, control systems, or semiconductor models.

Basically, the mathematical model of a power component (e.g., line, cable, transformer, rotating machine) for electromagnetic transient analysis must represent the effects of electromagnetic fields and losses [1]:

1. Electromagnetic field effects are, in general, represented using a circuit approach: magnetic field effects are represented by means of inductors and coupling between them, while electric fields effects are replaced by capacitors. In increased precision models, such as distributed-parameter transmission lines, parameters cannot be lumped, and mathematical models are based on solving differential equations with matrix coupling.

2. Losses can be caused in windings, cores, or insulations. Other sources of losses are corona in overhead lines or screens and sheaths in insulated cables. As for electromagnetic field effects, they are represented using a circuit approach. In many situations, losses cannot be separated from electromagnetic fields: skin effect is caused by the magnetic field constrained in windings, and produces frequency-dependent winding losses; core losses depend on the peak magnetic flux and the frequency of this field; corona losses are caused when the electric field exceeds the inception corona voltage; insulation losses are caused by the electric field and show an almost linear behavior. Approaches that can be used to represent losses would include a resistor, with either linear or nonlinear behavior, a hysteresis cycle, or a combination of various types of circuit elements. More sophisticated loss models must include frequency dependence.

The parameters used to represent electromagnetic field effects and losses can be deduced using the following techniques [1]:

- Techniques based on geometry, for instance a numerical solution aimed at solving the partial-differential equations of the electromagnetic fields developed within the component and based on the finite element method (FEM), a technique that can be used with most components. However, more simple techniques have also been developed; for instance, an analytical solution based on a simplified geometry and the separation of the electric and magnetic field are used with lines and cables. Factory measurements can be needed to obtain material properties (i.e., resistivity, permeability, and permittivity), although very often these values can be also obtained from standards or manufacturer catalogues. If the behavior of the component is assumed linear, permeabilities are approximated by that of the vacuum. If the behavior of the component is nonlinear (i.e., made of ferromagnetic material), factory tests may be needed to obtain saturation curves and/or hysteresis cycles.
- Factory tests, mainly used with transformers and rotating machines. Tests developed with this purpose can be grouped as follows:
  - Steady-state tests, which can be classified into two groups: fixed frequency tests (no load and short circuit tests are frequently used) and variable frequency tests (frequency response tests).
  - Transient tests; e.g., those performed to obtain parameters of the equivalent electric circuits of a synchronous machine.

When parameter determination is based on factory tests, a data conversion procedure can be required; that is, in many cases, parameters to be specified in a given model are not directly provided by factory measurements.

Factory tests are usually performed according to standards. However, the tests defined by standards do not always provide all of the data needed for transient modeling, and there are some cases for which no standard has been proposed to date. This is applicable to both transformers and rotating machines, although the most significant case is related to the representation of three-phase core transformers in low- and midfrequency transients [6]. The simulation of the asymmetrical behavior that can be caused by some transients

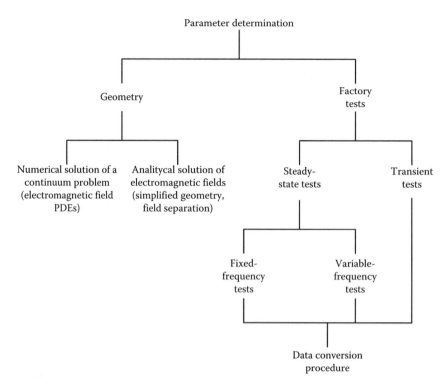

FIGURE 1.4
Classification of methods for parameter determination. (From Martinez, J.A. et al., *IEEE Power Energy Mag.*, 3, 16, 2005. With permission.)

must be based on models for which no standard has been yet developed, although several tests have been proposed in the specialized literature.

Figure 1.4 shows a flowchart on the parameter determination approaches presented above.

Some common approaches followed to obtain the mathematical model for lines and synchronous machines when using time-domain simulation tools will clarify the above discussion.

Parameters and/or mathematical models needed for representing an overhead line in simulation packages are obtained by means of a "supporting function" which is available in most EMTP-like tools and can be named, for the sake of generality, as line constants (LC) routine [7] (see Chapter 2). LC routine users must enter the physical parameters of the line and select the desired model. The following data must be inputted: (*x, y*) coordinates of each conductor and shield wire, bundle spacing, and orientations, sag of phase conductors and shield wires, phase and circuit designation of each conductor, phase rotation at transposition structures, physical dimensions of each conductor, DC resistance of each conductor (including shield wires) and ground resistivity of the ground return path. Note that all the above information except conductor resistances and ground resistivity is from geometric line dimensions. The following models can be created: lumped-parameter model or nominal pi-circuit model, at the specified frequency; constant distributed-parameter model, at the specified frequency; frequency-dependent distributed-parameter model, fitted for a given frequency range (see Figure 1.5). In addition, the following information

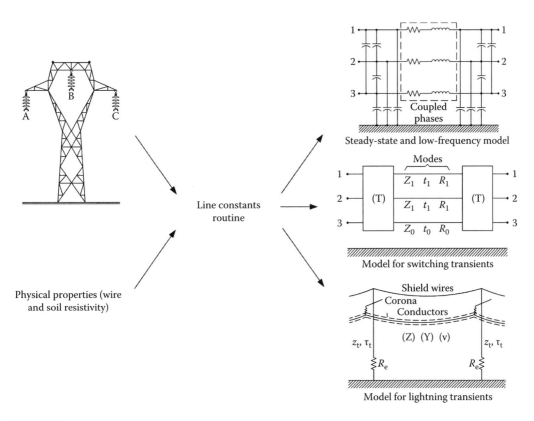

FIGURE 1.5
Application of a line constants routine to obtain overhead line models. (From Martinez, J.A. et al., *IEEE Power Energy Mag.*, 3, 16, 2005. With permission.)

usually becomes available: the capacitance or the susceptance matrix; the series impedance matrix; resistance, inductance, and capacitance per unit length for zero and positive sequences, at a given frequency or for the specified frequency range; surge impedance, attenuation, propagation velocity, and wavelength for zero and positive sequences, at a given frequency or for a specified frequency range. Line matrices can be provided for the system of physical conductors, the system of equivalent phase conductors, or with symmetrical components of the equivalent phase conductors. Note that the model created by the LC routine is representing only phase conductors and, if required, shield wires, while in some simulations the representation of other parts is also needed; for instance, towers, insulators, grounding impedances, and corona models are needed when calculating lightning overvoltages in overhead transmission lines.

The conversion procedures that have been proposed for the determination of the electrical parameters to be specified in the equivalent circuits of a synchronous machine use data from several sources; e.g., short circuit tests or standstill frequency response (SSFR) tests [8]. The diagram shown in Figure 1.6 illustrates the determination of the electrical parameters of a synchronous machine from SSFR tests. Low-voltage frequency response tests at standstill are becoming a widely used alternative to short circuit tests due to their advantages: they can be performed either in the factory or in site at a relatively low cost, equivalent circuits of high order can be derived, identification of field responses is possible. The equivalent circuits depicted in Figure 1.6 are valid to represent a synchronous machine in

Standstill frequency response tests (IEEE std. 115)

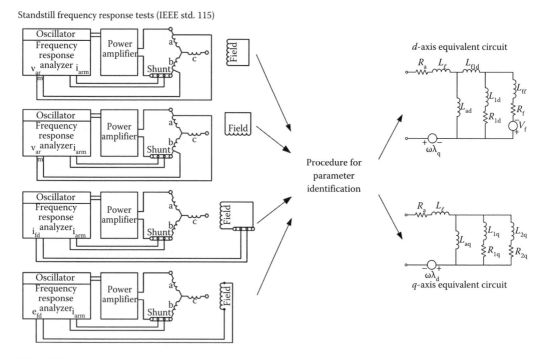

FIGURE 1.6
Determination of the electrical parameters of a synchronous machine from SSFR tests. (From Martinez, J.A. et al., *IEEE Power Energy Mag.*, 3, 16, 2005. With permission.)

low-frequency transients; e.g., transient stability studies. However, the information provided by SSFR tests can be also used to obtain more complex equivalent circuits.

There are two important exceptions which are worth mentioning: circuit breakers and arresters. The approaches usually followed for the determination of parameters to be specified in the most common models proposed for these two components are discussed below.

- The currently available approach for accurate analysis of a circuit breaker performance during fault clearing is based on using arc equations. Such black-box model equations are able to represent, with sufficient precision, the dynamic conductance of the arc and predict thermal reignition [9]. Although some typical parameters are available in the literature, there are no general purpose methods for arc model parameter evaluation. In some case studies, it becomes compulsory to perform or require specific laboratory tests for establishing needed parameters. The detailed arc model is useful for predicting arc quenching and arc instability. In the case of transient recovery voltage studies, a simple ideal switch model can be sufficient.

- The metal oxide surge arrester is modeled using a set of nonlinear exponential equations. The original data is deduced from arrester geometry and manufacturer's data [10]. Techniques based on factory measurements for determination of parameters have also been developed. A combination of arrester equations with inductances and resistances can be used to derive an accurate arrester model for both switching and lightning surges.

1.4   Scope of the Book

This book presents techniques and methods for parameter determination of six basic power components: overhead line, insulated cable, transformer, synchronous machine, surge arrester, and circuit breaker.

The role that each of these components plays in a power system is different, so it is their design, functioning, and behavior, as well as the approaches to be considered when determining parameters that are adequate for transient models. Synchronous machines are the most important components in generation plants; lines, cables, and transformers transmit and distribute the electrical energy; surge arresters and circuit breakers protect against overvoltages and overcurrents, respectively, although circuit breakers can make and break currents under both normal and abnormal (e.g., faulted) conditions, and are the origin of switching transient caused by opening and closing operations.

There are, however, some similarities between some of these components: lines and cables are both aimed at playing the same role in distribution and transmission networks; transformers and rotating machines are made of windings whose behavior during high-frequency transients (several hundreds of kHz and above) is very similar. Consequently, similar approach can be applied to determine their parameters for transient models: the most common approach for parameter determination of lines and cables at any frequency range is based on the geometry arrangement of conductors and insulation; transformer and rotating machine parameters for high-frequency models are also obtained from the geometry of windings, magnetic cores, and insulation.

A single chapter has been dedicated to any of the aforementioned components. Although there are some exceptions, the organization is similar in all the chapters: guidelines for representation of the component in transient simulations, procedures for parameter determination, and some practical examples are common features of most chapters. It is important to keep in mind that although this is a book related to transient analysis, and some transient cases are analyzed in all chapters, most examples are mainly addressed to emphasize the techniques and conversion procedures that can be used for parameter determination.

The various approaches that can and often must be used to represent a power component, when considering transients with different frequency ranges, do not require a parameter specification with the same accuracy and detail. The literature on modeling guidelines can help users of transient tools to choose the most adequate representation and advice about the parameters that can be of paramount importance, see Section 1.2.

Another important aspect is the usefulness of international standards, basically those developed by IEEE and IEC. As discussed in the previous section, parameter to be specified in transient models can be obtained from factory tests. Setups and measurements to be performed in these tests are specified in standards, although present standards do not cover the testing setups and the requirements that are needed to obtain parameters for any frequency range. In addition, it is important to make a distinction between test procedures established in standards, the determination of characteristic values derived from those measurements and the calculation/estimation of parameters to be specified in mathematical models, and for which a data conversion procedure can be needed. All these aspects, when required, are covered in the book.

Table 1.3 lists some IEEE and IEC standards that can be useful to understand either factory tests or the guidelines to be used for constructing electromagnetic transient models.

Readers are encouraged to consult standards related to the power components analyzed in this book; they are a very valuable source of information, regularly revised and

TABLE 1.3

Some International Standards

Component	IEEE Standards	IEC Standards
Overhead line	IEEE Std. 738-1993, IEEE Standard for Calculating the Current–Temperature Relationship of Bare Overhead Conductors.	IEC 61089, Round wire concentric lay overhead electrical stranded conductors, 1991.
	IEEE Std. 1243-1997, IEEE Guide for Improving the Lightning Performance of Transmission Lines.	IEC 61597, Overhead electrical conductors—Calculation methods for stranded bare conductors, 1997.
	IEEE Std. 1410–1997, IEEE Guide for Improving the Lightning Performance of Electric Power Overhead Distribution Lines.	
Insulated cable	IEEE Std. 575-1988, IEEE Guide for the Application of Sheath-Bonding Methods for Single-Conductor Cables and the Calculation of Induced Voltages and Currents in Cable Sheaths.	IEC 60141-X, Tests on oil-filled and gas-pressure cables and their accessories, 1980.
	IEEE Std. 635–1989, IEEE Guide for Selection and Design of Aluminum Sheaths for Power Cables.	IEC 60228, Conductors of insulated cables, 2004.
	IEEE Std. 848–1996, IEEE Standard Procedure for the Determination of the Ampacity Derating of Fire-Protected Cables.	IEC 60287-X, Electric cables. Calculation of the current rating, 2001.
	IEEE Std. 844-2000, IEEE Recommended Practice for Electrical Impedance, Induction, and Skin Effect Heating of Pipelines and Vessels.	IEC 60840, Power cables with extruded insulation and their accessories for rated voltages above 30 kV (Um = 36 kV) up to 150 kV (Um = 170 kV)—Test methods and requirements, 2004.
Transformer	IEEE Std. C57.12.00-2000, IEEE Standard General Requirements for Liquid-Immersed Distribution, Power, and Regulating Transformers.	IEC 60076-X, Power Transformers, 2004.
	IEEE Std. C57.12.01-1998, IEEE Standard General Requirements for Dry-Type Distribution and Power Transformers Including Those with Solid-Cast and/or Resin-Encapsulated Windings.	IEC 60905, Loading guide for dry-type power transformers, 1987.
	IEEE Std. C57.12.90-1999, IEEE Standard Test Code for Liquid-Immersed Distribution, Power, and Regulating Transformers and IEEE Guide for Short-Circuit Testing of Distribution and Power Transformers.	IEC 60354, Loading guide for oil-immersed power transformers, 1991.

Component	IEEE standards	IEC standards
	IEEE Std. C57.12.91-2001, IEEE Standard Test Code for Dry-Type Distribution and Power Transformers.	
	IEEE Std. C57.123-2002, IEEE Guide for Transformer Loss Measurement.	
Synchronous machine	IEEE Std. 1110-1991, IEEE Guide for Synchronous Generator Modeling Practices in Stability Studies.	IEC 60034-4, Rotating electrical machines—Part 4: Methods for determining synchronous machine quantities from tests, 1985.
	IEEE Std. 115-1995, IEEE Guide: Test Procedures for Synchronous Machines.	
Metal oxide surge arrester	IEEE Std. C62.11-1999, IEEE Standard for Metal-Oxide Surge Arresters for Alternating Current Power Circuits (>1kV).	IEC 60099-4, Surge arresters—Part 4: Metal-oxide surge arresters without gaps for a.c. systems, 2004.
	IEEE Std. C62.22-1997, IEEE Guide for the Application of Metal-Oxide Surge Arresters for Alternating-Current Systems.	
Circuit breaker	IEEE Std. C37.04-1999, IEEE Standard Rating Structure for AC High-Voltage Circuit Breakers.	IEC 60427, Synthetic testing of high-voltage alternating current circuit-breakers, 2000.
	IEEE Std. C37.09-1999, IEEE Standard Test Procedure for AC High-Voltage Circuit Breakers Rated on a Symmetrical Current Basis.	IEC 60694, Common specifications for high-voltage switchgear and controlgear standards, 2002.
	ANSI/IEEE C37.081-1981, IEEE Guide for Synthetic Fault Testing of AC High-Voltage Circuit Breakers Rated on a Symmetrical Current Basis.	IEC 61233, High-voltage alternating current circuit-breakers—Inductive load switching, 1994.
	IEEE Std. C37.083-1999, IEEE Guide for Synthetic Capacitive Current Switching Tests of AC High-Voltage Circuit Breakers.	IEC 61633, High-voltage alternating current circuit-breakers—Guide for short circuit and switching test procedures for metal-enclosed and dead tank circuit-breakers, 1995.

Source: Martinez, J.A. et al., *IEEE Power Energy Mag.,* 3, 16, 2005. With permission.

TABLE 1.4

Parameter Determination of Power Components in Present Simulation Tools

Component	Parameter Determination	Status in Current Simulation Tools
Overhead line	If only the model of the conductors and shield wires is required, parameters are determined (as a function of frequency) from the line geometry and from the physical properties of the conductors and wires. However, line models in fast-front transient simulations must include insulators, towers, footing impedances, or even corona effect. A different approach can be used for each part, although models based on geometry and physical properties of each part have been proposed.	Overhead line parameters are usually obtained by means of a LINE CONSTANTS supporting routine, which is implemented in most transients simulation tools [7].
Insulated cable	Parameters are determined (as a function of frequency) from the cable geometry and from the physical properties of the different parts (conductor, insulation, sheath, armor) of the cable.	Insulated cable parameters are usually obtained by means of a CABLE CONSTANTS supporting routine, which is implemented in most transients simulation tools [11].
Transformer	Parameters for low- and midfrequency models are usually derived from nameplate data and the anhysteretic curve provided by the manufacturer. These values are obtained from standard tests. There are, however, transformer models whose parameters cannot be derived from this standard information. Users can also consider the possibility of estimating parameters from transformer geometry and physical properties (e.g., magnetic saturation). High-frequency unsaturated models may be derived from frequency response tests.	A supporting routine is usually available in most transient tools to obtain an unsaturated transformer model from nameplate data [6]. Fitting routines have been also implemented in some tools to obtain unsaturated high-frequency models from frequency response data.
Rotating machine	Electrical parameters to be specified in the low- and midfrequency model of a synchronous machine are derived from standard tests (e.g., open-circuit tests, short circuit tests, SSFR tests). Parameters for high-frequency transient models (e.g., voltage distribution in stator winding caused by a steep-fronted wave) are derived from machine geometry and physical properties.	A supporting routine is usually available in most transients tools to estimate parameters of a low-frequency synchronous machine model from values obtained in standard open circuit and short circuit tests. Fitting routines have been also implemented in some tools to obtain an unsaturated model from a SSFR.
Metal oxide surge arrester	Parameters of a standard model, acceptable for a wide range of frequencies, can be derived from manufacturer's data and the arrester geometry.	No arrester model has been implemented in any transient tool, although users can create a module based on standard models. A simple iterative procedure is usually needed to fit parameters.
Circuit breaker	Parameters to be specified in black-box models, suitable for representing dynamic arc during a thermal breakdown, can be derived from circuit tests using an evaluation or fitting procedure.	Ideal statistical switch models have been implemented in most transient tools. However, a dynamic arc model is only available in very few tools, although users can create modules to represent Mayr- and Cassie-type models.

Source: Martinez, J.A., *IEEE PES General Meeting*, Tampa, 2007. With permission.

updated. It is finally worth mentioning that some standard tests or measurements that can be important for the purpose of this book have not been detailed or even mentioned in any chapter. They are basically aimed at determining physical properties of some materials whose values can be crucial for the transient behavior of some components. For instance, no details are provided about the standard procedure that can be followed for estimating ground resistivity values, whose knowledge is of paramount importance for computing overvoltages caused by lightning.

Sophisticated numerical techniques (e.g., the FEM) have not been included in the book, but they are referred in some chapters. Although the FEM can be used for calculation of parameters and it has been extensively applied to most power components [12–19], it is a time-consuming approach that cannot be used in the transient analysis of many real world applications. There has been, however, a steady progress in both software and hardware, and some important experience is already available [20]. The increasing capabilities of simulation tools and computers are a challenge for developers, and more sophisticated and rigorous models will be developed and implemented at lower costs.

Other techniques applied in the estimation or identification of system and component parameters [21], as well as more modern procedures, such as those based on genetic algorithms [22,23], or ant colony optimization [24], are out of the scope of this book.

The book includes two appendices. The first appendix provides an update of techniques for identification of linear systems from frequency responses; these techniques can be useful in parameter determination and for fitting frequency-dependent models of several components analyzed in this book (e.g., lines, cables, transformers, and synchronous machines). The second appendix reviews simulation tools for electromagnetic transient analysis, considering both off-line and online (real time) tools, and discusses limitations and topics for practical simulation needs. Table 1.4 summarizes the capabilities of present simulation tools to obtain models considering the different approaches for parameter determination.

References

1. J.A. Martinez, J. Mahseredjian, and R.A. Walling, Parameter determination for modeling system transients, *IEEE Power Energy Magazine*, 3(5), 16–28, September/October 2005.
2. CIGRE WG 33.02, *Guidelines for Representation of Network Elements when Calculating Transients*, CIGRE Brochure 39, 1990.
3. IEC TR 60071-4, *Insulation Co-ordination—Part 4: Computational Guide to Insulation Co-ordination and Modeling of Electrical Networks*, IEC, 2004.
4. A. Gole, J.A. Martinez-Velasco, and A. Keri (eds.), *Modeling and Analysis of Power System Transients Using Digital Programs*, IEEE Special Publication TP-133-0, IEEE Catalog No. 99TP133-0, 1999.
5. J.A. Martinez, Parameter determination for power systems transients, *IEEE PES General Meeting*, Tampa, June 24–28, 2007.
6. J.A. Martinez, R. Walling, B. Mork, J. Martin-Arnedo, and D. Durbak, Parameter determination for modeling systems transients. Part III: Transformers, *IEEE Transactions on Power Delivery*, 20(3), 2051–2062, July 2005.
7. J.A. Martinez, B. Gustavsen, and D. Durbak, Parameter determination for modeling systems transients. Part I: Overhead lines, *IEEE Transactions on Power Delivery*, 20(3), 2038–2044, July 2005.

8. J.A. Martinez, B. Johnson, and C. Grande-Morán, Parameter determination for modeling systems transients. Part III: Rotating Machines, *IEEE Transactions on Power Delivery*, 20(3), 2063–2072, July 2005.
9. J.A. Martinez, J. Mahseredjian, and B. Khodabakhchian, Parameter determination for modeling systems transients. Part VI: Circuit breakers, *IEEE Transactions on Power Delivery*, 20(3), 2079–2085, July 2005.
10. J.A. Martinez and D. Durbak, Parameter determination for modeling systems transients. Part V: Surge arresters, *IEEE Transactions on Power Delivery*, 20(3), 2073–2078, July 2005.
11. B. Gustavsen, J.A. Martinez, and D. Durbak, Parameter determination for modeling systems transients. Part II: Insulated cables, *IEEE Transactions on Power Delivery*, 20(3), 2045–2050, July 2005.
12. F. Piriou and A. Razek, Calculation of saturated inductances for numerical simulation of synchronous machines, *IEEE Transactions on Magnetics*, 19(6), 2628–2631, November 1983.
13. M.P. Krefta and O. Wasynezuk, A finite element based state model of solid rotor synchronous machines, *IEEE Transactions on Energy Conversion*, 2(1), 21–30, March 1987.
14. K. Shima, K. Ide, M. Takahashi, Y. Yoshinari, and M. Nitobe, Calculation of leakage inductances of a salient-pole synchronous machine using finite elements, *IEEE Transactions on Energy Conversion*, 14(4), 1156–1161, December 1999.
15. K.J. Meessen, P. Thelin, J. Soulard, and E.A. Lomonova, Inductance calculations of permanent-magnet synchronous machines including flux change and self- and cross-saturations, *IEEE Transactions on Magnetics*, 44(10), 2324–2331, October 2008.
16. O. Moreau, R. Michel, T. Chevalier, G. Meunier, M. Joan, and J.B. Delcroix, 3-D high frequency computation of transformer R–L parameters, *IEEE Transactions on Magnetics*, 41(5), 1364–1367, May 2005.
17. S.V. Kulkarni, J.C. Olivares, R. Escarela-Perez, V.K. Lakhiani, and J. Turowski, Evaluation of eddy current losses in the cover plates of distribution transformers, *IEE Proceedings—Science, Measurement and Technology*, 151(5), 313–318, September 2004.
18. Y. Yin and H.W. Dommel, Calculation of frequency-dependent impedances of underground power cables with finite element method, *IEEE Transactions on Magnetics*, 25(4), 3025–3027, July 1989.
19. H. Nam O, T.R. Blackburn, and B.T. Phung, Modeling propagation characteristics of power cables with finite element techniques and ATP, *AUPEC 2007*, December 9–12, 2007.
20. B. Asghari, V. Dinavahi, M. Rioual, J.A. Martinez, and R. Iravani, Interfacing techniques for electromagnetic field and circuit simulation programs, *IEEE Transactions on Power Delivery*, 24(2), 939–950, April 2009.
21. A. van den Bos, *Parameter Estimation for Scientists and Engineers*, Wiley, Chichester, 2007.
22. R. Escarela-Perez, T. Niewierowicz, and E. Campero-Littlewood, Synchronous machine parameters from frequency-response finite-element simulations and genetic algorithms, *IEEE Transactions on Energy Conversion*, 16(2), 198–203, June 2001.
23. B. Abdelhadi, A. Benoudjit, and N. Nait-Said, Application of genetic algorithm with a novel adaptive scheme for the identification of induction machine parameters, *IEEE Transactions on Energy Conversion*, 20(2), 284–291, June 2005.
24. L. Sun, P. Qu, Q. Huang, and P. Ju, Parameter identification of synchronous generator by using ant colony optimization algorithm, *2nd IEEE Conference on Industrial Electronics and Applications*, ICIEA 2007, Heilongjiang, China, pp. 2834–2838, May 23–25, 2007.

2

Overhead Lines

Juan A. Martinez-Velasco, Abner I. Ramirez, and Marisol Dávila

CONTENTS

2.1 Introduction 18
2.2 Phase Conductors and Shield Wires 20
    2.2.1 Line Equations 21
    2.2.2 Calculation of Line Parameters 23
        2.2.2.1 Shunt Capacitance Matrix 23
        2.2.2.2 Series Impedance Matrix 24
    2.2.3 Solution of Line Equations 28
    2.2.4 Solution Techniques 29
        2.2.4.1 Modal-Domain Techniques 29
        2.2.4.2 Phase-Domain Techniques 31
        2.2.4.3 Alternate Techniques 33
    2.2.5 Data Input and Output 33
2.3 Corona Effect 43
    2.3.1 Introduction 43
    2.3.2 Corona Models 45
2.4 Transmission Line Towers 50
2.5 Transmission Line Grounding 61
    2.5.1 Introduction 61
    2.5.2 Grounding Impedance 66
        2.5.2.1 Low-Frequency Models 66
        2.5.2.2 High-Frequency Models 67
        2.5.2.3 Discussion 68
    2.5.3 Low-Frequency Models of Grounding Systems 68
        2.5.3.1 Compact Grounding Systems 68
        2.5.3.2 Extended Grounding Systems 71
        2.5.3.3 Grounding Resistance in Nonhomogeneous Soils 72
    2.5.4 High-Frequency Models of Grounding Systems 74
        2.5.4.1 Distributed-Parameter Grounding Model 74
        2.5.4.2 Lumped-Parameter Grounding Model 77
        2.5.4.3 Discussion 78
    2.5.5 Treatment of Soil Ionization 81
    2.5.6 Grounding Design 87
2.6 Transmission Line Insulation 89
    2.6.1 Introduction 89
    2.6.2 Definitions 90
        2.6.2.1 Standard Waveshapes 90

        2.6.2.2 Basic Impulse Insulation Levels 91
        2.6.2.3 Statistical/Conventional Insulation Levels........................... 91
    2.6.3 Fundamentals of Discharge Mechanisms... 93
        2.6.3.1 Description of the Phenomena... 93
        2.6.3.2 Physical–Mathematical Models 93
    2.6.4 Dielectric Strength for Switching Surges... 98
        2.6.4.1 Introduction... 98
        2.6.4.2 Switching Impulse Strength 102
        2.6.4.3 Phase-to-Phase Strength 104
    2.6.5 Dielectric Strength under Lightning Overvoltages............................. 111
        2.6.5.1 Introduction... 111
        2.6.5.2 Lightning Impulse Strength 112
        2.6.5.3 Conclusions... 115
    2.6.6 Dielectric Strength for Power-Frequency Voltage 123
    2.6.7 Atmospheric Effects.. 125
References... 128

2.1 Introduction

Results derived from electromagnetic transients (EMT) simulations can be of vital importance for overhead line design. Although the selection of an adequate line model is required in many transient studies (e.g., power quality, protection, or secondary arc studies), it is probably in overvoltage calculations where adequate and accurate line models are crucial.

Voltage stresses to be considered in overhead line design are [1–3]

1. Normal power-frequency voltage in the presence of contamination
2. Temporary (low-frequency) overvoltages, produced by faults, load rejection, or ferroresonance
3. Slow front overvoltages, as produced by switching or disconnecting operations.
4. Fast front overvoltages, generally caused by lightning flashes

For some of the required specifications, only one of these stresses is of major importance. For example, lightning will dictate the location and number of shield wires and the design of tower grounding. The arrester rating is determined by temporary overvoltages, while the type of insulators will be dictated by the contamination. However, in other specifications, two or more of the overvoltages must be considered. For example, switching overvoltages, lightning, or contamination may dictate the strike distances and insulator string length. In transmission lines, contamination may determine the insulator string creepage length, which may be longer than that obtained from switching or lightning overvoltages.

In general, switching surges are important only for voltages of 345 kV and above; for lower voltages, lightning dictates larger clearances and insulator lengths than switching overvoltages do. However, this may not be always true for compact designs [3].

As a rule of thumb, distribution overhead line design is based on lightning stresses. By default, it is assumed that a distribution line flashovers every time it is impacted by a lightning stroke. In addition, the selected distribution line insulation level is usually the highest one of standardized levels. Except when calculating overvoltages caused by nearby strokes to ground, there is little to do with EMT simulation of distribution lines in insulation coordination studies, although those simulations can be very important in other studies (e.g., power quality).

The rest of this chapter has been organized bearing in mind the models and parameters needed in the calculation of voltage stresses that can be required for transmission line design.

As mentioned in Chapter 1, the simulation of transient phenomena may require a representation of network components valid for a frequency range that varies from DC to several MHz. Although an accurate and wideband representation of a transmission line is not impossible, it is more advisable to use and develop models appropriate for a specific range of frequencies. Each range of frequencies will correspond to a particular transient phenomenon (e.g., models for low-frequency oscillations will be adequate for calculation of temporary overvoltages).

Two types of time-domain models have been developed for overhead lines: lumped- and distributed-parameter models. The appropriate selection of a model depends on the line length and the highest frequency involved in the phenomenon.

Lumped-parameter line models represent transmission systems by lumped R, L, G, and C elements whose values are calculated at a single frequency. These models, known as pi-models, are adequate for steady-state calculations, although they can also be used for transient simulations in the neighborhood of the frequency at which the parameters were evaluated. The most accurate models for transient calculations are those that take into account the distributed nature of the line parameters [4–6]. Two categories can be distinguished for these models: constant parameters and frequency-dependent parameters.

The number of spans and the different hardware of a transmission line, as well as the models required to represent each part (conductors and shield wires, towers, grounding, and insulation), depend on the voltage stress cause. The following rules summarize the modeling guidelines to be followed in each case (Section 1.1):

1. In power-frequency and temporary overvoltage calculations, the whole transmission line length must be included in the model, but only the representation of phase conductors is needed. A multiphase model with lumped and constant parameters, including conductor asymmetry, will generally suffice. For transients with a frequency range above 1 kHz, a frequency-dependent model could be needed to account for the ground propagation mode. Corona effect can be also important if phase conductor voltages exceed the corona inception voltage.

2. In switching overvoltage calculations, a multiphase distributed-parameter model of the whole transmission line length, including conductor asymmetry, is in general required. As for temporary overvoltages, the frequency dependence of parameters is important for the ground propagation mode, and only phase conductors need to be represented.

3. The calculation of lightning-caused overvoltages requires a more detailed model, in which towers, footing impedances, insulators, and tower clearances, in addition to phase conductors and shield wires, are represented. However, only a few spans at both sides of the point of impact must be considered in the line model. Since lightning is a fast-front transient phenomenon, a multiphase model with distributed parameters, including conductor asymmetry and corona effect, is required for the representation of each span.

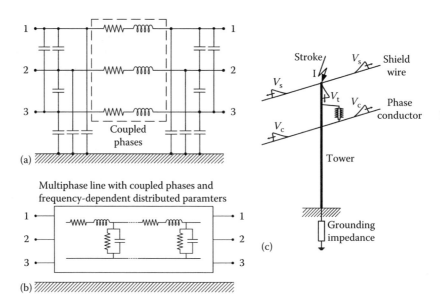

FIGURE 2.1
Line models for different ranges of frequency. (a) Steady-state and low-frequency transients, (b) switching (slow-front) transients, and (c) lightning (fast-front) transients.

Note that the length extent of an overhead line that must be included in a model depends on the type of transient to be analyzed, or, more specifically, on the range of frequencies involved in the transient process. As a rule of thumb, the lower the frequencies, the more length of line to be represented. For low- and mid-frequency transients, the whole line length is included in the model. For fast and very fast transients, a few line spans will usually suffice. These guidelines are illustrated in Figure 2.1 and summarized in Table 2.1, which provide modeling guidelines for overhead lines proposed by CIGRE [4], IEEE [5], and IEC [6].

The rest of the chapter is dedicated to analyzing the models to be used in the representation of each part of a transmission line, as well as the parameters to be specified or calculated for each part.

2.2 Phase Conductors and Shield Wires

Presently, overhead line parameters are calculated using supporting routines available in most EMT programs. The parameters to be calculated depend on the line and ground model to be applied, but they invariably involve the series impedance (longitudinal field effects) and the shunt capacitance (transversal field effects) of the line.

This section deals with, among other aspects, data input that is required for a proper modeling of overhead lines in transient simulations. To fully understand the meaning of the data input/output, the main theory of line modeling is introduced in this chapter. The concepts include the description of the time-domain and frequency-domain line equations, the calculation of the line parameters, and the description of the adopted techniques for solving the line equations.

TABLE 2.1

Modeling Guidelines for Overhead Lines

Topic	Low-Frequency Transients	Slow-Front Transients	Fast-Front Transients	Very Fast-Front Transients
Representation of transposed lines	Lumped-parameter multiphase PI circuit	Distributed-parameter multiphase model	Distributed-parameter multiphase model	Distributed-parameter single-phase model
Line asymmetry	Important	Capacitive and inductive asymmetries are important, except for statistical studies, for which they are negligible	Negligible for single-phase simulations, otherwise important	Negligible
Frequency-dependent parameters	Important	Important	Important	Important
Corona effect	Important if phase conductor voltages can exceed the corona inception voltage	Negligible	Very important	Negligible
Supports	Not important	Not important	Very important	Depends on the cause of transient
Grounding	Not important	Not important	Very important	Depends on the cause of transient
Insulators	Not included, unless flashovers are to be simulated			

Source: CIGRE WG 33.02, Guidelines for Representation of Network Elements when Calculating Transients, CIGRE Brochure no. 39, 1990; Gole, A. et al., eds., Modeling and Analysis of Power System Transients Using Digital Programs, IEEE Special Publication TP-133-0, IEEE Catalog No. 99TP133-0, 1999; IEC TR 60071-4, Insulation co-ordination—Part 4: Computational guide to insulation co-ordination and modelling of electrical networks, 2004.

The major objective of this section is that readers be able to either fully understand the input/output data from commercial EMT programs or build their own computational code. Once the corresponding data are understood, a proper interpretation of the transient results can be given by the user.

2.2.1 Line Equations

Figure 2.2 shows the frame and the equivalent circuit of a differential section of a single-phase overhead line.

Assuming that the line conductors are parallel to the ground and uniformly distributed, the time-domain equations of a single-conductor line can be expressed as follows:

$$-\frac{\partial v(x,t)}{\partial x} = Ri(x,t) + L\frac{\partial i(x,t)}{\partial t} \qquad (2.1a)$$

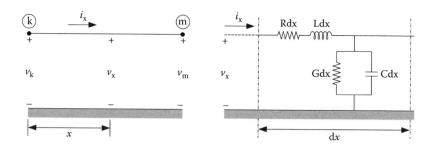

FIGURE 2.2
Single-conductor overhead line.

$$-\frac{\partial i(x,t)}{\partial x} = Gv(x,t) + C\frac{\partial v(x,t)}{\partial t} \qquad (2.1b)$$

where
  $v(x, t)$ and $i(x, t)$ are respectively the voltage and the current
  $R, L, G,$ and C are the line parameters expressed in per unit length

Figure 2.3 depicts a differential section of a three-phase unshielded overhead line illustrating the couplings among series inductances and among shunt capacitances. The time-domain equations of a multiconductor line can be expressed as follows:

$$-\frac{\partial \mathbf{v}(x,t)}{\partial x} = \mathbf{R}i(x,t) + \mathbf{L}\frac{\partial \mathbf{i}(x,t)}{\partial t} \qquad (2.2a)$$

$$-\frac{\partial \mathbf{i}(x,t)}{\partial x} = \mathbf{G}v(x,t) + \mathbf{C}\frac{\partial \mathbf{v}(x,t)}{\partial t} \qquad (2.2b)$$

where
  $\mathbf{v}(x,t)$ and $\mathbf{i}(x,t)$ are respectively the voltage and the current vectors
  $\mathbf{R}, \mathbf{L}, \mathbf{G},$ and $\mathbf{C}$ are the line parameter matrices expressed in per unit length

For a more accurate modeling of the line, these parameters are considered frequency dependent, although $\mathbf{C}$ can be assumed constant, and $\mathbf{G}$ can usually be neglected.

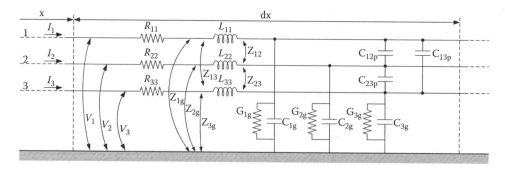

FIGURE 2.3
Differential section of a three-phase overhead line.

Advanced models can consider an additional distance dependence of the line parameters (nonuniform line) [7], the effect of induced voltages due to distributed sources caused by nearby lightning (illuminated line) [8], and the dependence of the line capacitance with respect to the voltage (nonlinear line, due to corona effect) [9,10]. This chapter presents in detail the case of the frequency-dependent uniform line, being the nonlinear model roughly mentioned.

Given the frequency dependence of the series parameters, the approach to the solution of the line equations, even in transient calculations, is performed in the frequency domain. The behavior of a multiconductor overhead line is described in the frequency domain by two matrix equations:

$$-\frac{d\mathbf{V}_x(\omega)}{dx} = \mathbf{Z}(\omega)\,\mathbf{I}_x(\omega) \tag{2.3a}$$

$$-\frac{d\mathbf{I}_x(\omega)}{dx} = \mathbf{Y}(\omega)\,\mathbf{V}_x(\omega) \tag{2.3b}$$

where $\mathbf{Z}(\omega)$ and $\mathbf{Y}(\omega)$ are respectively the series impedance and the shunt admittance matrices per unit length.

The series impedance matrix of an overhead line can be decomposed as follows:

$$\mathbf{Z}(\omega) = \mathbf{R}(\omega) + j\omega\mathbf{L}(\omega) \tag{2.4}$$

where $\mathbf{Z}$ is a complex and symmetric matrix, whose elements are frequency dependent.

Most EMT programs are capable of calculating $\mathbf{R}$ and $\mathbf{L}$ taking into account the skin effect in conductors and on the ground. This is achieved by using either Carson's ground impedance [11] or Schelkunoff's surface impedance formulas for cylindrical conductors [12]. Other approaches base the calculations on closed form approximations [13,14]. Refs. [15,16] provide a description of the procedures.

The shunt admittance can be expressed as follows:

$$\mathbf{Y}(\omega) = \mathbf{G} + j\omega\mathbf{C} \tag{2.5}$$

where the elements of $\mathbf{G}$ may be associated with currents leaking into the ground through insulator strings, which can mainly occur with polluted insulators. Their values can usually be neglected for most studies; however, under a corona effect, conductance values can be significant. That is, under noncorona conditions, with clean insulators and dry weather, conductances can be neglected. As for $\mathbf{C}$ elements, they are not frequency dependent within the frequency range that is of concern for overhead line design.

The main formulas used by commercial programs for calculating line parameters (impedance and admittance) in per unit length and suitable for computer implementation are presented in this section.

2.2.2 Calculation of Line Parameters

2.2.2.1 Shunt Capacitance Matrix

The capacitance matrix is only a function of the physical geometry of the conductors. Consider a configuration of n arbitrary wires in the air over a perfectly conducting ground.

Assuming the ground as a perfect conductor allows the application of the image method, as shown in Figure 2.4. The potential vector of the conductors with respect to the ground due to the charges on all of them is

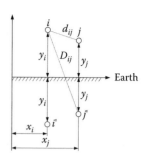

$$\mathbf{v} = \mathbf{P}\mathbf{q} \tag{2.6}$$

where
  $\mathbf{v}$ is the vector of voltages applied to the conductors
  $\mathbf{q}$ is the vector of electrical charges needed to produce these voltages
  $\mathbf{P}$ is the matrix of potential coefficients whose elements are given by

FIGURE 2.4
Application of the method of images.

$$\mathbf{P} = \frac{1}{2\pi\varepsilon_0} \begin{bmatrix} \ln\dfrac{D_{11}}{r_1} & \cdots & \ln\dfrac{D_{1n}}{d_{1n}} \\ \vdots & \ddots & \vdots \\ \ln\dfrac{D_{n1}}{d_{n1}} & \cdots & \ln\dfrac{D_{nn}}{r_n} \end{bmatrix} \tag{2.7}$$

where
  ε_0 is the permittivity of free space
  r_i is the radius of the ith conductor and (see Figure 2.4)

$$D_{ij} = \sqrt{(x_i - x_j)^2 + (y_i + y_j)^2} \tag{2.8a}$$

$$d_{ij} = \sqrt{(x_i - x_j)^2 + (y_i - y_j)^2} \tag{2.8b}$$

When calculating electrical parameters of transmission lines with bundled conductors r_i may be substituted by the geometric mean radius of the bundle:

$$R_{eq,i} = \sqrt[n]{n\, r_i\, (r_b)^{n-1}} \tag{2.9}$$

n is the number of conductors
r_b is the radius of the bundle
  Finally, the capacitance matrix is calculated by inverting the matrix of potential coefficients

$$\mathbf{C} = \mathbf{P}^{-1} \tag{2.10}$$

2.2.2.2 Series Impedance Matrix

The series or longitudinal impedance matrix is computed from the geometric and electric characteristics of the transmission line. In general, it can be decomposed into two terms:

$$\mathbf{Z} = \mathbf{Z}_{ext} + \mathbf{Z}_{int} \tag{2.11}$$

where $\mathbf{Z}_{ext}$ and $\mathbf{Z}_{int}$ are respectively the external and the internal series impedance matrix.

The external impedance accounts for the magnetic field exterior to the conductor and comprises the contributions of the magnetic field in the air ($\mathbf{Z}_g$) and the field penetrating the earth ($\mathbf{Z}_e$).

2.2.2.2.1 External Series Impedance Matrix

The contribution of the Earth return path is a very important component of the series impedance matrix. Carson reported the earliest solution of the problem of a thin wire above the earth [11] in the form of an integral, which can be expressed as a series [11,17–23].

The calculation of the electrical parameters of multiconductor lines is presently performed by using the complex image method [24], which consists in replacing the lossy ground by a perfect conductive line at a complex depth. Deri et al. extended this idea to the case of multilayer ground return [14], showing that the results from this method are valid from very low frequencies up to several MHz.

All the solutions provided in those works are valid when the conductors can be considered as thin wires. For practical purposes it can be said that the so-called thin wire approximation is valid when $(r/2h)\ln(2h/r) \ll 1$, being r the conductor radius and h the conductor height [25].

Consider again a configuration of n arbitrary wires in the air over a lossy ground. Using the complex image method, (see Figure 2.5), the external impedance matrix can be written as follows

$$\mathbf{Z}_{ext} = \frac{j\omega\mu_0}{2\pi} \begin{bmatrix} \ln\dfrac{D'_{11}}{r_1} & \cdots & \ln\dfrac{D'_{1n}}{d_{1n}} \\ \vdots & \ddots & \vdots \\ \ln\dfrac{D'_{n1}}{d_{n1}} & \cdots & \ln\dfrac{D'_{nn}}{r_n} \end{bmatrix} \tag{2.12}$$

where

$$D'_{ij} = \sqrt{\left(y_i + y_j + 2p\right)^2 + \left(x_i - x_j\right)^2} \tag{2.13}$$

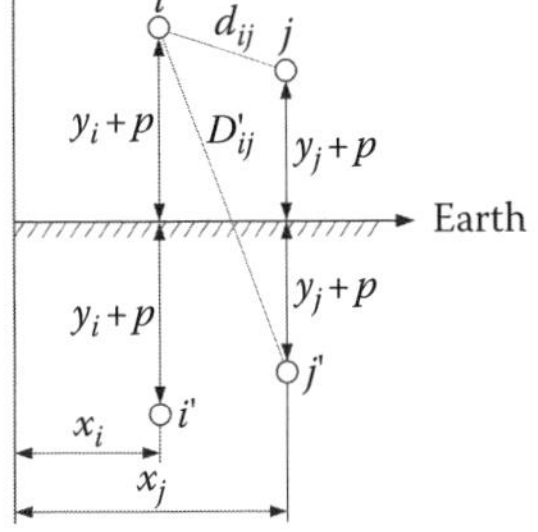

and the complex depth p is given by

$$p = \sqrt{\frac{1}{j\omega\mu_e(\sigma_e + j\omega\varepsilon_e)}} \tag{2.14}$$

where σ_e, μ_e, and ε_e are the ground conductivity (S/m), permeability (H/m), and permittivity (F/m), respectively. Note that the traditional definition of the complex depth is

$$p = \sqrt{\frac{1}{j\omega\mu_e\sigma_e}} \tag{2.15}$$

FIGURE 2.5
Geometry of the complex images.

Defining p as in Equation 2.14 makes the Earth impedance of a single conductor calculated with the complex image method equal to that calculated with a simple and accurate expression given by Sunde [26].

Multiplying each element of Equation 2.12 by D_{ij}/D_{ij}, the external impedance can be cast in terms of the geometrical impedance, $\mathbf{Z}_g$, and the Earth return impedance, $\mathbf{Z}_e$:

$$\mathbf{Z}_{\text{ext}} = \mathbf{Z}_g + \mathbf{Z}_e \tag{2.16}$$

where

$$\mathbf{Z}_g = \frac{j\omega\mu_0}{2\pi}
\begin{bmatrix}
\ln\dfrac{D_{11}}{r_1} & \cdots & \ln\dfrac{D_{1n}}{d_{1n}} \\
\vdots & \ddots & \vdots \\
\ln\dfrac{D_{n1}}{d_{n1}} & \cdots & \ln\dfrac{D_{nn}}{r_n}
\end{bmatrix} \tag{2.17}$$

$$\mathbf{Z}_e = \frac{j\omega\mu_0}{2\pi}
\begin{bmatrix}
\ln\dfrac{D'_{11}}{D_{11}} & \cdots & \ln\dfrac{D'_{1n}}{D_{1n}} \\
\vdots & \ddots & \vdots \\
\ln\dfrac{D'_{n1}}{D_{n1}} & \cdots & \ln\dfrac{D'_{nn}}{D_{nn}}
\end{bmatrix} \tag{2.18}$$

2.2.2.2.2 Internal Series Impedance

When the wires are not perfect conductors, the total tangential electric field in the wires is not zero; that is, there is a penetration of the electric field into the conductor. This phenomenon is taken into account by adding the internal impedance. The internal impedance of a round wire is found from the total current in the wire and the electric field intensity at the surface (surface impedance) [27]:

$$Z_{\text{int}} = \frac{E\big|_{\text{surface}}}{I_{\text{total}}} = -\frac{Z_{\text{cw}}}{2\pi r_c} \frac{I_0(\gamma_c r_c)}{I_1(\gamma_c r_c)} \tag{2.19}$$

where
$I_0(.)$ and $I_1(.)$ are modified Bessel functions
Z_{cw} is the wave impedance in the conductor given by

$$Z_{\text{cw}} = \sqrt{\frac{j\omega\mu_c}{\sigma_c + j\omega\varepsilon_c}} \tag{2.20}$$

γ_c is the propagation constant in the conducting material

$$\gamma_c = \sqrt{j\omega\mu_c(\sigma_c + j\omega\varepsilon_c)} \tag{2.21}$$

The conductivity, permittivity, permeability, and the radius of the conductor are denoted by σ_c, ε_c, μ_c, and r_c, respectively.

For low frequencies $|\gamma_c r_c| \ll 1$; using series expansions of the Bessel functions the internal impedance is [27]

$$Z_{\text{LF}} \approx R_{\text{dc}} \left[1 + \frac{1}{48} \left(\frac{r_c}{\delta} \right)^2 \right] + j \frac{\omega \mu}{8\pi} \tag{2.22}$$

where

$$\delta = \sqrt{\frac{2}{\omega \mu_c \sigma_c}} \tag{2.23}$$

is the skin depth in the conductor. The second term in the brackets is a correction useful for r_c/δ as large as unity, while the imaginary term corresponds to a low-frequency internal inductance and R_{dc} is the direct current resistance given by

$$R_{\text{dc}} = \frac{1}{\pi r_c^2 \sigma_c} \tag{2.24}$$

At very high frequencies $|\gamma_c r_c| \gg 1$, and using asymptotic expressions of the Bessel functions, the internal impedance becomes

$$Z_{\text{HF}} \approx \frac{1}{2\pi r_c p_c \sigma_c} \tag{2.25}$$

where p_c is the complex penetration depth for the conductor and is given by

$$p_c = \sqrt{\frac{1}{j\omega \mu_c \sigma_c}} \tag{2.26}$$

Note that Equation 2.25 can be interpreted as the complex resistance of an annulus defined by the conductor perimeter and the complex penetration depth. Using different approximations for different frequency ranges produces discontinuities in the calculated impedance. To avoid such discontinuities the following expression can be used for the whole frequency range:

$$Z_{\text{int}} = \sqrt{R_{\text{dc}}^2 + Z_{\text{HF}}^2} \tag{2.27}$$

For the case of bundled conductors Z_{int} must be divided by the number of conductors in the bundle. Finally, the internal impedance matrix for a multiconductor line with n phases is defined as follows:

$$\mathbf{Z}_{\text{int}} = \text{diag} \left(Z_{\text{int},1}, Z_{\text{int},2}, \ldots, Z_{\text{int},n} \right) \tag{2.28}$$

Galloway et al. [17] and Gary [24] provided formulas for the internal impedance, which take into account the stranding of real power conductors. Results from these formulas differ from those obtained using Equation 2.25 only by a multiplicative constant. Being the internal impedance a small part of the total impedance, using Equation 2.27 provides results within measurement errors.

2.2.3 Solution of Line Equations

The general solution of the line equations in the frequency domain can be expressed as follows:

$$\mathbf{I}_x(\omega) = e^{-\Gamma(\omega)x}\mathbf{I}_f(\omega) + e^{+\Gamma(\omega)x}\mathbf{I}_b(\omega) \tag{2.29a}$$

$$\mathbf{V}_x(\omega) = \mathbf{Y}_c^{-1}(\omega)[e^{-\Gamma(\omega)x}\mathbf{I}_f(\omega) - e^{+\Gamma(\omega)x}\mathbf{I}_b(\omega)] \tag{2.29b}$$

where
$\mathbf{I}_f(\omega)$ and $\mathbf{I}_b(\omega)$ are the vectors of forward and backward traveling wave currents at $x = 0$
$\Gamma(\omega)$ is the propagation constant matrix
$\mathbf{Y}_c(\omega)$ is the characteristic admittance matrix given by

$$\Gamma(\omega) = \sqrt{\mathbf{YZ}} \tag{2.30}$$

$$\mathbf{Y}_c(\omega) = \sqrt{(\mathbf{YZ})^{-1}\mathbf{Y}} \tag{2.31}$$

$\mathbf{I}_f(\omega)$ and $\mathbf{I}_b(\omega)$ can be deduced from the boundary conditions of the line. Considering the frame shown in Figure 2.6, the solution at line ends can be formulated as follows:

$$\mathbf{I}_k(\omega) = \mathbf{Y}_c(\omega)\mathbf{V}_k(\omega) - \mathbf{H}(\omega)\left[\mathbf{Y}_c(\omega)\mathbf{V}_m(\omega) + \mathbf{I}_m(\omega)\right] \tag{2.32a}$$

$$\mathbf{I}_m(\omega) = \mathbf{Y}_c(\omega)\mathbf{V}_m(\omega) - \mathbf{H}(\omega)\left[\mathbf{Y}_c(\omega)\mathbf{V}_k(\omega) + \mathbf{I}_k(\omega)\right] \tag{2.32b}$$

where $\mathbf{H} = \exp(-\Gamma\ell)$, ℓ being the length of the line.
Transforming Equations 2.32 into the time domain gives

$$\mathbf{i}_k(t) = \mathbf{y}_c(t) * \mathbf{v}_k(t) - \mathbf{h}(t) * \left\{\mathbf{y}_c(t) * \mathbf{v}_m(t) + \mathbf{i}_m(t)\right\} \tag{2.33a}$$

$$\mathbf{i}_m(t) = \mathbf{y}_c(t) * \mathbf{v}_m(t) - \mathbf{h}(t) * \left\{\mathbf{y}_c(t) * \mathbf{v}_k(t) + \mathbf{i}_k(t)\right\} \tag{2.33b}$$

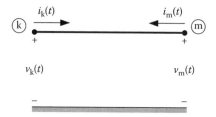

FIGURE 2.6
Reference frame.

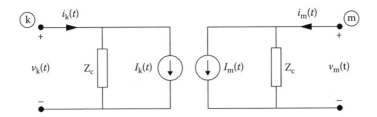

FIGURE 2.7
Equivalent circuit for time-domain simulations.

where the symbol "*" indicates convolution and $x(t) = F^{-1}\{X(\omega)\}$ is the inverse Fourier transform.

These equations suggest that an overhead line can be represented at each end by a multi-terminal admittance paralleled by a multiterminal current source, as shown in Figure 2.7.

The implementation of this equivalent circuit requires the synthesis of an electrical network to represent the multiterminal admittance. In addition, the current source values have to be updated at every time step during the time-domain calculation. Both tasks are not straightforward, and many approaches have been developed to cope with this problem, as described in Section 2.2.4.

2.2.4 Solution Techniques

The techniques developed to solve the equations of a multiconductor frequency-dependent overhead line can be classified into two main categories: modal-domain techniques and phase-domain techniques. An overview of the main approaches is presented in the following [28].

2.2.4.1 Modal-Domain Techniques

Overhead line equations can be solved by introducing a new reference frame:

$$\mathbf{V}_{ph} = \mathbf{T}_v \mathbf{V}_m \tag{2.34a}$$

$$\mathbf{I}_{ph} = \mathbf{T}_i \mathbf{I}_m \tag{2.34b}$$

where the subscripts "ph" and "m" refer to the original phase quantities and the new modal quantities. Matrices $\mathbf{T}_v$ and $\mathbf{T}_i$ are calculated through an eigenvalue/eigenvector problem such that the products $\mathbf{YZ}$ and $\mathbf{ZY}$ are diagonalized

$$\mathbf{T}_v^{-1} \mathbf{ZY} \mathbf{T}_v = \Lambda \tag{2.35a}$$

$$\mathbf{T}_i^{-1} \mathbf{YZ} \mathbf{T}_i = \Lambda \tag{2.35b}$$

Λ being a diagonal matrix.

Thus, the line equations in modal quantities become

$$-\frac{d\mathbf{V}_m}{dx} = \mathbf{T}_v^{-1} \mathbf{Z} \mathbf{T}_i \mathbf{I}_m \tag{2.36a}$$

$$-\frac{\mathrm{d}\mathbf{I}_\mathrm{m}}{\mathrm{d}x} = \mathbf{T}_i^{-1}\mathbf{YT}_v\mathbf{V}_\mathrm{m} \qquad (2.36\mathrm{b})$$

It can be proved that $[\mathbf{T}_v]^{-1} = [\mathbf{T}_i]^\mathrm{T}$ (superscript T denotes transposed) and that the products $\mathbf{T}_v^{-1}\mathbf{ZT}_i\ (=\mathbf{Z}_\mathrm{m})$ and $\mathbf{T}_i^{-1}\mathbf{YT}_v\ (=\mathbf{Y}_\mathrm{m})$ are diagonal [15,16].

The solution of a line in modal quantities can be then expressed in a similar manner as in Equation 2.32. The solution in time domain is obtained again by using convolution, as in Equation 2.33.

However, since both $\mathbf{T}_v$ and $\mathbf{T}_i$ are frequency dependent, a new convolution is needed to obtain line variables in phase quantities:

$$\mathbf{v}_\mathrm{ph}(t) = \mathbf{T}_v(t) * \mathbf{v}_\mathrm{m}(t) \qquad (2.37\mathrm{a})$$

$$\mathbf{i}_\mathrm{ph}(t) = \mathbf{T}_i(t) * \mathbf{i}_\mathrm{m}(t) \qquad (2.37\mathrm{b})$$

The procedure to solve the equations of a multiconductor frequency-dependent overhead line in the time domain involves in each time step the following:

1. Transformation from phase-domain terminal voltages to modal domain.
2. Solution of the line equations using modal quantities and calculation of (past history) current sources.
3. Transformation of current sources to phase-domain quantities.

Figure 2.8 shows a schematic diagram of the solution of overhead line equations in the modal domain.

Two approaches have been used for the solution of the line equations in modal quantities: constant and frequency-dependent transformation matrices.

1. The modal decomposition is made by using a constant real transformation matrix $\mathbf{T}$ calculated at a user-specified frequency, being the imaginary part usually discarded. This has been the traditional approach for many years. It has an obvious advantage, as it simplifies the problem of passing from modal quantities to phase quantities and reduces the number of convolutions to be calculated in the time domain, since $\mathbf{T}_v$ and $\mathbf{T}_i$ are real and constant. Differences between methods in the time-domain implementation, based on this approach, differ from the way in which the characteristic admittance function $\mathbf{Y}_\mathrm{c}$ and the propagation function $\mathbf{H}$

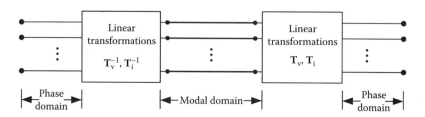

FIGURE 2.8
Transformations between phase-domain and modal-domain quantities.

of each mode are represented. The characteristic admittance function is in general very smooth and can be easily synthesized with *RC* networks. To evaluate the convolution that involves the propagation function, several alternatives have been proposed: weighting functions [29], exponential recursive convolution [30,31], linear recursive convolution [32], and modified recursive convolution [33,34]. The work presented in [35] uses the constant Clarke's transformation matrix for passing from model domain to phase domain, and represents the frequency dependence of uncoupled line modes by a cascade of synthetic π-circuits.

2. The frequency dependence of the modal transformation matrix can be very significant for some untransposed multicircuit lines. An accurate time-domain solution using a modal-domain technique requires then frequency-dependent transformation matrices. This can, in principle, be achieved by carrying out the transformation between modal- and phase-domain quantities as a time-domain convolution, with modal parameters and transformation matrix elements fitted with rational functions [36–38]. Although working for cables, see [36], it has been found that for overhead lines, the elements of the transformation matrix cannot be always accurately fitted with stable poles only [38]. This problem is overcome by the phase-domain approaches.

2.2.4.2 Phase-Domain Techniques

Some problems associated with frequency-dependent transformation matrices could be avoided by performing the transient calculation of an overhead line directly with phase quantities. A summary of the main approaches is presented in the following.

1. Phase-domain numerical convolution: Initial phase-domain techniques were based on a direct numerical convolution in the time domain [39,40]. However, these approaches are time consuming in simulations involving many time steps. This drawback was partially solved in [41] by applying linear recursive convolution to the tail portion of the impulse responses.

2. z-Domain approaches: An efficient approach is based on the use of two-sided recursions (TSR), as presented in [42]. The basic input–output in the frequency domain is usually expressed as follows:

$$\mathbf{y}(s) = \mathbf{H}(s)\mathbf{u}(s) \tag{2.38}$$

Taking into account the rational approximation of $\mathbf{H}(s)$, Equation 2.38 becomes

$$\mathbf{y}(s) = \mathbf{D}^{-1}(s)\mathbf{N}(s)\mathbf{u}(s) \tag{2.39}$$

$\mathbf{D}(s)$ and $\mathbf{N}(s)$ being polynomial matrices. From this equation one can derive

$$\mathbf{D}(s)\mathbf{y}(s) = \mathbf{N}(s)\mathbf{u}(s) \tag{2.40}$$

This relation can be solved in the time domain using two convolutions:

$$\sum_{k=0}^{n} D_k y_{r-k} = \sum_{k=0}^{n} N_k u_{r-k} \tag{2.41}$$

The identification of both side coefficients can be made using a frequency-domain fitting [42].

A more powerful implementation of the TSR, known as ARMA (Auto-Regressive Moving Average) model, was presented in [43,44] by explicitly introducing modal time delays in Equation 2.41.

3. s-Domain approaches: A third approach is based on s-domain fitting with rational functions and recursive convolutions in the time domain. Two main aspects are issued: how to obtain the symmetric admittance matrix, **Y**, and how to update the current source vectors. These tasks imply the fitting of $\mathbf{Y}_c(\omega)$ and $\mathbf{H}(\omega)$. The elements of $\mathbf{Y}_c(\omega)$ are smooth functions and can be easily fitted. However, the fitting of $\mathbf{H}(\omega)$ is more difficult since its elements may contain different time delays from individual modal contributions; in particular, the time delay of the ground mode differs from those of the aerial modes. Some works consider a single time delay for each element of $\mathbf{H}(\omega)$ [45,46]. However, a very high order fitting can be necessary for the propagation matrix in the case of lines with a high ground resistivity, as an oscillating behavior can result in the frequency domain due to the uncompensated parts of the time delays. This problem can be solved by including modal time delays in the phase domain. Several line models have been developed on this basis, using polar decomposition [47], expanding $\mathbf{H}(\omega)$ as a linear combination of the natural propagation modes with idempotent coefficient matrices [48], or calculating unknown residues once the poles and time delays have been precalculated from the modes, in the so-called universal line model (ULM) [49]. For a discussion on the advantages and limitations of these models see [50].

4. Nonhomogeneous models: Overhead line parameters associated with external electromagnetic fields are frequency independent. That is, the series impedance matrix **Z** is a full matrix that can be split up as follows:

$$\mathbf{Z}(\omega) = \mathbf{Z}_{\text{loss}}(\omega) + j\omega\mathbf{L}_{\text{ext}} \qquad (2.42)$$

where

$$\mathbf{Z}_{\text{loss}} = \mathbf{R} + j\omega\Delta\mathbf{L} \qquad (2.43)$$

Elements of $\mathbf{L}_{\text{ext}}$ are frequency independent and related to the external flux, while elements of **R** and $\Delta\mathbf{L}$ are frequency dependent and related to the internal flux. Finally, the elements of the shunt admittance matrix, $\mathbf{Y}(\omega) = j\omega\mathbf{C}$, depend on the capacitances, which can be assumed frequency independent. Taking into account this behavior, frequency-dependent effects can be separated, and a line section can be represented as shown in Figure 2.9 [48,51].

Modeling $\mathbf{Z}_{\text{loss}}$ as lumped has advantages, since their elements can be synthesized in phase quantities, and limitations, since a line has to be divided into sections to reproduce the distributed nature of parameters.

Some models based on a phase-domain approach have been implemented in some EMT programs [43,49,50].

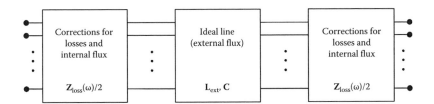

FIGURE 2.9
Section of a nonhomogeneous line model.

2.2.4.3 Alternate Techniques

1. Finite differences models: In this type of models the set of partial differential equations (Equation 2.1) are converted to an equivalent set of ordinary differential equations. This new set is discretized with respect to the distance and time by finite differences and solved sequentially along the time [52]. It has been shown that these models have advantages over those described above when the line has to be discretized, for instance in the presence of incident external fields or/and corona effect.

2. Frequency-domain solution: The line equations are solved in the frequency domain directly and going to the time domain when the solution is found. This is done through the use of Fast Fourier transform (FFT)-based routines such as the numerical Laplace transform [53].

2.2.5 Data Input and Output

Users of EMT programs obtain overhead line parameters by means of a dedicated supporting routine which is usually denoted "line constants" (LC) [15]. In addition, several routines are presently implemented in transients programs to create line models considering different approaches [31,34,43,49]. This section describes the most basic input requirements of LC-type routines. It is followed by an example that investigates the sensitivity of line parameters (R, L, and C) to variations in the representation of an overhead line, and shows the influence that some parameters can have on the transient response.

LC routine users enter the physical parameters of the line and select the desired type of line model. This routine allows users to request the following models:

- Lumped-parameter equivalent or nominal pi-circuits, at the specified frequency
- Constant distributed-parameter model, at the specified frequency
- Frequency-dependent distributed-parameter model, fitted for a given frequency range

In order to develop line models for transient simulations, the following input data must be available:

- (x, y) Coordinates of each conductor and shield wire
- Bundle spacing, orientations
- Sag of phase conductors and shield wires

- Phase and circuit designation of each conductor
- Phase rotation at transposition structures
- Physical dimensions of each conductor
- DC resistance of each conductor and shield wire (or resistivity)
- Ground resistivity of the ground return path

Other information, such as segmented grounds, can be important.

Note that all the above information, except conductor resistances and ground resistivity, come from geometric line dimensions.

The following information can be usually provided by the routine:

- The capacitance or the susceptance matrix
- The series impedance matrix
- Resistance, inductance, and capacitance per unit length for zero and positive sequences, at a given frequency or for a specified frequency range
- Surge impedance, attenuation, propagation velocity, and wavelength for zero and positive sequences, at a given frequency or for a specified frequency range

Line matrices can be provided for the system of physical conductors, the system of equivalent phase conductors, or symmetrical components of the equivalent phase conductors.

The following example is included to illustrate

- Proper input of physical parameters
- Examination of LC output
- Benchmarking impedances Z_0, Z_1/Z_2
- Benchmarking for frequency response
- Application considerations

Example 2.1

1. Test Line

   Figure 2.10 shows the geometry of the 345 kV transmission line studied in this example (distances in meters). Conductor data for this line are presented in Table 2.2.

2. Sensitivity Analysis of Line Parameters

   A parametric study of sequence parameters was performed. To obtain the frequency dependence of the resistance and the inductance of conductors, users can assume either a solid conductor or a hollow conductor and apply the skin effect correction. Skin effect entails that the highest current density is at the conductor surface. To include skin effect for hollow conductors in an LC routine, users must specify the ratio T/D, being T the thickness and D the diameter of the conductor. For the results shown in the following a solid conductor is considered.

   The studies presented in the following are aimed at determining the sensitivity of line parameters with respect to frequency, ground resistivity, skin effect, and line geometry:

   - Figure 2.11 shows the dependency of the series parameters (R, L) with respect to ground resistivity and frequency.

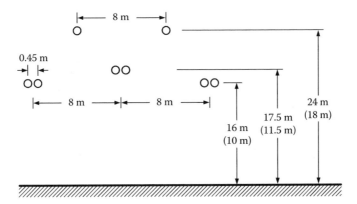

FIGURE 2.10
Example 2.1: 345 kV single-circuit overhead line configuration (values between brackets are midspan heights).

TABLE 2.2

Example 2.1: Conductor Characteristics

	Diameter (cm)	DC Resistance (Ω/km)
Phase conductors	3.0	0.0664
Shield wire conductors	1.0	1.4260

- Figure 2.12 shows the dependency of the series parameters (R, L) with respect to frequency, using the average height of the lowest conductors above ground as a parameter. These results were deduced by assuming a ground resistivity of 100 Ω m. The average height is calculated here by $h = h_m + h_s/3$ where h_m and h_s correspond to the conductor height at midspan and to the sag, respectively [15].
- Figure 2.13 shows the dependence of capacitances with respect to the average height of the lower conductor. Since capacitances are not frequency dependent within the range of frequencies considered in this analysis, frequency is not used as a parameter in this case.

All calculations were performed by assuming full transposition of phase conductors. One can deduce from these plots the conclusions listed in the following.

- The dependence of the resistance with respect to frequency can be significant, and it is particularly important for the zero-sequence resistance at high frequencies, but differences between values obtained with several ground resistivities are not very significant in this example below 5 kHz.
- Inductance values are also frequency dependent, but their dependence is very different for positive- and zero-sequence values. The positive-sequence inductance does not show large variation along the whole range of frequencies. However, for the zero-sequence inductance the frequency dependence is much larger; on the other hand, there are no significant differences with different ground resistivity values.
- When the skin effect is included in the calculation of line parameters, differences obtained by assuming either a solid or a hollow conductor are small, and negligible for frequencies below 5 kHz.
- When the average height of conductors is varied (between 12 and 32 m for the lower conductors), the variation of the inductance values is rather small, less than 2%, in

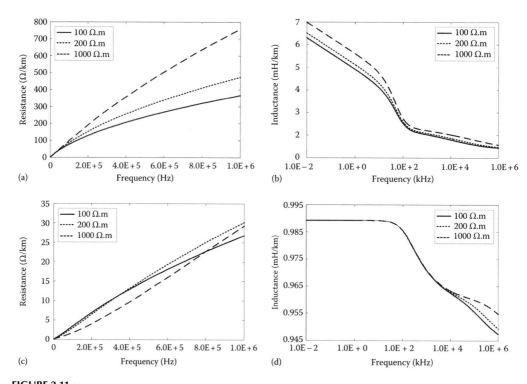

FIGURE 2.11
Example 2.1: Relationship between overhead line parameters and ground resistivity. (a) Zero-sequence resistance (Ω/km), (b) zero-sequence inductance (mH/km), (c) positive-sequence resistance (Ω/km), (d) positive-sequence inductance (mH/km).

the whole range of frequencies. However, the variation is more important for resistance values; in fact, the positive-sequence resistance can vary more than a 50% at high frequencies.

- The variation of the capacitance per unit length along a line span, see Figure 2.13, is very small.

From these results one can conclude that

- Not much accuracy is required to specify line geometry since a rather small variation in parameters is obtained for large variations in distances between conductors and heights above ground.
- Since accurate frequency-dependent models are not required when simulating low- and mid-frequency transients (below 10 kHz), the value of the ground resistivity is not critical.

Except for very short lines, the distributed nature of line parameters must be considered, and a rather accurate specification of the ground resistivity can be required when simulating high frequency transients, as shown in the following.

3. Transient Behavior
   The test line (assumed here as 80 km long) was used to illustrate the effect that losses, frequency dependence of parameters, and the value of the ground resistivity can have on some simple transients.
   Results depicted in Figures 2.14 and 2.15 show the propagation of a step voltage on one of the outer phases of the line when this step is applied to the three-phase conductors (zero-sequence energization). Calculations presented in Figure 2.14 were performed by assuming

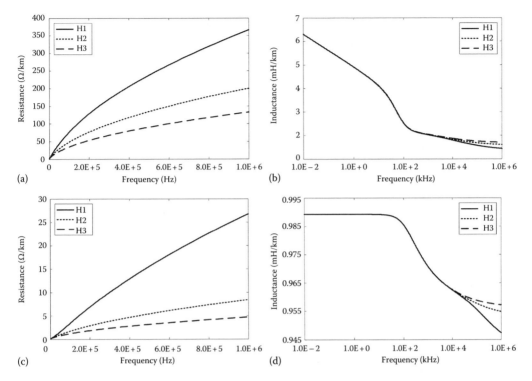

FIGURE 2.12
Example 2.1: Relationship between overhead line parameters and conductor heights. (H1 = 12 m; H2 = 22 m; H3 = 32 m). (a) Zero-sequence resistance (Ω/km), (b) zero-sequence inductance (mH/km), (c) positive-sequence resistance (Ω/km), (d) positive-sequence inductance (mH/km).

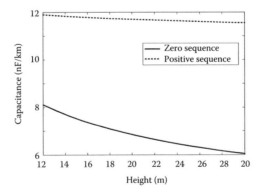

FIGURE 2.13
Example 2.1: Relationship between capacitances and conductor heights.

a constant distributed-parameter line model and calculating line parameters at power frequency (60 Hz) and at 5 kHz. It is obvious that the propagation takes place without too much distortion in both cases, but the attenuation is quite significant when parameters are calculated at 5 kHz. In addition, the propagation velocity is faster with a low ground resistivity. When the frequency dependence of line parameters is included in the transient simulation, as shown in Figure 2.15, the propagation is made with noticeable distortion of the wave front. The velocity of propagation decreases again as the ground resistivity is increased. This effect

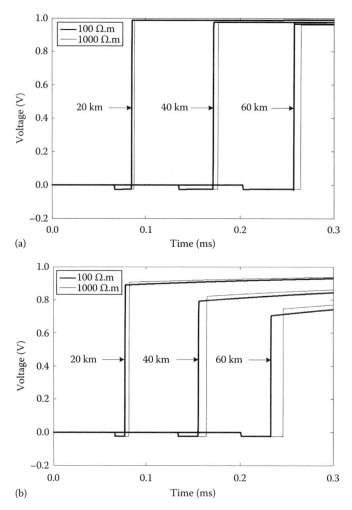

FIGURE 2.14
Example 2.1: Zero-sequence energization of an 80 km untransposed overhead line (Constant distributed-parameter model, Source = 1 V step). (a) Parameters calculated at 50 Hz, (b) parameters calculated at 5 kHz.

is due to the increase of the inductance with respect to the ground resistivity, as shown in Figures 2.11 and 2.12.

Figures 2.16 and 2.17 demonstrate again the effect of the ground resistivity on transient simulations during zero-sequence energizations. These simulations were performed by using a frequency-dependent parameter model of the line. All plots present the receiving end voltage on one of the outer phases when the line is open. Both figures show that increasing the soil resistivity by a factor of 10 leads to a noticeable reduction of the dominant frequency.

It is obvious that a significant attenuation can be obtained in wave propagation, even for short distances, when overhead line parameters are calculated taking into account their frequency dependence, and it is very evident when compared to the propagation that is obtained if this dependency is not included in calculations. However, since the highest frequency transients in overhead lines usually involve the simulation of a few sections (spans) of the line, a very accurate representation of this effect is not usually needed. Accordingly, the IEEE TF on Fast Front Transients proposes to obtain line parameters at a constant frequency, between 400 and 500 kHz, for simulating lightning overvoltages [54].

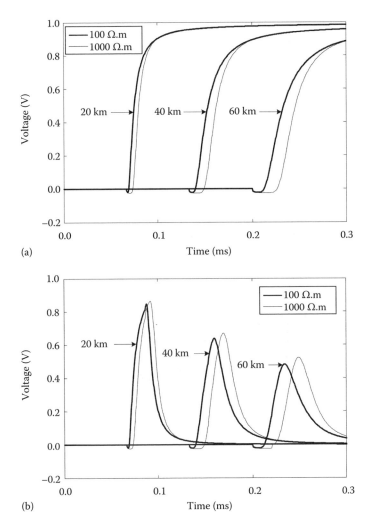

FIGURE 2.15
Example 2.1: Zero-sequence energization of an 80 km untransposed overhead line (Frequency-dependent distributed-parameter model). (a) Source = 1 V step, (b) source = 1 V, 20 μs pulse.

If simulations are performed without ground wires, the impact of the increased ground resistivity becomes much stronger.

Zero-sequence resistance increases with increasing ground resistivity, while zero-sequence inductance exhibits the opposite trend; this dependence being much smaller for positive-sequence quantities. The influence of this effect on attenuation and velocity of propagation is not negligible. Therefore some care is needed to specify the ground resistivity when high values of this parameter are possible.

As a further experiment, the test line is used to illustrate the effect of ground resistivity and frequency dependence of the line parameters when it is sequentially energized from a three-phase balanced sinusoidal 60 Hz source with an internal resistance of 0.001 Ω and having the receiving end open. Phases *a*, *b*, and *c* are closed at 3, 5, and 8 ms, respectively. Figure 2.18a shows the voltage of phase *a* using a frequency-dependent line model with different ground resistivity. Figure 2.18b presents again the voltage of phase *a* obtained by means of both a constant-parameter (at 5 kHz) and a frequency-dependent line model. The effect of ground

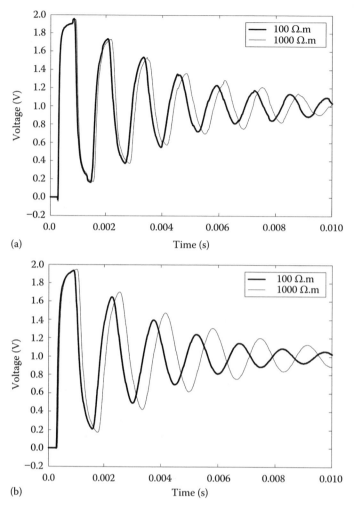

FIGURE 2.16
Example 2.1: Zero-sequence energization of an 80 km untransposed overhead line (Source = 1 V step). (a) With ground wires, (b) without ground wires.

wires on a sequential closure is illustrated in Figure 2.18c for the test line represented by a frequency-dependent line model.

4. Discussion [55]

The above simulations were performed without including corona effect. This effect can have a strong influence on the propagation of waves when the phase conductor voltage exceeds the so-called corona inception voltage, see Section 2.3. Corona causes additional attenuation and distortion, mainly on the wave front and above the inception voltage, so a noncorona model will provide conservative results. Some programs allow users to include this effect in transient simulations. Several approaches can be considered. The simplest one includes corona effect from line geometry, although some models also consider the air density factor and even an irregularity factor. In fact, corona is a very complex phenomenon whose accurate representation should be based on a distributed-hysteresis behavior. Perhaps the most important study for which corona can have a strong influence is the determination of incoming surges in

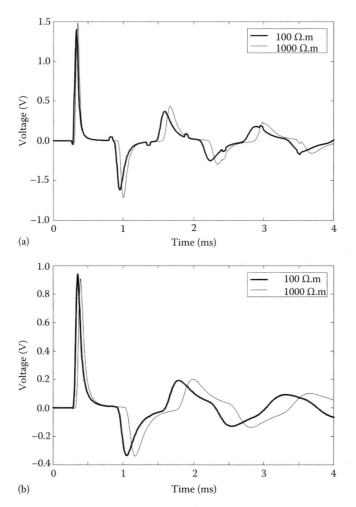

FIGURE 2.17
Example 2.1: Zero-sequence energization of an 80 km untransposed overhead line (Source = 1 V, 50 μs pulse). (a) With ground wires, (b) without ground wires.

substations [3]. When an accurate representation of the corona effect is possible, then additional input parameters are required for a full characterization of the model [9,10].

The concept of nonuniform line is used to deal with line geometries where the longitudinal variation of line parameters can be significant. Examples of this type of line are lines crossing rivers or entering substations. In such cases, nonuniformities can be very important for surge propagation [7]. When corona effect (which represents a distributed nonlinearity) is included or a nonuniform line model (where the line parameters are distance dependent) is assumed, the line has to be subdivided. Thus finite differences based methods are preferred. On the other hand, the line can be subdivided into uniform line subsegments to represent such phenomena (nonlinearity and nonuniformity); however, the resultant model is prone to numerical oscillations [10].

Line configurations more complex than the one used in this example must be often simulated. In all cases, the input data to be specified for these lines are similar to that required for the test line. And the main conclusions from a transient study would be similar to those derived in this example.

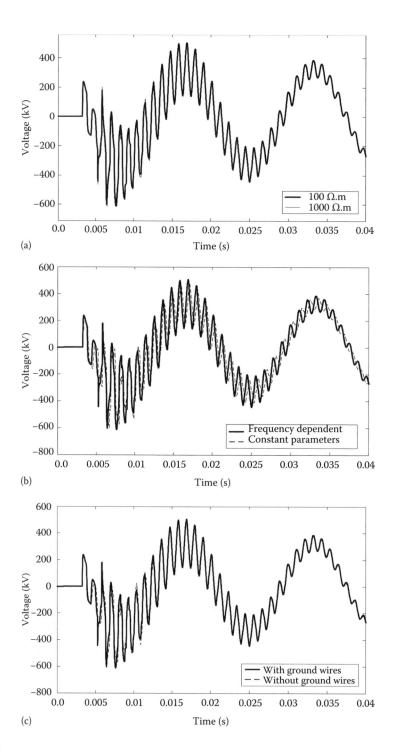

FIGURE 2.18
Example 2.1: Sequential closure of an 80 km untransposed overhead line (60 Hz balanced source). (a) Frequency-dependent model with different ground resistivity, (b) constant-parameters versus frequency-dependent model, (c) effect of ground wires.

Phase-conductor resistances depend on temperature. This effect can add a nonnegligible increase to the resistance value. It can be easily included by specifying the correct value of conductor resistances.

From the above results one can conclude that, when only phase conductors and shield wires are to be included in the line model, the line parameters can be calculated from the line geometry, as well as from physical properties of phase conductors, shield wires, and ground. A great accuracy is not usually required when specifying input values if the goal is to duplicate low frequency and slow front transients, but more care is needed, mainly with the ground resistivity value, if the goal is to simulate fast transients.

2.3 Corona Effect

2.3.1 Introduction

When the voltage of a conductor reaches a critical value (v_c) and the electrical field in its neighborhood is higher than the dielectric strength of the air, ionization is produced around the conductor. As a consequence, there will be storage and movement of charges in the ionized region, which can be viewed as an increase of the conductor radius and consequently of the capacitance to ground. This phenomenon is known as corona effect.

The increase in capacitance results in both a decrease in the velocity of propagation and a decrease in the surge impedance. The decrease in velocity causes distortion of the surge voltage during propagation; that is, the wave front is pushed back and the steepness of the surge is decreased. Depending on the tail of the initial surge, the crest voltage is also decreased. This effect is illustrated in Figure 2.19.

In practice, the corona inception voltage is a statistical quantity and a function of the voltage steepness.

A model of the corona effect based on microscopic processes of the phenomenon is very complicated and impractical for transient analysis in transmission lines. An example of this kind of model was developed in [56]. In propagation analysis, it is common to use models based on a macroscopic description, specifically models based on charge–voltage curves (q–v curves). Figure 2.20 illustrates the approximate behavior of the line capacitance with respect to the change of voltage. The models proposed in the literature can be classified into two groups: static and dynamic. Static models are those in which the corona capacitance is only a function of the voltage, $C_c = f(v)$. When the capacitance is a function of the voltage and its derivatives, $C_c = f(v, \partial v/\partial t, ...)$, the models are dynamic.

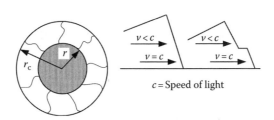

FIGURE 2.19
Corona effect on wave propagation. (From Hileman, A.R., *Insulation Coordination for Power Systems*, Marcel Dekker, New York, 1999. With permission.)

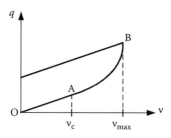

FIGURE 2.20
Charge versus voltage (q–v) curve (v_c = corona inception voltage).

Models based on q–v curves can be classified as piecewise linear, parabolic, polynomial, and dynamic. In the first, q–v curves are approximated by straight lines. In the parabolic (polynomial) models, corona capacitance is approximated by a generalized parabola (polynomial) with the power not necessarily equal to 2. Dynamic models take into account the fact that the charge depends on the voltage and on the rate of change of the voltage.

In general, all methods need to calculate beforehand the corona inception voltage. The empirical formula proposed by Peek is widely used to calculate the critical electrical field (in kV/cm) around a conductor [57]:

$$E_c = gm\delta p\left(1+\frac{0.308}{\sqrt{\delta r}}\right)$$

(2.44)

where
  $g = 30\,\text{kV/cm}$ is the critical strength of air in an uniform electric field
  m is the surface irregularity factor of the conductor, commonly set to 0.75
  δ is the relative air density
  p is the voltage polarity factor, equal to 1.0 for negative polarity and 0.5 for positive
  r is the conductor radius, in cm

When the corona effect appears in the line, the capacitance becomes a function of voltage and/or derivatives of voltage, and Equation 2.1b becomes

$$-\frac{\partial i(x,t)}{\partial x} = Gv(x,t)+C_c(v,\partial v/\partial t,\ldots)\frac{\partial v(x,t)}{\partial t}$$

(2.45)

To solve this nonlinear propagation problem, most of the methods that have been proposed resort to one of the following techniques:

1. Representing lines by means of cascaded pi-circuits with nonlinear shunt branches.

2. Subdividing frequency-dependent lines in linear subsections with nonlinear shunt branches at each junction.

3. Applying finite differences methods to the line equations.

From the results in the literature it can be concluded that the last technique yields better results since it lacks the numerical oscillations, which are present in other methods [52].

2.3.2 Corona Models

Some characteristic models proposed in the literature for representing corona are briefly described in this section. The corresponding equations contain the basic parameters needed for simulations/measurements.

A. Piecewise models

In these models, the corona capacitance (i.e., the one corresponding to the nonlinear segment from A to B in Figure 2.20) is represented by a straight line, see for instance [58]. The corresponding equation is

$$C_c = C_A + K(v - v_c) \tag{2.46}$$

where C_A is the value of capacitance at point A, while factor K (in pF/kV m) is related to the conductor height h and the polarity of the voltage, and it is included by

$$K = 0.1/h^2 \text{ (positive)} \tag{2.47a}$$

$$K = 0.05/h^2 \text{ (negative)} \tag{2.47b}$$

B. Parabolic models

For these cases, the segment A–B in Figure 2.20 is represented by means of a parabolic expression as a better approximation of the q–v curve. For instance, the following model is proposed in [59]

$$C_c = \begin{cases} C_g & v \le v_c, \quad \partial v/\partial t > 0 \\ C_g p \left(\dfrac{v}{v_c} \right)^{p-1} & v > v_c, \quad \partial v/\partial t > 0 \\ C_g & \partial v/\partial t \le 0 \end{cases} \tag{2.48}$$

where
  C_g corresponds to the geometrical capacitance
  p represents a parameter whose value depends on the polarity of the voltage, the conductor radius r in cm, and the number of conductors in bundle n as shown in Table 2.3

C. Polynomial models

An alternate option to represent the nonlinear variation of the capacitance is to approximate the q–v curve by a polynomial such as [60]:

$$q = b_0 + b_1 v + b_2 v^2 + b_3 v^3 \tag{2.49}$$

TABLE 2.3

Parameter for Corona Models

Configuration	Positive Polarity	Negative Polarity
Single conductor	$k = 0.22r + 1.2$	$k = 0.07r + 1.12$
Bundle	$k = 1.52 - 0.15\ln(n)$	$k = 1.28 - 0.08\ln(n)$

Source: Gray, C. et al., Distorsion and attenuation of travelling waves caused by transient corona, *CIGRE Report*, Study Committee 33: Overvoltages and Insulation Coordination, 1989. With permission.

where the coefficients b_0, b_1, b_2, and b_3 are obtained from the fitting of experimental data. Although this approximation of the q–v curve is more accurate than piecewise or parabolic models, the disadvantage of the polynomial approximation is the availability of measurements.

D. Dynamic models

A major consideration in this type of models is that the corona capacitance is not only the function of the voltage but of its derivatives with respect to time as well. Thus, a dynamic model can be described by

$$q = f(v, \partial v/\partial t, \ldots) \tag{2.50}$$

Despite the complexity of these models, some comprehensive formulations have been described in the literature. In [57,61] for example, the difference in time to calculate the charge $q(t)$ and its generating voltage is taken into account. This means that, in measuring the charge $q(t)$ produced by $v(t)$, the voltage has changed to $v(t + T_c)$, where T_c is the period of time to produce a new charge. The total charge in $(t + T_c)$ corresponds to the sum of the charge of a conductor of radius r, which has no delay, and the charge in a cylinder of radius r_{cor}, corresponding to the conductor with corona. Mathematically, this can be expressed as

$$q(t + T_c) = 2\pi\varepsilon\alpha\, E_c r_{cor}(t)\frac{2h - r_{cor}(t)}{2h} + C_g[v(t + T_c) - v(t)] \tag{2.51}$$

where α, equal to 0.9, is based on the experimental and theoretical results, according to which the electromagnetic field in the conductor surface is reduced to approximately 90% in corona conditions [57,61]. The radius r_{cor} is calculated iteratively with

$$r_{cor} = \left(r + \frac{v}{\alpha E_c}\right)\bigg/\left[1 + \frac{2h - r_{cor}}{2h}\ln\left(\frac{2h - r_{cor}}{r_{cor}}\right)\right] \tag{2.52}$$

Therefore, the corona capacitance can be expressed as

$$C_c(t + T_c) = \frac{\Delta q(t)}{\Delta v(t)} = \frac{q(t + T_c) - q(t + T_c - \Delta t)}{v(t + T_c) - v(t + T_c - \Delta t)} \tag{2.53}$$

Note that from Figure 2.20, as the voltage decreases (i.e., $\partial v/\partial t < 0$), the capacitance takes again its geometrical value C_g.

Corona effect adds some extra damping during wave propagation that cannot be justified by the increase in capacitance and a decrease in propagation velocity. This damping is a consequence of the corona losses caused by the transversal current due to ionization around line conductors when the voltage on the conductor exceeds the corona inception voltage. This damping can be represented in a line model as a shunt conductance.

The single-phase corona model can be extended to the case of several parallel conductors above the ground plane considering that the charge–voltage relationship is given by a matrix of Maxwell potential coefficients, see Equation 2.6. Considering that conductor "*i*" enters in corona, assuming that the spatial corona charge is uniformly distributed in a cylindrical shell around the conductor, and that its equivalent radius is still much smaller than the distance between conductors and the distance to ground, only the self-potential coefficients P_{ii} in Equation 2.6 become variable with their radii depending on voltage v_i. If in addition to conductor "*i*" there are other conductors entering in corona, their corresponding self-potential coefficients in Equation 2.6 become functions of their own voltages. Then, the extension of Equation 2.6 to the case of more coronating conductors is straightforward. Thus, the elements of matrix $\mathbf{C} = \mathbf{P}^{-1}$, which have the dimensions of capacitance per unit of length, will become dependent on the voltages of the coronating conductors.

The following examples illustrate the effect of corona on the propagation of waves considering both a single-phase and a three-phase line, and using different modeling approaches. In all cases, results were obtained by applying a finite-differences method (method of characteristics) [52,63,64].

Example 2.2

This example has been taken from a practical experiment reported in [62]. The test line consists of three aluminum cable steel reinforced (ACSR) conductors in horizontal layout with equal radii of $r = 2.54\,\text{cm}$ and medium height $h = 18.9\,\text{m}$, see Figure 2.21. The distance between adjacent conductors is $d = 9.75\,\text{m}$. The line length is 2185 m. The earth resistivity is $\rho_e = 100\,\Omega\,\text{m}$.

Assume that only one phase is energized and the conductor is represented as a frequency-dependent single-phase transmission line. At the sending end a double exponential $v(t) = AK$

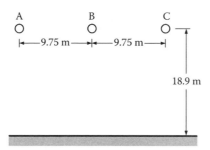

FIGURE 2.21
Example 2.2: Test line configuration.

$(e^{-t/t_m} - e^{-t/t_i})$, with $A = 1560\,\text{kV}$, $K = -1.2663$, $t_m = 0.35\,\mu\text{s}$, and $t_i = 6\,\mu\text{s}$, is applied. The corona capacitance is represented by a polynomial model as in Equation 2.49 with $b_0 = -1.7832$, $b_1 = 0.6509$, $b_2 = 9.0659 \times 10^{-4}$, and $b_3 = 1.5734 \times 10^{-7}$. The corona inception voltage has been taken equal to 450 kV.

Figure 2.22 depicts the propagation of the impinging wave without and with corona effect included in the line model. This example shows that above the corona inception voltage, the front of the surge is pushed back, being the degree of this push back dependent on the traveled distance.

In the next experiment the same impulse is injected at the sending end of the central conductor (phase B), while the line is represented by a three-phase model. In addition, the line is terminated in a resistance $R = 400\,\Omega$ located at the central phase, whereas the lateral conductors are left open as shown in Figure 2.23. The corona inception voltage is determined at 450 kV by means of Peek's formula [57].

Figure 2.24 provides the waveforms calculated at various points along the line with and without frequency dependence of electrical parameters included. The plots show respectively the waveforms for the energized center phase and the waveforms measured at either of the lateral phases (A or C). Note that a common timescale has been used in both figures for all waveforms. One can deduce from these results that the attenuation by including both frequency dependence and corona is more pronounced for the induced voltages.

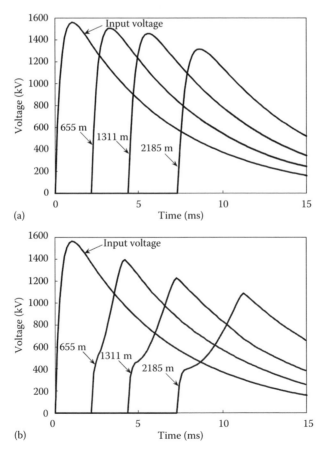

FIGURE 2.22
Example 2.2: Wave propagation in a single-phase line. (a) Without corona effect, (b) with corona effect.

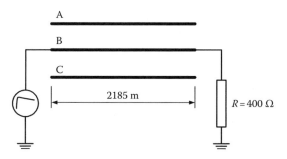

FIGURE 2.23
Example 2.2: Test experiment.

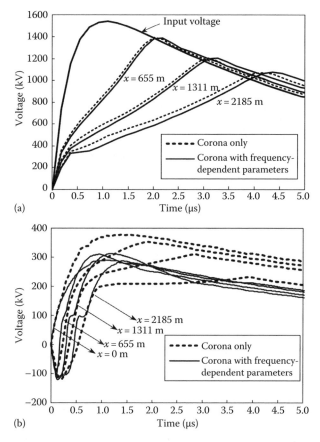

FIGURE 2.24
Example 2.2: Simulation results with a three-phase line model. (a) Energized phase response, (b) induced waveforms.

A very severe test for corona models is the application of a truncated tail wave. In such case, the involved discontinuity may cause numerical oscillations in some methods while energy passes from the wave front to the wave tail. Figure 2.25 shows the simulation results with a truncated tail wave.

For more details on single-phase and three-phase corona models see Refs. [63,64], respectively.

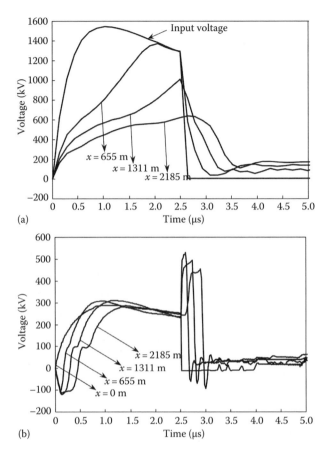

FIGURE 2.25
Example 2.2: Simulation results. (a) Energized phase response, (b) induced waveforms.

2.4   Transmission Line Towers

Although the lightning response of a transmission line tower is an electromagnetic phenomenon [65], the representation of a tower is usually made in circuit terms; that is, the tower is represented by means of several line sections and circuit elements that are assembled taking into account the tower structure [66–73]. There are some reasons to use this approach: such representation can be implemented in general-purpose simulation tools (e.g., EMTP-like programs), and it is easy to understand by the practical engineer.

Due to the fast-front times associated to lightning stroke currents, most tower models assume that the tower response is dominated by the transverse electromagnetic mode (TEM) wave and neglect other types of radiations. By introducing this simplification, a tower can be represented as either (1) an inductance connecting shield wires to ground, (2) a constant impedance transmission line, (3) a variable impedance transmission line, or (4) a radiating structure [74].

Several tower models have been developed over the years and can be categorized in several ways, see [65,75]. Basically, two approaches have been applied [76]: models developed using a theoretical approach [66,67,71–73] and models based on an experimental work [70]. The simplest representation is a lossless distributed-parameter transmission line, characterized by a surge impedance and a travel time.

The first models were deduced by assuming a vertical leader channel that hits the tower top. In fact, the response of a tower to horizontal stroke currents (i.e., the return stroke hits midway between towers) is different from the response to vertical stroke currents (i.e., the return stroke hits at the tower top). In addition, the surge impedance of the tower varies as the wave travels from top to ground. To cope with this behavior, some corrections were introduced into the first models and more complicated models have been developed: They are based on nonuniform transmission lines, [77,78], or on a combination of lumped- and distributed-parameter circuit elements [70–73]. The latter approach is also motivated by the fact that in many cases it is important to obtain the lightning overvoltages across insulators located at different heights above ground; this is particularly important when two or more transmission lines with different voltage levels are sharing the same tower.

The models based on a constant-parameter circuit representation can be classified into three groups: The tower is represented as a single vertical lossless line, as a multiconductor vertical line, or as a multistory model. A short summary of each approach is provided in the following [76].

a. Single vertical lossless line models

The first models were developed by using electromagnetic field theory, representing the tower by means of simple geometric forms, and assuming a vertical stroke to the tower top. Wagner and Hileman used a cylindrical model and concluded that the tower impedance varies as the wave travels down to the ground [66]. Sargent and Darveniza used a conical model and suggested a modified form for the cylindrical model [67]. Chisholm et al. proposed a modified equation for the above models in front of a horizontal stroke current and recommended a new model for waisted towers [68]. These models have been implemented in the FLASH program [79]. The waist model was recommended by CIGRE [74], although the version presently implemented in the FLASH program is a modified one [80].

The surge propagation velocity along tower elements can be assumed that of the light; however, the multiple paths of the lattice structure and the crossarms introduce some time delays; consequently the time for a complete reflection from ground is longer than that obtained from a travel time whose value is the tower height divided by the speed of light. Therefore, the propagation velocity in some of the above models was reduced to include this effect in the tower response.

The overvoltages caused when using the above models should be the same between terminals of all insulator strings, since these models do not distinguish between line phases. In fact, some differences result due to the different coupling between the shield wires and the phase conductors located at different heights above ground. The effect of crossarms was analyzed in [69] and found that they behave as short stub lines with open-circuit ends. Experimental results showed that travel times in crossarms are longer than those derived by assuming a propagation velocity equal to that of light. On the other hand, the incorporation of line sections representing crossarms reduces slightly the tower impedance. In general, the net effect is not significant. Experimental studies of reflections from tower bases showed that the initial reflection differed from that predicted by assuming a lumped-circuit representation of the grounding impedance. This could be justified by including transient ground-plane impedance [81]. This effect may be incorporated as an additional inductance in the grounding impedance [81–83] (Section 2.5.4.2).

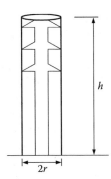

FIGURE 2.26
Cylindrical tower shape.

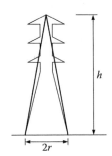

FIGURE 2.27
Conical tower shape.

The surge impedance, measured in Ω, for the most common tower shapes can be calculated, according to the above references, by means of the following expressions:

- Cylindrical tower (Figure 2.26)

$$Z = 60 \cdot \left(\ln\left(2\sqrt{2}\, \frac{h}{r} \right) - 1 \right) \qquad (2.54)$$

where
  h is the tower height, in m
  r is the tower base radius, in m

- Conical tower (Figure 2.27)

$$Z = 60 \cdot \ln\left(\sqrt{2} \sqrt{\left(\frac{h}{r} \right)^2 + 1} \right) \qquad (2.55)$$

where
  h is the tower height, in m
  r is the tower base radius, in m

Since $h > r$, this expression can be usually approximated by

$$Z \approx 60 \cdot \ln\left(\sqrt{2}\, \frac{h}{r} \right) \qquad (2.56)$$

The above expressions were derived by assuming a vertical path of the lightning stroke. When the stroke hits somewhere in midspan, the lightning current approaches the tower following a horizontal path. Chisholm et al. recommended a different expression for the average surge impedance [69].

- Cylindrical tower

$$Z = 60 \cdot \left(\ln\left(\cot \frac{\theta}{2} \right) - 1 \right) \qquad (2.57)$$

- Conical tower

$$Z = 60 \cdot \ln\left(\cot \frac{\theta}{2} \right) \qquad (2.58)$$

where

$$\theta = \tan^{-1} \frac{r}{h} \qquad (2.59)$$

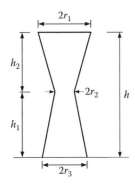

FIGURE 2.28
Waist tower shape.

Chisholm et al. proposed also an expression for the surge impedance of a general transmission-tower shape using the geometry shown in Figure 2.28, see [69]. The surge impedance would be obtained by defining a weighted average of the tower radius

$$r_{av} = \frac{r_1 h_2 + r_2 h + r_3 h_1}{h} \qquad (h = h_1 + h_2) \qquad (2.60)$$

and using the expression for a conical tower over a ground plane (Equation 2.58),

where

$$\theta = \tan^{-1}\left(\frac{r_{av}}{h}\right) \qquad (2.61)$$

h_1 is the height from base to waist, in m
h_2 is the height from waist to top, in m
r_1 is the tower top radius, in m
r_2 is the tower radius at waist, in m
r_3 is the tower base radius, in m

The new approach was implemented into FLASH as a new model for representing waisted tower shapes. The formula for the calculation of this surge impedance was further modified. The expression presently implemented into FLASH is the following one [75]:

$$Z = \sqrt{\frac{\pi}{4}} 60 \cdot \left(\ln\left(\cot\frac{\theta}{2} \right) - \ln\sqrt{2} \right) \qquad (2.62)$$

where θ is obtained according to Equations 2.60 and 2.61.

Table 2.4 presents a summary of tower models presently implemented in the FLASH program [79,82,84]. Note that a fourth configuration, the H-frame has been included in the table.

b. Multiconductor vertical line models
Each segment of the tower between crossarms is represented as a multiconductor vertical line, which can be reduced to a single conductor. The tower model is then a single-phase line whose section increases from top to ground, as shown in Figure 2.29. This representation has been analyzed in several Refs. [71–73], using each one a different approach to obtain the parameters of each section.

The modified model presented in [72] included the effect of bracings (represented by lossless lines in parallel to the main legs) and crossarms (represented as lossless line branched at junction points). As an example, the development of this model, proposed by Hara and Yamamoto, is presented in the following paragraphs.

TABLE 2.4

Tower Models

Tower Waveshape	Diagram	Surge Impedance—Travel Time
Cylindrical	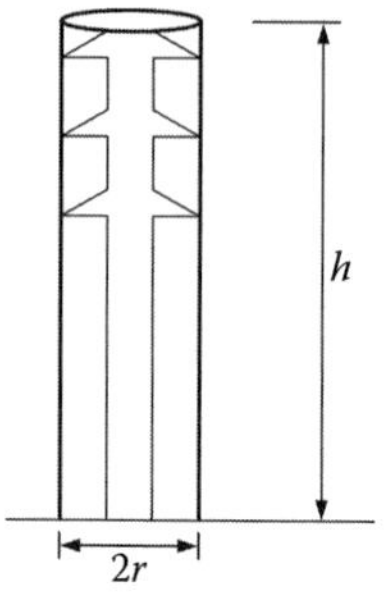	$Z = 60 \cdot \left(\ln\left(2\sqrt{2}\, \dfrac{h}{r} \right) - 1 \right)$ $t = \dfrac{h}{0.85 \cdot c}$
Conical	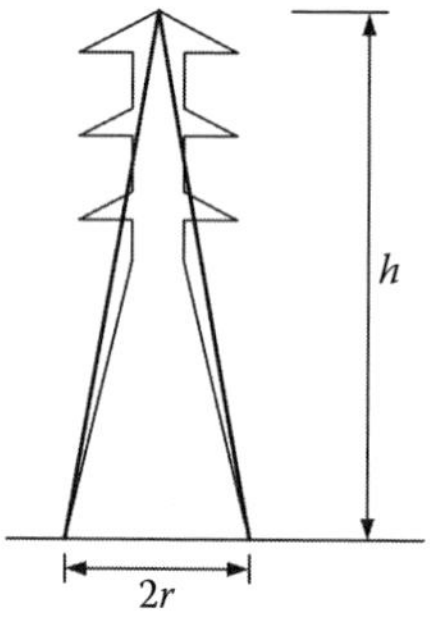	$Z = 60 \cdot \ln\left(\sqrt{2} \sqrt{ \left(\dfrac{h}{r} \right)^2 + 1 } \right)$ $t = \dfrac{h}{c}$
Waist	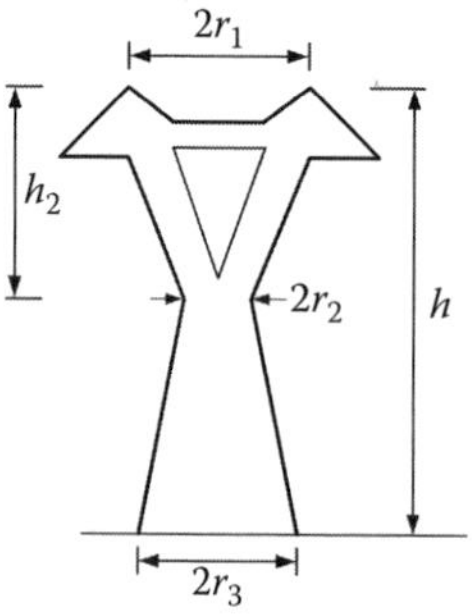	$Z = \sqrt{\dfrac{\pi}{4}}\, 60 \cdot \left(\ln\left(\cot \dfrac{\tan^{-1}(r/h)}{2} \right) - \ln\sqrt{2} \right)$ $r = \dfrac{r_1 h_2 + r_2 h + r_3 h_1}{h} \qquad (h = h_1 + h_2)$ $t = \dfrac{h}{0.85 \cdot c}$
H-frame	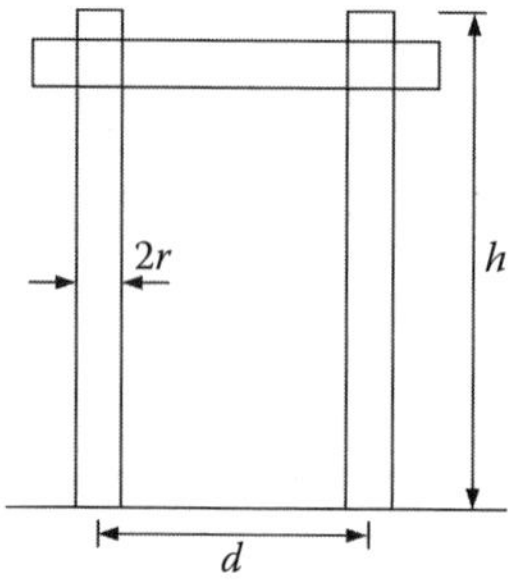	$Z_1 = 60 \cdot \ln\left(2\sqrt{2}\, \dfrac{h}{r} \right) - 60$ $Z_2 = \dfrac{d \cdot 60 \cdot \ln\left(2 \dfrac{h}{r} \right) + h \cdot Z_1}{h + d}$ $Z = \dfrac{Z_1 \cdot Z_2}{Z_1 + Z_2}$ $t = \dfrac{1}{c \cdot Z} \dfrac{h \cdot Z_1 \cdot (d + h) \cdot Z_2}{h \cdot Z_1 + (d + h) \cdot Z_2}$

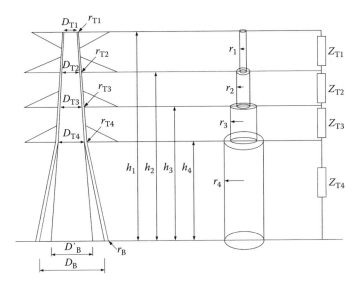

FIGURE 2.29
Multiconductor tower model.

1. The following empirical equation is used for the surge impedance of a cylinder [71]:

$$Z_T = 60 \cdot \left(\ln \left(2\sqrt{2}\, \frac{h}{r} \right) - 2 \right) \qquad (2.63)$$

where r and h are the radius and height of the cylinder, respectively. Note that this expression is different from Equation 2.54.

2. The total impedance of n parallel cylinders is given by the following expression:

$$Z_n = \frac{1}{n} \cdot \left(Z_{11} + Z_{12} + \cdots + Z_{1n} + Z_{22} + \cdots + Z_{2n} + \cdots \right) \qquad (2.64)$$

where
n is the number of conductors
Z_{kk} is the self-surge impedance of the kth conductor
Z_{km} is the mutual surge impedance between the kth and the mth conductors

3. Based on Equation 2.63, the values of these impedances are obtained as follows:

$$Z_{kk} = 60 \cdot \left(\ln \left(\sqrt{2}\, \frac{2h}{r} \right) - 2 \right) \qquad (2.65a)$$

$$Z_{km} = 60 \cdot \left(\ln \left(\sqrt{2}\, \frac{2h}{d_{km}} \right) - 2 \right) \qquad (2.65b)$$

where
   h is the conductor height
   r is the conductor radius
   d_{km} is the distance between the kth and the mth conductors

4. The surge impedance for vertical parallel multiconductor systems consisting of two or more conductors is obtained as follows:

$$Z_n = 60 \cdot \left(\ln \left(\sqrt{2} \, \frac{2h}{r_{eq}} \right) - 2 \right) \tag{2.66}$$

where r_{eq} is the equivalent radius, which is given by

$$r_{eq} = \begin{cases} r^{1/2} \cdot D^{1/2} & n = 2 \\ r^{1/3} \cdot D^{2/3} & n = 3 \\ 2^{1/8} \cdot r^{1/4} \cdot D^{3/4} & n = 4 \end{cases} \tag{2.67}$$

where
   r is the radius of the conductors
   D is the distance between two neighbor conductors

5. For a geometry as that depicted in Figure 2.30, the expressions of the surge impedances are still valid, and they may be used after replacing r and D by the following results:

$$r = \sqrt[3]{r_T \cdot r_B^2} \tag{2.68}$$

$$D = \sqrt[3]{D_T \cdot D_B^2} \tag{2.69}$$

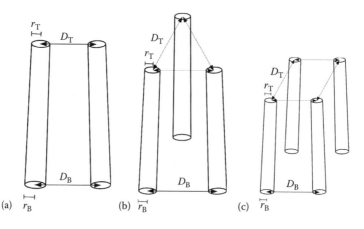

FIGURE 2.30
Different configurations of conductor systems. (a) Two conductors, (b) three conductors, (c) four conductors.

where

r_T and r_B are the radii at the top and the base of the conductors, respectively
D_T and D_B are the distances between two adjacent conductors at the top and the base, respectively

6. The tower model shown in Figure 2.29 can be divided into several sections (four in this case), being the surge impedance of each section obtained as follows:

$$Z_{Tk} = 60 \cdot \left(\ln \left(\sqrt{2}\, \frac{2h_k}{r_{ek}} \right) - 2 \right) \quad (k = 1,\ 2,\ 3,\ 4) \tag{2.70}$$

where

$$r_{ek} = 2^{1/8} \cdot \left(\sqrt[3]{r_{Tk} \cdot r_B^2} \right)^{1/4} \cdot \left(\sqrt[3]{D_{Tk} \cdot D_B^2} \right)^{3/4} \quad (k = 1,\ 2,\ 3,\ 4) \tag{2.71}$$

in which h_k, r_{Tk}, D_{Tk}, r_B, and D_B are from the geometry depicted in Figures 2.29 and 2.30.

According to the authors, these equations are applicable to towers fabricated with tubular components. When the tower is constructed with angle sections, the values of r_T and r_B (see Figure 2.30) must be replaced by half of the side length of the angle section.

7. The measured results show that the surge impedance of conductors is reduced about 10% by adding the bracings to the main legs. This performance is represented by adding in parallel to the line that represents each tower section another line of the same length and with the following surge impedance:

$$Z_{Lk} = 9 \cdot Z_{Tk} \tag{2.72}$$

8. Finally, the crossarms are represented by line sections branched at the junction points, being the corresponding surge impedances obtained from the expression of a conventional horizontal conductor:

$$Z_{Ak} = 60 \cdot \ln \frac{2h_{Ak}}{r_{Ak}} \quad (k = 1,\ 2,\ 3,\ 4) \tag{2.73}$$

where h_{Ak} and r_{Ak} are respectively the height and the equivalent radius of the kth crossarm.

The full model of the transmission tower would be that shown in Figure 2.31. See Ref. [72] for more details.

c. Multistory model

It is composed of four sections that represent the tower sections between crossarms [70]. Each section consists of a lossless line in series with a parallel R–L circuit, included for attenuation of the traveling waves, see Figure 2.32.

The parameters of this model were deduced from experimental results. The values of the parameters, and the model itself, have been revised in more recent years [75]. The approach was originally developed for representing towers of ultra high voltage (UHV) transmission lines.

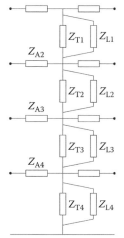

FIGURE 2.31
Equivalent model of a transmission tower.

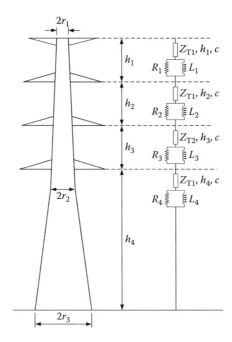

FIGURE 2.32
Multistory model of a transmission tower. (From Ishii, M. et al., *IEEE Trans. Power Deliv.*, 6, 1327, July 1991; Yamada, T. et al., *IEEE Trans. Power Deliv.*, 10, 393, January 1995. With permission.)

The surge impedance of each line section is obtained as in the previous model, while the propagation velocity is that of light. Without including the representation of the crossarms, the multistory model is as shown in the figure. The damping resistances and inductances are deduced according to the following equations [70,85]:

$$R_i = \frac{-2 \cdot Z_{T1} \cdot \ln\sqrt{\gamma}}{h_1 + h_2 + h_3} h_i \quad (i = 1,\ 2,\ 3) \tag{2.74}$$

$$R_4 = -2 \cdot Z_{T2} \cdot \ln\sqrt{\gamma} \tag{2.75}$$

$$L_i = \alpha \cdot R_i \cdot \frac{2h}{c} \quad (i = 1,\ 2,\ 3,\ 4) \tag{2.76}$$

where
    Z_{T1} is the surge impedance of the three upper tower sections
    Z_{T2} is the surge impedance of the lower tower section
    h_i is the height of each tower section
    γ is the attenuation coefficient
    α is the damping coefficient

The attenuation coefficient is between 0.7 and 0.8, while unity has been the value usually chosen for the damping coefficient [70,85].

    A further experimental investigation found that an adequate calculation for both Z_{T1} and Z_{T2} could be based on Jordan's formula:

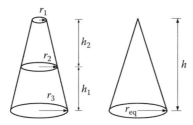

FIGURE 2.33
Geometry for equivalent radius calculation.

$$Z = 60\left[\ln\left(\frac{h}{r_{eq}}\right) - 1\right]$$ (2.77)

where
  h is the tower height
  r_{eq} is the equivalent radius obtained from the geometry shown in Figure 2.33 and given by [85]

$$r_{eq} = \frac{r_1 h_2 + r_2 h + r_3 h_1}{2h} \quad (h = h_1 + h_2)$$ (2.78)

The propagation velocity is that of the light.

A study presented in [86] concluded that the multistory model with surge impedance values originally proposed in [70] was not adequate for representing towers of lower voltage transmission lines. According to this study, the tower model for shorter towers can be simpler than that assumed by the multistory model; that is, a single lossless line for each tower section, whose surge impedance was calculated from Equations 2.77 and 2.78 would suffice.

Antenna theory approach might be used for an accurate computation of lightning voltages, and for validation and improvement of simpler circuit and transmission-line models, as discussed in [65]. This reference also discussed possible improvements, as well as the use of nonuniform lines for transmission tower modeling.

Example 2.3

Figure 2.34 shows the tower design of the lines analyzed in this example. Main characteristics of phase conductors and shield wires are presented in Table 2.5. The aim is to simulate and compare the maximum lightning overvoltages that will occur across the insulator strings of each test line, considering different lightning stroke waveshapes.

The following tower models are used with each test line:

- Towers of the test line 1 are represented using the conical and the waisted model, as it is presently implemented into FLASH [79,80] and as proposed by CIGRE [74].
- Towers of the test line 2 are represented using the conical model implemented into FLASH [67], the waisted model recommended by CIGRE [74], and the multistory model, whose parameters were calculated according to [85].

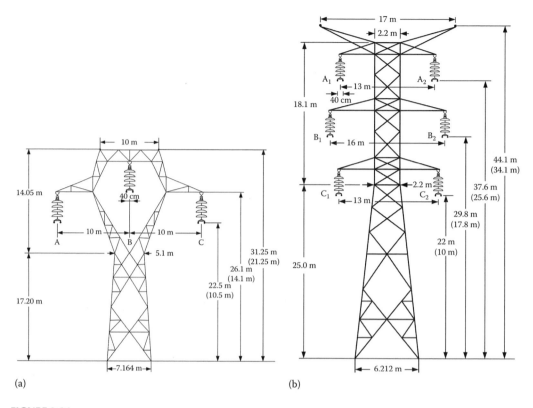

FIGURE 2.34
Example 2.3: 400 kV line configurations (values within parenthesis are midspan heights). (a) Test line 1, (b) test line 2.

TABLE 2.5

Example 2.3: Characteristics of Wires and Conductors

	Type	Diameter (mm)	DC Resistance (Ω/km)
Phase conductors	CURLEW	31.63	0.05501
Shield wires	94S	12.60	0.642

The models of both test lines are created considering the following values and modeling guidelines:

- Each line is represented by means of three 400 m spans plus a 3 km section as line termination at each side of the point of impact. Each line section is represented as a constant distributed-parameter model whose values are calculated at 400 kHz. Soil resistivity, needed to obtain the electrical parameters, is 200 Ω m.
- The insulators are represented as open switches since the main goal is to compare the overvoltages that will be caused in both lines with different models of the towers.
- The grounding impedance of all towers of both lines is represented as a constant resistance.
- The lightning stroke is represented as an ideal current source (infinite parallel impedance) with a double ramp waveform, defined by the peak current magnitude, I_{100}, the time-to-crest (also known as rise time or front time), t_f, and the tail time, t_h, which is the time, measured along the tail, needed to reach 50% of the peak value.
- Power-frequency voltages at phase conductors are neglected.

Figure 2.35 shows the equivalent circuit that results for any line after following the guidelines detailed above. Tables 2.6 and 2.7 summarize the different approaches applied to represent each transmission tower, as well as the resulting parameters. The tower models used in simulations have some minor differences with respect to those depicted in tables since small resistors have been added to separate phases and shield wires when the tower is represented as a single-phase lossless line.

Some simulation results derived with the two test lines are shown in Figure 2.36. These plots depict the overvoltages caused by two different return strokes across the insulator strings of the outer phase of the first test line and an upper phase of the second test line, respectively. One can observe that the differences between simulation results that correspond to one line depend on the tower model and the rise time of the return stroke current (t_f): As the rise time of the return stroke waveshape increases, the differences between the results derived with different tower models decrease, and the differences are more important between results obtained with the tower models used for the second line.

On the other hand, the results derived with the models tested in this example are similar to the two lightning current waveshapes, except when the multistory approach is applied, which provides much larger overvoltages. This is according to the conclusions of [86].

In fact the tower representation has less influence than the tower grounding as can be seen from the results derived with different grounding resistances. It is worth mentioning that a constant resistance is a very conservative approach to represent the grounding of a transmission line, as discussed in the following section.

2.5   Transmission Line Grounding

2.5.1   Introduction

Power equipment is in general connected to earth through a ground connection of sufficiently low impedance and having sufficient current-carrying capacity to prevent the buildup of voltages that may damage the equipment or result in hazard to persons [83,87–90]. Grounding affects the power-frequency voltages of unfaulted phases or the voltages caused by lightning, and influences the choice of surge protection (see Chapter 6).

According to EN-50341 [88], the grounding design of a power line has to

1. Ensure mechanical strength and corrosion resistance
2. Withstand, from a thermal point of view, the highest fault current
3. Avoid damage to properties and equipment
4. Ensure personal safety with regard to voltages that appear during ground faults
5. Achieve a certain reliability of the line

When an overhead line is constructed with two or more different voltage levels, all these requirements should be met for each voltage level.

The grounding requirements of a high voltage power line depends on characteristics such as the type of neutral point design (insulated, resonant grounding, and low resistance neutral), the type of supports (supports with or without built-in disconnectors or transformers), the material used for supports (steel, reinforced concrete, and wood), or support sites (normal or particularly exposed) [89].

Overhead line supports of nonconducting material need not be grounded, although poles of distribution lines are grounded if ground wires are installed to improve the lightning

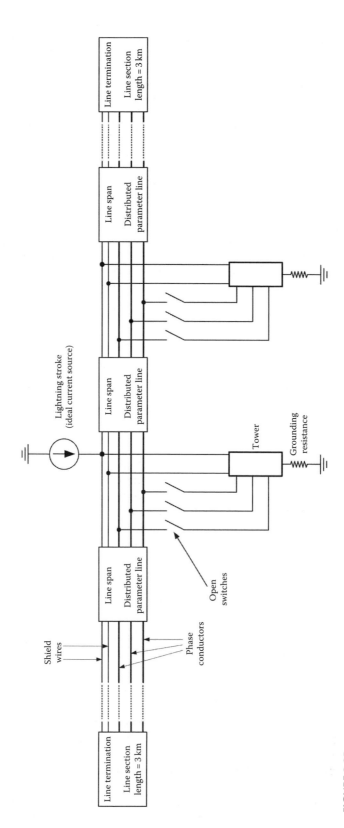

FIGURE 2.35
Example 2.3: Line model used in lightning overvoltage calculations.

TABLE 2.6

Example 2.3: Parameters of the Tower—Test Line 1

Geometry	Equivalent Circuit	Parameters

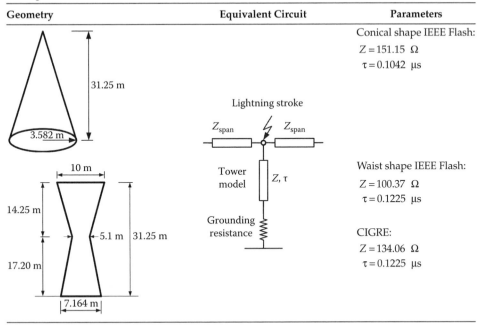

Conical shape IEEE Flash:

$Z = 151.15 \ \Omega$

$\tau = 0.1042 \ \mu s$

Waist shape IEEE Flash:

$Z = 100.37 \ \Omega$

$\tau = 0.1225 \ \mu s$

CIGRE:

$Z = 134.06 \ \Omega$

$\tau = 0.1225 \ \mu s$

performance of the line. Supports of conducting material are in principle grounded by their footings, but additional measures may be required. Ground wires, if used, are connected directly at the support top. Grounding of metallic supports may be done by burying the structure into the earth, but supplementary grounding is required when this design does not provide satisfactory impedance.

A grounding system is generally composed of one or more horizontal, vertical, or inclined electrodes, buried or driven into the soil. For wood poles, one or more ground rods are buried near the pole, and a down lead, running vertically from the top of the pole along its length, is connected to these rods.

Ground electrodes for overhead lines can be classified into three groups [83]:

- Surface-earth electrodes, which usually consist of strip wires or bands arranged as radial, ring, or meshed electrodes (or a combination of these) buried at depths of up to approximately 1 m.
- Deep-earth electrodes, which can pass through various soil layers.
- Foundation electrodes, which are a part of the civil construction.

In practice, it can be necessary to apply a combination of deep-earth (e.g., vertical ground rod) and surface electrodes to achieve a good resistance and safe surface potential gradients. Foundation electrodes are often ignored, but they can offer additional areas of contact with soil, which can improve both ground resistance and surface potential gradients. On the other hand, deep-earth electrodes can be advantageous where shallow upper layers have high resistivity or (when vertically driven) in places with a limited surface.

TABLE 2.7

Example 2.3: Parameters of the Tower—Test Line 2

Geometry	Equivalent Circuit	Parameters

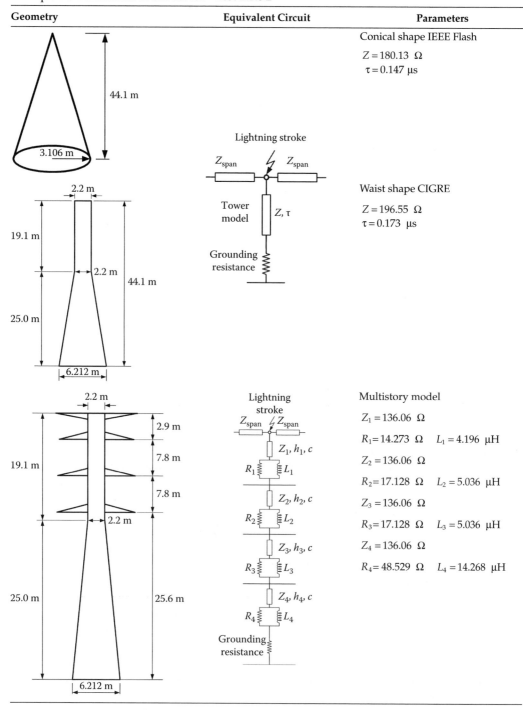

Conical shape IEEE Flash

$Z = 180.13\ \Omega$
$\tau = 0.147\ \mu s$

Waist shape CIGRE

$Z = 196.55\ \Omega$
$\tau = 0.173\ \mu s$

Multistory model

$Z_1 = 136.06\ \Omega$
$R_1 = 14.273\ \Omega \quad L_1 = 4.196\ \mu H$
$Z_2 = 136.06\ \Omega$
$R_2 = 17.128\ \Omega \quad L_2 = 5.036\ \mu H$
$Z_3 = 136.06\ \Omega$
$R_3 = 17.128\ \Omega \quad L_3 = 5.036\ \mu H$
$Z_4 = 136.06\ \Omega$
$R_4 = 48.529\ \Omega \quad L_4 = 14.268\ \mu H$

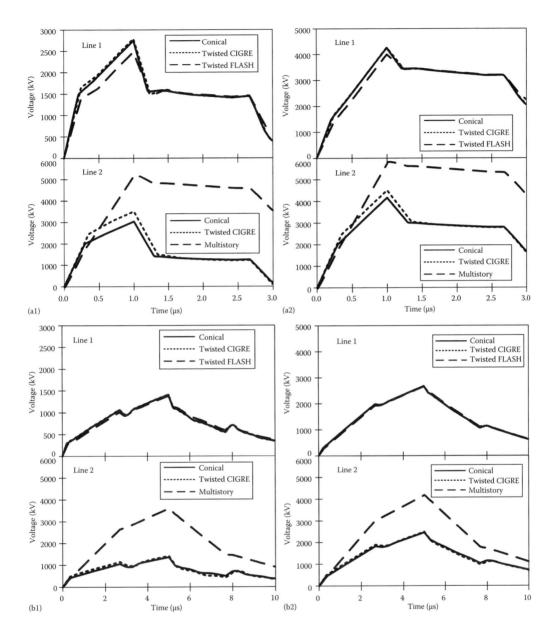

FIGURE 2.36
Example 2.3: Tower model performance—lightning current $I_{max} = 120\,kA$, $t_h = 77.5\,\mu s$. (a) $t_f = 1\,\mu s$, (b) $t_f = 5\,\mu s$.

High voltages can be generated on grounded parts of a power line support when either a ground wire or a phase conductor is struck by lightning. If lightning strikes a tower or a ground wire, the discharge should be then safely led to the earth and dissipated there. The purpose of grounding for protection against lightning is to bypass the energy of the lightning discharge safely to the ground; that is, most of the energy of the lightning discharge should be dissipated into the ground without raising the voltage of the protected system. The tower grounding impedance depends on the area of the tower steel (or grounding conductor) in contact with the earth, and on the resistivity of the earth. The latter is not constant, fluctuates over time, and is a function of soil type, moisture content, temperature, current magnitude, and waveshape.

Sections 2.5.2 to 2.5.5 discuss the representation of the grounding impedance as a function of the discharge current waveform, taking into account soil ionization and the fact that real soils are nonhomogenous and soil resistivity varies with depth. The design of grounding system for a power line is out of the scope of this chapter; however, some general considerations about the selection of grounding electrodes for improving the lightning performance of overhead lines are included in Section 2.5.6.

2.5.2 Grounding Impedance

When current is discharged into the soil through a ground electrode, potential gradients are set up as a result of the conduction of current through the soil. The grounding impedance of a power line is given by the relationship between the potential rise of the electrode and the current discharged into the ground. The representation of the grounding impedance depends on the frequency range of the discharged current. Grounding models can be classified into two groups: low- and high-frequency models. In practice, they correspond respectively to power-frequency and to lightning stroke discharged currents. The following paragraphs summarize the approaches developed to represent grounding arrangements of overhead lines.

2.5.2.1 Low-Frequency Models

The complex power-frequency impedance of a ground system can be expressed as follows [83]:

$$Z = (R_L + R_E + R_C) + j(X_E + X_L) \tag{2.79}$$

where
  R_L and X_L are respectively the resistance and the reactance of the ground electrode leads
  R_C is the contact resistance between the surface of the electrode and the surrounding soil, due to the imperfect contact between the soil and the surface of the electrode
  R_E is the dissipation resistance of the soil surrounding the electrode (i.e., the resistance of the earth between the electrode surface and a remote earth)
  X_E is the reactance of the current paths in the soil

The series resistance of the metallic conductors that make up the ground electrodes and leads is typically much lower than the contact resistance and the resistance of the surrounding soil. On the other hand, the reactance of the metallic conductors and the leads is much higher than the reactance of the current dissipated in the soil. Equation 2.79 can therefore be simplified to

$$Z = (R_L + R_E) + jX_L \tag{2.80}$$

The contact resistance can typically be neglected when the soil has settled around the electrode conductors. At power frequencies, the reactance of the grounding conductors and leads becomes negligible compared to the dissipation resistance. Therefore, at low (power) frequencies, the grounding impedance can be represented by the dissipation resistance:

$$Z \approx R_E \qquad (2.81)$$

2.5.2.2 High-Frequency Models

For high-frequency phenomena such as lightning, several aspects determine the magnitude and shape of the transient voltage that appear at the tower base. They include the surge impedance of buried wires, the surge impedance of the ground plane, and soil ionization [83].

During lightning discharges the effective impedance of a buried horizontal wire is not constant. Such behavior can be explained by considering surge propagation along the buried conductor, and can be summarized as follows [91]:

- The effective impedance is initially equal to the surge impedance of the buried wire and it reduces in a few microseconds to a level that corresponds to the leakage resistance.

- The transition from the initial surge impedance to the final leakage resistance is accomplished in a few round-trip travel times along the wire. Although multivelocity waves will exist, the only one of importance is slow and travels at approximately 30% of the velocity of light.

- The surge impedance rises abruptly in less than $1\,\mu s$, and increases at a slow rate thereafter; the leakage resistance is initially a very high value, but it decreases as the reflection of the traveling waves builds up the voltage along the conductor, being its final value equal to the low-frequency resistance. Resistance quickly dominates as a current wave propagates along the conductor.

Counterpoises of more than 60–90 m in length are not justifiable. Several short parallel-connected counterpoises are a more efficient way to reduce the ground electrode surge impedance.

Chisholm and Janischewskyj reported from field measurements that the initial reflection coefficient defined between the impedance of the tower and the ground impedance was approximately 0.6, even for a test geometry designed to obtain a reflection coefficient of unity [81]. This imperfect reflection, also observed by other authors [92], can be justified if the surge impedance of the ground plane is taken into account, and its influence is included in the ground impedance model by adding an inductance that is a function of tower height [81], see Section 2.5.4.2.

Another factor is soil ionization which leads to an additional decay in the electrode resistance when high currents are discharged into the soil. Under lightning surge conditions and some power-frequency fault conditions, the high current density in the soil increases the electric field strength up to values that cause electrical discharges in the soil that surrounds the electrode. The plasma of the discharges has a resistance lower than that of the surrounding soil, so there is an apparent decrease of the ground resistivity in areas where the ionization occurs. Since ionization occurs mainly near the electrode where the current density is highest, it increases the effective size of the electrode and results in a reduction in the electrode resistance.

Electrical breakdown occurs in the soil at an average surface ionization gradient of approximately 300–400 kV/m. In some soil types, the ionization gradient can be as high as 1000 kV/m [93]. The threshold level and intensity of the ionization are especially high when the soil is dry or when it has a high resistivity. Depending on the electrode configuration, ionization can take place under impulse currents as low as 1 kA.

The soil ionization around typical grounding electrodes approaches a hemispheric shape for very high levels of current. The maximum extent of this hemispheric zone of ionization rarely exceeds 10 m for typical lightning surge currents of up to 200 kA. This means that the ionization will usually occur in the upper layer of the soil. In terms of the electrode resistance, ionization reduces the contact resistance term without modifying the geometric resistance of the electrode. The effect of soil ionization is more pronounced on small, compact electrodes, and it can be negligible for large grounding foundations.

2.5.2.3   Discussion

From the above analysis it can be concluded that

- For low frequencies the ground impedance may be represented as a pure resistance and remains constant and equal to its DC resistance. At high frequencies, the behavior is frequency dependent [83,94,95].

- Some effects can affect the lightning performance of a power line: There is a residual inductance in any grounding arrangement that adds some stress to line insulation, while soil ionization and capacitive displacement currents tend to reduce the apparent impedance. Electrode inductance increases the grounding impedance, being this effect more significant for lines with low grounding impedance. The ionization effects are important for many soil types, and decrease resistance by increasing the effective radius of the electrodes. The capacitive displacement current is only important in areas where soil resistivities are greater than 10 kΩ m.

Although much work has been done on the behavior of the ground impedance under lightning discharge, there is no consensus on how to apply this knowledge to the representation and the design of actual electrode systems, and the power-frequency grounding resistance is often used to predict the lightning performance of power lines.

The grounding system of a power line support can be broadly classified as compact (concentrated) or extended (distributed) [95,96]. A compact grounding can be represented by lumped-circuit elements. In an extended grounding system the travel time of the electromagnetic fields along the electrodes is comparable with that along the support itself; this generally applies to grounding systems with physical dimensions exceeding 20 m.

Sections 2.5.3 and 2.5.4 provide a theoretical and practical base for the analysis of power line ground systems taking into account their frequency-dependent behavior, and show how to treat soil ionization.

2.5.3   Low-Frequency Models of Grounding Systems

2.5.3.1   Compact Grounding Systems

One of the simplest grounding configurations is a hemispherical electrode, see Figure 2.37. Assuming uniform soil resistivity, a current flowing from the hemisphere into the ground produces a current density in the surrounding soil given by the following expression:

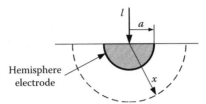

FIGURE 2.37
Hemisphere electrode.

$$J = \frac{I}{2\pi x^2} \tag{2.82}$$

where
  J is the current density
  I is the total current
  x is the distance from the center of the electrode

The electric field strength that such current density produces in the soil is

$$E = \rho J = \frac{\rho I}{2\pi x^2} \tag{2.83}$$

where ρ is the soil resistivity.
  The voltage at any distance x is

$$V = \int_a^x E \, dx = \frac{\rho I}{2\pi} \int_a^x \frac{1}{x^2} \, dx = \frac{\rho I}{2\pi} \left(\frac{1}{a} - \frac{1}{x} \right) \tag{2.84}$$

where a is the radius of the electrode.
  The total voltage between the electrode and a far distance point ($x \cong \infty$) is then

$$V = \frac{\rho I}{2\pi a} \tag{2.85}$$

Finally, the total resistance of the electrode is obtained as follows:

$$R = \frac{V}{I} = \frac{\rho}{2\pi a} \tag{2.86}$$

This is the resistance experienced by current flowing through the entire surrounding space. Most of this resistance is encountered in the region immediately around the electrode. From Equation 2.84, 50% of the total resistance is contained in $x \le 2a$, and 90% is contained in $x \le 10a$.

The rod is the most common type of ground electrode. Ground rods are generally made of galvanized steel, 2.5–3 m in length, less than 2.5 cm in diameter, and driven vertically down from the earth's surface. When the length of the ground rod is much greater than its radius, the low-current, low-frequency impedance of a single ground rod is approximated by a resistance whose value may be obtained from the following expression:

$$R_0 = \frac{\rho}{2\pi\ell}\left(\ln\frac{4\ell}{a} - 1\right) \tag{2.87}$$

where
  ℓ is the length of the buried rod, in m
  a is the radius of the rod, in m
  ρ is the soil resistivity, in Ω m

Some discrepancies can be found in the literature related to the expression given by some authors for the resistance of a driven ground rod; they are mainly due to the different approaches used to derive the above expression [26,81,83,97–100].

R_0 decreases as either the buried length or the radius of the rod increase, but it does not decrease directly with length, so an increase in length above a certain limit will not significantly reduce the resistance.

Ground resistance can be reduced by connecting several rods in parallel. The resistance is inversely proportional to the number of parallel rods, provided the spacing between rods is large compared to their length. If the spacing is short and the n ground rods are arranged on a circle of diameter D, then Equation 2.87 is still valid if the radius a is replaced by an equivalent radius r_{eq}[26]:

$$R_0 = \frac{\rho}{2\pi\ell}\left(\ln\frac{4\ell}{r_{eq}} - 1\right) \tag{2.88}$$

where

$$r_{eq} = \left(na\left(\frac{D}{2}\right)^{n-1}\right)^{1/n} \tag{2.89}$$

That is, when the rods are closely spaced compared with their length, the whole ground arrangement behaves as one rod with a larger apparent diameter and a small reduction in resistance. As the rod spacing increases, the combined resistance decreases. When the spacing between adjacent rods, arranged on a circle of diameter D, is equal to or longer than the length of rods, the combined resistance of n ground rods can be approximated as follows [26]:

$$R_0 = \frac{1}{n}\frac{\rho}{2\pi\ell}\left(\ln\frac{4\ell}{a} - 1 + \frac{\ell}{D}\sum_{m=1}^{n-1}\frac{1}{\sin\frac{m\pi}{n}}\right) \tag{2.90}$$

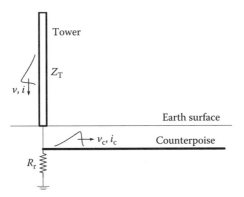

FIGURE 2.38
Wave propagation along a counterpoise. (Adapted from Hileman, A.R., *Insulation Coordination for Power Systems*, Marcel Dekker, New York, 1999.)

The proximity effect between the ground rods tends to increase the combined resistance, thus diminishing the advantage of multiple rods.

2.5.3.2 Extended Grounding Systems

The contact area of a grounding system with the earth can be increased by installing a counterpoise, which is a conductor buried in the ground at a depth of about 1 m. Common arrangements include one or more radial wires extending out from each tower base, single or multiple continuous wires from tower to tower, or combinations of radial and continuous wires.

Counterpoises may be also combined with driven rods. In such case, a wave traveling down the tower impinges on the combination of the vertically driven rod and the counterpoise, and results in a wave which travels out along the counterpoises at about 1/3 of the speed of light, see Figure 2.38 [3]. The current wave initially meets the surge impedance of the conductor, whose value decreases with time and reaches a steady-state value when the current is distributed through the counterpoise length. After a few round-trip travel times, the impedance is reduced to the total leakage resistance of the counterpoise (Section 2.5.2.2).

When the length of the conductor is much greater than its burial depth, the low-current low-frequency resistance of a horizontal conductor buried in the soil can be given by [26]

$$R_0 = \frac{\rho}{\pi \ell}\left(\ln \frac{2\ell}{\sqrt{2ad}} - 1 \right) \tag{2.91}$$

where
ℓ is the length of the buried conductor, in m
a is the conductor radius, in m
d is the burial depth of the conductor, in m
ρ is the soil resistivity, in Ω m

The steady-state resistance is not greatly influenced by either a or d.

Several short wires, arranged radially, may be more effective than a single long wire even if the total length and contact resistance of both arrangements are the same. The initial surge impedance of several wires is lower and the steady-state contact resistance is reached faster. The low-current low-frequency resistance of n radial conductors is [26]

$$R_0 = \frac{\rho}{n\pi\ell}\left(\ln\frac{2\ell}{\sqrt{2ad}} - 1 + \sum_{m=1}^{n-1}\ln\frac{1+\sin\dfrac{\pi m}{n}}{\sin\dfrac{\pi m}{n}} \right) \qquad (2.92)$$

As n becomes very large, the last term between brackets approaches $1.22n$ [95], and in the limit

$$R_0 \rightarrow \frac{\rho}{2.57\ell} \qquad (2.93)$$

Table 2.8 shows the resistance of several simple electrodes. It is important to remember that several different equations have been proposed for the calculation of ground electrodes, and each one gives a slightly different result. See for instance Ref. [83].

2.5.3.3 Grounding Resistance in Nonhomogeneous Soils

The expressions presented for calculating the grounding resistance use only one resistivity value, since a homogeneous soil is considered. Several methods have been developed to deal with nonhomogeneous soils. One of the methods consists in stratifying the soil in two layers, which could represent the effect of multiple layers. The first layer reaches a depth d and is characterized by a resistivity ρ_1, while the second layer has an infinite depth and a resistivity ρ_2. For a great majority of lines, it is possible to determine a two-layer soil structure that can represent typical soils for grounding purposes. In some cases it might not be possible to define a two-layer soil structure, then a three-layer soil model must be considered.

The computation of the grounding resistance depends on the type of ground electrode and the depth of the first layer with respect to the length of the grounding electrode. For a rod driven into the upper layer only, the ground resistance is deduced as follows [89,100]:

$$R = \frac{\rho_1}{2\pi\ell}\left(\ln\left(\frac{4\ell}{a}-1\right) + \sum_{n=1}^{\infty}\frac{\Gamma^n}{2}\ln\frac{nd/\ell+1}{nd/\ell-1} \right) \qquad (2.94)$$

where

$$\Gamma = \frac{\rho_2 - \rho_1}{\rho_2 + \rho_1} \qquad (2.95)$$

is the reflection coefficient.

The first term is the resistance of a rod having length ℓ driven into soil of resistivity ρ_1, and the second term represents the additional resistance due to the second layer.

If the rod penetrates both layers, then the ground resistance is deduced as follows [89]:

$$R = F\cdot\left(R_1 + R_a\right) = \frac{\rho_1}{2\pi\ell}\frac{1+\Gamma}{1-\Gamma+2\Gamma d/\ell}\left(\ln\frac{2\ell}{a} + \sum_{n=1}^{\infty}\Gamma^n\ln\frac{nd/\ell+1}{(2n-2)d/\ell-1} \right) \qquad (2.96)$$

TABLE 2.8

Ground Resistance of Elementary Electrodes

Electrode	Schematic Representation	Dimensions	Grounding Resistance
Hemisphere		Radius a	$R_0 = \dfrac{\rho}{2\pi a}$
Single vertical rod		Length ℓ Radius a	$R_0 = \dfrac{\rho}{2\pi\ell}\left(\ln\dfrac{4\ell}{a} - 1\right)$
n vertical rods		Length ℓ Radius a Circle diameter D $D\sin\dfrac{\pi}{n} > \ell$	$R_0 = \dfrac{1}{n}\dfrac{\rho}{2\pi\ell}\ln\left(\dfrac{4\ell}{a} - 1 + \dfrac{\ell}{D}\sum_{m=1}^{n-1}\dfrac{1}{\sin\dfrac{m\pi}{n}}\right)$
Single buried horizontal wire		Length ℓ Radius a Depth d	$R_0 = \dfrac{\rho}{\pi\ell}\left(\ln\dfrac{2\ell}{\sqrt{2ad}} - 1\right)$
n buried horizontal radial wires		Length ℓ Radius a Depth d	$R_0 = \dfrac{\rho}{n\pi\ell}\left(\ln\dfrac{2\ell}{\sqrt{2ad}} - 1 + \sum_{m=1}^{n-1}\ln\dfrac{1+\sin\dfrac{\pi m}{n}}{\sin\dfrac{\pi m}{n}}\right)$
Ring		Diameter D Ring radius a (independent of depth)	$R_0 = \dfrac{\rho}{\pi^2 D}\left(\ln\dfrac{\pi D}{a} - 1\right)$
Round plate		Diameter D (zero burial depth)	$R_0 = \dfrac{\rho}{4D}$

where

    F is the penetration factor given by

$$F = \frac{1+\Gamma}{1-\Gamma+2\Gamma d/\ell} \tag{2.97}$$

    R_1 is the resistance of the rod driven in uniform soil of resistivity ρ_1
    R_a is the additional resistance due to the second layer

A simpler two-layer soil treatment is appropriate when $\rho_2 \gg \rho_1$; that is, when the reflection coefficient approaches unity. Under this condition, the resistance of a single hemisphere of radius a in a layer soil with resistivity ρ_1 and thickness d may be approximated by the following expression [96,101]:

$$R = \frac{\rho_1}{2\pi a}\left(1 + 2\sum_{n=1}^{\infty}\frac{\Gamma^n}{\sqrt{1+(2nd/a)^2}}\right) \tag{2.98}$$

For a reflection coefficient $\Gamma = 1$, the series in Equation 2.98 becomes the harmonic series and converges slowly. Loyka proposes the following approximation [96,102]:

$$R \approx \frac{\rho_1}{2\pi a}\left(1 + \frac{a}{d}\ln\left(\frac{50000}{1+\sqrt{1+(a/2d)^2}}\right)\right) \tag{2.99}$$

This approximation is accurate enough for a wide range of transmission line grounding applications, and proves that a finite upper layer depth has a strong influence on the electrode resistance [96].

2.5.4 High-Frequency Models of Grounding Systems

Voltage and current waves are dissipated in a counterpoise while traveling along its length. A model based on a resistive behavior is not an accurate representation if the injected current has significant high-frequency content. Several approaches have been proposed to represent line grounding when a lightning discharge impinges the electrode system. Although all of them account for a frequency-dependent behavior, different equivalent circuits have been proposed.

A didactic description of the behavior of grounding systems when subjected to lightning currents has been presented in Ref. [103].

2.5.4.1 Distributed-Parameter Grounding Model

The electrode is represented as the series–parallel equivalent circuit shown in Figure 2.39, where R is the parallel resistance, L is the series inductance, and C the parallel capacitance to ground per unit length.

The equations of this circuit are similar to the equations of a line with distributed parameters and can be expressed as follows:

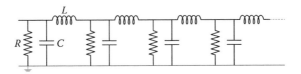

FIGURE 2.39
Equivalent circuit of a ground electrode at high frequencies.

$$\frac{\partial V(x,s)}{\partial x} = -sLI(x,s)$$

$$\frac{\partial I(x,s)}{\partial x} = -(G+sC)V(x,s) \qquad (G = 1/R) \tag{2.100}$$

To take into account soil ionization a nonlinear resistance might be included in the model instead of a constant one. However, soil ionization is not instantaneous and soil resistivity decreases with a time constant of about $2\,\mu s$. This value is rather large compared with the front times associated with fast-front lightning strokes, e.g., most negative subsequent strokes. Moreover if soil ionization occurs, it will always cause a reduction of the ground potential rise. Therefore, ignoring this phenomenon always gives conservative results. The following analysis considers only an equivalent circuit with constant and linear parameters.

The expressions for calculation of R, L, and C in per unit length are [26]

- For vertical conductors:

$$R = \frac{\rho}{2\pi} A_1 \quad L = \frac{\mu_o}{2\pi} A_1 \quad C = \frac{2\pi\varepsilon}{A_1} \quad A_1 = \ln\frac{4\ell}{a} - 1 \quad (\ell \gg a) \tag{2.101}$$

- For horizontal conductors:

$$R = \frac{\rho}{\pi} A_2 \quad L = \frac{\mu_o}{2\pi} A_2 \quad C = \frac{\pi\varepsilon}{A_2} \quad A_2 = \ln\frac{2\ell}{\sqrt{2ad}} - 1 \quad (\ell \gg a, \ell \gg d) \tag{2.102}$$

where
ℓ is the conductor length, in m
a is the radius of the conductor, in m
d is the burial depth, in m
ρ is the soil resistivity (supposed to be homogeneous) in Ω m
μ_o is the vacuum permittivity ($= 4\pi \times 10^{-7}\,H/m$)
ε is the soil permittivity (typical value: $\varepsilon = 10\varepsilon_o$, with $\varepsilon_o = 8.85 \times 10^{-12}\,F/m$)

Note that the above calculations of parameters in per unit length depend on the electrode length. Although different authors diverge in determining these values of R, L, and C, the discrepancies remain usually small with respect to the approximations due to seasonal variations of soil resistivity or lack of soil homogeneity.

If the circuit in Figure 2.39 is energized at one end and the other end is open, the input impedance $Z(j\omega)$ can be computed as follows [27]:

$$Z(\omega) = Z_c \coth \gamma\ell \tag{2.103}$$

where
  Z_c is the surge impedance
  γ is the propagation constant

From Equations 2.101 and 2.102, the following results are derived [104]:

- For vertical conductors:

$$Z_c = \frac{A_1}{2\pi} \sqrt{\frac{j\omega\mu_o\rho}{(1+j\omega\varepsilon\rho)}} \qquad \gamma = \sqrt{\frac{j\omega\mu_o}{\rho}(1+j\omega\varepsilon\rho)} \tag{2.104}$$

- For horizontal conductors:

$$Z_c = \frac{A_2}{\pi} \sqrt{\frac{j\omega\mu_o\rho}{2(1+j\omega\varepsilon\rho)}} \qquad \gamma = \sqrt{\frac{j\omega\mu_o}{2\rho}(1+j\omega\varepsilon\rho)} \tag{2.105}$$

The ground impedance is complex and frequency dependent. When the harmonic impedance $Z(\omega)$ of a ground electrode is known, the transient voltage developed at the feeding point $v(t)$ as a response to an injected current pulse $i(t)$ can be evaluated as follows:

$$v(t) = L^{-1}\left\{Z(\omega) \cdot L\left[i(t)\right]\right\} \tag{2.106}$$

where
  $i(t)$ is the injected current pulse
  L and L^{-1} are respectively the Laplace (or Fourier) and the inverse Laplace (or inverse Fourier) transforms

Several parameters may be used to characterize the transient behavior of ground electrodes. One is the impulse impedance $z_0(t)$, defined as the ratio between the instantaneous values of the voltage and the current at the feeding point [105]:

$$z_0(t) = \frac{v(t)}{i(t)} \tag{2.107}$$

Another way to characterize the transient response is the conventional impedance Z, which is defined as the ratio between the peak values of $v(t)$ and $i(t)$:

$$Z = \frac{\max\left[v(t)\right]}{\max\left[i(t)\right]} = \frac{V_{max}}{I_{max}} \tag{2.108}$$

Another parameter is the *impulse efficiency* defined as the ratio between the conventional impedance and the low-frequency resistance, Z/R_0 [106]. It is worth noting that a lower value of the impulse efficiency means better impulse performance.

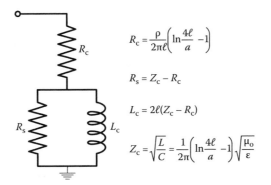

$$R_c = \frac{\rho}{2\pi\ell}\left(\ln\frac{4\ell}{a} - 1\right)$$

$$R_s = Z_c - R_c$$

$$L_c = 2\ell(Z_c - R_c)$$

$$Z_c = \sqrt{\frac{L}{C}} = \frac{1}{2\pi}\left(\ln\frac{4\ell}{a} - 1\right)\sqrt{\frac{\mu_o}{\varepsilon}}$$

FIGURE 2.40
Equivalent circuit of a counterpoise.

2.5.4.2 Lumped-Parameter Grounding Model

Based on experimental results, Bewley suggested that the lightning behavior of counterpoise electrodes could be represented by the simple equivalent circuit presented in Figure 2.40 [91], where

- R_c represents the counterpoise leakage resistance.
- R_s is a resistor selected so that the high-frequency impedance of the circuit corresponds to the surge impedance of the counterpoise (Z_c).
- L_c is the inductor responsible for the transition from the surge impedance to the low-frequency impedance; its value is dependent on the length of the counterpoise.

Chisholm and Janischewskyj suggest from their experimental results that the apparent impedance of the ground plane as a function of time can be obtained as follows [81]:

$$Z = 60\frac{h}{ct} \quad \left(t > t_t = \frac{h}{c}\right) \tag{2.109}$$

where
h is the height of the tower
c is the velocity of light (=300 m/μs)
t is time
t_t is the surge travel time of the tower

It is possible to represent the initial surge response of a perfectly grounded tower as an inductance that is a function of tower height. The value of this inductance can be approximated by means of the following expression:

$$L \approx 60 t_t \ln\left(\frac{t_f}{t_t}\right) \tag{2.110}$$

where t_f is the front time.

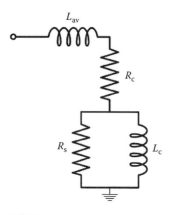

FIGURE 2.41

Modified equivalent circuit of a counterpoise.

This equivalent inductance is valid for the normal case, in which the front time t_f is much greater than the tower travel time t_t. Consequently, the modified equivalent circuit of the counterpoise is that shown in Figure 2.41. Longer front times give higher values of average inductance, but the voltage rise from the inductance at the crest current is actually lower.

2.5.4.3 Discussion

A comparison between the performance of three frequency-dependent models for a vertically driven ground rod was presented in [107] a lumped-parameter circuit (see Figure 2.42), a distributed-parameter circuit (see Figure 2.39), and a more rigorous approach based on antenna theory. The main conclusions of the study can be summarized as follows:

- The application of the lumped-parameter circuit model is limited to cases where the length of the rod is less than one-tenth the wavelength in earth, which limits the frequency range of the validity of this model to low frequencies. This approach can be used for a preliminary analysis, but keeping in mind that it greatly overestimates the ground rod impedance at high frequencies.

- The approximate distributed-parameter circuit reduces the overestimation of the ground rod impedance at high frequencies in comparison with the *RLC* circuit.

- The best fit to ground rod impedance was achieved by means of a model based on antenna theory, with a nonuniform distributed-parameter model, whose parameters can be deduced by curve matching.

The following example analyzes the behavior and validity of different models for a ground rod.

Example 2.4

Consider a ground rod with a length of 9 m and a radius of 10 mm, vertically driven into a homogeneous soil whose resistivity and relative permittivity are respectively 100 Ω m and 10. The goal of this example is to analyze the transient response and the frequency-domain response of this ground rod considering three different modeling approaches: a constant resistance, a high-frequency lumped-parameter circuit (Figure 2.42) and a high-frequency distributed-parameter circuit (Figure 2.39).

Table 2.9 shows the parameters calculated for each model. To obtain the high-frequency distributed-parameter representation, the circuit has been built with sections that correspond to 0.5 m of grounding rod.

Figure 2.43 depicts the frequency response of the two high-frequency models. The plots show the relationship of the complex impedance modulus, $|Z(j\omega)|$, and the low-frequency resistance value, R_0, as well as the corresponding phase angles, as a function of the input current frequency.

It can be seen that the impedance is frequency independent and equal to the low-frequency resistance up to a frequency of about 100 kHz. For higher frequencies, the lumped-parameter model exhibits inductive behavior whereas the distributed-parameter model becomes capacitive

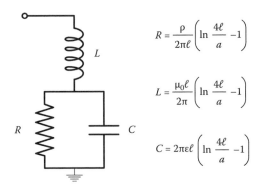

FIGURE 2.42
High-frequency lumped equivalent circuit of a ground electrode.

for frequencies close to 1 MHz. The frequency until which the electrode shows a resistive behavior increases as the dimension of the electrode decreases and the soil resistivity increases [108]. A resistive–capacitive behavior can be advantageous since the whole impedance is equal or smaller than the low-frequency resistance. Usually capacitive behavior is typical for electrodes with small dimensions buried in highly resistive soils; otherwise, the grounding electrode behavior is mostly inductive [109].

Figure 2.44 shows the transient response of the ground rod for two input current pulses with different front time.

- For a slow front-time current pulse (left plots in Figure 2.44), the transient voltage response at the feeding point is very similar with the two high-frequency models and very close to the response that would result with a constant resistance model, although the transient impedance may reach very high values with both high-frequency models. Since such current pulse does not have significant frequency content above 100 kHz, except during the front time, the voltage response is not substantially influenced by the inductive part of the harmonic impedance (Figure 2.43), although the voltage response precedes in both cases the current pulse during the current front time, as in a typical inductive behavior. The transient impedance starts from a high value, but it settles down to the low-frequency resistance value during the rise of the current.
- The transient voltages for a faster front-time current pulse (right plots in Figure 2.44) show some differences with respect to those obtained with the previous current pulse. Since the new current pulse has significant frequency content above 100 kHz, the new response exhibits very different behavior, as predicted from the frequency responses. The transient impedance in both models starts again from high values and settles down to a value close to the low-frequency resistance value after the current peak. The voltage response in both models is oscillatory and reaches its peak value before the current pulse peak, as for a typical inductive behavior. However, the peak value of the transient voltage is very different with the two high-frequency models. The transient voltage obtained with the lumped-parameter *RLC* model reaches a large peak value, while this value is a little bit larger than the value obtained with a constant-resistance model if the distributed-parameter model is used. That is, the *RLC* circuit overestimates the ground rod impedance at high frequencies, and the approximate distributed-parameter circuit reduces the overestimation, as mentioned in the previous discussion (Section 2.5.4.3).

When the ground electrode exhibits inductive behavior at high frequencies and the current pulse has enough high frequency content, it can be concluded that voltage peak will occur

TABLE 2.9

Example 2.4: Parameters of Grounding Models

Grounding Model	Parameters

$$R_0 = \frac{\rho}{2\pi\ell}\left(\ln\frac{4\ell}{a}-1\right) = 12.71\ \Omega$$

$$R = \frac{\rho}{2\pi\ell}\left(\ln\frac{4\ell}{a}-1\right) = 12.71\ \Omega$$

$$L = \frac{\mu_0}{2\pi}\ell\left(\ln\frac{4\ell}{a}-1\right) = 12.94\ \mu H$$

$$C = 2\pi\varepsilon\ell\left(\ln\frac{4\ell}{a}-1\right) = 35.98\ nF$$

$$R_i = \frac{\rho}{2\pi}\frac{n}{\ell}\left(\ln\frac{4\ell}{a}-1\right) = 228.82\ \Omega$$

$$L_i = \frac{\mu_0}{2\pi}\frac{\ell}{n}\left(\ln\frac{4\ell}{a}-1\right) = 0.72\ \mu H$$

$$C_i = 2\pi\varepsilon\frac{\ell}{n}\left(\ln\frac{4\ell}{a}-1\right) = 2.0\ nF$$

Note: Rod length = 9 m; rod radius = 10 mm; soil resistivity = 100 Ω m; relative soil permittivity = 10; n = 18.

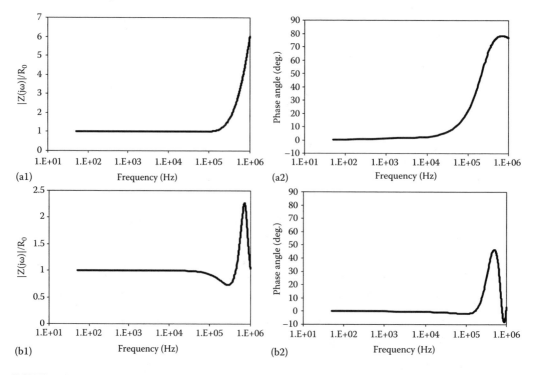

FIGURE 2.43

Example 2.4: Frequency response of the ground rod. (a) High-frequency lumped-parameter model, (b) high-frequency distributed-parameter model.

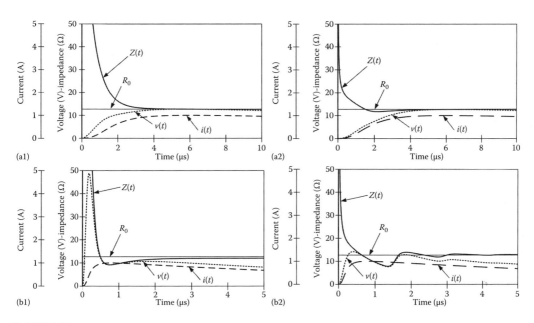

FIGURE 2.44
Example 2.4: Transient response of the ground rod. (a) High-frequency lumped-parameter model, (b) High-frequency distributed-parameter model.

during the rise of the current pulse; however, after few microseconds, the transient process is finished and the transient impedance settles down to the low-frequency resistance. An important result of this analysis is that the use of the impulse impedance, as defined in Equation 2.107 can mislead to wrong conclusions; that is, although the impulse impedance can exhibit very high values, the transient voltage peak of an accurate high-frequency model for a ground rod might not be much higher than the voltage peak obtained with a constant-resistance model.

Interestingly, the use of the conventional impedance, as defined in Equation 2.108, may be more appropriate for this application. The values of this impedance for the cases analyzed in this example were as follows:

- Slow front-time current pulse:

  Constant resistance model $Z = 12.71\,\Omega$
  High-frequency lumped-parameter model $Z = 12.68\,\Omega$
  High-frequency distributed-parameter model $Z = 12.67\,\Omega$

- Fast front-time current pulse:

  Constant resistance model $Z = 12.71\,\Omega$
  High-frequency lumped-parameter model $Z = 48.82\,\Omega$
  High-frequency distributed-parameter model $Z = 14.17\,\Omega$

2.5.5 Treatment of Soil Ionization

The resistance of a ground electrode may decrease due to ionization of the soil. When a current is injected to the electrode, an ionization process will occur in regions around the electrode if a critical field gradient is exceeded. In those regions, low ohmic discharge channels are formed, being the resistance of the ionized zone reduced to a negligible value. The ground resistance of an electrode remains at the value determined by the electrode geometry and the soil resistivity until ionization breakdown is reached; after breakdown, the resistance varies. This soil breakdown can be viewed as an increase of the geometry of the electrode. The transient electric fields needed to ionize small volumes of soil, or to flashover across the soil surface, are typically between 100 and 1000 kV/m [93,110].

For a vertically driven rod, this process may be represented with the simplified model of Figure 2.45, which also shows the final area. As the ionization progresses, the shape of the zone becomes more spherical; that is, when the gradient exceeds a critical value E_0, breakdown of soil occurs and the ground rod can be modeled as a hemisphere electrode.

After ionization the current density J and the voltage gradient E at a distance x from the rod for an injected current I are given by Equations 2.82 and 2.83. The ionization zone is described by the critical field strength E_0 at which the radius is equal to r. As shown above, the equations for a hemisphere electrode having a radius a are (Figure 2.46)

$$R_0 = \frac{\rho}{2\pi a} \quad J = \frac{I}{2\pi x^2} \quad E = \frac{\rho I}{2\pi x^2} \tag{2.111}$$

Assuming the soil resistivity is zero within the ionization zone means that the perfectly conducting hemisphere radius has expanded to a radius r as defined by setting $E = E_0$ in Equation 2.111. Replacing a with this new radius, the impulse resistance results in the following equation:

$$R_i = \sqrt{\frac{E_0 \rho}{2\pi I}} \tag{2.112}$$

which can be also expressed as

$$\ln R_i = \frac{1}{2}\left(\ln \frac{E_0 \rho}{2\pi} - \ln I\right) \tag{2.113}$$

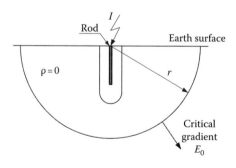

FIGURE 2.45
After soil ionization a rod becomes a hemisphere.

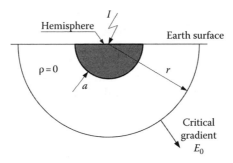

FIGURE 2.46
A hemisphere electrode before and after soil ionization.

This relationship does not exist until there is sufficient current to produce the critical gradient E_0 at the surface of the sphere. To determine this current I_g, set $E = E_0$, and $I = I_g$ in Equation 2.111. Then

$$I_g = \frac{2\pi a^2 E_0}{\rho} = \frac{1}{2\pi} \frac{\rho E_0}{R_0^2}$$ (2.114)

Upon substitution of Equation 2.114 into Equation 2.112, the following expression results:

$$R_i = R_0 \sqrt{\frac{I_g}{I}}$$ (2.115)

The impulse resistance is inversely proportional to the reciprocal of the square root of the current, but on log-to-log plot, the impulse resistance as a function of the current is a straight line, as shown in Figure 2.47. The low-current resistance value R_0 is maintained until the current exceeds I_g, after which the resistance is given by Equation 2.115.

This result may be adapted for ground rods noting that for high currents, rods act as spheres. A simplification for rod electrodes should account for some aspects (i.e., they have the low-current resistance for currents close to zero, they approach the square-root dependence for very high current values and they approximate the log dependence between these two extremes). A possible expression is [74,111]

$$R_i = \frac{R_0}{\sqrt{1 + I/I_g}}$$ (2.116)

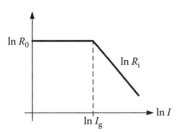

FIGURE 2.47
Impulse resistance of a hemisphere electrode including soil ionization.

As the current increases, a point is reached where the ionized zone is approximately spherical and R_i decreases as illustrated in Figure 2.48. The limit current I_g is given by Equation 2.114, which states that the final decrease of the resistance is determined by the field E_0, the soil resistivity, and the resistance R_0 representing the physical dimensions.

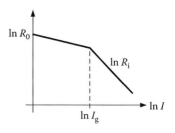

The proposed simplification strictly applies to ground electrodes with small extensions. A maximum effective length of 30 m is estimated. In unfavorable soil conditions in which counterpoises with large extensions are required, the simplification is not accurate enough.

Korsuncev used similarity analysis to relate parameters and variables of compact ground electrodes [93,110,112]. By using D to denote the diameter of the hemisphere, Equation 2.112 may be rearranged as follows:

FIGURE 2.48
Impulse resistance of a rod electrode including soil ionization.

$$\frac{R_i D}{\rho} = \sqrt{\frac{E_0 D^2}{2\pi\rho I}}$$

(2.117)

Using the following notations

$$\Pi_1 = \frac{R_i D}{\rho} \qquad \Pi_2 = \frac{\rho I}{E_0 D^2}$$

(2.118)

Equation 2.117 results in

$$\ln \Pi_1 = \frac{1}{2}\left(\ln\frac{1}{2\pi} - \ln \Pi_2\right)$$

(2.119)

The two variables Π_1 and Π_2 are dimensionless and form the similarity relationship.

The method can be applied to ground electrodes with more complex shapes by using a characteristic dimension. Korsuncev defined this characteristic dimension s as the distance from the geometric center of the electrode on the surface of the earth to the most distant point of the electrode [110].

Oettle suggested an alternative definition [93]:

$$s = \sqrt{h_1^2 + h_2^2 + d^2}$$

(2.120)

where
d is the depth of the electrode from the ground surface
h_1 is the largest horizontal dimension of the electrode
h_2 is the horizontal dimension that lies perpendicular to the largest dimension, see Figure 2.49

Chisholm proposes four relevant dimensions for this analysis [96]:

- A, the surface area in contact with soil
- s, the three-dimensional distance from the center of the electrode to its furthest point

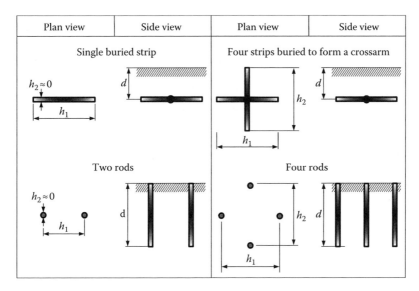

FIGURE 2.49
Characteristic dimensions of ground electrodes. (From Oettle, E.E., *IEEE Trans. Power Deliv.*, 3, 2020, October 1988. With permission.)

- g, the geometric radius of the electrode
- L, the total length of wire

Table 2.10 shows the expressions to be used for these dimensions in some common ground arrangements [96]. See also Table 2.8.

A generalized relationship, based on a large sample of experimental data, has been proposed as a function of two dimensionless parameters, Π_1 and Π_2, for compact electrodes

$$\Pi_1 = \frac{R_i s}{\rho} \qquad \Pi_2 = \frac{\rho I}{E_0 s^2} \qquad \Pi_1 = f(\Pi_2) \tag{2.121}$$

where s is the three-dimensional distance from the center to the furthest point on the electrode.

Figure 2.50 shows the normalized current dependence. It should be noted that the full range of Π_2 has not been obtained by varying the current amplitude for one ground electrode, but by varying the length of the diameter of the electrodes. Thus the data on the left correspond to extended electrodes and that on the right to compact electrodes.

The generalized curve can be divided into two regions: (1) geometry-dependent region and (2) geometry-independent region. The ground electrode in the geometry-dependent region is the low-current power-frequency ground resistance, which can be calculated as detailed above.

Chisholm and Janischewskyj developed a simple equation for the power-frequency ground resistance of compact ground electrodes [81]:

$$\Pi_1^0 = 0.4517 + \frac{1}{2\pi} \ln\left(\frac{s^2}{A}\right) \tag{2.122}$$

TABLE 2.10

Characteristic Dimensions of Ground Electrodes

Electrode	Surface Area	3-D Distance	Geometric Radius	Wire Length
Hemisphere (Radius a)	$A = 2\pi a^2$	$s = a$	$g = \sqrt{3}a$	$L = \infty$
Single vertical rod (Length ℓ, radius a)	$A = 2\pi a\ell$	$s \approx \ell$	$g \approx \ell$	$L = \infty$
Single buried horizontal wire (Burial depth h, length ℓ)	–	–	–	$L = h + \ell$
Ring (Burial depth h, radius a)	–	–	–	$L = 2\pi a$
Disk on surface (Thickness t, radius a)	$A \approx \pi a^2$	$s \approx a$	$g \approx \sqrt{2}a$	$L = \infty$

Source: Chisholm, W.A., *Power Systems*, Chapter 12, Grigsby, L.L., ed., CRC Press, Boca Raton, FL, 2007.

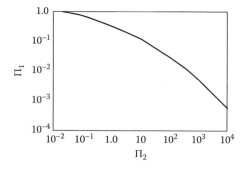

FIGURE 2.50
Similarity relationship of the generalized impulse resistance. (From Popolansky, F., Generalization of model measurements results of impulse characteristics of concentrated earths, *CIGRE Session*, Paris, Paper 33–80, 1980. With permission.)

where
   Π_1^0 is the geometry-dependent unionized part of the Korsuncev curve
   s is the characteristic distance of the ground electrode
   A is its surface area

In the geometry-independent region, soil ionization and the associated nonlinearity of the ground resistance set in. Chisholm and Janischewskyj have proposed a similar equation based on a least-square fit of the Korsuncev curve for Π_2 [81]

$$\Pi_1 = 0.2631\Pi_2^{-0.3082} \tag{2.123}$$

The derivation of this equation was based on equations for calculating the ground electrode resistances some of which are different from those used in this chapter.
   The procedure for calculation of the ionized electrode resistance may be the following one [96]:

1. A value of Π_1^0 is obtained by means of Equation 2.122. This unionized value ranges from 0.159 (for a hemisphere) to 1.07 for a 3 m long, 0.01 m radius cylinder.
2. A value of Π_2 is calculated from Equation 2.121.

3. A value of Π_1 is computed using Equation 2.123. This value represents the fully ionized sphere with the gradient of E_0 at the injected current.

4. If the value of Π_1 is not greater than Π_1^0, then there is not enough current to ionize the grounding, and the unionized value Π_1^0 is used in Equation 2.121 to obtain the resistance. Otherwise, the ionized value from Equation 2.123 is used. Ionization caused by high lightning currents can significantly reduce the low-current resistance.

Equation 2.123 covers a small region of the Π_1–Π_2 plot. To cover the entire region, the plot can be piecewise linearized as follows [95]:

$$
\begin{aligned}
&\text{For } \Pi_2 \leq 5 &&\Pi_1 = 0.2965\Pi_2^{-0.2867} \\[2mm]
&\text{For } 5 \leq \Pi_2 \leq 50 &&\Pi_1 = 0.4602\Pi_2^{-0.6009} \\[2mm]
&\text{For } 50 \leq \Pi_2 \leq 500 &&\Pi_1 = 0.9534\Pi_2^{-0.7536} \\[2mm]
&\text{For } 500 \leq \Pi_2 &&\Pi_1 = 1.8862\Pi_2^{-0.8693}
\end{aligned}
\tag{2.124}
$$

With expanded electrodes, ionization effects will tend to reduce the contact resistance without altering the surface area A or the characteristic dimension s, which tend to reduce the influence of ionization.

Some aspects must be kept in mind when using this approach to obtain the impulse impedance of concentrated ground electrodes:

- The Korsuncev curve suggests that all ground electrodes act like hemispheres at high values of Π_2. This could be due to the fact that all results have been obtained for electrodes with limited dimensions, which may not represent all practical ground electrodes.

- The Korsuncev curve does not show any dependence on the rate of rise of the impulse current. It is well known that the soil has an ionization time delay which can offset the results for steep-front lightning currents [98].

In addition, the critical breakdown voltage gradient of soil needs to be known accurately; E_0 is difficult to define, and can be quite different in the field from that measured in the laboratory. E_0 in the laboratory is calculated when the soil between the electrodes breaks down completely. In the field, the soil breakdown is incomplete. Even in the laboratory, E_0 would be a function of the soil resistivity and the electrode configuration.

2.5.6 Grounding Design

A complete line design will specify the types and sizes of ground electrodes needed to achieve the required grounding impedance. The electrode sizes and shapes will depend on the range of soil conductivities. The inputs to the design process are the prospective lightning surge current, the soil resistivity, and the dimensions of either the tower base, or a trial electrode. In addition, dimensioning of ground electrodes should be made taking into account other considerations such as corrosion and mechanical strength, thermal strength, or human safety [88].

As a first approach, the low-frequency geometric resistance of the electrode is computed following the procedure proposed in Section 2.5.5. If the electrode is compact (e.g., a driven rod), the ionization effects should be included, which will reduce resistance under high-current conditions. If the electrode is extended (e.g., a counterpoise), ionization will also reduce the contact resistance, but this effect can be neglected and the resistance can be obtained from electrode geometry.

For adequate transient performance, it is recommendable to select electrode sizes that do not rely on ionization. This can be achieved by using the following process [96]:

- Obtain the shape coefficient of the electrode Π_1^0 from Equation 2.122.
- Establish the value of Π_2^0 that will just cause ionization.

$$\Pi_2^0 = 0.0131 \left(\Pi_1^0 \right)^{-3.24} \tag{2.125}$$

- The dimension of the electrode s needed to prevent ionization with current I and voltage gradient E_0 in soil resistivity ρ is given by

$$s = \sqrt{\frac{\rho I}{E_0 \Pi_2^0}} \tag{2.126}$$

The fraction of surge current absorbed by a given electrode will change as its size is adjusted, making the design process iterative.

Some practical recommendations where a low impedance ground is needed to limit lightning-caused overvoltages in tall structures (e.g., transmission towers) are summarized in the following paragraphs [3].

- *Ground Rods.* For a constant-resistivity soil, the driven depth of rods should be from 2 to 6 m. Since soil resistivity may vary with depth, rods are frequently driven to greater depths. Multiple rods decrease the resistance, but due to mutual effects between the rods the benefit decreases as more rods are added. Thus three to five rods, spaced 3 or more meters apart, is a usual limit. The diameter of the rod is not an influential factor, and any mechanically suitable diameter is acceptable from an electrical viewpoint.

- *Counterpoises.* For a constant soil resistivity, counterpoise length should be limited to about 50 m. Additional counterpoises decrease the resistance, but spacings should be in the range of about 10 m. The depth of burial is usually about 1 m.

- For mechanical and electrical reasons, the minimum cross sections shall be copper = 16 mm^2 and steel = 50 mm^2 [88]. Aluminum is not recommended since this material can vanish in a few years [3].

- Ground rods are generally employed for low-resistivity soils while counterpoises are generally used in high-resistivity soils.

- Surface arcing may transfer lightning currents to unprotected facilities. Therefore, the size of ground electrodes for lightning protection must be selected with sufficient area and multiplicity to limit ionization, mainly in areas where transferred lightning potentials could be dangerous to adjacent objects or systems. For a vertically driven rod, the final ionized diameter D can be derived from Equation 2.111 and results

$$D = \sqrt{\frac{2\rho I}{\pi E_0}} \qquad (2.127)$$

Ionized diameters can range from 5 to 10 m. To obtain the maximum effectiveness of parallel rods, spacings should be increased to approximately 5 m.

2.6 Transmission Line Insulation

2.6.1 Introduction

Power line insulation is of external type (i.e., exposed to atmospheric conditions) and self-restoring (i.e., it recovers its insulating properties after a disruptive discharge), and it must be selected to meet the stresses produced by power-frequency voltage and any type of overvoltage.

Insulation strength is expressed in terms of withstand voltage, a quantity determined by tests conducted under specified conditions with specified waveshapes. The same insulation may have different withstand voltages for different voltage waveshapes; that is, insulation withstand strength depends greatly on the waveshape of the applied voltage. If the value of withstand strength measured during tests is to be a fair measure of the insulation behavior, the waveshape of the test voltage should well represent the waveshape of the actual stress on the insulation. However, overvoltages may have a wide range of waveshapes, so rather than attempting to determine the withstand strength for each stress by test, it is common practice to assign a specific testing waveshape and duration of its application to each category of overvoltage.

A plot of withstand strength versus waveshape is known as a volt–time curve. The parameter of waveshape generally used is the so-called front time of the testing impulse. Such a curve shows only the withstand strength of the insulation when exposed to standard laboratory-generated test voltages.

Power line insulation must be chosen after a careful study of both transient and power-frequency voltage stresses and the corresponding strengths of each insulation element. Transient stresses may be divided into lightning and switching surges. As a rule of thumb, one can consider that lightning overvoltages dictate the insulation level at distribution and subtransmission levels, and switching overvoltages control the design at transmission levels. Nevertheless, it is advisable to consider both types of transients.

Power line insulation must be sufficient to ensure a reliable operation; however, a design that assumes no insulation failures is extremely conservative and expensive. In addition, insulation design for transient voltages is complicated by the great variety of transient and meteorological conditions that an overhead line experiences, and by statistical fluctuations in the insulation strength itself. Transients appear in all shapes and with different amplitudes. Each type of transient may occur with any meteorological condition, which affects the flashover strength differently. The larger transients occur more rarely than the smaller ones, and the greatest possible transients may have an extremely low probability of occurring in actual service. Therefore, a proper balance must be reached between the reliability and the cost of insulation. Sophisticated design methods may assess the relative frequencies of occurrence of all the combinations of electrical and meteorological events, evaluate how overall performance would change as the choice of insulation is changed, and determine how additional insulation investment would affect the outage rate. These methods explore the problem in detail, but they require a systematic evaluation of meteorological and transient voltage conditions.

This section covers the aspects needed to understand how power line insulation behaves under any type of stress, waveshape, and meteorological condition. It presents a short description of the phenomena that occur during breakdown caused by either switching or lightning stresses, a summary of mathematical models aimed at representing the behavior of power line insulation, and procedures to obtain the dielectric strength for power-frequency voltage, switching and lightning impulses, at any atmospheric condition. A summary of standard concepts needed to describe external and self-restoring insulation is also presented.

The section is mostly based on Refs. [3,113,114]. For more details and further information, readers are encouraged to consult these references. Other basic references related to the main topics of this chapter are [1,2,74,115].

2.6.2 Definitions

2.6.2.1 Standard Waveshapes

The standard lightning and switching impulse waveshapes are described by their time-to-crest and their time-to-half value measured on the tail, see Figure 2.51.

- Lightning impulse waveshape: The (virtual) time-to-crest or front time, t_f, can be defined by the equation

$$t_f = 1.67 \left(t_{90} - t_{30} \right)$$

(2.128)

where
t_{90} is the actual time to 90% of the crest voltage
t_{30} is the actual time to 30% of crest voltage

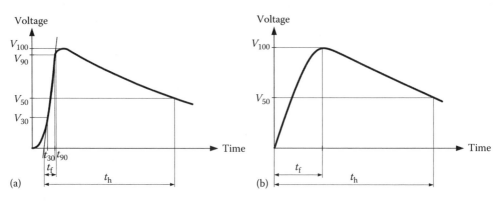

FIGURE 2.51
Standard impulse waveshapes. (a) Lightning impulse waveshape, (b) switching impulse waveshape.

The time-to-half value or tail time, t_h, is the time between the virtual origin (the point at which the line between t_{30} and t_{90} intersects the zero voltage) and the point at which the voltage decreases to 50% of the crest value. In general, the wave-shape is denoted as a t_f/t_h impulse. The standard lightning impulse waveshape is 1.2/50 μs.

- Switching impulse waveshape: The time-to-crest or front time is measured from the actual time zero to the actual crest of the impulse, while the tail is defined as the time-to-half value, and is also measured from the actual time zero. The wave-shape is denoted in the same manner as for the lightning impulse. The standard switching impulse waveshape is 250/2500 μs.

2.6.2.2 Basic Impulse Insulation Levels

- Basic lightning impulse insulation level (BIL) [116]: It is the electrical strength expressed in terms of the crest value of the standard lightning impulse, at standard dry atmospheric conditions. The BIL may be either statistical or conventional. The "statistical BIL" is the crest value of standard lightning impulse for which the insulation exhibits a 90% probability of withstand; that is, a 10% probability of failure. The "conventional BIL" is the crest value of a standard lightning impulse for which the insulation does not exhibit disruptive discharge when subjected to a specific number of applications of this impulse. The statistical BIL is applicable only to self-restoring insulations, whereas the conventional BIL is applicable to non-self-restoring insulations. In IEC Std 60071 [117], the BIL is defined in the same way but known as the lightning impulse withstand voltage.
- Basic switching impulse insulation level (BSL) [116]: It is the electrical strength expressed in terms of the crest value of a standard switching impulse, at standard wet atmospheric conditions. The BSL may be either statistical or conventional. As with the BIL, the statistical BSL is applicable only to self-restoring insulations, while the conventional BSL is applicable to non-self-restoring insulations. The statistical BSL is the crest value of a standard switching impulse for which the insulation exhibits a 90% probability of withstand, a 10% probability of failure. The conventional BSL is the crest value of a standard switching impulse for which the insulation does not exhibit disruptive discharge when subjected to a specific number of applications of this impulse. In IEC Std 60071 [117], the BSL is called the switching impulse withstand voltage and the definition is the same.

IEEE and IEC standards recommend values for both BILs and BSLs that equipment manufacturers are encouraged to use [116,117].

2.6.2.3 Statistical/Conventional Insulation Levels

In general, the strength characteristic of self-restoring insulation may be represented by a cumulative normal of Gaussian distribution [1,3]. Under this assumption, the probability of flashover for a specified voltage V is given by the following expression:

$$F(V) = \frac{1}{\sqrt{2\pi}\sigma_f} \int_{-\infty}^{V} e^{-\frac{1}{2}\left(\frac{V-\mu}{\sigma_f}\right)^2} dV \tag{2.129}$$

where
  μ is the mean value
  σ_f is the standard deviation

In a normalized form, this expression can be written as follows:

$$F(V) = \frac{1}{\sqrt{2\pi}} \int_{-\infty}^{Z} e^{-\frac{Z^2}{2}} \, dV \tag{2.130}$$

where

$$Z = \frac{V - \mu}{\sigma_f} \tag{2.131}$$

The mean of this distribution is known as the critical flashover voltage or CFO (in IEC it is known as the U_{50} or 50% flashover voltage). The insulation exhibits a 50% probability of flashover when the CFO is applied; that is, half the impulses flashover. From the definition of BIL and BSL (at the 10% point) the following equations result:

$$BIL = CFO\left(1 - 1.28 \frac{\sigma_f}{CFO}\right) \tag{2.132a}$$

$$BSL = CFO\left(1 - 1.28 \frac{\sigma_f}{CFO}\right) \tag{2.132b}$$

where σ in per unit of the CFO is called the coefficient of variation.

Assuming a normal distribution, insulation strength is fully specified by providing the values of CFO and σ_f. An alternate method is to provide the values CFO and σ_f/CFO.

The assumption of a normal or Gaussian distribution is adequate for most practical applications, but some researches have shown that the probability distributions can sometimes be far from normal (Gaussian) [118]. In addition, the Gaussian distribution is unbounded to the right and the left; that is, it is defined between $+\infty$ and $-\infty$. A limit of $-\infty$ indicates that there exists a probability of flashover for a voltage equal to zero, which is physically impossible. Since the Gaussian distribution is valid to at least four standard deviations below the CFO, it is reasonable to believe that there exists a nonzero voltage for which the probability of flashover is zero. A distribution with this property is the Weibull whose cumulative distribution function is

$$F(V) = 1 - e^{-\left(\frac{V - \alpha_0}{\alpha}\right)^{\beta}} \qquad \infty > V > \alpha_0 \tag{2.133}$$

In IEC 60071, the value of β is rounded to 5, and the distribution is approximated by the following expression [2]:

$$F(V) = 1 - e^{-(\ln 2)\left(\frac{Z}{4} + 1\right)^5} = 1 - 0.5^{\left(\frac{Z}{4} + 1\right)^5} \tag{2.134}$$

where Z is defined as in Equation 2.131, with $\mu = $ CFO.

2.6.3   Fundamentals of Discharge Mechanisms

2.6.3.1   *Description of the Phenomena*

The behavior of air insulation depends on various factors, such as the type and polarity of the applied voltage, electric field distribution, gap length, and atmospheric conditions. The dielectric breakdown of a gas is characterized by the formation of free charges of opposite sign due to ionization of molecules by collision with free electrons accelerated by the electric field. This phenomenon develops in successive phases.

1. The first phase is the first corona, which develops in the region of high electric field in the proximity of the electrodes, in the form of thin filaments, called streamers, propagating only for one part of the gap.
2. The second phase is the leader phase, characterized by the formation and the elongation of a channel, more ionized than a streamer, which propagates with leader corona development from its tip. Depending on the value and shape of the applied voltage and to the gap length, the leader can either stop or reach the opposite electrode.
3. Once either first corona streamers or leader corona streamers have reached the opposite electrode, the third phase develops. The leader channel elongates at increasing velocity bridging the whole gap. At this point the channel becomes highly ionized and the electrodes are short circuited.

In the case of rod–rod geometry, the phenomenon develops from both electrodes, and the last phase occurs when filaments of opposite polarity meet inside the gap.

The first corona phase and the leader phase strongly depend on the polarity of the applied voltage. In almost all practical geometries, the breakdown voltage with positive polarity is lower than that with negative polarity, and the minimum breakdown voltage, with the same gap length and impulse shape, belongs to the case of rod–plane gap (i.e., a geometry in which only one region of the gap is stressed).

The different phases detailed above are clearly defined and detectable in the case of impulse voltages of long-duration front. By reducing the time-to-crest, some overlapping of the various phases can occur, and some phases can be absent. In the case of lightning impulses, because of the fast decrease of the applied voltage during the tail, the leader cannot develop as in the case of switching impulses. To reach breakdown, the voltage has to be increased very much. Due to the very high rate of rise of the voltage during the front, first corona is almost immediately followed by a succession of pulses which launch streamers into the gap. Before a leader has time to develop significantly, the streamers reach the opposite electrode, causing a condition in the gap similar to that occurring at the final jump; that is, there is a leader-propagation phase. Breakdown voltage corresponds to the average field along the streamer zone multiplied by the gap length. In the case of short-front impulse voltage, the influence of the tail duration may be very important.

2.6.3.2   *Physical–Mathematical Models*

They do not describe physical phenomena, although they are based on the characteristics of the various phases of the discharge mechanism. Since the influence of these phases depends on the electrode geometry and voltage-impulse shape, a good modeling of the dominant phase is of primary importance. Some approaches developed to represent the behavior of air insulation under lightning and switching impulses are summarized in the following.

1. Lightning impulse strength models

The wide variety of lightning stroke characteristics, together with the modification effects that the line components introduce, stresses line insulation with a diversity of impulse-voltage waveshapes. While insulation coordination is based on impulse strength determined for standard impulse voltages (see Section 2.6.2), it is important to be able to evaluate insulation performance when stressed by nonstandard lightning impulses. The following paragraphs summarize the main characteristics of these models and provide recommendations for the model to be used. The proposed approaches for predicting the dielectric strength of air gaps under lightning overvoltages usually allow the calculation of both the minimum breakdown voltage and the time-to-breakdown.

A. Voltage–time curves

They give the dependence of the peak voltage of the specific impulse shape on the time-to-breakdown, see Figure 2.52. Volt-time (or time-lag) curves are determined experimentally for a specific gap or for an insulator string, and may be represented with empirical equations, applicable only within the range of parameters covered experimentally [119]. In practice, measurements can be affected by several factors: impulse front shape, front times of the applied standard lightning impulse, gap distance and gap geometry, polarity, and internal impedance of the impulse generator (due to the predischarge currents in the gap).

There are special cases when the use of these curves can be advantageous. The IEEE simplified backflash method takes advantage of the insensitivity of the time-to-breakdown to waveshapes of some insulator strings by performing overvoltage analysis at a fixed time of 2 μs [84]. At times-to-breakdown shorter than 3 μs, the nonstandard impulse strength can be more that 10% higher than the standard impulse strength. Therefore, more

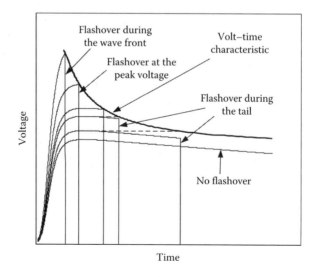

FIGURE 2.52
Volt–time characteristic.

representative descriptions must be used for this type of application. The standard volt–time curves do not apply to multiple flashover studies and their accuracy might be poor when long times-to-breakdown or low probability flashovers (i.e., taking place on the tail of the surge) are being studied.

B. Integration methods
Their aim is to predict insulation performance as a function of one or more significant parameters of the nonstandard voltage waveshape [120–128]. Their common basic assumptions are that

- There is a minimum voltage V_0 that must be exceeded before any breakdown can start or continue.
- The time-to-breakdown is a function of both the magnitude and the time duration of the applied voltage above the minimum voltage V_0.
- There exists a unique set of constants associated with breakdown for each insulator configuration.

In the most general formulation, different weights can be given to the effects of voltage magnitude and time. The dielectric breakdown of the insulation is obtained then from the following equation:

$$D = \int_{t_0}^{t_c} \left(v(t) - V_0 \right)^n \mathrm{d}t \tag{2.135}$$

where
t_0 is the time after which the voltage $v(t)$ is higher than the required minimum voltage V_0 (also known as reference voltage)
t_c is the time-to-breakdown

The constant D is known as disruptive effect constant.
Different values for V_0, n, and D have been proposed, but each proposal refers to a particular set of results. If $n = 1$, the method is know as the equal-area law, see Figure 2.53.

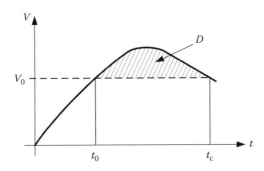

FIGURE 2.53
Integration method.

Although easy to use, these methods can be applied to specific geometries and voltage shapes only. This law has been applied in several typical gap configurations of overhead lines and substations, and fits the test results better than other simplifications; however, for nonuniform gaps, the value of V_0 that gives the best conformity between test results and the equal-area law does not coincide with results derived from measurements.

C. Physical models

They consider the different phases and their dependence on the applied voltage and compute the time-to-breakdown as the time that is necessary for completion of all the phases of the discharge process [74,127,129–131].

As detailed above, when the applied voltage exceeds the corona inception voltage, streamers propagate and cross the gap after a certain time if the voltage remains high enough. The streamer propagation is accompanied by current impulses of significant magnitude. Only when the streamers have crossed the gap, the leaders can start their development. Usually the leader velocity increases exponentially. Both the steamer and the leader phase can develop from only one or from both electrodes. When the leader has crossed the gap, or when the two leaders meet, the breakdown occurs.

Most studies have been performed with double-exponential type impulses. Although the presence of oscillations or other abrupt changes in the applied voltage does not change the overall behavior of the breakdown process, they can disturb the leader propagation, causing discontinuous breakdown development, as the voltage can fall below that necessary for leader propagation; the result is a discharge where the leader propagates and stops at each cycle, as modulated by the presence of the oscillations.

The time-to-breakdown can be expressed as the sum of three components:

$$t = t_i + t_s + t_\ell \tag{2.136}$$

where

t_i is the corona inception voltage

t_s is the time the streamers need to cross the gap or to meet the streamers from the opposite electrode

t_ℓ is the leader propagation time

- Corona phase: Corona inception occurs at time t_i when the applied voltage has reached a convenient value, which depends on electrode geometry, gap distance, and rate of rise of the applied voltage. This voltage can be computed and t_i derived if the impulse shape of the voltage is known. In the case of gaps with nonuniform field distribution (most of the practical air insulations), the inception voltage is below the breakdown voltage, and this time can be usually neglected without introducing large errors.

- Streamer propagation phase: It starts at corona inception and is completed when the gap is fully crossed by streamers. The corresponding time depends on the applied voltage; the minimum value necessary for streamers to cover the whole gap is not far from the CFO when standard lightning impulses are applied. At this voltage, the time the streamers need to cross the gap is maximum, and it decreases as voltage is raised. This time is almost independent of voltage polarity, electrode configuration, and gap clearance, but it is strongly dependent on the ratio E/E_{50}, where E is the average field in the gap at the applied voltage V and E_{50} is the average field at CFO. In almost all the proposed models, it is, however, assumed that the streamer phase is completed when the applied voltage has reached a value which gives an average field in the gap equal to E_{50}. The duration of the streamer phase, t_s, can be estimated as follows [74]:

$$\frac{1}{t_s} = 1.25\frac{E}{E_{50}} - 0.95 \tag{2.137}$$

For geometries different from the rod–plane configuration, where two streamers are present, the resulting velocity of the streamer has to be interpreted as the average velocity of an equivalent streamer that would cross over the same gap in the same time but starting from one electrode only.

• Leader phase: The length of the leader at the end of t_s is a small percentage of the total gap length. The propagation of the leader is described in terms of the instantaneous values of its velocity. The leader propagation time is generally calculated from its velocity, which depends on the applied voltage and the leader length. Various expressions have been proposed for the leader velocity. The experimental evidence shows that the leader velocity increases proportionally to the length of the gap which is not yet covered by the leader, and can be described by the following equation:

$$\frac{d\ell_\ell}{dt} = g\left[v(t), d_g\right]\left(\frac{v(t)}{d_g - \ell_\ell} - V_0\right) \tag{2.138}$$

where
g and V_0 are some functions
d_g is the gap length
ℓ_ℓ is the leader length
$v(t)$ is the actual (absolute value) voltage in the gap, which does not generally correspond to the theoretical value, due to the current flowing into the circuit.

For all configurations, the following equation has been proposed for calculation of the leader velocity [130]

$$\frac{d\ell_\ell}{dt} = 170 d_g\left[\frac{v(t)}{d_g - \ell_\ell} - E_0\right]e^{\frac{0.0015v(t)}{d_g}} \tag{2.139}$$

A simpler approach had been proposed in [131]

$$\frac{d\ell}{dt} = kv(t)\left[\frac{v(t)}{d_g - \ell_\ell} - E_0\right] \tag{2.140}$$

The constants k and E_0 have been found to be dependent of the gap configuration and insulator type, see Table 2.11.

• Leader current: During the discharge development a current flows in the circuit so that the voltage applied at the gap is not the voltage from the unloaded generating

TABLE 2.11

Values for Factors k and E_0

Configuration	Polarity	$k(m^2/V^2/s)$	$E_0(kV/m)$
Air gaps, post and	+	$0.8 \cdot 10^{-6}$	600
long-rod insulators	−	$1.0 \cdot 10^{-6}$	670
Cap and pin insulators	+	$1.2 \cdot 10^{-6}$	520
	−	$1.3 \cdot 10^{-6}$	600

Source: CIGRE WG 33-01, Guide to Procedures for Estimating the Lightning Performance of Transmission Lines, CIGRE Brochure no. 63, 1991. With permission.

circuit. The value $v(t)$ has to be the actual voltage applied to the gap, due to the voltage drop caused by current flowing during leader propagation. The actual applied voltage can be determined when the circuit characteristics and the current flowing during the leader propagation are known.

The physical model is valid for a large variety of impulse shapes and can be used in the evaluation of dielectric strength of a variety of geometries. The *leader progression model* (LPM) shown in Equation 2.140 has proved to have adequate accuracy for most calculations. The integration methods have comparable accuracies but more restricted application in relation to waveshapes. The empirical methods can give a good accuracy when they are used within their validity limits (i.e., when specific data are used for a specific insulator or gap, together with a careful application of the model). The use of volt–time curves works well in the short time-to-breakdown domain (2–6 μs). Based on the above, it can be concluded that no single approach alone can be recommended for all applications.

2. Switching impulse strength models

In the case of switching impulses, leader inception and propagation are the most important phases for the assessment of the gap strength. Although it is a prerequisite to leader inception, first corona has little influence on the gap strength; on the other hand, breakdown cannot be avoided once the final jump phase is reached. Several models aimed at calculating the critical strength of simple configurations have been proposed to account for leader inception and propagation on long air gaps under positive switching impulses. Most works deal with impulses having critical time-to-crest, and with rod–plane and conductor–plane gaps. Few models extend to other configurations such as rod–rod, conductor–rod, and conductor–tower leg. A thorough review of switching impulse strength models was presented in the CIGRE Brochure 72 [114]. Unlike lightning impulse strength models, these models are not usually part of calculations performed in transmission line design. Readers interested in this subject are referred to the CIGRE publication.

2.6.4 Dielectric Strength for Switching Surges

2.6.4.1 Introduction

There are some important reasons for the analysis of switching surges in transmission line design. Surge waveshapes may fall in the region of times-to-crest, which, for positive polarity, are critical for the insulation strength of typical line air gaps; consequently, the strength can be lower than for lightning impulses and for power-frequency voltages. On the other hand, the increase in switching surge strength with gap spacing is less than proportional, whereas the increase in lightning-impulse strength is linear.

The variables that can affect the switching surge strength can be classified as follows [132]:

- Geometrical: The strike distances (i.e., air clearances); the insulator arrangement (V-string, I-string, or strain string); the size and configuration of the tower; I-string swing angle; the phase location (center or outer); number of insulator units.
- Electrical: Waveshape.
- Meteorological: Relative air density; humidity (relative and absolute); precipitation.

A summary of the influence that several factors have on the switching-impulse strength of overhead lines is summarized in the following paragraphs:

- Waveshape: Switching surge waveshapes are practically infinite in their variety. The standard waveshape used to simulate switching surges is a double exponential wave, characterized by the *time-to-crest* and the *time-to-half value*, see Figure 2.51. The insulation strength is a function of these parameters, particularly the time-to-crest. Good agreement of data obtained for different waveshapes is achieved by using a criterion that accounts for the differences in the shapes of the impulses. This criterion is based on the analysis of the time interval defined as the active part of the impulse front, i.e., the part of the impulse when the flashover process essentially takes place. Such an approach is valid when a flashover occurs before or near the crest of a switching impulse. Tests have been made to define the portion of the wave that is important for the insulation strength. The inception and propagation of the leader occurs at voltages between 60%–75% and 100% of the crest value of impulses corresponding to a 50% probability of flashover. It is generally agreed that the waveshape below 70% of the crest value does not influence the value of the flashover voltage. Thus, the active portion of the waveform was defined as that between the 70% point and the crest of the impulse. Two waves having the same time for the active portion produce practically the same results, even if their times-to-crest are different.

  Waves representative of switching surges encountered on electric power systems have times-to-crest ranging from 50 to 2000 μs or more [133,134]. The flashover voltage of a gap depends on the rate of rise of voltage and on the electrode spacing. These dependences are evident in the existence of the so-called U-curves. For fixed electrode spacing, the flashover voltage, plotted as a function of impulse time-to-crest, passes through a minimum; that is, there exists a wave front that produces minimum insulation strength, the critical wave front (CWF), also known as critical time-to-crest. There is a different U-curve for each electrode spacing, with the minimum occurring at larger times-to-crest as the spacing increases. Examples for rod–plane gaps are shown in Figure 2.54.

  For waves shorter than critical, flashovers generally occur after crest; therefore, the insulation strength is partially influenced also by the time-to-half value.
- Polarity: The switching-surge flashover strength is greatly affected by the polarity of the surge. For most practical configurations, the grounded electrode is either the ground plane, or massive structures, or a combination of the two. In general, the dimensions of the energized electrode are much smaller, so the electric field

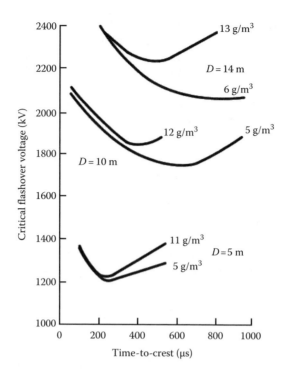

FIGURE 2.54
Impulse voltages of various times-to-crest applied to rod–plane gaps. (From Büsch, W., *IEEE Trans. Power Apparatus Syst.*, 97, 2086, November/December 1978. With permission.)

in the gap between the electrodes is very nonuniform with the highest intensity at the energized electrode. For most practical configurations, the positive-polarity switching-surge strength is lower than the negative-polarity strength because of the particular mechanism of switching-surge flashover. The more nonuniform the field is, the more pronounced the polarity effect. Phase-to-ground insulation of lines and stations designed to withstand positive-polarity surges is minimally affected by negative-polarity surges. Although positive- and negative-polarity surges occur statistically with the same frequency and amplitude, it is common practice in the design of phase-to-ground insulation to disregard switching surges of negative polarity.

- Geometry: The breakdown process in large gaps is greatly influenced by the gap geometry. If the line electrode is of small dimensions, the presence of large ground planes is particularly harmful and reduces flashover strength on positive-polarity switching surges. Consequently, the physical size of a tower can adversely influence the breakdown strength of the insulation, particularly if the tower approximates a ground plane. Corona shields and bundle conductor configurations also affect insulator flashover strength, sometimes lowering the flashover strength while improving the radio-noise performance.

- Insulator length: For dry conditions, as the insulator length increases, the CFO increases until the insulator length is equal to the strike distance. If the insulator length is less than the strike distance, flashovers will occur across the insulators, and thus the insulator string limits the tower strength. But if the strike distance is less than the insulator length, flashovers will occur across the air strike distance, and the strike distance is limiting. For wet conditions, this saturation point increases to a level where the insulator length is 1.05–1.10 times the strike distance. To obtain the maximum CFO within a tower window, or for a fixed strike distance, the insulator length should be 5% to 10% greater than the strike distance. The strike distance S to be used is the shortest of the three distances (see Figure 2.55): S_1 to the upper truss, S_2 to the tower side, and $S_I/1.05$, where S_I is the insulator string length. For practical designs, the insulator string length is the control variable.

Wet conditions degrade insulators' CFO more than that of the air. While dry tests show consistent results, tests under wet conditions are extremely variable. For insulators in vertical position, water cascades down them too much. Only when the string is moved more than 20° from the vertical position (a V-string is normally at a 45° angle) water drips off each insulator and test results become consistent. I-strings do not have lower insulation strength than V-strings; however, the CFO of I-strings is difficult to measure for practical rain conditions.

- Phase position: The CFO of the outside phase with V-string insulator strings is usually about 8% greater than that of the center phase, since there exist only one tower side [136–138].

- Meteorology: The main weather parameters for switching surges are relative air density, absolute humidity, rain, and wind. Air density and humidity may change insulator-flashover strength, and correction curves must be provided for both (see Section 2.6.7). Rain reduces the switching-surge strength of insulator strings, but it does not affect the strength of air gaps. For an insulator length/strike distance ratio of 1.05–1.10, which represents the design condition, wet conditions decrease the dry CFO. For application purposes, a reduction value of 4% is suggested.

The rest of this section is dedicated to present a general procedure aimed at estimating the dielectric strength of power line insulation under switching impulses and analyzing the various factors that affect its behavior. The section also includes a discussion about the characterization of phase-to-phase strength. The first subsection introduces the concept of gap factor.

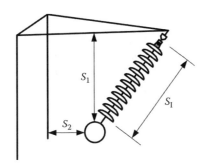

FIGURE 2.55
Outside arm strike distances.

2.6.4.2 Switching Impulse Strength

A practical design procedure of the switching-surge strength of various tower and line configurations for a variety of geometrical, electrical, and meteorological conditions must be based on a set of standard parameters for the gap in question.

A two-step procedure may be applied: (1) obtain a CFO versus gap distance curve, and (2) supply auxiliary correction factors, curves, or formulae that describe how strength has to be corrected to the standard parameters, as parameters are varied in their range.

The insulation strength characteristic is of random nature and may be approximated by a cumulative normal (Gaussian) distribution characterized by a mean (the CFO) and a standard deviation (σ_f). The standard deviation is usually given in per unit or percentage of the CFO, and is known as the coefficient of variation. The value of the standard deviation depends on waveshape, polarity, geometry, and weather conditions. The standard deviation in per unit of the CFO (σ_f/CFO) tends to increase with increased strike distance. For nonuniform gaps and waveshapes having critical time-to-crest, it averages about 4%–5% of CFO, independently of the voltage level. For practical purposes, the lower limit of the insulation strength is fixed at about 4 standard deviations below the CFO. An average value of 5% is normally used for both wet and dry conditions. For dry conditions, the average value of σ_f/CFO is 4.3% and for wet conditions, it is 4.9%; σ_f/CFO also varies with the wave front [139], corresponding the CWF to a wave front of about 400 µs.

Corrections to be considered in the estimation of the dielectric strength at nonstandard atmospheric conditions are analyzed in Section 2.6.7.

The flashover voltage of an air gap depends on the spatial extent of first corona, which depends on the electric field in the vicinity of the electrodes, and hence on their geometry. Paris and Cortina proposed a general approach to the estimation of the positive polarity CFO for different gap configurations [140], based on the fact that CFO curves versus gap spacing had basically the same shape. They also noted that the rod–plane gap has the lowest CFO.

The gap factor is defined as the ratio of the flashover voltage for a given electrode configuration to that for the rod–plane gap, which is taken as a reference. Paris and Cortina proposed the following general equation for positive polarity and dry conditions:

$$\text{CFO} = 500 k_g S^{0.6} \tag{2.141}$$

where
S is the gap spacing or strike distance in meters with the CFO in kV
The parameter k_g is the gap factor and is equal to 1.0 for a rod–plane gap

For other gap configurations, this factor increases to a maximum of about 1.9 for a conductor-to-rod gap. Since Paris and Cortina used a 250 µs front, which is not the CWF for all strike distances, Equation 2.141 is not applicable to the minimum strength. Paris et al. [141] proposed the gap factors shown in Table 2.12; see also [3,137].

Gallet et al. developed an alternate expression of the CFO for the CWF [142]:

$$\text{CFO} = k_g \frac{3400}{1 + (8/S)} \tag{2.142}$$

which is only valid for positive polarity and is normally applied only for dry conditions. To use this equation the gap factors shown in Table 2.12 should be applied. Using

Equation 2.142, the CFO for a rod–plane gap approaches 3400 kV as S approaches infinity, which is not correct. In general, this equation is assumed valid for gap spacing in the range of about 15 m.

With further study, it became obvious that the gap factor was not simply a specific number but did vary with the specific parameters of the gap configuration.

Other equations for rod–plane gaps with a different range of validity have been proposed, see for instance [143].

TABLE 2.12

Gap Factors

Electrode Configuration	Diagram	Paris et al. [137,141]	Gallet [12]
Rod–plane	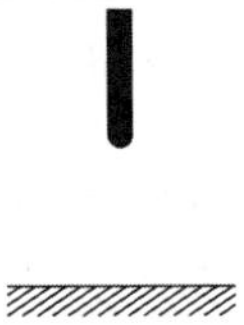	1.00	1.00
Rod–rod	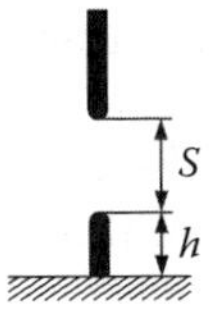	1.30 ($h = 6$)	$1 + 0.6\dfrac{h}{h+S}$ or $\dfrac{h}{h+S}e^{0.5}$
Conductor–plane	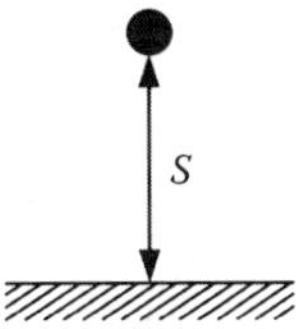	1.15	1.10
Conductor–rod	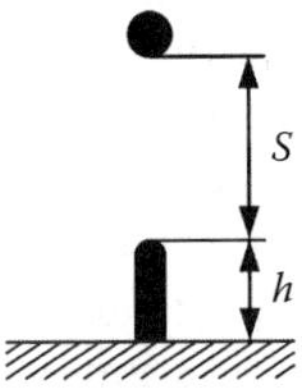	1.65 ($h = 3$ m) 1.90 ($h = 6$ m)	$1.1 + 1.4\left(\dfrac{h}{h+S}\right)^{1.62}$ or $1.1\dfrac{h}{h+S}e^{0.7}$
Conductor–structure	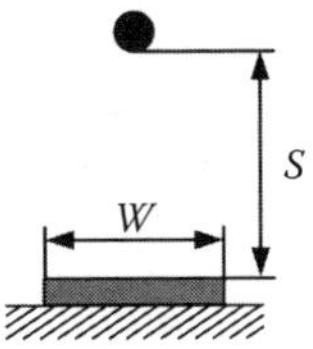	1.30	$1.1 + \dfrac{0.3}{1+\dfrac{W}{S}}$
Rod–structure	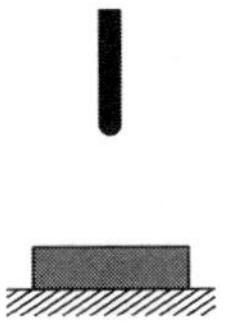	1.05	1.05

For rod–plane gaps up to 25 m, IEC 6007-2 recommends obtaining the positive polarity CFO by means of the following expression [2]:

$$CFO = 1080k_g \ln(0.46S + 1)$$

(2.143)

According to IEC 60071-2, the following expression provides a better approximation for standard switching impulses [2]:

$$CFO = 500k_g S^{0.6}$$

(2.144)

Both expressions are applicable at sea level; CFO is in kV crest and gap spacing in m.

In general, these approaches give similar results with practical gap spacing. The gap factors of some actual configurations are shown in Table 2.13, which includes the diagram of the configuration, the expression to be used for calculation of the gap factor, the range of applicability of this expression, and typical values.

Breakdown depends on a combined effect of time-to-crest of voltage and electric field distribution. The U-curve (see Figure 2.54) of any electrode system and that of the rod–plane gap are not actually related by a simple proportionality, since the gap factor varies also with the waveshape. Experimental results have shown significant variations with time-to-crest over a range of switching impulses [144]. For a vertical rod–rod gap, the gap factor depends on the height of the tip of the grounded rod above ground [145]. The gap factor increases with the height of the gap above ground as a consequence of the effect of the ground plane on the electric field distribution in the gap. The proximity of the ground plane to a rod–plane gap has a much smaller effect than in the case of the rod–rod gap [146].

A procedure for calculation of the CFO for positive-polarity switching impulses of critical waveshape is proposed in [132].

2.6.4.3 *Phase-to-Phase Strength*

Switching surges on transmission lines are applied not only to the insulation existing between each phase and ground but also to the insulation between phases. The stress appearing between two phases is more complex than that between one phase and ground because of the varying proportions of surges that may appear on each phase. Phase-to-phase insulation strength is a function not only of the total voltage between phases but also of the relative proportion between the two phase-to-ground voltages. For example, the system of Figure 2.56a may be considered as the combination of three phase-to-ground insulations (A–G, B–G, and C–G) and two phase-to-phase insulations (A–B and B–C); it is generally assumed that the strength of the insulation between electrodes A and B does not depend on the voltage on electrode C, and the phase-to-phase insulation system between A and C is generally not considered. However, this simplification is not possible for insulation systems such as that shown in Figure 2.56b, in which the strength of the insulation between A and B is affected by the voltage on C.

The number of variables needed to characterize phase-to-phase stresses is much larger than for the phase-to-ground switching surges. An example of switching surges between phases is shown in Figure 2.57. The voltage between phases A and B is the difference of the voltage between phase A and ground, V_A, and the voltage between phase B and ground, V_B. Parameters necessary to characterize this combination of surges are the crest voltages

TABLE 2.13

Gap Factors of Actual Configurations

Configuration	Gap Factor	Typical Value
Conductor: crossarm 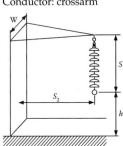	$$k_g = 1.45 + 0.015\left(\frac{h}{S_1} - 6\right) + 0.35\left(e^{\frac{8W}{S_1}} - 0.2\right)$$ $$+ 0.135\left(\frac{S_2}{S_1} - 1.5\right)$$ Applicable in the range: $S_1 = 2\text{--}10\,\text{m}$; $S_2/S_1 = 1\text{--}2$ $W/S_1 = 0.01\text{--}1.0$; $h/S_1 = 2\text{--}10$	$k_g = 1.45$
Conductor: window 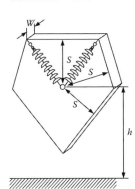	$$k_g = 1.25 + 0.005\left(\frac{h}{S} - 6\right) + 0.25\left(e^{\frac{8W}{S}} - 0.2\right)$$ Applicable in the range: $S = 2\text{--}10\,\text{m}$ $W/S = 0.01\text{--}1.0$ $h/S = 2\text{--}10$	$k_g = 1.25$
Conductor: lower structure 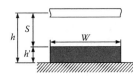	$$k_g = 1.15 + 0.81\left(\frac{h'}{h}\right)^{1.167} + 0.02\left(\frac{h'}{S}\right)$$ $$- A\left[1.209\left(\frac{h'}{h}\right)^{1.167} + 0.03\left(\frac{h'}{h}\right)\right]\left(0.67 - e^{-\frac{2W}{S}}\right)$$ ($A = 0$, if $W/S < 0.2$, otherwise $A = 1$) Applicable in the range: $S = 2\text{--}10$; $W/S = 0\text{--}\infty$; $h'/h = 0\text{--}1$	$k_g = 1.45$ for conductor–plane to 1.50 or more
Conductor: lateral structure 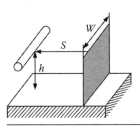	$$k_g = 1.45 + 0.024\left(\frac{h}{S} - 6\right) + 0.35\left(e^{-\frac{8W}{S}} - 0.2\right)$$ Applicable in the range: $S = 2\text{--}10\,\text{m}$, $W/S = 0.1\text{--}1.0$, $h/S = 2\text{--}10$	$k_g = 1.45$

(continued)

TABLE 2.13 (continued)

Gap Factors of Actual Configurations

Configuration	Gap Factor	Typical Value
Rod–rod structure 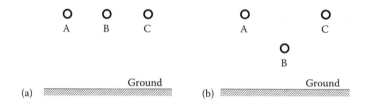	$k_{g1} = 1.35 - 0.1\dfrac{h'}{h} - \left(\dfrac{S_1}{h} - 0.5\right)$ $k_{g2} = 1.00 + 0.6\dfrac{h'}{h} - 1.093 A\dfrac{h'}{h}\left(0.549 - e^{\frac{3W}{S_2}}\right)$ ($A = 0$ if $W/S_2 < 0.2$, otherwise $A = 1$) Applicable in the range: k_{g1} $S_1 = 2$–10 m, $S_1/h = 0.1$–0.8, $S_1 < S_2$ k_{g2} $S_2 = 2$–10 m, $W/S_2 = 0$–∞, $S_2 < S_1$	$k_{g1} = 1.30$ $k_{g2} = 1 + 0.6\dfrac{h'}{h}$

Source: B. Hutzler, B. et al., Strength under switching overvoltages in reference ambient conditions, in *CIGRE WG 33.01, Guide for the Evaluation of the Dielectric Strength of External Insulation*, Chapter 4, CIGRE Technical Brochure no. 72, 1992. With permission.

FIGURE 2.56
Different phase-to-phase configurations. (a) Two phase-to-phase insulations, (b) three phase-to-phase insulations.

of the phase-to-ground surges, V_A and V_B; the crest voltage of the surge between phases, V_{AB}; the interval ΔT between phase-to-ground crest voltages; the times above 70%, t_A, t_B, and t_{AB}, which characterize the waveshapes; the voltages, V^+ and V^-, on the two phases at the instant of the maximum phase-to-phase voltage. The shape of the negative-voltage wave has a negligible influence on the flashover voltage between phases; consequently, phase-to-phase research may be performed using positive and negative waves with identical shapes.

The time-to-crest of the positive component, however, is important. Minimum flashover voltages for rod–rod gaps occur for times-to-crest between 150 and 300 μs. For conductor-to-conductor gaps the effect of the time-to-crest is less pronounced.

Experimental results with long waves (equivalent time-to-crest ≈ 1100 μs) are a better representation of the overvoltages occurring on actual systems. It has been found that flashover occurs around the crest of the waves and that the time interval between the crests of the negative and the positive surges has little effect, provided the components V^+ and V^- at the instant of maximum phase-to-phase voltage do not change [147].

Only the highest peak value of the overvoltage between phases should be considered. Subsequent peaks, even if they have about the same magnitude, do not substantially change the flashover strength of the insulation.

Phase-to-phase surges can be reproduced with the impulses shown in Figure 2.58. Hence, the strength becomes a function only of the phase-to-phase voltage V_{tot} and of the proportions of its components, which may be characterized by the ratio $\alpha = V^-/V_{tot}$. The

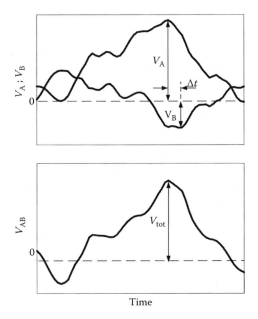

FIGURE 2.57
Example of phase-to-phase switching surge.

CFO between phases can be plotted versus the parameter α and the geometrical variables of the insulation.

The main conclusions derived from test measurements of phase-to-phase flashovers can be summarized as follows [132]:

- The lowest phase-to-phase voltage for the same gap spacing is obtained when a positive voltage is applied to one conductor with the other grounded, while the highest voltage is obtained when a negative voltage is applied to one conductor with the other grounded.

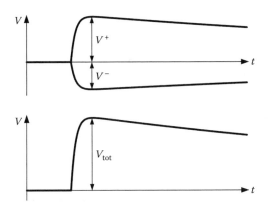

FIGURE 2.58
Phase-to-ground and phase-to-phase impulses.

- For either α = 0 or α = 0.33, the effect of the height above ground is important. In fact, proximity of the ground reduces the positive-polarity strength (α = 0) and increases the negative-polarity strength (α = 1). For α = 0.5, the CFO is almost unaffected by the height of the conductors above ground.

- The presence of an insulator string between phases does not appreciably alter the phase-to-phase switching-surge strength. This presence does not influence the CFO with fair weather, but some reduction (5%–10%) is recommended to account for foul weather.

- With a configuration similar to that shown in Figure 2.56b, the third phase does not appreciably influence the phase-to-phase strength, whether the phase is at ground potential or at floating potential. In practice, this type of stress is not of great concern because the highest phase-to-phase voltages will usually occur with α = 0.5 (e.g., a large positive voltage on phase B and a large negative voltage on phase C), and they will be accompanied by low voltages on the third phase (i.e., phase A) [148].

Some procedures have been developed to deal with phase-to-phase design [1–3,132]. For a thorough discussion on phase-to-phase insulation strength characterization see Chapter 4 of [3].

The following example presents the selection of the insulator string striking distance from the probability distribution of switching overvoltages using a simplified method.

Example 2.5

Figure 2.59 shows the conductor geometry of the test line, a 50 Hz, 400 kV transmission line. The objective of this example is to select the striking distance of insulator strings for this line from the

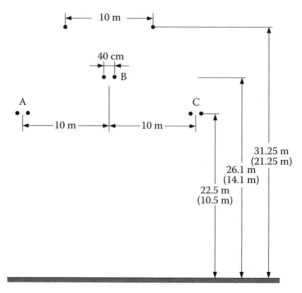

FIGURE 2.59
Example 2.5: Configuration of conductors and wires of the test line.

probability distribution of statistical switching overvoltages. Although lightning and contamination requirements should be also considered, switching overvoltages are of concern for a 400 kV line and they may dictate the insulation level.

This test line was analyzed in Example 2.3. Characteristics of phase conductors and shield wires are provided in Table 2.5. For the case under study the line length is 200 km, and the line is assumed to be at sea level, so standard atmospheric conditions will be used (see Section 2.6.7). To perform the calculations, the source side will be represented by a network equivalent to a short-circuit capacity of 10,000 MVA and assuming the following ratios: $Z_1 = Z_2$, $X_1/R_1 = 12.0$, $X_1/X_0 = 1.3$, $X_0/R_0 = 8.0$.

The transmission line is represented by means of a nontransposed, constant, and distributed-parameter model, with parameters calculated at power frequency. Although a more rigorous approach should be based on a frequency-dependent line model [4,149], the model used in this study will provide conservative overvoltage values since the dependence with respect to the frequency increases the conductor resistance and damping.

The design will be based on a flashover rate of 1/100; i.e., it is accepted one flashover every 100 switching operations. With such a risk of failure, a conservative probabilistic approach is [3]

$$V_3 = E_2 \qquad (2.145)$$

where
V_3 is the insulation strength that is 3 standard deviations below the CFO

$$V_3 = \text{CFO} - 3\sigma_f = \text{CFO} \cdot \left(1 - 3\frac{\sigma_f}{\text{CFO}}\right) = \text{CFO} \cdot (1 - 3\sigma) \qquad (2.146)$$

E_2 is known as the statistical switching overvoltage; that is, 2% of switching overvoltages equal or exceed E_2

A value of 5% of the standard deviation in per unit of the CFO will be assumed for the insulation strength, $\sigma/\text{CFO} = 0.05$.

The first step is then to determine the probability distribution of switching overvoltages. For an introduction on the methodology used in this example see Example 7.10.

It is well known, as a consequence of multimodal wave propagation, that the overvoltages that can occur at the open end of an overhead transmission line during energizing may be greater than 2 per unit. However, the most onerous scenario corresponds to a reclosing operation; that is, a line energization with trapped charge. The presence of the trapped charge can be due to a phase-to-ground fault. As a consequence of the fault, the breakers at both line ends open to clear the fault; and a charge of about 1 per unit is trapped on the unfaulted phases. After closing the breakers at one end, the magnitude of the resulting voltages at the open end may be above 3 per unit.

Reclosing overvoltages can be reduced by preinserting resistors. First, the auxiliary contacts of the preinsertion resistors close; after a time interval of about half-cycle of the power frequency the main contacts close.

The three scenarios (energizing, reclosing, and preinsertion of resistors) are analyzed to decide which one should be used to achieve a reasonable striking distance.

The simulation of the test line is in all scenarios made with following common features:

- It is assumed that only phase-to-ground overvoltages are of concern, and only the highest peak value of the three overvoltages is collected from each run.
- The energizations are performed over the entire range of a cycle and assuming that the three poles are independent. The closing time of each pole is randomly varied according to a normal (Gaussian) probability distribution, with a standard deviation of 2.5 ms.
- The aiming time is chosen following the same method applied in Example 7.10.

Reclosing is analyzed by assuming that a 1 per unit voltage is trapped on each phase, and two different values of preinsertion resistances, 200 and 400 Ω, are used.

TABLE 2.14

Example 2.5: Statistical Distribution of Phase-to-Ground Voltages

Case	Mean Value (pu)	Standard Deviation (pu)	E_2 (pu)
Energizing	2.301	0.268	2.851
Reclosing	3.190	0.666	4.558
Preinsertion resistors: $200\,\Omega$	1.842	0.124	2.097
Preinsertion resistors: $400\,\Omega$	1.541	0.032	1.607

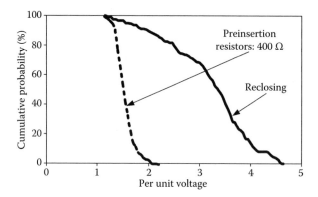

FIGURE 2.60
Example 2.5: Peak voltage distribution—normal distribution.

Table 2.14 shows the characteristic parameters that result for each probability density function. The value of the statistical-switching overvoltage E_2 is deduced by using the following expression:

$$E_2 = \mu_S + 2.054\sigma_S \tag{2.147}$$

where μ_S and σ_S are respectively the mean and the standard deviation of the probability density function of switching overvoltage stresses.

Figure 2.60 depicts the peak voltage distributions obtained when reclosing the line and when using a preinsertion resistor of $400\,\Omega$.

Since standard atmospheric conditions are assumed, the required strike distance will be obtained from Equation 2.142. The strike distance, in meters, will be then

$$S = \frac{8}{k_g \dfrac{3400}{\text{CFO}} - 1} \tag{2.148}$$

The gap factor for this test line has to be obtained from the expression specific for a conductor–crossarm configuration, see Table 2.13. However, this calculation cannot be performed since the gap factor is a function of the desired strike distance. Therefore, an iterative procedure is necessary. The typical value of 1.45 will be initially used. This can be checked once the strike distance was determined, and the whole procedure repeated until a good match is achieved.

Using the results obtained from line reclosing (i.e., the most onerous case), it results

$$E_2 = 4.558 \cdot \sqrt{2}\, \frac{400}{\sqrt{3}} = 1488.6\,\text{kV}$$

Since σ_f is assumed to be 5% of the CFO, and $V_3 = E_2$

$$\text{CFO} = \frac{E_2}{1-3\sigma} = \frac{1488.6}{1-3\cdot 0.05} = 1751.3\,\text{kV}$$

Then

$$S = \frac{8}{1.45\dfrac{3400}{1751.3} - 1} = 4.41\,\text{m}$$

This is a rather long distance for a 400 kV line. Therefore, the calculations are repeated with the results derived from the other scenarios. Table 2.15 presents the striking distances derived from each case.

The striking distance that results from all cases, except reclosing, can be implemented for a transmission line with a rated voltage of 400 kV. However, it is important to emphasize that for this voltage level lightning and contamination requirements must be also considered. It is also worth mentioning that a more rigorous method may be applied following standard procedures, see for instance [1,2].

As for the striking distance to be selected, that deduced from the energizing scenario could be considered only when the line reclosing is made after a time span long enough to remove most of the trapped charge. Otherwise, preinsertion resistors should be used.

2.6.5 Dielectric Strength under Lightning Overvoltages

2.6.5.1 Introduction

The lightning impulse strength characteristic of power line insulation can be also modeled as a cumulative normal (Gaussian) distribution. IEC suggests using a Weibull distribution [2]. For application purposes, the following information is needed for any given impulse, gap, and geometry:

- The CFO (or U_{50}) and the conventional deviation σ_f/CFO.
- The volt–time characteristics, which relate the maximum voltage reached before breakdown, V_B, to the time-to-breakdown, T_B, see Figure 2.52.

The standard deviation for lightning impulses is smaller than that for switching impulses, usually in the range of 1% or 3% of the CFO, although values as high as 3.6% have been

TABLE 2.15

Example 2.5: Selected Striking Distances

Case	E_2 (kV)	CFO (kV)	Striking Distance (m)
Energizing	931.13	1095.5	2.29
Reclosing	1488.6	1751.3	4.41
Preinsertion resistors: 200 Ω	684.88	805.74	1.56
Preinsertion resistors: 400 Ω	524.84	617.46	1.15

obtained for specific cases. Very often, the lightning impulse insulation strength is characterized by a single value, the CFO or the BIL. That is, voltages applied to the insulation that are below the BIL or the CFO are assumed to have a zero probability of flashover or failure; alternately, applied voltages that are greater than the CFO or the BIL are assumed to have a 100% probability of flashover or failure.

In general, the CFO curve as a function of strike distance is linear, and the CFO can be given by a single value of gradient at the critical flashover voltage, or a CFO gradient, in terms of kV/m. In addition, the CFO is primarily a function of the wave tail, and the front is only of importance when considering very short wave tails. Section 2.6.5.2 discusses the behavior of power line insulation under both standard and nonstandard lightning impulses.

2.6.5.2 Lightning Impulse Strength

Although lightning strokes to overhead lines may lead to traveling surges with a shape similar to the standard impulse, in most actual conditions lightning overvoltages are generated with shapes substantially different. Therefore, the lightning impulse strength of power lines must be analyzed for both standard and nonstandard impulses.

A. Standard lightning impulses

- CFO: The CFO for rod–plane configuration with positive polarity is much lower than that with negative polarity. When plotted against the gap clearance, the CFO is nonlinear with negative polarity while it is linear with positive one. Figure 2.61 shows a generalization of the influence of the gap geometry on the CFO [74]. Figure 2.61a refers to rod–plane configuration and gives the average gradient in the gap ($E_{50} = U_{50}/d$) under positive and negative polarities as a function of the gap clearance. With positive polarity, E_{50} is about 525 kV/m and independent of the clearance. With negative polarity, E_{50} is higher than that with positive polarity and decreases when the gap clearance increases. Figure 2.61b refers to various electrode configurations and gives the value of the CFO gradient in per unit of the rod–plane E_{50} as a function of the gap factor k_g of the configuration. It is evident that E_{50} with positive polarity tends to increase as k_g increases, while with negative polarity the trend is the opposite one.

  Table 2.16 shows a summary of expressions proposed to estimate the CFO for lightning impulses. As per Figure 2.61, the negative polarity CFO for large gap factors may be less than that for positive polarity. However, gap factors greater than 1.4 seldom occur in practical configurations found in transmission lines.

  The CFO for conductor–crossarm configuration without insulators is independent of polarity; however, polarity affects the CFO when insulators are in the gap, although CFO with positive polarity is very close to that with negative polarity, and it is higher than that with rod–plane under positive polarity. In general, the presence of insulators reduces breakdown voltage in the case of negative polarity. The influence of cap and pin insulators is reduced when the stress on the first insulator at both extremities of the string is reduced using shielding rings. It is also reduced for more practical configurations, with insulators at both extremities less stressed than in the case of rod–plane gap. For some configurations, like conductor–upper structure and even conductor–crossarm (see Table 2.13), the strength is close to that of air gaps. For other configurations, and especially when

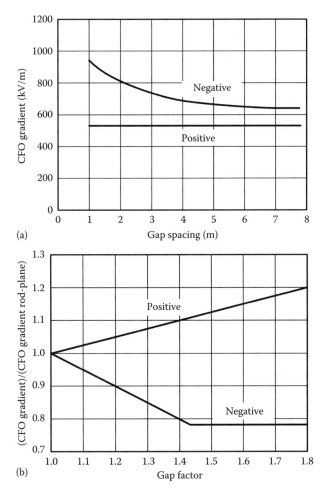

FIGURE 2.61
Average gradient at CFO in a rod–plane gap: (a) as a function of the gap spacing; (b) as a function of the gap factor.

TABLE 2.16

CFO for Lightning Impulses

Lightning Impulse	CFO Estimation
Positive polarity	Paris-Cortina: $CFO^+ = 383 + 147k_g$
	CIGRE TB 72: $CFO^+ = 394 + 131k_g$
	CIGRE TB 72: $CFO^+ = 354 + 154k_g$
	IEC 60071: $CFO^+ = 530S(0.74 + 0.26k_g)$ with spacings up to 10 m
Negative polarity[a]	IEC 60071: $CFO^- = 950S^{0.8}(1.5 - 0.5k_g)$ for k_g from 1 to 1.44
	$CFO^- = 741S^{0.8}$ for k_g greater than 1.44

[a] Present version of IEC 60071–2 recommends a general formula ($CFO^- = 700S$) to estimate the breakdown strength for negative polarity impulses [2], see Example 2.6.

TABLE 2.17

Lightning Impulse CFOs for Practical Gap Configurations with and without Insulators

Gap Configuration	Diagram	Positive Polarity CFO (kV/m)		Negative Polarity CFO (kV/m)	
		Without Insulators	With Insulators	Without Insulators	With Insulators
Outside arm		600	500	600	595
		625	520	625	620
Conductor: upper structure		575	560	625	610
Conductor: upper rod		655	500*	595	585

Source: Hileman, A.R., *Insulation Coordination for Power Systems*, Marcel Dekker, New York, 1999. With permission.
Note: The CFO gradient as a function of the distance is non linear. Value given for 4 meters.

large clearances are involved, testing is advised to get accurate results. Table 2.17 shows ranges of the CFO gradient as presented in [3] and obtained from the literature [129,140,150], for some configurations that can be encountered in transmission lines. The dielectric performance depends on insulator type (capacitance between units, distance between metal parts along the insulator set). A lower influence is to be expected for insulators with few metal parts (e.g., post insulators, long rod, and composite). The influence of rain on the flashover voltage is secondary as in the case of air gaps.

- Standard deviation: For air gaps, it is higher for negative impulses than for positive impulses. It can be affected by the presence of insulators, reaching a maximum of 5% to 9% in connection with the cases that exhibit the largest influence of the insulators on CFO. In the other cases a value close to that of air gaps is applicable. The standard deviation for positive polarity is about 1.0% of the CFO, and increases to 3.6% of the CFO for negative polarity.

- Volt–time curves: They vary significantly with gap configuration. As the gap configuration approaches a uniform field gap, the upturn of the volt–time curve becomes less pronounced: the curve becomes flatter. Oppositely, as the gap configuration approaches a more nonuniform field gap, the upturn at short times becomes greater. The following equation may be used to approximate volt–time curves from about 2–11 µs:

$$V_B = \left(0.58 + \frac{1.39}{\sqrt{t}}\right)\text{CFO} \tag{2.149}$$

where
  V_B is the breakdown crest voltage
  t is the time to breakdown or flashover

Suggested values for breakdown voltage are [3]

- Tower insulation: Breakdown voltage at $2\,\mu s = 1.67 \cdot \text{CFO}$.
  Breakdown voltage at $3\,\mu s = 1.38 \cdot \text{CFO}$.
- Porcelain insulations: Breakdown voltage at $2\,\mu s = 1.32 \div 148 \cdot \text{CFO}$.

Breakdown voltage at $3\,\mu s = 1.22 \div 1.31 \cdot \text{CFO}$.

B. Nonstandard impulses

Experimental results show that V_B increases when the tail duration decreases. If the voltage is increased so that breakdown takes place around the crest, differences in impulse shape tend to be less significant and all curves converge; minor differences can be due to circuit interaction and slight modifications associated with wave front parameters. Due to the variety of impulses and configurations, the available results cannot be generalized.

The following expression quantifies the effect of the time-to-half value on the CFO:

$$\text{CFO}_{NS} = \left(0.977 + \frac{2.82}{t_h}\right)\text{CFO}_S \tag{2.150}$$

where
  CFO_S is the CFO for the standard $1.2/50\,\mu s$ waveshape
  CFO_{NS} is the nonstandard CFO for the tail time t_h in µs

From this equation one can deduce that the ratio $\text{CFO}_{NS}/\text{CFO}_S$ increases as the tail time decreases. The equation is valid for times-to-crest between 0.5 and 5 µs and for tail time constants between 10 and 100 µs. For the standard tail of 50 µs, the ratio is 1.016.

2.6.5.3 Conclusions

Experimental information is, in general, sufficient to estimate the dielectric strength of practical configurations for standard lightning impulses. Once the configuration and the gap distance are fixed, the gap factor can be obtained. The CFO is estimated by entering this value in Figure 2.61. In general, experimental results are also sufficient to obtain indications about the volt–time curve.

In the case of nonstandard impulses, however, experimental data cannot be generalized, and suitable models have to be used to predict insulation behavior. Various approaches have been proposed to evaluate the time-to-breakdown under both standard and nonstandard lightning impulses, see Section 2.6.3.2.

The CFO gradients of tower insulation (wet), for either the center or the outside phase, can be approximately obtained from Table 2.17.

The application of models for fast-front transients of the different parts of an overhead transmission line is presented in the following example, whose main goal is to analyze the lightning performance of an actual transmission line. A discussion of the various options available in transients tools is included to justify the line model chosen in this example.

Example 2.6

Figure 2.62 shows the tower design of the test transmission line, a 400 kV line, with two conductors per phase and two shield wires, whose characteristics are presented in Table 2.5. This tower was analyzed in Example 2.3.

The goal of this example is to estimate the lightning performance of this line. This analysis will be performed by using up to three different approaches to represent insulator strings: (1) insulator strings are modeled as open switches, the goal is to obtain overvoltages caused by lightning strokes; (2) insulator strings are modeled as open switches controlled by the voltage generated across them, when the voltage exceeds the BIL of any insulator string, the line flashovers; (3) insulator strings are modeled by means of the LPM, as detailed in Section 2.6.3.2. The average span length is 400 m and the striking distance of insulator strings is 3.212 m. The calculations will be made by assuming that the line is located at the sea level.

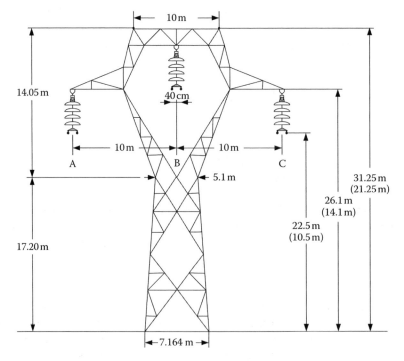

FIGURE 2.62
Example 2.6: 400 kV line configuration (values in brackets are midspan heights).

1. Modeling guidelines

The models used to represent the different parts of a transmission line in lightning overvoltage calculations are detailed in the following paragraphs [4–6,54].

  a. The shield wires and phase conductors are modeled by including four spans at each side of the point of impact. Each span is represented as a multiphase untransposed frequency-dependent and distributed-parameter line section. Corona effect is not included.

  b. A line termination at each side of the above model is needed to avoid reflections that could affect the simulated overvoltages caused around the point of impact. Each termination is represented by means of a long enough section whose parameters are also calculated as for line spans.

  c. A tower is represented as an ideal single conductor distributed-parameter line, see Example 2.3. The approach used in this example is the twisted model recommended by CIGRE [74], see Section 2.4. This model can suffice for single circuits with towers shorter than 50 m [3].

  d. The grounding impedance is represented as a nonlinear resistance whose value is approximated by the expression recommended in standards (see Section 2.5.5) [1,2]:

$$R = \frac{R_0}{\sqrt{1+I/I_g}} \qquad \left(I_g = \frac{E_0 \rho}{2\pi R_0^2} \right)$$

where
  R_0 is the footing resistance at low current and low frequency
  I is the stroke current through the resistance
  I_g is the limiting current to initiate sufficient soil ionization
  ρ is the soil resistivity (Ω m)
  E_0 is the soil ionization gradient (400 kV/m in this example, [151])

  e. The options considered for the representation of insulator strings have been mentioned above. When the representation of insulator strings relies on the application of the LPM, the leader propagation is obtained by means of the following equation (see Section 2.6.3.2):

$$\frac{d\ell}{dt} = k_\ell V(t) \left[\frac{V(t)}{g-\ell} - E_{\ell 0} \right]$$

where
  $V(t)$ is the voltage across the gap
  g is the gap length
  ℓ is the leader length
  $E_{\ell 0}$ is the critical leader inception gradient
  k_l is a leader coefficient

When insulator strings are represented as voltage-controlled switches, the CFO for negative polarity strokes is calculated according to the expression proposed by IEC 60071–2 [2]

$$CFO^- = 700S \tag{2.151}$$

where S is the striking distance of the insulator strings. With this model, a flashover occurs if the lightning overvoltage exceeds the lightning insulation withstand voltage.

  f. Phase voltages at the instant at which the lightning stroke hits the line are included. For statistical calculations, phase voltage magnitudes are deduced by randomly determining the phase voltage reference angle and considering a uniform distribution between 0° and 360°.

g. A lightning stroke is represented as an ideal current source (infinite parallel impedance). In statistical calculations, stroke parameters are randomly determined according to the distribution density functions recommended in the literature [152–154]. See below for more details.

The model that results from the application of these guidelines is similar to that presented in Figure 2.35, which depicts the case that corresponds to an impact on a tower. When the stroke hits a shield wire or a phase conductor midway between two towers, the model has to be modified; the line span model has to be split up into two sections and the current source that represents the lightning stroke moved to the new impact position.

The study of this example has been divided into two parts. The first part summarizes the main results of a sensitivity study aimed at analyzing the influence that some line parameters have on the overvoltages originated across insulator strings. The second part presents the application of the Monte Carlo method to assess the lightning performance of this line.

2. Sensitivity study

The goal is to simulate overvoltages caused by strokes to towers (backflashover) and phase conductors (shielding failure), and determine the influence that some parameters have on the peak voltages. All the calculations presented in this example have been performed by representing insulator strings as open switches and grounding impedances of line towers as constant resistances. As discussed in Section 2.5, this grounding model is not always realistic and provides conservative results; however, it will facilitate the analysis of the results since the full line model becomes linear.

The lightning stroke current is assumed of negative polarity and represented as a concave waveform, with no discontinuity at $t = 0$, see Figure 2.63. The mathematical expression used in this example is the so-called Heidler model, which is given by [155]

$$i(t) = \frac{I_\mathrm{p}}{\eta} \frac{k^n}{1+k^n} e^{-t/\tau_2} \qquad (2.152)$$

where
  I_p is the peak current
  η is a correction factor of the peak current
  n is the current steepness factor, $k = t/\tau_1$
  τ_1 and τ_2 are time constants determining current rise and decay time, respectively

The value selected for parameter n, to be specified in Equation 2.152, is 5 in all simulations performed for this example.

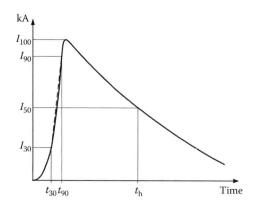

FIGURE 2.63
The lightning stroke current represented as a concave waveform.

A. Stroke to a tower

Since the line model used in this study is linear, the relationship between the peak stroke current and the voltage across insulator strings is also linear; however, this relationship is more complex with respect to other values, such as the rise time of the lightning return stroke current and the grounding resistance. Figure 2.64a shows the voltage that results across the insulator string of the outer phase of the test line. Figure 2.64b and c depict the relationships with respect to the two parameters mentioned above. One can easily deduce that both parameters have a strong influence: the greater the grounding resistance value and the shorter the rise time, the higher the overvoltages. These results were derived without including power-frequency voltages.

B. Strokes to a phase conductor

For the same stroke peak current, overvoltages originated by strokes to phase conductors will be much higher than those originated by strokes to towers or shield wires. Figure 2.65a shows the overvoltage originated by a stroke to the outer phase, including the power-frequency voltage that is present at the moment the stroke hits the line. A new parametric study was made to deduce the influence of the stroke peak current and the voltage angle, using the phase angle of the outer phase (phase A) to which the lightning stroke impacts as a reference. Figure 2.65b and c present the results obtained by considering the worst case from each simulation. The plots show very high voltages, but it is worth noting that shield wires will prevent strokes with a peak current higher than 20 kA from reaching phase conductors, as it is shown in the subsequent section. The influence of the grounding resistance when the lightning stroke hits a phase conductor is negligible.

The conclusions from these results are straightforward. The overvoltage caused by a lightning stroke to a phase conductor increases linearly with the peak current magnitude and depends on the phase angle at the moment it hits the conductor.

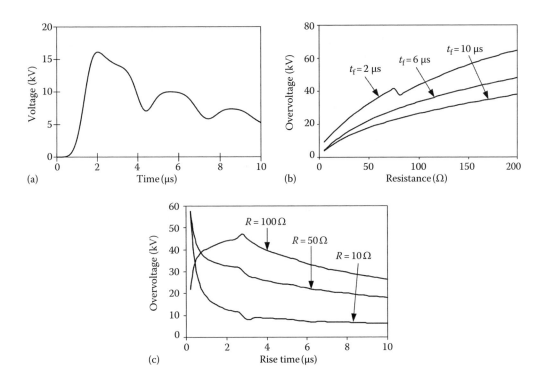

FIGURE 2.64

Example 2.6: Overvoltages caused by strokes to a tower. (a) Insulator string overvoltage: outer phase ($I_p = 1\,kA$, $R_{tower} = 20\,\Omega$); (b) insulator string overvoltage as a function of the grounding resistance ($I_p = 1\,kA$); (c) insulator string overvoltage as a function of the rise time of the lightning return stroke ($I_p = 1\,kA$).

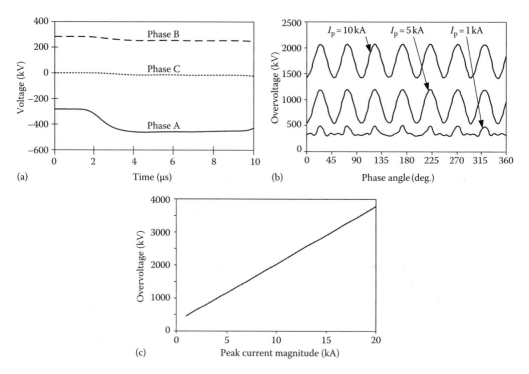

FIGURE 2.65
Example 2.6: Overvoltages caused by strokes to phase conductors. (a) Insulator string overvoltage: outer phase ($I_p = 1\,\text{kA}$, $R_{\text{tower}} = 20\,\Omega$); (b) insulator string overvoltage as a function of the reference phase angle; (c) insulator string overvoltage as a function of the peak current magnitude.

3. Statistical calculation of lightning overvoltages

A. General procedure
The main aspects of the Monte Carlo procedure used in this example are discussed in the following:
  a. The values of the random parameters are generated at every run according to the probability distribution function assumed for each one. The calculation of random values includes the parameters of the lightning stroke (peak current, rise time, tail time, and location of the vertical channel), the phase conductor voltages, the grounding resistance and the insulator strength.
  b. The determination of the point of impact requires of a method for discriminating strokes to line conductors from those to ground. It is important to distinguish return strokes to shield wires from those to phase conductors. This step will be based on the application of the electrogeometric model recommended by Brown–Whitehead [156]

$$r_c = 7.1I^{0.75} \tag{2.153a}$$

$$r_g - 6.4I^{0.75} \tag{2.153b}$$

where
  r_c is the striking distance to both phase conductors and shield wires
  r_g is the striking distance to earth
  I is the peak current magnitude of the lighting return stroke current

With this model, only return strokes with a peak current magnitude below 20 kA will reach phase conductors.

c. The overvoltage calculations can be performed once the point of impact of the randomly generated stroke has been determined. Overvoltages originated by nearby strokes to ground are neglected.

d. Voltages across insulator strings are continuously monitored. When the goal of the procedure is to obtain the lightning flashover rate, every time a flashover is produced the run is stopped, the counter is increased and the flashover rate is updated.

e. The entire simulation is stopped when the convergence of the Monte Carlo method is achieved or the specified maximum number of iterations is reached. The convergence is checked by comparing the probability density function of all variables to their theoretical functions, so the procedure is stopped when they match within the maximum error or the maximum number of runs is reached [157].

B. Lightning stroke waveform and parameters

Most lightning flashes are of negative polarity, and they can consist of multiple strokes, although only the first and the second strokes are of concern for transmission insulation levels. The incidence of positive flashes increases during winter, although their percentage is usually below 10% [152,153,158,159]. Very rarely positive flashes have more than one stroke [153]. Only single-stroke negative polarity flashes are considered in this study.

The stroke waveform is the same that was used in the first part of this example; see Figure 2.63 and Equation 2.152. Main parameters used in statistical calculations to define the waveform of a lightning stroke are the peak current magnitude, I_p (or I_{100}), the front time, t_f (=1.67 $(t_{90}-t_{30})$), and the tail time, t_h, see Figure 2.63. Since these values cannot be directly defined in the Heidler model, a conversion procedure [160] is performed to derive the parameters to be specified in Equation 2.152 from the stroke parameters, which are randomly calculated, as detailed in the following.

The statistical variation of the lightning stroke parameters is usually approximated by a lognormal distribution, with the following probability density function [153]:

$$p(x) = \frac{1}{\sqrt{2\pi}x\sigma_{\ln x}} \exp\left[-0.5\left(\frac{\ln x - \ln x_m}{\sigma_{\ln x}}\right)^2\right] \qquad (2.154)$$

where
$\sigma_{\ln x}$ is the standard deviation of $\ln x$
x_m is the median value of x

A joint probability density function of two stroke parameters, $p(x,y)$, may be considered by introducing a coefficient of correlation ρ_c. If x and y are independently distributed, then $\rho_c = 0$, and $p(x,y) = p(x)p(y)$.

C. Random parameters

The following probability distributions have been assumed for each random value:

- Insulator string parameters are determined according to a Weibull distribution. The mean value and the standard deviation of E_{l0} are 570 kV/m and 5%, respectively. The value of the leader coefficient is $k_l = 1.3E-6\,m^2/(V^2s)$ [74]. The value of the average gradient at the CFO is assumed to be the same as that E_{l0}.
- Stroke parameters are determined assuming a lognormal distribution for all of them. Table 2.18 shows the values used for each parameter. In addition, parameters are independently distributed.
- The phase conductor reference angle has a uniform distribution, between 0° and 360°.

TABLE 2.18

Example 2.6: Statistical Parameters of the
Return Stroke

Parameter	x	$\sigma_{\text{ln}x}$
I_{100}, kA	34.0	0.740
t_f, µs	2.0	0.494
t_h, µs	77.5	0.577

Source: IEEE TF on Parameters of Lightning
Strokes, *IEEE Trans. Power Deliv.*, 20,
346, January 2005. With permission.

- The grounding impedance has a normal distribution with a mean value $R_m = 50\,\Omega$ and a standard deviation $\sigma = 5\,\Omega$. Remember that the grounding resistance model accounts for soil ionization effects, so parameter R_m is the mean value of the low-current and low-frequency resistance, R_0. The value of the soil resistivity is $200\,\Omega$ m.
- The stroke location, before the application of the electrogeometric model, is estimated by assuming a vertical path and a uniform ground distribution of the leader.

The line model has been implemented considering that only flashovers across insulator strings can occur.

D. Simulation results

To check the convergence of the Monte Carlo procedure summarized above, the resulting and the theoretical statistical distributions are compared at 10%, 30%, 50%, 70%, and 90% of each cumulative distribution function. More than 10,000 runs are needed to match them within an error margin of 10%. For an error margin of 5% no less than 30,000 runs were needed. All the studies have been performed by executing a maximum of 40,000 runs [157].

Figures 2.66 and 2.67 show the results obtained with each of the scenarios considered in this example. From these results one can observe that the range of peak current magnitudes that cause backflashover (stroke to a shield wire or to a tower) is different from the range of values that cause shielding failure flashover (strokes to phase conductors). It is also evident that after 40,000 runs the statistical distribution of the overvoltages caused by backflashovers is well defined, which is not the case for overvoltages caused by shielding failures, indicating that even more runs are needed to obtain an accurate enough distribution of this type of overvoltages. The same conclusion is valid for the probabilities of flashover derived from each study.

The lightning flashover rate of the test line, assuming a ground flash density $N_g = 1$ fl/km² year, is 0.845 per 100 km-year when the insulator strings are modeled as voltage-controlled switches, and 0.347 per km-year when they are represented by means of the LPM. Although the parameters used in both approaches are those recommended by IEC [2] and CIGRE [74], the difference between rates is significant.

Theoretically, an LPM representation of the line insulation is more rigorous, provided that the parameters used to represent the leader propagation are accurate enough. It is interesting to note that the range of values that cause flashover with each insulation representation is different and exhibit a different trend: when the LPM is used, the range of values that cause flashover is narrower than when the insulation is modeled as a voltage-controlled switch, while the trend is opposite in case of shielding failure flashover. This later performance means that, when the LPM is used, there can be flashovers caused by lightning strokes with lower peak current magnitudes. Remember that with this modeling approach a flashover can occur during the tail of the lightning current; that is, the insulation can flashover after passing the peak value of the lightning overvoltage, which is not possible with the other modeling approach.

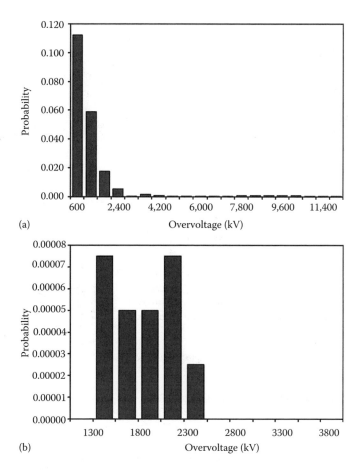

FIGURE 2.66

Example 2.6: Distribution of lightning overvoltages. Insulator strings are represented as open switches. (a) Overvoltages caused by impacts on a shield wire (including impacts on line towers); (b) overvoltages caused by impacts on phase conductors (shielding failures).

2.6.6 Dielectric Strength for Power-Frequency Voltage

External insulation design of overhead transmission lines in the extra high voltage (EHV) and UHV range is most often determined by switching overvoltage requirements. Very rarely the design is controlled by lightning overvoltages. Generally, insulation designed on the basis of these requirements is also sufficient for power-frequency voltages because in clean and wet conditions the strength is usually more than twice the stress. There have been, however, many cases of power-frequency flashovers on transmission lines without evidence of transient overvoltages. These flashovers usually take place in wet weather conditions, and are caused by contamination of the insulator surfaces. For transmission lines, the insulation strength under power-frequency voltage for clean or noncontaminated conditions is seldom a determinant factor for insulator design or for strike distance selection; it is the performance of external insulation under contaminated conditions that dictates the insulation design.

Contamination flashovers on transmission systems are initiated by particles deposited on the insulators. These particles may be of natural origin or generated by pollution, which can be caused by industrial, agricultural, or construction activities. A common natural

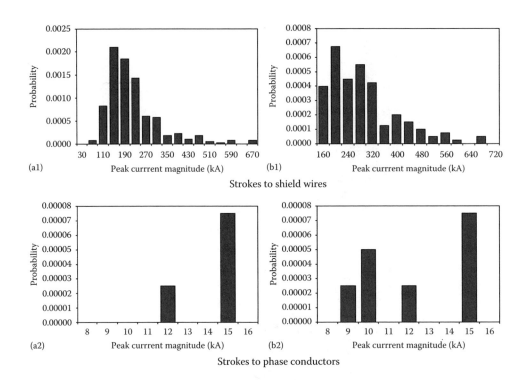

FIGURE 2.67

Example 2.6: Distribution of stroke currents that caused flashovers. (a) Voltage controlled switch model; (b) leader progression model (LPM).

deposit is sea salt, which causes a severe contamination in coastal areas. In industrial or agricultural areas, a great variety of contaminants may reduce insulation strength. These deposits do not decrease the insulation strength when the insulators are dry. The loss of strength is caused by the combination of contaminants and moisture, which is always necessary to produce a conductive layer on contaminated insulator surfaces.

Insulator contamination is a serious threat to reliable transmission-line operation, since several successive flashovers may occur, which may force to deenergize the line. This is particularly common for transmission lines in heavily contaminated areas during periods of wet weather. After these wet periods, when the insulator surfaces dry out, their insulation strength is restored. However, the possibility of flashover remains unless insulator surface is cleaned by either natural or artificial means. Consequently, transmission lines experiencing contamination flashovers are likely to have repeated outages [161].

The flashover strength of a contaminated insulator string is a nonlinear function of the string length, and flashover strength saturates at higher voltage levels. Contamination can become the most important factor in selecting conductor-to-tower clearances if switching overvoltages are controlled at a low level for EHV and UHV levels.

Rain may substantially reduce the insulation strength dependent on the rate of rainfall, the conductivity, and the insulator configuration (V-string, I-string, or horizontal). For I-strings or vertical insulator strings, this decrease may approach 30%. The effect of rain on air gaps is negligible.

IEC Std 60071 [2] provides an approximate equation for the power-frequency CFO (or U_{50}) of a rod–plane gap:

$$CFO_{PF-RP} = \sqrt{2} \cdot 750 \cdot \ln\left(1 + 0.55 S^{1.2}\right) \tag{2.155}$$

where S is the strike distance, in m.

The CFO peak value under power-frequency voltage is about 20%–30% higher than the corresponding value under positive switching impulse at critical front time.

For gaps between 1 and 2 m, the above expression is a conservative approach. For gaps longer than 2 m, the strength (under dry conditions) could be evaluated as follows [2,162]:

$$CFO_{PF} = CFO_{PF-RP} \cdot \left(1.35 k_g - 0.35 k_g^2\right) \tag{2.156}$$

where k_g is the gap factor.

By using this expression, the resulting power-frequency CFO can be greater than the switching impulse, positive-polarity CFO, for 3 m gaps and a gap factor larger than 1 [3].

Insulators may significantly reduce the flashover voltage of the same gap but without insulators, especially in conditions of high humidity [2,162].

Several countermeasures can be applied to avoid or mitigate breakdowns caused by power-frequency voltages under contaminated conditions [161]: overinsulation by increasing the number of insulators, application of fog-type insulators, application of silicone grease, washing and cleaning of deenergized insulators, and use of semiconducting glazed insulators.

Although several models have been developed and tested to represent the performance of contaminated insulators under power frequency, they are not frequently used to predict flashover. A comprehensive survey of surface discharge models is presented in [163].

2.6.7 Atmospheric Effects

The flashover voltage of external insulation depends on atmospheric conditions. BILs and BSLs are specified for standard atmospheric conditions. However, laboratory atmospheric conditions are rarely standard, and correlation factors are needed to determine the crest impulse voltage that should be applied so that the BIL and BSL will be valid for standard conditions. It is necessary, therefore, to establish procedures by which it is possible to estimate flashover voltages at any atmospheric condition from those measured at another condition.

Usually, the flashover voltage for a given path in air is raised by an increase either in air density or the humidity [164]. Denoting the voltage measured under nonstandard atmospheric conditions as V_A and the voltage for standard conditions as V_S, the suggested relationship is [1,3]

$$V_A = \delta^m H_c^w V_S \tag{2.157}$$

where
   δ is the relative air density
   H_c is the humidity correction factor
   m and w are constants dependent on the factor G_0, which is defined as follows:

$$G_0 = \frac{CFO_S}{500S} \tag{2.158}$$

where
  S is the strike distance in m
  CFO_S is the CFO under standard conditions

The values of m and w may be obtained from Table 2.19.

The standard atmospheric condition assumes an absolute humidity of 11 g moisture content per cubic meter of air, 760 torr and 293 K (20°C).

Density effects: Breakdown voltages can be assumed to vary linearly with the density of the air, which is determined by temperature and pressure. The relative air density is defined as

$$\delta = \frac{PT_0}{P_0 T} \tag{2.159}$$

where
  P_0 and T_0 are the standard pressure and temperature with the temperature in degrees Kelvin
  P and T are the ambient pressure and temperature

Over the normal range of temperatures encountered in practice, the density correction factor is satisfactory. Values and curves of flashover voltages presented in the literature are usually described as uncorrected, for which $\delta = 1$.

Since the relative air density is a function of pressure and temperature, it is also a function of altitude. At any specific altitude, the air pressure and the temperature, and thus the relative air density, are not constant but vary with time. Therefore, the relative air density can be approximated by a Gaussian distribution [165]. The mean value of the relative air density is related to the altitude by the expression [1]

$$\delta = e^{-A/8.6} \tag{2.160}$$

where A is the altitude, in km.

Humidity effects: The effects of humidity on flashover are quite complex, but in general humidity has its strongest influence on the positive pre-breakdown discharge; it does not

TABLE 2.19

Values of m and w

G_0	m	w
$G_0 < 0.2$	0	0
$0.2 < G_0 < 1.0$	$m = w = 1.25\, G_0\,(G_0 - 0.2)$	
$1.0 < G_0 < 1.2$	1	1
$1.2 < G_0 < 2.0$	1	$w = 1.25\,(2.2 - G_0)\,(2 - G_0)$
$G_0 > 2.0$	1	0

exhibit a significant effect on the leader gradients, although it increases the leader velocity, and has not much influence on the negative flashover under lightning impulse. For wet or simulated rain conditions, $H_c = 1.0$.

Other atmospheric effects: Overhead lines must contend with atmospheric pollution, rain, ice, and snow. All of these can have a significant effect on insulators. For further information see Refs. [166–168].

Improved correction factors were suggested in [169].

The conversion procedure can be used for impulse, alternating, and direct voltage measurements. It is also satisfactory for normal variations of pressure at altitudes up to 2000 m, although at greater altitudes (lower pressures) the linearity breaks down and a modified procedure must be used [170].

By definition, Equation 2.157 can be applied to correct CFO, BIL, and BSL. That is,

$$\text{CFO}_A = \delta^m H_c^w \text{CFO}_S$$

$$\text{BIL}_A = \delta^m H_c^w \text{BIL}_S$$

$$\text{BSL}_A = \delta^m H_c^w \text{BSL}_S \tag{2.161}$$

Table 2.20 shows a summary of results for both lightning and switching impulses.

IEC Std 60071.2 proposes to account for nonstandard atmospheric conditions by means of the following relationship [2]:

$$V_A = K_a V_S \tag{2.162}$$

being the atmospheric correction factor K_a calculated from

$$K_a = e^{-m(A/8.15)} \tag{2.163}$$

where $m = 1$ for lightning impulse withstand voltages, while it depends on various parameters (type of insulation, coordination switching impulse withstand voltage) for switching impulse withstand voltages.

TABLE 2.20

Correction for Nonstandard Atmospheric Conditions

Impulse Type	Atmospheric Conditions	Correction
Switching impulse	Wet/rain conditions	$V_A = \delta^m V_S$
	($H_c = 1$)	$\text{CFO}_A = \delta^m \text{CFO}_S$
		$\text{BSL}_A = \delta^m \text{BSL}_S$
	G_0 is between 0.2 and 1.0	
	$m = w = 1.25 G_0 (G_0 - 0.2)$	
Lightning impulse	Wet/rain conditions	$V_A = \delta V_S$
	($H_c = 1$)	$\text{CFO}_A = \delta \text{CFO}_S$
		$\text{BIL}_A = \delta \text{BIL}_S$
	G_0 is between 1.0 and 1.2	
	($m = 1, w = 1$)	

Standard correction factors for relative air density and humidity generally used for EHV design are not applicable to UHV tower design [132]. Tests on different gaps with varying atmospheric conditions show that a change of the relative air density and humidity had a smaller influence on the switching-surge flashover voltage at UHV spacings than at EHV spacings. Similarly, tests on various gaps in the UHV range show that the wet value of switching flashover voltage is very close to the dry value. The spread in flashover data in the total range of EHV and UHV spacings can be reduced if standard atmospheric correction factors are applied to the air-gap clearance instead of to the flashover voltage. Because of the saturated flashover characteristics of UHV gap configurations, this approach produces successively reduced effects of the atmospheric correction on the voltage as the spacings are increased. This method of correction has its experimental justification in several works [171–173].

References

1. IEEE Std 1313.2-1999, IEEE Guide for the Application of Insulation Coordination.
2. IEC 60071-2, Insulation co-ordination, Part 2: Application Guide, 1996.
3. A.R. Hileman, *Insulation Coordination for Power Systems*, New York: Marcel Dekker, 1999.
4. CIGRE WG 33.02, Guidelines for Representation of Network Elements when Calculating Transients, CIGRE Brochure no. 39, 1990.
5. A. Gole, J.A. Martinez-Velasco, and A. Keri (Eds.), Modeling and Analysis of Power System Transients Using Digital Programs, IEEE Special Publication TP-133-0, IEEE Catalog No. 99TP133-0, 1999.
6. IEC TR 60071-4, Insulation co-ordination—Part 4: Computational guide to insulation co-ordination and modelling of electrical networks, 2004.
7. A.I. Ramirez, A. Semlyen, and R. Iravani, Modeling nonuniform transmission lines for time domain simulation of electromagnetic transients, *IEEE Transactions on Power Delivery*, 18(3), 968–974, July 2003.
8. A. Ramirez, J.L. Naredo, and P. Moreno, Full frequency dependent line model for electromagnetic transient simulation including lumped and distributed sources, *IEEE Transactions on Power Delivery*, 20(1), 292–299, January 2005.
9. P.S. Maruvada, D.H. Nguyen, and H. Hamadani-Zadeh, Studies on modeling corona attenuation of dynamic overvoltages, *IEEE Transactions on Power Delivery*, 4(2), 1441–1449, April 1989.
10. S. Carneiro and J.R. Martí, Evaluation of corona and line models in electromagnetic transients simulations, *IEEE Transactions on Power Delivery*, 6(1), 334–342, January 1991.
11. J.R. Carson, Wave propagation in overhead wires with ground return, *Bell System Technical Journal*, 5, 539–554, October 1926.
12. S.A. Schelkunoff, The electromagnetic theory of coaxial transmission lines and cylindrical shields, *Bell System Technical Journal*, 13, 532–579, 1934.
13. L.M. Wedepohl and D.J. Wilcox, Transient analysis of underground power transmission system; system-model and wave propagation characteristics, *Proceedings of the IEE*, 120(2), 252–259, February 1973.
14. A. Deri, G. Tevan, A. Semlyen, and A. Castanheira, The complex ground return plane. A simplified model for homogeneous and multi-layer earth return, *IEEE Transactions on Power Apparatus and Systems*, 100(8), 3686–3693, August 1981.
15. H.W. Dommel, *EMTP Theory Book*, Portland, Oregon: Bonneville Power Administration, 1996.
16. J.A. Martinez-Velasco and J.R. Martí, Electromagnetic transients analysis, in *Electric Energy Systems*, Chapter 12, A. Gómez-Expósito, A. Conejo, and C. Cañizares (Eds.), Boca Raton, FL: CRC Press, 2008.

17. R.H. Galloway, W.B. Shorrocks, and L.M. Wedepohl, Calculation of electrical parameters for short and long polyphase transmission lines, *Proceedings of the IEE*, 111(12), 2051–2059, December 1964.

18. R.E. Judkins and D.E. Nordell, Electromagnetic effects of overhead transmission lines practical problems, safeguards and methods of calculation, *IEEE Transactions on Power Apparatus and Systems*, 93, 901–902, May/June 1974.

19. W.H. Wise, Propagation of HF currents in ground return circuits, *Proceedings of the IEE*, 22, 522–527, 1934.

20. H.W. Dommel, Discussion of Electromagnetic effects of overhead transmission lines practical problems, safeguards and methods of calculation, *IEEE Transactions on Power Apparatus and Systems*, 93(3), 900–901, May/June 1974.

21. J.R. Wait and K.P. Spies, On the image representation of the quasi-static fields of a line current source above the ground, *Canadian Journal of Physics*, 47, 2731–2733, December 1969.

22. J.R. Wait, Theory of wave propagation along a thin wire parallel to an interface, *Radio Science*, 7(6), 675–679, June 1972.

23. L.M. Wedepohl and A.E. Efthymiadis, Wave propagation in transmission lines over lossy ground: A new, complete field solution, *Proceedings of the IEE*, 125(6), 505–510, June 1978.

24. C. Gary, Approche complète de la propagation multifilaire en haute fréquence par utilization des matrices complexes, *EDF Bulletin de la Direction des Ètudes et Reserches-Serie B*, no. 3/4, pp. 5–20, 1976.

25. R.L. Pogorzelski and D.C. Chang, On the validity of the thin wire approximation in analysis of wave propagation along a wire over ground, *Radio Science*, 12(5), 699–707, September/October 1977.

26. E.D. Sunde, *Earth Conduction Effects in Transmission Systems*, New York: Dover, 1968.

27. S. Ramo, J.R. Whinnery, and T. van Duzer, *Fields and Waves in Communication Electronics*, 3rd edn., New York: Wiley, 1994.

28. J.A. Martinez-Velasco and B. Gustavsen, Overview of overhead line models and their representation in digital simulations, *4th IPST*, Rio de Janeiro, Brazil, June 24–28, 2001.

29. W. Scott Meyer and H.W. Dommel, Numerical modeling of frequency dependent transmission-line parameters in an electromagnetic transients program, *IEEE Transactions on Power Apparatus Systems*, 939(5), 1401–1409, September/October 1974.

30. A. Semlyen and A. Dabuleanu, Fast and accurate switching transient calculations on transmission lines with ground return using recursive convolutions, *IEEE Transactions on Power Apparatus and Systems*, 94(2), 561–571, March/April 1975.

31. A. Semlyen, Contributions to the theory of calculation of electromagnetic transients on transmission lines with frequency dependent parameters, *IEEE Transactions on Power Apparatus and Systems*, 100(2), 848–856, February 1981.

32. A. Ametani, A highly efficient method for calculating transmission line transients, *IEEE Transactions on Power Apparatus and Systems*, 95(5), 1545–1551, September/October 1976.

33. J.F. Hauer, State-space modeling of transmission line dynamics via nonlinear optimization, *IEEE Transactions on Power Apparatus and Systems*, 100(12), 4918–4924, December 1981.

34. J.R. Marti, Accurate modeling of frequency-dependent transmission lines in electromagnetic transient simulations, *IEEE Transactions on Power Apparatus and Systems*, 101(1), 147–155, January 1982.

35. M.C. Tavares, J. Pissolato, and C.M. Portela, Mode domain multiphase transmission line model—Use in transients analysis, *IEEE Transactions on Power Delivery*, 14(4), 1533–1544, October 1999.

36. L. Marti, Simulation of transients in underground cables with frequency-dependent modal transformation matrices, *IEEE Transactions on Power Delivery*, 3(3), 1099–1110, July 1988.

37. L.M. Wedepohl, H.V. Nguyen, and G.D. Irwin, Frequency-dependent transformation matrices for untransposed transmission lines using Newton-Raphson method, *IEEE Transactions on Power Systems*, 11(3), 1538–1546, August 1996.

38. B. Gustavsen and A. Semlyen, Simulation of transmission line transients using vector fitting and modal decomposition, *IEEE Transactions on Power Delivery*, 13(2), 605–614, April 1998.

39. H. Nakanishi and A. Ametani, Transient calculation of a transmission line using superposition law, *IEE Proceedings*, 133, Pt. C, no. 5, 263–269, July 1986.

40. R. Mahmutcehajic, S. Babic, R. Garcanovic, and S. Carsimamovic, Digital simulation of electromagnetic wave propagation in a multiconductor transmission system using the superposition principle and Hartley transform, *IEEE Transactions on Power Delivery*, 8(3), 1377–1385, July 1993.

41. B. Gustavsen, J. Sletbak, and T. Henriksen, Calculation of electromagnetic transients in transmission cables and lines taking frequency dependent effects accurately into account, *IEEE Transactions on Power Delivery*, 10(2), 1076–1084, April 1995.

42. G. Angelidis and A. Semlyen, Direct phase-domain calculation of transmission line transients using two-sided recursions, *IEEE Transactions on Power Delivery*, 10(2), 941–949, April 1995.

43. T. Noda, N. Nagaoka, and A. Ametani, Phase domain modeling of frequency-dependent transmission lines by means of an ARMA model, *IEEE Transactions on Power Delivery*, 11(1), 401–411, January 1996.

44. T. Noda, N. Nagaoka, and A. Ametani, Further improvements to a phase-domain ARMA line model in terms of convolution, steady-state initialization, and stability, *IEEE Transactions on Power Delivery*, 12(3), 1327–1334, July 1997.

45. H.V. Nguyen, H.W. Dommel, and J.R. Marti, Direct phase-domain modeling of frequency-dependent overhead transmission lines, *IEEE Transactions on Power Delivery*, 12(3), 1335–1342, July 1997.

46. B. Gustavsen and A. Semlyen, Combined phase and modal domain calculation of transmission line transients based on vector fitting, *IEEE Transactions on Power Delivery*, 13(2), 596–604, April 1998.

47. B. Gustavsen and A. Semlyen, Calculation of transmission line transients using polar decomposition, *IEEE Transactions on Power Delivery*, 13(3), 855–862, July 1998.

48. F. Castellanos, J.R. Marti, and F. Marcano, Phase-domain multiphase transmission line models, *Electrical Power and Energy Systems*, 19(4), 241–248, May 1997.

49. A. Morched, B. Gustavsen, and M. Tartibi, A universal model for accurate calculation of electromagnetic transients on overhead lines and underground cables, *IEEE Transactions on Power Delivery*, 14(3), 1032–1038, July 1999.

50. B. Gustavsen, G. Irwin, R. Mangelrød, D. Brandt, and K. Kent, Transmission line models for the simulation of interaction phenomena between parallel AC and DC overhead lines, *3rd IPST*, Budapest, Hungary, June 20–24, 1999.

51. F. Castellanos and J.R. Marti, Full frequency-dependent phase-domain transmission line model, *IEEE Transactions on Power Systems*, 12(3), 1331–1339, August 1997.

52. J.L. Naredo, A.C. Soudack, and J.R. Martí, Simulation of transients on transmission lines with corona via the method of characteristics, *IEE Proceedings. Generation, Transmission and Distribution*, 142(1), 81–87, January 1995.

53. P. Moreno and A. Ramirez, Implementation of the numerical Laplace transform: A review, *IEEE Transactions on Power Delivery*, 23(4), 2599–2609, October 2008.

54. IEEE Task Force on Fast Front Transients (A. Imece, Chairman), Modeling guidelines for fast transients, *IEEE Transactions on Power Delivery*, 11(1), 493–506, January 1996.

55. J.A. Martinez, B. Gustavsen, and D. Durbak, Parameter determination for modeling systems transients. Part I: Overhead lines, *IEEE Transactions on Power Delivery*, 20(3), 2038–2044, July 2005.

56. M. Abdel-Salam and E. Keith Stanek, Mathematical-physical model of corona from surges on high-voltage lines, *IEEE Transactions on Industry Applications*, 23(3), 481–489, May/June 1987.

57. X.-R. Li, O.P. Malik, and Z.-D. Zhao, A practical mathematical model of corona for calculation of transients on transmission lines, *IEEE Transactions on Power Delivery*, 4(2), 1145–1152, April 1989.

58. H.J. Köster and K.H. Weck, Attenuation of travelling waves by impulse corona, *CIGRE Report*, Study Committee 33: Overvoltages and Insulation Coordination, 1981.

59. C. Gary, D. Cristescu, and G. Dragan, Distorsion and attenuation of travelling waves caused by transient corona, *CIGRE Report*, Study Committee 33: Overvoltages and Insulation Coordination, 1989.

60. A. Ramirez, Electromagnetic transients in transmission line considering frequency-dependent parameters and corona effect, (in Spanish), M.A.Sc. thesis, Cinvestav, Guadalajara, Mexico, 1998.

61. X.-R. Li, O.P. Malik, and Z.-D. Zhao, Computation of transmission lines transients including corona effects, *IEEE Transactions on Power Delivery*, 4(3), 1816–1821, July 1989.

62. C.F. Wagner, I.W. Gross, and B.L. Lloyd, High-voltage impulse test on transmission lines, *AIEE Transactions*, 73, Pt. III-A, 196–210, April 1954.
63. A. Ramirez, J.L. Naredo, P. Moreno, and L. Guardado, Electromagnetic transients in overhead lines considering frequency dependence and corona effect via the method of characteristics, *Electrical Power and Energy Systems*, 23, 179–188, 2001.
64. M. Davila, J.L. Naredo, and P. Moreno, A characteristics model for electromagnetic transient analysis of multiconductor lines with corona and skin effect, *6th IPST*, Montréal, Quebec, Canada, June 19–23, 2005.
65. L. Grcev and F. Rachidi, On tower impedance for transient analysis, *IEEE Transactions on Power Delivery*, 19(3), 1238–1244, July 2004.
66. C.F. Wagner and A.R. Hileman, A new approach to the calculation of the lightning performance of transmission lines—Part III, *AIEE Transactions Part III*, 79(3), 589–603, October 1960.
67. M.A. Sargent and M. Darveniza, Tower surge impedance, *IEEE Transactions on Power Apparatus and Systems*, 88(3), 680–687, May 1969.
68. W.A. Chisholm, Y.L. Chow, and K.D. Srivastava, Lightning surge response of transmission towers, *IEEE Transactions on Power Apparatus and Systems*, 102(9), 3232–3242, September 1983.
69. W.A. Chisholm, Y.L. Chow, and K.D. Srivastava, Travel time of transmission towers, *IEEE Transactions on Power Apparatus and Systems*, 104(10), 2922–2928, October 1985.
70. M. Ishii, T. Kawamura, T. Kouno, E. Ohsaki, K. Shiokawa, K. Murotani, and T. Higuchi, Multistory transmission tower model for lightning surge analysis, *IEEE Transactions on Power Delivery*, 6(3), 1327–1335, July 1991.
71. A. Ametani, Y. Kasai, J. Sawada, A. Mochizuki, and T. Yamada, Frequency-dependent impedance of vertical conductors and a multi-conductor tower model, *IEE Proceedings Generation, Transmission and Distribution*, 141(4), 339–345, July 1994.
72. T. Hara and O. Yamamoto, Modelling of a transmission tower for lightning surge analysis, *IEE Proceedings Generation, Transmission and Distribution*, 143(3), 283–289, May 1996.
73. J.A. Gutierrez, P. Moreno, J.L. Naredo, J.L. Bermudez, M. Paolone, C.A. Nucci, and F. Rachidi, Nonuniform transmission tower model for lightning transient studies, *IEEE Transactions on Power Delivery*, 19(2), 490–496, April 2004.
74. CIGRE WG 33-01, Guide to Procedures for Estimating the Lightning Performance of Transmission Lines, CIGRE Brochure no. 63, 1991.
75. Y. Baba and M. Ishii, Numerical electromagnetic field analysis on measuring methods of tower surge response, *IEEE Transactions on Power Delivery*, 14(2), 630–635, April 1999.
76. J.A. Martinez and F. Castro-Aranda, Tower modeling for lightning analysis of overhead transmission lines, *IEEE PES General Meeting*, San Francisco, June 2005.
77. M.E. Almeida and M.T. Correia de Barros, Tower modelling for lightning surge analysis using electro-magnetic transients program, *IEE Proceedings Generation, Transmission and Distribution*, 141(6), 637–639, November 1994.
78. C. Menemenlis and Z.T. Chun, Wave propagation of nonuniform lines, *IEEE Transactions on Power Apparatus and Systems*, 101(4), 833–839, April 1982.
79. IEEE Std 1243-1997, IEEE Guide for Improving the Lightning Performance of Transmission Lines, 1997.
80. Y. Baba and M. Ishii, Numerical electromagnetic field analysis on lightning surge response of tower with shield wire, *IEEE Transactions on Power Delivery*, 15(3), 1010–1015, July 2000.
81. W.A. Chisholm and W. Janischewskyj, Lightning surge response of ground electrodes, *IEEE Transactions on Power Delivery*, 14(2), 1329–1337, April 1989.
82. IEEE WG on Lightning Performance of Transmission Lines, Estimating lightning performance of transmission lines II: Updates to analytical models, *IEEE Transactions on Power Delivery*, 8(3), 1254–1267, July 1993.
83. Guide for Transmission Line Grounding: A Roadmap for Design, Testing and Remediation: Part I—Theory Book, EPRI Report 1013594, Palo Alto, CA, November 2006.

84. IEEE WG on Lightning Performance of Transmission Lines, A simplified method for estimating lightning performance of transmission lines, *IEEE Transactions on Power Apparatus and Systems*, 104(4), 919–932, April 1985.

85. T. Yamada, A. Mochizuki, J. Sawada, E. Zaima, T. Kawamura, A. Ametani, M. Ishii, and S. Kato, Experimental evaluation of a UHV tower model for lightning surge analysis, *IEEE Transactions on Power Delivery*, 10(1), 393–402, January 1995.

86. T. Ito, T. Ueda, H. Watanabe, T. Funabashi, and A. Ametani, Lightning flashover on 77-kV systems: Observed voltage bias effects and analysis, *IEEE Transactions on Power Delivery*, 18(2), 545–550, April 2003.

87. ANSI/IEEE Std 80-2000, IEEE Guide for Safety in AC Substation Grounding.

88. EN 50341-1, Overhead electrical lines exceeding AC 45 kV. Part I: Requirements—Common specifications, CENELEC, October 2001.

89. F. Kiesling, P. Nefzger, J.F. Nolasco, and U. Kaintzyk, *Overhead Power Lines*, Berlin, Germany: Springer, 2003.

90. H. Griffiths and N. Pilling, Earthing, in *Advances in High Voltage Engineering*, Chapter 8, A. Haddad and D. Warne (Eds.), London: The Institution of Electrical Engineers, 2004.

91. L.V. Bewley, *Traveling Waves on Transmission Systems*, 2nd edn., New York: Wiley, 1951.

92. J.L. Bermúdez, Lightning currents and electromagnetic fields associated with return strokes to elevated strike objects, École Polytechnique Fédérale de Lausanne, Ph.D. thesis no. 2741, Lausanne, Switzerland, 2003.

93. E.E. Oettle, A new general estimation curve for predicting the impulse impedance of concentrated earth electrodes, *IEEE Transactions on Power Delivery*, 3(43), 2020–2029, October 1988.

94. CIGRE WG C4.4.02, Protection of MV and LV Networks Against Lightning—Part I Common Topics, CIGRE Brochure no. 287, February 2006.

95. P. Chowdhuri, *Electromagnetic Transients in Power Systems*, Chichester: RSP-Wiley, 1996.

96. W.A. Chisholm, Transmission line transients—Grounding, in *Power Systems*, Chapter 12, L.L. Grigsby (Ed.), Boca Raton, FL: CRC Press, 2007.

97. H.B. Dwight, Calculation of resistances to ground, *Electrical Engineering*, 55, 1319–1328, 1936.

98. A.C. Liew and M. Darveniza, Dynamic model of impulse characteristics of concentrated earths, *IEE Proceedings*, 121(2), 123–135, February 1974.

99. R. Rudenberg, Grounding principles and practices I—Fundamental considerations on ground currents, *Electrical Engineering*, 64, 1–13, January 1945.

100. G.F. Tagg, *Earth Resistivity*, London: Newnes, 1964.

101. G.G. Keller and F.C. Frischknecht, *Electrical Methods in Geophysical Prospecting*, New York: Pergamon, 1982.

102. S.L. Loyka, A simple formula for the ground resistance calculation, *IEEE Transactions on Electromagnetic Compatibility*, 41(2), 152–154, May 1999.

103. S. Visacro, A comprehensive approach to the grounding response to lightning currents, *IEEE Transactions on Power Delivery*, 22(1), 381–386, January 2007.

104. L. Grcev and V. Arnautovski, Comparison between simulation and measurement of frequency dependent and transient characteristics of power transmission line grounding, *24th International Conference on Lightning Protection*, Birmingham, U.K., 524–529, 1998.

105. C. Mazzetti and G.M. Veca, Impulse behavior of ground electrodes, *IEEE Transactions on Power Apparatus and Systems*, 102(9), 3148–3156, September 1983.

106. L. Grcev, Impulse efficiency of ground electrodes, *IEEE Transactions on Power Delivery*, 24(1), 441–451, January 2009.

107. L. Grcev and M. Popov, On high-frequency circuit equivalents of a vertical ground rod, *IEEE Transactions on Power Delivery*, 20(2), 1598–1603, April 2005.

108. C. Gary, L'impédance de terre des conducteurs enterrés horizontalement, *International Conference on Lightning and Mountains*, Chamonix, France, 1994.

109. L. Grcev, Improved earthing system design practices for reduction of transient voltages, *CIGRE Session*, Paris, France, Paper 36–302, 1998.

110. A.V. Korsuncev, Application of the theory of similarity to calculation of impulse characteristics of concentrated electrodes, *Elektrichestvo*, no. 5, 31–35, 1958.

111. K.H. Weck, Remarks to the current dependence of tower footing resistances, *CIGRE Session*, Paris, France, Paper 33–85, 1988.

112. F. Popolansky, Generalization of model measurements results of impulse characteristics of concentrated earths, *CIGRE Session*, Paris, France, Paper 33–80, 1980.

113. *Transmission Line Reference Book*. 345 kV and Above, 2nd edn., EPRI, Palo Alto, CA, 1982.

114. CIGRE WG 33.01, Guide for the Evaluation of the Dielectric Strength of External Insulation, Technical Brochure no. 72, 1992.

115. N.L. Allen, Fundamental aspects of air breakdown, in *High Voltage Engineering and Testing*, Chapter 20, H.M. Ryan (Ed.), London: The Institution of Electrical Engineers, 2001.

116. IEEE Std 1313.1-1996, IEEE Standard for Insulation Coordination—Definitions, Principles and Rules.

117. IEC 60071-1, Insulation Co-ordination, Part 1: Definitions, principles and rules, 2006.

118. C. Menemenlis and G. Harbec, Particularities of air insulation behaviour, *IEEE Transactions on Power Apparatus and Systems*, 95(6), 1814–1821, November/December 1976.

119. IEC 60060-1, 1989, High-voltage test techniques—Part 1: General definitions and test requirements.

120. R.L. Witzke and T.J. Bliss, Surge protection of cable connected equipment, *AIEE Transactions*, 69, 527–542, 1950.

121. R.L. Witzke and T.J. Bliss, Co-ordination of lightning arrester location with transformer insulation level, *AIEE Transactions*, 69, 964–975, 1950.

122. A.A. Akopian, V.P. Larionov, and A.S. Torosian, On impulse discharge voltage across high insulation as related to the shape of the voltage wave, *CIGRE Session*, Paris, France, Paper 411, 1954.

123. A.R. Jones, Evaluation of the integration method for analysis of non-standard surge voltages, *AIEE Transactions*, 73, 984–990, 1954.

124. S. Rusck, Effect of non-standard surge voltages on insulation, *CIGRE Session*, Paris, France, Paper 403, 1958.

125. R.O. Caldwell and M. Darveniza, Experimental and analytical studies of the effect of non-standard waveshapes on the impulse strength of external insulation, *IEEE Transactions on Power Apparatus and Systems*, 92(4), 1420–1428, July/August 1973.

126. T. Suzuki and K. Miyake, Experimental study of the breakdown voltage time characteristics of large air-gaps with lightning impulses, *IEEE Transactions on Power Apparatus and Systems*, 96(1), 227–233, January/February 1977.

127. K. Alstad, J. Huse, H.M. Paulsen, A. Schei, H. Wold, T. Henriksen, and A. Rein, Lightning impulse flashover criterion for overhead line insulation, *Proc. ISH*, Milan, Italy, Paper 42.19, 1979.

128. M. Darveniza and A.E. Vlastos, The generalized integration method for predicting impulse volt-time characteristics for non-standard wave shapes—A theoretical basis, *IEEE Transactions on Electrical Insulation*, 23(3), 373–381, June 1988.

129. G. Baldo, B. Hutzler, A. Pigini, and E. Garbagnati, Dielectric strength under fast front overvoltages in reference ambient conditions, in *CIGRE WG 33.01, Guide for the Evaluation of the Dielectric Strength of External Insulation*, Chapter 5, CIGRE Technical Brochure no. 72, 1992.

130. A. Pigini, G. Rizzi, E. Garbagnati, A. Porrino, G. Baldo, and G. Pesavento, Performance of large air gaps under lightning overvoltages: Experimental study and analysis of accuracy of predetermination methods, *IEEE Transactions on Power Delivery*, 4(2), 1379–1392, April 1989.

131. K.H. Weck, Lightning performance of substations, *CIGRE Conference*, SC 33, Rio de Janeiro, Brazil, 1981.

132. K.J. Lloyd and L.E. Zaffanella, Insulation for Switching Impulses, Chapter 11 of *Transmission Line Reference Book*. 345 kV and Above, 2nd edn., EPRI, 1982.

133. AIEE Committee Report, Switching surges; Pt. I, Phase-to-ground voltages, *AIEE Transactions on Power Apparatus and Systems*, 80, Pt. III, 240–261, June 1961.

134. IEEE Committee Report. Switching Surges: Pt. II, Selection of typical waves for insulation coordination, *IEEE Transactions on Power Apparatus and Systems*, 85(5), 1091–1097, October 1966.
135. W. Büsch, Air humidity, an important factor of UHV design, *IEEE Transactions on Power Apparatus and Systems*, 97(6), 2086–2093, November/December 1978.
136. W.C. Guyker, A.R. Hileman, and J.F. Wittibschlager, Full scale tower insulation test for APS 500 kV system, *IEEE Transactions on Power Apparatus and Systems*, 85(6), 614–623, June 1966.
137. J.K. Dillard and A.R. Hileman, UHV transmission tower insulation tests, *IEEE Transactions on Power Apparatus and Systems*, 89(8), 1772–1784, November/December 1970.
138. A.W. Atwood Jr., A.R. Hileman, J.W. Skooglund, and J.F. Wittibschlager, Switching surge tests on simulated and full scale EHV tower insulation systems, *IEEE Transactions on Power Apparatus and Systems*, 84(4), 293–303, April 1965.
139. C. Menemenlis and G. Harbec, Coefficient of variation of the positive-impulse breakdown of long air-gaps, *IEEE Transactions on Power Apparatus and Systems*, 93(3), 916–927, May/June 1974.
140. L. Paris and R. Cortina, Switching and lightning impulse characteristics of large air gaps and long insulator strings, *IEEE Transactions on Power Apparatus and Systems*, 87(4), 947–957, April 1968.
141. L. Paris., A. Taschini, K. H. Schneider, and K. H. Week, Phase-to-ground and phase-to-phase air clearances in substations, *Electra*, no. 29, 29–44, July 1973.
142. G. Gallet, G. LeRoy, R. Lacey, and I. Kromel, General expression for positive switching impulse strength valid up to extra long air gaps, *IEEE Transactions on Power Apparatus and Systems*, 94(6), 1989–1973, November/December 1975.
143. A. Pigini, G. Rizzi, and R. Brambilla, Switching impulse strength of very large air gaps, *3rd ISH*, Milan, Italy, 1979.
144. B. Hutzler, E. Garbagnati, E. Lemke, and A. Pigini, Strength under switching overvoltages in reference ambient conditions, in *CIGRE WG 33.01, Guide for the Evaluation of the Dielectric Strength of External Insulation*, Chapter 4, CIGRE Technical Brochure no. 72, 1992.
145. W. Diesendorf, *Insulation Coordination in High Voltage Electric Power Systems*, London: Butterworth, 1974.
146. W.G. Standring, D.N. Browning, R.C. Hughes, and W.J. Roberts, Impulse flashover of air gaps and insulators in the voltage range 1–2.5 MV, *Proceedings of the IEE*, 110(6), 1082–1088, November 1963.
147. G. Gallet, B. Hutzler, and J. Riu, Analysis of the switching impulse strength of phase-to-phase air gaps, *IEEE Transactions on Power Apparatus and Systems*, 97(2), 485–494, March/April 1978.
148. CIGRE Work Group, Phase-to-phase insulation coordination, *Electra, no.* 64, 137–236, May 1979.
149. D.W. Durbak, A.M. Gole, E.H. Camm, M. Marz, R.C. Degeneff, R.P. O'Leary, R. Natarajan, J.A. Martinez-Velasco, K.C. Lee, A. Morched, R. Shanahan, E.R. Pratico, G.C. Thomann, B. Shperling, A.J.F. Keri, D.A. Woodford, L. Rugeles, V. Rashkes, and A. Sharshar, Modeling guidelines for switching transients, in *Modeling and Analysis of System Transients using Digital Systems*, Chapter 4, A. Gole, J.A. Martinez-Velasco and A.J.F. Keri (Eds.), IEEE Special Publication, TP-133-0, 1998.
150. G.N. Aleksandrov, Y.A. Gerasimov, and P.V. Gorbunov, The electric strength of long insulator strings by lightning impulses, *7th ISH*, Dresden, Germany, 1991.
151. A.M. Mousa, The soil ionization gradient associated with discharge of high currents into concentrated electrodes, *IEEE Transactions on Power Delivery*, 9(3), 1669–1677, July 1994.
152. V.A. Rakov and M.A. Uman, *Lightning: Physics and Effects*, New York: Cambridge University Press, 2003.
153. IEEE TF on Parameters of Lightning Strokes, Parameters of lightning strokes: A review, *IEEE Transactions on Power Delivery*, 20(1), 346–358, January 2005.
154. R.B. Anderson and A.J. Eriksson, Lightning parameters for engineering applications, *Electra*, no. 69, 65–102, March 1980.

155. F. Heidler, J.M. Cvetic, and B.V. Stanic, Calculation of lightning current parameters, *IEEE Transactions on Power Delivery*, 14(2), 399–404, April 1999.
156. G.W. Brown and E.R. Whitehead, Field and analytical studies of transmission line shielding: Part II, *IEEE Transactions on Power Apparatus and Systems*, 88(3), 617–626, May 1969.
157. J.A. Martinez and F. Castro-Aranda, Lightning performance analysis of overhead transmission lines using the EMTP, *IEEE Transactions on Power Delivery*, 20(3), 2200–2210, July 2005.
158. S. Yokoyama, K. Miyake, T. Suzuki, and S. Kanao, Winter lightning on Japan sea coast— Development of measuring system on progressing feature of lightning discharge, *IEEE Transactions on Power Delivery*, 5(3), 1418–1425, July 1990.
159. A. Asakawa, K. Miyake, S. Yokoyama, T. Shindo, T. Yokota, T. Sakai, and M. Ishii, Two types of lightning discharge to a high stack on the coast of the sea of Japan in winter, *IEEE Transactions on Power Delivery*, 12(3), 1222–1231, July 1997.
160. J.A. Martinez, F. Castro-Aranda, and O.P. Hevia, Generación aleatoria de los parámetros del rayo en el cálculo de sobretensiones atmosféricas, (in Spanish), *ALTAE 2003*, San José, Costa Rica, August 18–23, 2003.
161. K.J. Lloyd and H.M. Schneider, Insulation for Power Frequency Voltage, Chapter 10 of *Transmission Line Reference Book. 345 kV and Above*, EPRI, 2nd edition, 1982.
162. A. Pigini, L. Thione, and F. Rizk, Dielectric strength under AC and DC voltages, in *CIGRE WG 33.01, Guide for the Evaluation of the Dielectric Strength of External Insulation*, Chapter 6, CIGRE Technical Brochure no. 72, 1992.
163. F. Amarh, Electric transmission line flashover prediction system, Ph.D. thesis, Arizona State University, PSERC Publication 01–16, May 2001.
164. Y. Aihara, T. Harada, Y. Aoshima, and Y. Ito, Impulse flashover characteristics of long air gaps and atmospheric correction, *IEEE Transactions on Power Apparatus and Systems*, 97(2), 342–348, March/April 1978.
165. A.R. Hileman, Weather and its effect on air insulation specifications, *IEEE Transactions on Power Apparatus and Systems*, 103(10), 3104–3116, October 1984.
166. E. Lemke, A. Fracchia, E. Garbagnati, and A. Pigini, Performance of contaminated insulators under transient overvoltages. Survey of experimental data, in *CIGRE WG 33.01, Guide for the Evaluation of the Dielectric Strength of External Insulation*, Chapter 9, CIGRE Technical Brochure no. 72, 1992.
167. T. Kawamura and K. Naito, Influence of ice and snow, in *CIGRE WG 33.01, Guide for the Evaluation of the Dielectric Strength of External Insulation*, Chapter 10, CIGRE Technical Brochure no. 72, 1992.
168. J. Fonseca, K. Sadursky, A. Britten, M. Moreno, and J. Van Name, Influence of high temperature and combustion particles (presence of fires), in *CIGRE WG 33.01, Guide for the Evaluation of the Dielectric Strength of External Insulation*, Chapter 11, CIGRE Technical Brochure no. 72, 1992.
169. C. Menemenlis, G. Carrara, and P. J. Lambeth, Application of insulators to withstand switching surges I: Switching impulse insulation strength, *IEEE Transactions on Power Delivery*, 4(1), 545–560, January 1989.
170. M. Moreno, A. Pigini, and F. Rizk, Influence of air density on the dielectric strength of air insulation, in *CIGRE WG 33.01, Guide for the Evaluation of the Dielectric Strength of External Insulation*, Chapter 7, CIGRE Technical Brochure no. 72, 1992.
171. T.A. Phillips, A.F. Rohlfs, L.M. Robertson, and R.L. Thomson, The influence of air density on the electrical strength of transmission line insulation, *IEEE Transactions on Power Apparatus and Systems*, 86(4), 948–961, August 1967.
172. T. Harada, Y. Aihara, and Y. Aoshima, Influence of humidity on lightning and switching flashover voltages, *IEEE Transactions on Power Apparatus and Systems*, 90(3), 1433–1442, July/August 1971.
173. J. Kucera and V. Fiklik, Correction for switching impulse flashover voltages for air humidity, *IEEE Transactions on Power Apparatus and Systems*, 89(3), 441–447, March 1970.

3

Insulated Cables

Bjørn Gustavsen, Taku Noda, José L. Naredo, Felipe A. Uribe,
and Juan A. Martinez-Velasco

CONTENTS

3.1 Introduction 138
3.2 Material Properties 139
    3.2.1 Conductive Materials 139
    3.2.2 Insulating Materials 140
    3.2.3 Semiconducting Materials 141
3.3 Cable Designs 142
    3.3.1 Single-Core Self-Contained Cables 142
    3.3.2 Three-Phase Self-Contained Cables 142
    3.3.3 Pipe-Type Cables 143
3.4 Calculation of Cable Parameters 143
    3.4.1 Coaxial Cables 144
        3.4.1.1 Series Impedance Matrix 144
        3.4.1.2 Shunt Admittance Matrix 148
        3.4.1.3 Approximate Parameter Calculations 149
        3.4.1.4 Accuracy of Ground-Return Impedance Evaluation Methods 150
    3.4.2 Pipe-Type Cables 153
        3.4.2.1 Series Impedance Matrix 153
        3.4.2.2 Shunt Admittance Matrix 155
    3.4.3 Cables Laid in a Tunnel or a Trench 155
    3.4.4 Spiral Effect of a Wire Sheath 156
3.5 Grounding of Sheaths and Armors 156
3.6 Sensitivity of Transients to Cable Parameters and Cable Design 157
    3.6.1 Transient Response Sensitivity 157
    3.6.2 Ground-Return Impedance Sensitivity 158
3.7 Input Data Preparation 160
    3.7.1 Introduction 160
    3.7.2 Conversion Procedure 161
        3.7.2.1 Core 161
        3.7.2.2 Insulation and Semiconducting Screens 161
        3.7.2.3 Wire Screen 162
3.8 Cases Studies 162
3.9 Conclusions 173
References 174

3.1 Introduction

The representation of insulated cables for the simulation of electromagnetic transients requires the calculation of cable parameters from geometrical data and material properties [1,2], and the conversion of the cable parameters into a new set of parameters for usage by the transmission line/cable model.

Several line models have been implemented in commonly available Electromagnetic Transients Program (EMTP)-type programs which can represent the frequency dependence of cable systems [3–5]. All these models require the same type of input parameters, namely the series impedance matrix Z and the shunt admittance matrix Y. The basic equations used to represent insulated cables have the following form:

$$Z(\omega) = R(\omega) + j\omega L(\omega) \tag{3.1}$$

$$Y(\omega) = G(\omega) + j\omega C(\omega) \tag{3.2}$$

where R, L, G, and C are respectively the series resistance, series inductance, shunt conductance, and shunt capacitance per unit length of the cable system. These quantities are $(n \times n)$ matrices, n being the number of (parallel) conductors of the cable system. The variable ω reflects that these quantities are calculated as function of frequency.

Most EMTP-type programs have dedicated support routines for calculating cable parameters. These routines have very similar features, so hereinafter they will be given the generic name "Cable Constants" (CC). Z and Y are calculated by means of CC routines, using cable geometry and material properties as input parameters. In general, users must specify:

1. *Geometry*
   - Location of each conductor (x–y coordinates)
   - Inner and outer radii of each conductor
   - Burial depth of the cable system

2. *Material properties*
   - Resistivity, ρ, and relative permeability, μ_r, of all conductors (μ_r is unity for all nonmagnetic materials)
   - Resistivity and relative permeability of the surrounding medium, ρ, μ_r
   - Relative permittivity of each insulating material, ε_r

The calculation of Z and Y from the geometry and material properties follows similar steps for all CC routines [1,2,6]. The main challenge is the calculation of impedance which is based on computing surface impedances and transfer impedances of cylindrical metallic shields, as well as self and mutual ground-return impedances. CC routines differ in the actual expressions that are used in the calculation of these quantities. It is worth noting that these routines take the skin effect into account but neglect proximity effects. A procedure for including proximity effects is given in Yin and Dommel [7].

Sufficiently accurate input parameters are in general more difficult to obtain for cable systems than for overhead lines as the small geometrical distances make the cable parameters highly sensitive to errors in the specified geometry. In addition, it is not straightforward to represent certain features such as wire screens, semiconducting screens, armors, and lossy insulation materials; besides CC routines have some shortcomings in representing certain cable features.

A previous conversion procedure can be required in order to bring the available cable data into a form which can be used as input by a CC routine. This conversion is frequently needed because input cable data can have alternative representations, while CC routines only support one representation and CC routines do not consider certain cable features, such as semiconducting screens and wire screens. Nominal thickness of the various layers (insulation, semiconducting screens), as stated by manufacturers, can be smaller than the actual (design) thickness of the layers; therefore, the information on geometrical data from the manufacturer can be inaccurate from the viewpoint of cable parameter calculations.

This chapter has been organized as follows. The first part is dedicated to discuss about the properties of materials used in cables. The second part presents the most common cable designs. The third and the fourth parts deal, respectively, with the calculation of parameters for the most common types of high-voltage cable systems and the conversion procedure required for actual single-core (SC) coaxial cable systems. Some illustrative examples are detailed in the last part, which also includes a discussion on how to best use the available data to produce a reliable cable model, the effect of inaccurate data on a time-domain simulation and the shortcoming of CC routines in taking into account possible attenuation effects caused by the semiconducting screens. Guidelines on how to apply CC routines to both cables with extruded solid insulation (cross-linked polyethylene [XLPE], PE) and cables with oil-impregnated paper are also provided in this part of the chapter.

3.2 Material Properties

3.2.1 Conductive Materials

Table 3.1 shows appropriate values for the resistivity of some common conductor materials.

Stranded conductors need to be modeled as massive conductors. The resistivity should be increased with the inverse of the fill factor of the conductor surface so as to give the correct resistance of the conductor.

The resistivity of the surrounding ground depends strongly on the soil characteristics, ranging from about $1\,\Omega\cdot m$ (wet soil) to about $10\,k\Omega\cdot m$ (rock). The resistivity of sea water lies between 0.1 and $1\,\Omega\cdot m$.

Submarine cables are normally designed with a (magnetic) steel armor. The armor consists of a number of steel (round or square/flat) wires, or of steel tapes. In the case of a wired armor, the permeability depends on the wire diameter, the laying angle, and the intensity of the circumferential magnetic field.

Bianchi and Luoni [8] obtained curves for the permeability of round wire steel armors due to a magnetic field in the circumferential direction. Their calculations were based

TABLE 3.1

Resistivity of Conductive Materials

Material	Copper	Aluminum	Lead	Steel
$\rho\ (\Omega\cdot m)$	1.72E-8	2.83E-8	22E-8	18E-8

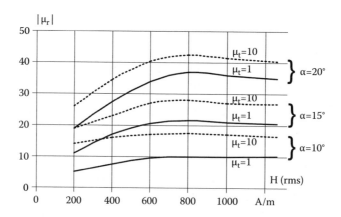

FIGURE 3.1
Circumferential permeability. Steel wire diameter = 5 mm. (From Bianchi, G. and Luoni, G., *IEEE Trans. Power Apparatus Syst.*, 95, 49, 1976. With permission.)

on measured permeability in the longitudinal direction of steel wires, and an assumption of the permeability in the perpendicular direction lying between 1 and 10. Figure 3.1 shows the permeability in the circumferential direction (magnitude) as a function of the circumferential magnetic field strength, for different lay angles (α) and different values of the permeability in the perpendicular direction (μ_t).

3.2.2 Insulating Materials

The relative permittivity of the cable main insulation can be obtained from the manufacturer. Table 3.2 shows typical values for common insulating materials at power frequency. XLPE is an extruded insulation while mass-impregnated and fluid-filled denote paper–oil-based insulations.

Most extruded insulations, including XLPE and PE, are practically lossless up to 1 MHz, whereas paper–oil-type insulations exhibit significant losses also at lower frequencies. The losses are associated with a complex, frequency-dependent permittivity.

$$\varepsilon_r(\omega) = \varepsilon_r'(\omega) - j\varepsilon_r''(\omega), \qquad \tan\delta(\omega) = \varepsilon_r''/\varepsilon_r' \tag{3.3}$$

tan δ being the insulation loss factor. At present, none of the available CC routines allow the entry of a frequency-dependent loss factor, so a constant value has to be specified. However, this leads to nonphysical frequency responses which cannot be accurately fitted by frequency-dependent transmission line models. Therefore, the loss-angle should instead be specified as zero.

Breien and Johansen [9] fitted a Cole–Cole model to the measured frequency response of insulation samples of a low-pressure fluid-filled cable, in the frequency range 10 kHz to 100 MHz. The permittivity is given as

TABLE 3.2

Relative Permittivity of Insulation Materials

Material	XLPE	Mass-Impregnated	Fluid-Filled
Permittivity	2.3	4.2	3.5

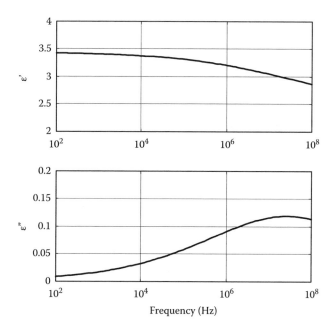

FIGURE 3.2
Complex permittivity of paper–oil insulation, according to Equation 3.4.

$$\varepsilon_r = 2.5 + \frac{0.94}{1 + (j\omega \cdot 6 \cdot 10^{-9})^{0.315}} \tag{3.4}$$

The permittivity at zero frequency is real-valued and equal to 3.44. According to Breien and Johansen [9], the frequency-dependent permittivity causes additional attenuation of pulses shorter than 5 µs. The frequency variation in Equation 3.4 is shown in Figure 3.2.

Equation 3.4 is a function of frequency $s = j\omega$, so its inclusion in a CC routine could be accomplished through an accurately fitted rational approximation. Therefore, CC routines should be modified to allow the permittivity to be specified in a Cole–Cole model, see Equation 3.4, with user-specified parameters. Section 3.8 shows the effect of insulation losses on a transient response.

3.2.3 Semiconducting Materials

The main insulation of high-voltage cables is always sandwiched between two semiconducting layers. This is the case for both extruded insulation and paper–oil insulation. The electric parameters of semiconducting screens can vary between wide limits; Table 3.3 gives indicative values for extruded insulation. The resistivity is required by norm to be smaller than 1E-3 Ω m. However, semiconducting layers can in most cases be taken into account by using a simplistic approach, as shown in Section 3.4.

TABLE 3.3

Parameters of Semiconducting Layers
(Extruded Insulation)

Resistivity (Ω·m)	<1E-3
Relative permittivity	>1000

3.3   Cable Designs

3.3.1   Single-Core Self-Contained Cables

SC cable systems comprise of three separate cables which are coaxial in nature, see Figure 3.3. The insulation system can be based on extruded insulation (e.g., XLPE) or oil-impregnated paper (fluid-filled or mass-impregnated). The core conductor can be hollow in the case of fluid-filled cables.

SC cables for high-voltage applications are always designed with a metallic sheath conductor, see Figure 3.3. The sheath conductor can be made of lead, corrugated aluminum, or copper wires. Such cables are also designed with an inner and an outer semiconducting screen, which are in contact with the core conductor and the sheath conductor, respectively. Submarine cables are normally designed with a steel armor to provide additional mechanical strength.

None of the available CC routines permit the user to directly specify the semiconducting layers. These must therefore be introduced by a modification of the input data. Semiconducting layers, which are in contact with a metallic conductor, can be taken into account by replacing the semiconductors with the insulating material of the main insulation, and increasing the permittivity of the total insulation so that the electric capacitance between the core and the sheath remains unchanged. The validity of this approach has been verified by measurements of up to at least 1 MHz [10]. An illustrative example is included in Section 3.8.1. For a rigorous treatment of semiconducting layers see Ametani et al. [11].

3.3.2   Three-Phase Self-Contained Cables

Three-phase cables essentially consist of three SC cables which are contained in a common shell. The insulation system of each SC cable can be based on extruded insulation or on paper–oil. Regarding the modeling in CC routines, most cable designs can be differentiated into the two designs shown in Figure 3.4:

Design #1: One metallic sheath for each SC cable, SC cables enclosed within metallic pipe (sheath/armor). This design can be directly modeled using the "pipe-type" representation available in some CC routines, where "pipe" denotes the common metallic enclosure. See Section 3.4.3.

Design #2: One metallic sheath for each SC cable, SC cables enclosed within insulating pipe. None of the present CC routines can directly deal with this type of design due to the common *insulating* enclosure. This limitation can be overcome in one of the following ways:

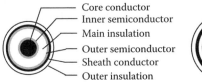

Core conductor
Inner semiconductor
Main insulation
Outer semiconductor
Sheath conductor
Outer insulation

Armor

FIGURE 3.3
SC XLPE cable, with and without armor.

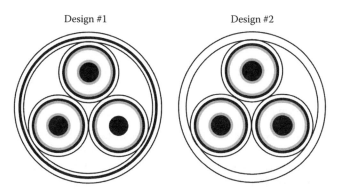

FIGURE 3.4
Three-phase cable designs.

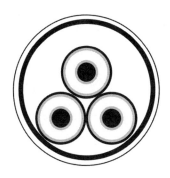

FIGURE 3.5
Pipe-type cable.

a. Place a very thin conductive conductor on the inside of the insulating pipe. The cable can then be represented as a pipe-type cable in a CC routine.

b. Place the three SC cables directly in earth (ignore the insulating pipe).

Both options should give reasonably accurate results when the sheath conductors are grounded at both ends. However, these approaches are not valid when we calculate induced sheath overvoltages.

The space between the SC cables and the enclosing pipe is for both designs filled by a composition of insulating materials; however, CC routines only permit to specify a homogenous material between sheaths and the metallic pipe. Fortunately, the representation of this medium is not very important, as explained in Section 3.6.

3.3.3   Pipe-Type Cables

Pipe-type cables consist of three SC paper cables that are laid asymmetrically within a steel pipe, which is filled with pressurized low viscosity oil or gas, see Figure 3.5. Each SC cable is fitted with a metallic sheath. The sheaths may touch each other. Most CC routines have an input template which is specifically dedicated to this cable design.

3.4   Calculation of Cable Parameters

Main procedures for calculating cable parameters are described in this section. Insulated cables come in such diversity of transversal geometries that a general CC routine seems impractical at present. This section focuses mostly on coaxial configurations. Other transversal geometries should be approximated to this or dealt with through auxiliary methods such as those based on finite element analysis [7] or on subdivision of conductors [12–17].

3.4.1 Coaxial Cables

3.4.1.1 Series Impedance Matrix

The series impedance matrix of a coaxial cable can be obtained by means of a two-step procedure. First, surface and transfer impedances of a hollow conductor are derived; then they are rearranged in the form of the series impedance matrix that can be used for describing traveling-wave propagation [1,18]. Figure 3.6 shows the cross section of a coaxial cable with the three conductors: core, metallic sheath, and armor. Some coaxial cables do not have armor. The materials of the insulation layers differ with cable types. The main insulation can be XLPE or oil-impregnated paper. Insulations A and B are sometimes called a bedding and a plastic sheath, respectively.

Consider a hollow conductor whose inner and outer radii are a and b, respectively. Figure 3.7 shows its cross section. The inner surface impedance Z_{aa} and the outer surface impedance Z_{bb}, both in per unit length (p.u.l.), are given by [18]

$$Z_{aa} = \frac{\rho m}{2\pi a} \frac{I_0(ma)K_1(mb) + K_0(ma)I_1(mb)}{I_1(mb)K_1(ma) - K_1(mb)I_1(ma)} \tag{3.5}$$

$$Z_{bb} = \frac{\rho m}{2\pi b} \frac{I_0(mb)K_1(ma) + K_0(mb)I_1(ma)}{I_1(mb)K_1(ma) - K_1(mb)I_1(ma)} \tag{3.6}$$

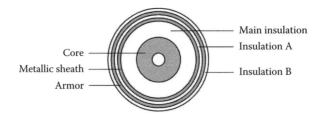

FIGURE 3.6
Cross section of a coaxial cable.

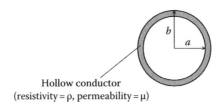

FIGURE 3.7
Cross section of a hollow conductor.

where

$$m = \sqrt{j\omega\mu/\rho} \qquad (3.7)$$

ρ and μ being the resistivity and the permeability of the conductor, respectively. $I_n(\cdot)$ and $K_n(\cdot)$ are the nth order modified Bessel functions of the first and the second kind, respectively.

The p.u.l. transfer impedance Z_{ab} from one surface to the other is calculated as follows [18]:

$$Z_{ab} = \frac{\rho}{2\pi ab} \frac{1}{I_1(mb)K_1(ma) - K_1(mb)I_1(ma)} \qquad (3.8)$$

Z_{aa} can be seen as the p.u.l. impedance of the hollow conductor for the current returning inside the conductor, while Z_{bb} is the p.u.l. impedance for the current returning outside the conductor.

In cases where cable cores are not hollow, only the outer surface impedance is required. Instead of Equation 3.6, one can use the following expression that is equivalent to Equation 2.19, see Chapter 2:

$$Z_{\text{CORE}} = \frac{\rho m}{2\pi b} \frac{I_0(mb)}{I_1(mb)} \qquad (3.9)$$

The impedance of an insulation layer between two hollow conductors, whose inner and outer radii are c and d, respectively, see Figure 3.8, is given by the following expression:

$$Z_i = \frac{j\omega\mu}{2\pi} \ln \frac{d}{c} \qquad (3.10)$$

where μ is the permeability of the insulation.

The ground-return impedance of an underground wire can be calculated by means of the following general expression [19,20]:

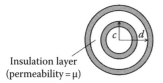

Insulation layer (permeability $= \mu$)

$$Z_g = \frac{\rho m^2}{2\pi} \left[K_0(mD_1) - K_0(mD_2) + \int_{-\infty}^{+\infty} \frac{e^{-Y\sqrt{\lambda^2 + m^2}}}{|\lambda| + \sqrt{\lambda^2 + m^2}} e^{j\lambda x} d\lambda \right] \qquad (3.11)$$

FIGURE 3.8
Insulation layer between two hollow conductors.

where
m is given by Equation 3.7
ρ is the resistivity of the ground soil

To obtain the p.u.l. mutual impedance of two wires, placed at depths of y_i and y_j with horizontal separation x_{ij}, substitute

$$D_1 = \sqrt{x^2 + (y_i - y_j)^2}, \quad D_2 = \sqrt{x^2 + Y^2}, \quad x = x_{ij}, \quad Y = y_i + y_j \tag{3.12}$$

into Equation 3.11.

The p.u.l. self-impedance of a wire placed at a depth of y, with radius r, is obtained by substituting

$$D_1 = r, \quad D_2 = \sqrt{r^2 + Y^2}, \quad x = r, \quad Y = 2y \tag{3.13}$$

into Equation 3.11.

One assumption made for Equation 3.11 is that the soils are homogeneous with constant earth resistivity. An extension to Equation 3.11 for soils with two layers of different resistivities can be found in Tsiamitros et al. [21]. The methodology presented in this reference can be extended with relative ease to include additional layers. Nevertheless, practical applications of these extensions are limited as it is usually very difficult to characterize real soils to the detail required by multilayer earth models.

Consider the coaxial cable shown in Figure 3.6. Assume that I_1 is the current flowing down the core and returning through the sheath, I_2 flows down the sheath and returns through the armor, and I_3 flows down on the armor and its return path is the external ground soil. If V_1, V_2, and V_3 are the voltage differences between the core and the sheath, between the sheath and the armor, and between the armor and the ground, respectively, the relationships between currents and voltages can be expressed as follows [6]:

$$-\frac{\partial}{\partial x}\begin{bmatrix} V_1 \\ V_2 \\ V_3 \end{bmatrix} = \begin{bmatrix} Z_{11} & Z_{12} & 0 \\ Z_{12} & Z_{22} & Z_{23} \\ 0 & Z_{23} & Z_{33} \end{bmatrix}\begin{bmatrix} I_1 \\ I_2 \\ I_3 \end{bmatrix} \tag{3.14}$$

where the matrix elements are given by

$$Z_{11} = Z_{bb(\text{core})} + Z_{i(\text{core-sheath})} + Z_{aa(\text{sheath})}$$

$$Z_{22} = Z_{bb(\text{sheath})} + Z_{i(\text{sheath-armor})} + Z_{aa(\text{armor})}$$

$$Z_{33} = Z_{bb(\text{armor})} + Z_{i(\text{armor-ground})} + Z_g$$

$$Z_{12} = -Z_{ab(\text{sheath})}$$

$$Z_{23} = -Z_{ab(\text{armor})}. \tag{3.15}$$

$Z_{aa \text{ (conductor)}}$, $Z_{bb(\text{conductor})}$, and $Z_{ab \text{ (conductor)}}$ are calculated by substituting the inner and outer radii of the conductor into Equations 3.5, 3.6, and 3.8; $Z_{i \text{ (insulator)}}$ is calculated by substituting the inner and outer radii of the designated insulator layer into Equation 3.10; Z_g is the self-ground-return impedance of the armor obtained from Equation 3.11.

An algebraic manipulation of Equation 3.14 using the relationship

$$V_1 = V_{\text{core}} - V_{\text{sheath}}, \quad V_2 = V_{\text{sheath}} - V_{\text{armor}}, \quad V_3 = V_{\text{armor}}$$

$$I_1 = I_{\text{core}}, \quad I_2 = I_{\text{core}} + I_{\text{sheath}}, \quad I_3 = I_{\text{core}} + I_{\text{sheath}} + I_{\text{armor}} \tag{3.16}$$

gives

$$-\frac{\partial}{\partial x}\begin{bmatrix} V_{\text{core}} \\ V_{\text{sheath}} \\ V_{\text{armor}} \end{bmatrix} = Z_{3\times3}\begin{bmatrix} I_{\text{core}} \\ I_{\text{sheath}} \\ I_{\text{armor}} \end{bmatrix} \tag{3.17}$$

where $Z_{3\times3}$ is the p.u.l. series impedance matrix of the coaxial cable shown in Figure 3.6 when a single coaxial cable is buried alone.

When more than two parallel coaxial cables are buried together, mutual couplings among the cables must be accounted for. The three-phase case is illustrated in the following paragraph. Among the circulating currents I_1, I_2, and I_3, only I_3 has mutual couplings between different cables. Using subscripts a, b, and c to denote the phases of the three cables, Equation 3.14 can be expanded into the following form [6]:

$$-\frac{\partial}{\partial x}\begin{bmatrix} V_{1a} \\ V_{2a} \\ V_{3a} \\ V_{1b} \\ V_{2b} \\ V_{3b} \\ V_{1c} \\ V_{2c} \\ V_{3c} \end{bmatrix} = \begin{bmatrix} Z_{11a} & Z_{12a} & 0 & 0 & 0 & 0 & 0 & 0 & 0 \\ Z_{12a} & Z_{22a} & Z_{23a} & 0 & 0 & 0 & 0 & 0 & 0 \\ 0 & Z_{23a} & Z_{33a} & 0 & 0 & Z_{g,ab} & 0 & 0 & Z_{g,ca} \\ 0 & 0 & 0 & Z_{11b} & Z_{12b} & 0 & 0 & 0 & 0 \\ 0 & 0 & 0 & Z_{12b} & Z_{22b} & Z_{23b} & 0 & 0 & 0 \\ 0 & 0 & Z_{g,ab} & 0 & Z_{23b} & Z_{33b} & 0 & 0 & Z_{g,bc} \\ 0 & 0 & 0 & 0 & 0 & 0 & Z_{11c} & Z_{12c} & 0 \\ 0 & 0 & 0 & 0 & 0 & 0 & Z_{12a} & Z_{22c} & Z_{23c} \\ 0 & 0 & Z_{g,ca} & 0 & 0 & Z_{g,bc} & 0 & Z_{23c} & Z_{33c} \end{bmatrix}\begin{bmatrix} I_{1a} \\ I_{2a} \\ I_{3a} \\ I_{1b} \\ I_{2b} \\ I_{3b} \\ I_{1c} \\ I_{2c} \\ I_{3c} \end{bmatrix} \tag{3.18}$$

where

$Z_{g,ab}$ is the mutual ground-return impedance between the armors of the phases a and b
$Z_{g,bc}$ and $Z_{g,ca}$ are the mutual ground-return impedances between b and c and between c and a, respectively

These mutual ground-return impedances can be obtained from Equation 3.11.

Using the relationship Equation 3.16 for each phase, an algebraic manipulation leads to the following final form:

$$-\frac{\partial}{\partial x}\begin{bmatrix} V_{\text{core},a} \\ V_{\text{sheath},a} \\ V_{\text{armor},a} \\ V_{\text{core},b} \\ V_{\text{sheath},b} \\ V_{\text{armor},b} \\ V_{\text{core},c} \\ V_{\text{sheath},c} \\ V_{\text{armor},c} \end{bmatrix} = Z_{9\times9}\begin{bmatrix} I_{\text{core},a} \\ I_{\text{sheath},a} \\ I_{\text{armor},a} \\ I_{\text{core},b} \\ I_{\text{sheath},b} \\ I_{\text{armor},b} \\ I_{\text{core},c} \\ I_{\text{sheath},c} \\ I_{\text{armor},c} \end{bmatrix} \tag{3.19}$$

where $Z_{9 \times 9}$ is the p.u.l. series impedance matrix of the three-phase coaxial cable.

A general and systematic method to convert the loop impedance matrix of cables into their series impedance matrix is presented in Noda [22].

3.4.1.2 Shunt Admittance Matrix

The p.u.l. capacitance of the insulation layer between the two hollow conductors shown in Figure 3.8 is given by

$$C_i = \frac{2\pi\varepsilon}{\ln \dfrac{d}{c}} \tag{3.20}$$

where ε is the permittivity of the insulation layer.

If the dielectric losses are ignored, the p.u.l. admittance is $Y_i = j\omega C_i$, and the relationship between currents and voltages can be expressed as follows:

$$-\frac{\partial}{\partial x}\begin{bmatrix} I_{\text{core}} \\ I_{\text{sheath}} \\ I_{\text{armor}} \end{bmatrix} = \begin{bmatrix} Y_1 & -Y_1 & 0 \\ -Y_1 & Y_1 + Y_2 & -Y_2 \\ 0 & -Y_2 & Y_2 + Y_3 \end{bmatrix}\begin{bmatrix} V_{\text{core}} \\ V_{\text{sheath}} \\ V_{\text{armor}} \end{bmatrix} \equiv Y_{3\times3}\begin{bmatrix} V_{\text{core}} \\ V_{\text{sheath}} \\ V_{\text{armor}} \end{bmatrix} \tag{3.21}$$

where $Y_{3\times3}$ is the p.u.l. shunt admittance matrix of the coaxial cable shown in Figure 3.6 when a single coaxial cable is buried alone.

If the dielectric losses are considered, a real part is added to Y_i, see Section 3.2.2 and Equation 3.3.

It is important to note that there are no electrostatic couplings between the cables, when more than two parallel coaxial cables are buried together. Thus, the p.u.l. shunt admittance matrix for a three-phase cable can be expressed as follows:

$$Y_{9\times9} = \begin{bmatrix} \begin{array}{ccc|ccc|ccc} Y_{1a} & -Y_{1a} & 0 & & & & & & \\ -Y_{1a} & Y_{1a} + Y_{2a} & -Y_{2a} & & 0 & & & 0 & \\ 0 & -Y_{2a} & Y_{2a} + Y_{3a} & & & & & & \\ \hline & & & Y_{1b} & -Y_{1b} & 0 & & & \\ & 0 & & -Y_{1b} & Y_{1b} + Y_{2b} & -Y_{2b} & & 0 & \\ & & & 0 & -Y_{2b} & Y_{2b} + Y_{3b} & & & \\ \hline & & & & & & Y_{1c} & -Y_{1c} & 0 \\ & 0 & & & 0 & & -Y_{1c} & Y_{1c} + Y_{2c} & -Y_{2c} \\ & & & & & & 0 & -Y_{2c} & Y_{2c} + Y_{3c} \end{array} \end{bmatrix} \tag{3.22}$$

where the subscripts *a*, *b*, and *c* denote the phases of the three cables.

The formulas shown above can be easily modified when an arbitrary number of coaxial conductors are present in each cable and an arbitrary number of parallel coaxial cables are buried together.

3.4.1.3 Approximate Parameter Calculations

The following approximations to Equations 3.5, 3.6, 3.8, and 3.9 have been proposed in [1]:

$$Z_{aa} = \frac{\rho m}{2\pi a}\coth\left[m(b-a)\right] - \frac{\rho}{2\pi a (a+b)} \quad \Omega/m \tag{3.23a}$$

$$Z_{bb} = \frac{\rho m}{2\pi b}\coth\left[m(b-a)\right] - \frac{\rho}{2\pi b (a+b)} \quad \Omega/m \tag{3.23b}$$

$$Z_{ab} = \frac{\rho m}{\pi(a+b)}\operatorname{cosech}\left[m(b-a)\right] \quad \Omega/m \tag{3.23c}$$

$$Z_{\text{CORE}} = \frac{\rho m}{2\pi b}\coth\left(0.777 mb\right) + \frac{0.356\rho}{\pi b^2} \quad \Omega/m \tag{3.23d}$$

Equations 3.23a through 3.23c provide a good accuracy within the following range of values for a and b:

$$\frac{(b-a)}{(b+a)} < \frac{1}{8}$$

Approximation (Equation 3.23d) has a maximum error of 4% at $b = 5/|m|$ [1].

Equations 3.23a through 3.23d are not necessarily less accurate than Equations 3.5, 3.6, 3.8, and 3.9. The latter ones assume cylindrical conductors excited by cylindrical EM field distributions. In practice, however, cable cores and armors often are made of strand layers, and the strands may have round or flattened transversal shapes. In addition, Equation 3.5, 3.6, 3.8, and 3.9 require evaluating modified Bessel functions given in the form of infinite series. Numerical methods to evaluate these special functions are well established for real arguments. Abramowitz and Stegun [23], for instance, provide methods that have been adopted in some CC routines used with EMT programs [6]; see also Noda [22]. Nevertheless, care should be taken as arguments in Equations 3.5 through 3.9 are complex.

Ground-return impedance given by Equation 3.11 involves three terms. The first two require evaluating modified Bessel function $K_0()$. The third term is an integral with no analytic-closed-form solution known as the Pollaczek Integral. Because of the difficulties encountered at solving this integral, CC routines often approximate it by the Carson Integral as follows [6]:

$$\int_{-\infty}^{+\infty} \frac{e^{-\gamma\sqrt{\lambda^2+m^2}}}{|\lambda|+\sqrt{\lambda^2+m^2}}e^{j\lambda x}d\lambda \cong \int_{-\infty}^{+\infty} \frac{e^{-\gamma|\lambda|}}{|\lambda|+\sqrt{\lambda^2+m^2}}e^{j\lambda x}d\lambda \tag{3.24}$$

Other approximations to Pollaczek Integral are available in the form of closed formulas. One of these is due to Saad, Gaba, and Giroux (SGG) and leads to the following expression for Z_g [24]:

$$Z_g = \frac{\rho m^2}{2\pi} \left[K_0 (mD_1) + \frac{2}{4 + (xm)^2} e^{-Ym} \right] \tag{3.25}$$

Another approximation is given by the following formula due to Wedepohl and Wilcox [1]:

$$Z_g = \frac{\rho m^2}{2\pi} \left[-\log\left(\frac{\gamma D_1}{2p} \right) + 0.5 - \frac{2Y}{3p} \right] \tag{3.26}$$

where γ is Euler's constant.

For mutual ground-return impedances, one should apply D_1, x, and Y as in Equation 3.12, while for self-ground-return impedances one should instead use Equation 3.13.

An additional approximation for calculating ground-return impedances, known as the Infinite Earth Model [6], is obtained by considering the ground as a hollow cylinder of infinite external radius. On taking $a = r$ in Equation 3.5 along with the limit of b that tends to infinity, the following expression is obtained for self Z_g:

$$Z_g = \frac{\rho m}{2\pi r} \frac{K_0 (mr)}{K_1 (mr)} \tag{3.27}$$

The expression corresponding to mutual Z_g is [6]:

$$Z_g = \frac{\rho}{2\pi r^2} \frac{K_0 (mD_1)}{\left[K_1 (mr) \right]^2} \tag{3.28}$$

where D_1 is given by Equation 3.12.

3.4.1.4 Accuracy of Ground-Return Impedance Evaluation Methods

Pollaczek integral in Equation 3.11 does not possess a closed-form solution, and its numerical integration cannot always be obtained reliably by general purpose integration routines. An integration method that guarantees its accurate solution is described in Uribe et al. [25]. This method, however, requires a considerable amount of computations. In consequence, it is not recommended here for its incorporation into CC routines. Instead, this method is adopted as reference to evaluate other methods as well as approximate formulas. It has been used, for instance, to assess solutions to Pollaczek Integral by the Gauss-Lobato quadrature routine [26], as well as by the Monte-Carlo quadrature method [27].

Pollaczek integral, in principle, is a function of five physical variables: three electrical ones, μ, ω, and σ, and two geometrical ones, Y and x. The latter variables are defined by Equations 3.12 and 3.13. Uribe et al. [25] have shown, however, that this dependency can be reduced to the following two dimensionless variables only:

$$\xi = Y|p| \tag{3.29}$$

$$\eta = x/D_2 \tag{3.30}$$

TABLE 3.4

Variation Ranges for Physical Variables

0.2	$\leq Y \leq$	2×10^2	(m)
2π	$\leq \omega \leq$	$2\pi \times 10^6$	(rad/s)
10^{-4}	$\leq \sigma \leq$	1	(S/m)
10^{-2}	$\leq x \leq$	10^3	(m)

where $p = 1/m$ is the complex thickness of the Skin Effect layer in the ground.

To assess approximate methods for calculating ground-return impedances, first, the following expression is adopted:

$$\epsilon_{\%} = \frac{\left| Z_{\text{g-REF}} - Z_{\text{g-APPROX}} \right|}{\left| Z_{\text{g-REF}} \right|} \times 100 \tag{3.31}$$

where

$\epsilon_{\%}$ is the percent relative error

$Z_{\text{g-REF}}$ is the ground-return impedance obtained from the reference method

$Z_{\text{g-APPROX}}$ is the ground-return impedance obtained with the method being assessed

Then, the two dimensionless variables ξ and η, defined by Equations 3.29 and 3.30, are introduced. By doing this, the error term $\epsilon_{\%}$ defined by Equation 3.31 becomes a function of these two dimensionless variables only. Next, the following ranges are established for ξ and η to encompass most practical cases:

$$10^{-4} \leq \xi \leq 10^2 \tag{3.32}$$

$$10^{-3} \leq \eta \leq 10^2 \tag{3.33}$$

The ranges have been obtained by considering the variations shown in Table 3.4 for physical variables D_2, x, ω, and σ. Magnetic permeability μ is considered here as constant and is equal to that of vacuum ($\mu_0 = 4\pi \times 10^{-7}$ H/m). Finally, the ranges defined by Equations 3.32 and 3.33, along with error formula (Equation 3.31), are used to produce error contour-maps, whose usefulness is illustrated in the following example.

Example 3.1

Consider a system of two cables with an external radius of $r = 0.04$ m, buried at a depth of $Y = 2$ m, with a horizontal distance of $x = 2$ m between them, the ground conductivity being $\sigma = 0.1$ S/m. Suppose that, for a particular study, the range of relevant frequencies is between 60 Hz and 200 kHz. For the self-impedance case, at a frequency of 60 Hz, Equations 3.13 and 3.29 result in $\xi = 0.0275$, while Equations 3.12 and 3.30 yield $\eta = 0.01$. These two values correspond to point $\mathbf{P_1}$ in Figures 3.9 and 3.10. As the frequency is varied continuously from 60 Hz to 200 kHz, trajectory $\mathbf{P_1}$–$\mathbf{P_2}$ is drawn in these two figures. For $\mathbf{P_2}$, at 200 kHz, $\xi = 1.589$ and η remains constant. This trajectory shows the global errors in the evaluation of Z_g as a particular approximation is employed. A comparison of errors between Carson Integral approximation and SGG formula for the self-impedance evaluation is provided in Figure 3.11a. One can observe here that for the particular case in consideration, the Carson approximation can produce errors above 10%, while those for the SGG formula are below 1%. Trajectory $\mathbf{P_3}$–$\mathbf{P_4}$ in Figures 3.9 and 3.10 corresponds to the values of ξ and η for the mutual impedance evaluation within the range of frequencies from 60 Hz to 200 kHz. Values of ξ for $\mathbf{P_3}$ and $\mathbf{P_4}$ are equal to those of $\mathbf{P_1}$ and $\mathbf{P_2}$, respectively, while η takes a constant value of 0.05. Figure 3.11b shows the error plots for the two methods being considered here. Note that the approximation by Carson produces an error greater than 10% at frequencies above 10 kHz, whereas the one for the SGG formula is less than 5% for the entire

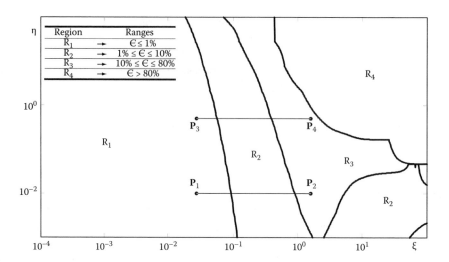

FIGURE 3.9
Example 3.1: Error map for evaluating Z_g by the Carson Integral Method.

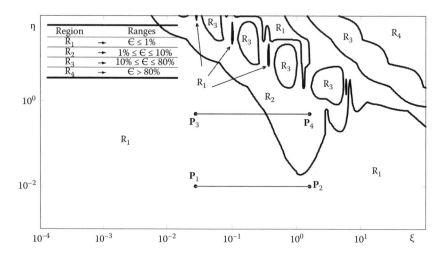

FIGURE 3.10
Example 3.1: Error map for the SGG formula.

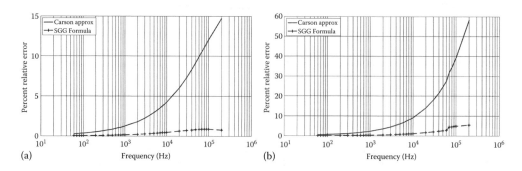

FIGURE 3.11
Example 3.1: (a) Plot of error vs. frequency for self-ground-return impedance. (b) Plot of error vs. frequency for mutual ground-return impedance.

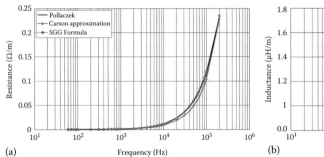
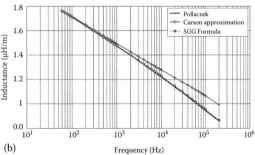

(a) Frequency (Hz)   (b) Frequency (Hz)

FIGURE 3.12
Example 3.1: (a) Plot of self-ground-return resistance vs. frequency. (b) Plot of self-ground-return inductance vs. frequency.

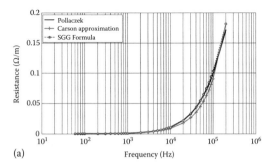
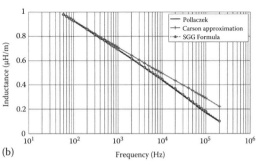

(a) Frequency (Hz)   (b) Frequency (Hz)

FIGURE 3.13
Example 3.1: (a) Plot of mutual ground-return resistance vs. frequency. (b) Plot of mutual ground-return inductance vs. frequency.

range of frequencies. Figure 3.12a and b provides the plots for the self-ground-return resistance and the self-ground-return inductance as obtained by the adopted integration method (labeled as Pollaczek in these figures), by the Carson approximation, and by the SGG formula. Figure 3.13a and b provides the corresponding plots for the mutual ground-return resistance and the mutual ground-return inductance.

3.4.2 Pipe-Type Cables

3.4.2.1 Series Impedance Matrix

Figure 3.14 shows the cross section of a pipe-type cable with a cradle configuration. Since the penetration depth into the pipe at 50/60 Hz is usually smaller than the pipe thickness, it is reasonable to assume that the pipe is the only return path and the ground-return current

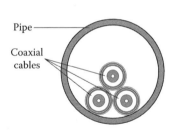

can be ignored. In this case, an infinite pipe thickness can be assumed. A technique to take into account the ground-return current is detailed in [2,6].

For each coaxial cable in the pipe, the impedance matrix for circulating currents given in Equation 3.14 can be used. The matrix elements are calculated using Equation 3.15, except for Z_{33}, which is replaced by

$$Z_{33} = Z_{bb(\text{armor})} + Z_{i(\text{armor–pipe})} + Z_{aa(\text{pipe})} \qquad (3.34)$$

FIGURE 3.14
Cross section of a pipe-type cable.

where $Z_{bb\,(\text{armor})}$ is obtained from Equation 3.6.

Since the conductor geometry of a pipe-type cable is not concentric with respect to the pipe center, the formula for $Z_{i(\text{armor–pipe})}$ is somewhat complicated compared with Equation 3.10:

$$Z_{i(\text{armor–pipe})} = \frac{j\omega\mu}{2\pi} \ln\left[\frac{R}{r}\left\{1-\left(\frac{d}{R}\right)^2\right\}\right] \tag{3.35}$$

where
μ is the permeability of the insulation between the armor and the pipe
R is the radius of the pipe
r is the radius of the armor of interest
d is the offset of the coaxial cable of interest from the pipe center

On the other hand, $Z_{aa(\text{pipe})}$ is calculated from the following expression:

$$Z_{aa(\text{pipe})} = \frac{j\omega\mu}{2\pi}\left[\frac{K_0(mR)}{mRK_1(mR)} + 2\sum_{n=1}^{\infty}\left(\frac{d}{R}\right)^{2n}\frac{K_n(mR)}{n\mu_r K_n(mR) - mRK_n'(mR)}\right] \tag{3.36}$$

where
m is given in Equation 3.7
$\mu = \mu_0\mu_r$ is the permeability of the pipe
$K_n'(\cdot)$ is the derivative of $K_n(\cdot)$

To take into account the mutual impedance among the coaxial cables in a pipe (the three-phase case is shown here), the 9 × 9 impedance matrix for circulating currents given in Equation 3.18 has to be built. Since an infinite pipe thickness is assumed, $Z_{g,ab}$, $Z_{g,bc}$, and $Z_{g,ca}$ are replaced by $Z_{p,ab}$, $Z_{p,bc}$, and $Z_{p,ca}$ (the subscript "p" designates a pipe) and they are deduced by substituting the phase indexes a, b, and c into i and j in the formula below.

$$Z_{p,ij} = \frac{j\omega\mu}{2\pi}\left[\ln\frac{R}{\sqrt{d_i^2 + d_j^2 - 2d_i d_j\cos\theta_{ij}}} + \mu_r\frac{K_0(mR)}{mRK_1(mR)}\right.$$
$$\left. + \sum_{n=1}^{\infty}\left(\frac{d_i d_j}{R^2}\right)^n\cos(n\theta_{ij})\left\{2\mu_r\frac{K_n(mR)}{n\mu_r K_n(mR) - mRK_n'(mR)} - \frac{1}{n}\right\}\right], \tag{3.37}$$

where
d_i is the offset of the i-phase coaxial cable from the pipe center
d_j is the offset of the j-phase coaxial cable from the pipe center
θ_{ij} is the angle that the i-phase and the j-phase cables make with respect to the pipe center

Equations 3.35, 3.36, and 3.37 are from Brown and Rocamora [28]. A method to take into account the saturation effect of a pipe wall was presented in Brown and Rocamora [29].

3.4.2.2 Shunt Admittance Matrix

The inverse of $Y_{3\times3}$ in Equation 3.21 multiplied by $j\omega$ gives the p.u.l. potential coefficient matrix of each coaxial cable in the pipe. If potential coefficients of phases *a*, *b*, and *c* are denoted as P_a, P_b, and P_c, the potential coefficient matrix of the whole cable system, including the pipe, is written in the form

$$P_{9\times9} = \begin{bmatrix} P_a + P_{aa} & P_{ab} & P_{ca} \\ P_{ab} & P_b + P_{bb} & P_{bc} \\ P_{ca} & P_{bc} & P_c + P_{cc} \end{bmatrix} \qquad (3.38)$$

where the submatrices P_{ab}, P_{bc}, and P_{ca} consist of nine identical elements which can be calculated by substituting the phase indexes *a*, *b*, and *c* into *i* and *j* in the following formulas [28]:

$$P_{ii} = \frac{1}{2\pi\varepsilon} \ln\left[\frac{R}{r_i}\left\{ 1 - \left(\frac{d_i}{R}\right)^2 \right\} \right] \qquad (3.39)$$

$$P_{ij} = \frac{1}{2\pi\varepsilon} \ln \frac{R}{\sqrt{d_i^2 + d_j^2 - 2d_id_j\cos\theta_{ij}}} \qquad (3.40)$$

where ε is the permittivity of the insulation between the armors and the pipe.

Finally, the p.u.l. shunt admittance matrix is calculated as follows:

$$Y_{9\times9} = j\omega P_{9\times9}^{-1} \qquad (3.41)$$

3.4.3 Cables Laid in a Tunnel or a Trench

It should be noted that theoretical formulas to calculate series impedance and shunt admittance matrices have been developed only for directly buried coaxial cables and pipe-type cables. However, it is common to install a cable in a tunnel or a trench whose cross section is not circular. In such situations, a cable can be reasonably approximated by a pipe-type cable. Figure 3.15 shows how to approximate a cable installed in a tunnel by means of a pipe-type cable. If the distance from the cable center to the closest wall is D_1 and the distance to the other side of the wall is D_2 as shown in Figure 3.15a, then the tunnel is replaced

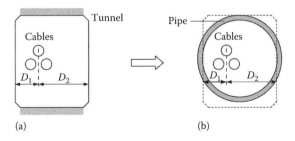

(a)   (b)

FIGURE 3.15 Approximation of a cable installed in a tunnel by a pipe-type cable.

by an equivalent pipe with diameter $D_1 + D_2$ as shown in Figure 3.15b. The cable is placed in the equivalent pipe so that the distances to the closest and the other side walls are kept equal. The resistivity of the pipe is set to that of the tunnel wall. If needed, reinforcing rods in the concrete wall of the tunnel are considered by reducing the resistivity. This approximation gives a close surge impedance value between the outermost conductor of a cable and the tunnel wall. A similar approximation may be used for a cable installed in a trench.

3.4.4 Spiral Effect of a Wire Sheath

For the metallic sheath, lower-voltage XLPE cables often use copper tapes, while higher-voltage cables use an aluminum layer or spiral wires with a stainless-steel layer. In the case of spiral wires with a stainless-steel layer, the sheath current mainly flows through the spiral wires due to the different conductivities of the spiral wires and the stainless-steel layer. This spiral current flow adds an additional inductance to the impedance between the core and the metallic sheath. It is proposed in [30] to add the following p.u.l. impedance to Equation 3.9:

$$\frac{j\omega\mu}{4\pi}\left(\frac{\pi d_{sm}}{p}\right)^2 \tag{3.42}$$

where
d_{sm} is the diameter of the wire sheath
p is the winding pitch

The winding pitch p is shown in Figure 3.16 for different winding types. Measured and calculated values of the surge impedances and the propagation velocities of six different cables are compared in [30], and it is shown that a good agreement is found when the additional impedance term in Equation 3.42 is added.

3.5 Grounding of Sheaths and Armors

CC routines require users to specify the grounding conditions of metallic sheaths and armors. Grounding a conductor in a CC routine means that this conductor will be assumed to be at ground potential at any point along the cable, which results in the conductor being eliminated from Z and Y and thus from the resulting cable model. The main advantage of this option over manual grounding (i.e., grounding is achieved by placing a very small resistor between the

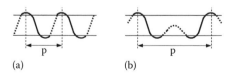

(a)   (b)

FIGURE 3.16
Winding pitch p for different winding types. (a) Helical type and (b) S-Z type.

conductor and ground) is a shorter simulation time. A disadvantage is that one loses the possibility of monitoring the current flowing through the eliminated conductor.

The armor can in nearly all situations be assumed to be at ground potential. In submarine cables, the armor is usually quite thick, thus preventing any high frequency flux to penetrate the armor, so no voltage drop will develop along it. Also, many submarine cables have a wet construction where the conductive sea water is allowed to penetrate the armor.

Sheath conductors are usually grounded at both cable ends. In this situation, the ideal grounding option usually applies because induced (transient) sheath voltages along the cables are in general very small compared to voltages on core conductors. Sheaths must, however, be included when high voltages can develop along them. This includes situations with a high ground potential rise at the cable grounding point (ground fault current, injection of lightning current), and cross-bonded cable systems. Their inclusion may also be needed in situations with the sheath grounded at one end through arresters. The cable sheaths must of course be included in any study of sheath overvoltages.

3.6 Sensitivity of Transients to Cable Parameters and Cable Design

3.6.1 Transient Response Sensitivity

All cable designs described in this chapter (SC, three-phase, pipe-type) are based on three SC cables having a core conductor and a sheath conductor. High frequency cable transients essentially propagate as decoupled coaxial waves between cores and sheaths [31,32], so the transient behavior of the cable is sensitive to the modeling of the core, main insulation, semiconductors, and the metallic sheath. The sensitivity of the coaxial wave can be summarized as follows:

1. Increasing the core resistivity increases the attenuation and slightly decreases propagation velocity.
2. Increasing the sheath resistivity (or decreasing the sheath thickness) increases the attenuation.
3. Increasing the insulation permittivity increases the cable capacitance. This decreases velocity and surge impedance.
4. With a fixed insulation thickness, adding semiconducting screens increases the inductance of the core–sheath loop without changing the capacitance. This decreases velocity and increases surge impedance.

Since the sheath conductors are normally grounded at both ends, the potential along this conductor is low as compared to that of the core conductor, even in transient conditions. As a result, the transients on phase conductors are insensitive to the specified properties of insulating materials external to the sheath.

The magnetic flux external to the sheath is small at frequencies above which the penetration depth δ is smaller than the sheath thickness.

$$\delta = \sqrt{\frac{2}{\omega \mu \sigma}} \qquad (3.43)$$

It follows that high frequency transients are not very sensitive to ground and to the conductors external to the sheaths. The shielding effect increases with decreasing resistance of the sheath.

Some care is needed when modeling armored SC cables at low and intermediate frequencies as the return path of each coaxial mode divides between the sheath and the armor. This makes the propagation characteristics sensitive to the modeling of the armor (and to the separation distance between the sheath and armor). The armor permeability now becomes an important parameter.

In studies of ground fault situations, a significant zero-sequence current at power frequency will flow in conductors external to the sheaths (armor, pipe) as the sheath will not shield the magnetic flux. In such situations, it is necessary to model the armor/pipe with care as they can strongly affect the zero-sequence impedance of the cable and thus the magnitude of the fault current.

3.6.2 Ground-Return Impedance Sensitivity

Often in practice, the physical parameters of a cable system under analysis are given with limited accuracy. A sensitivity analysis is thus convenient. The following example analyzes the effect of cable depth, horizontal distance between cables, and ground resistivities on the ground-return impedance.

Example 3.2

Consider two cables with external radii $r_1 = r_2 = 0.04\,$m, buried at depths $h_1 = h_2 = 1.0\,$m and separated at a horizontal distance $x = 0.5\,$m. The ground resistivity is considered at $\rho = 100\,\Omega \cdot$m and the frequency range of interest is between 20 Hz and 200 kHz.

First, the cable depths are changed to values $h_1 = h_2 = 0.5$, 1.0, 2.0, and 4.0 m. Figure 3.17a and b shows the effects of this variation on the self-ground-return resistance and on the self-ground-return inductance, respectively, while Figure 3.18a and b shows these effects on the mutual ground-return resistance and on the mutual ground-return inductance. In all these four figures, the parameter changes are below 10%.

Second, the cable depths are fixed at $h_1 = h_2 = 1.0\,$m and the horizontal distance is varied at values $x = 0.04$, 0.025, 0.5, 1.0, 2.0, and 4.0 m. Figure 3.19a and b shows the effects of these variations on the ground-return resistance and on the ground-return inductance, respectively. Note that distance $x = 0.04\,$m corresponds to self-ground-return impedance. Figure 3.19a shows that depth variations have little effect on ground-return resistance. Figure 3.19b shows that ground-return

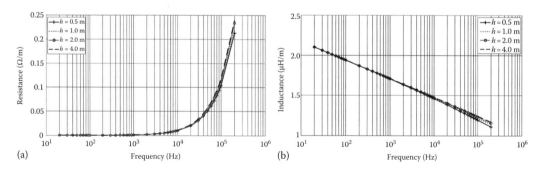

(a)

(b)

FIGURE 3.17
Example 3.2: (a) Effects of cable depth variations on self-ground-return resistance. (b) Effects of cab le depth variations on self-ground-return inductance.

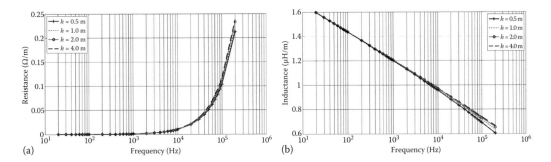

FIGURE 3.18
Example 3.2: (a) Effects of cable depth variations on mutual ground-return resistance. (b) Effects of cable depth variations on mutual ground-return inductance.

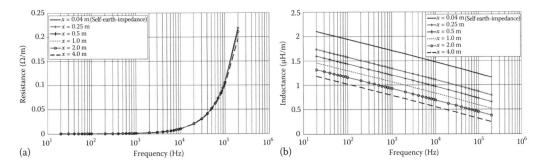

FIGURE 3.19
Example 3.2: (a) Effects of horizontal distance variations on ground-return resistance. (b) Effects of horizontal distance variations on ground-return inductance.

inductance is more sensitive to horizontal distance changes; still though, a 100% variation of this distance produces a change on the ground-return inductance of about a 15% only.

Finally, the horizontal distance between cables is fixed at $x = 0.5$ m and the ground resistivity is varied at values $\rho = 30, 100, 300, 1000, 3000,$ and $10,000\,\Omega\cdot$m. Figure 3.20a and b shows the effects of these variations on the self-ground-return resistance and on the self-ground-return inductance. Figure 3.21a and b shows these effects on the mutual ground-return resistance and on the mutual ground-return inductance. It can be observed in Figures 3.20a and 3.21a that resistivity

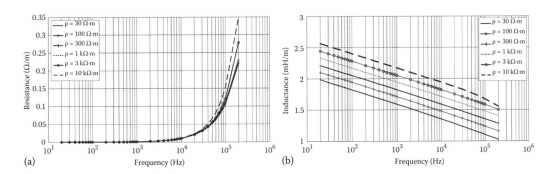

FIGURE 3.20
(a) Effects of ground resistivity variations on self-ground-return resistance. (b) Effects of ground resistivity variations on self-ground-return inductance.

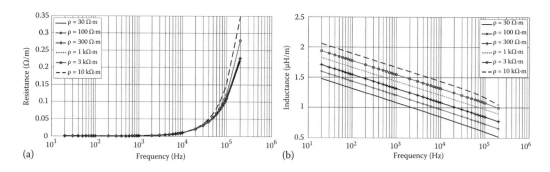

FIGURE 3.21
Example 3.2: (a) Effects of ground resistivity variations on mutual ground-return resistance. (b) Effects of ground resistivity variations on mutual ground-return inductance.

variations between 30 and 300 Ω·m have a very little decreasing effect on the ground-return resistance, while the variations from 300 to 10,000 Ω·m have a larger effect increasing by 20% its value as ρ is tripled. Figures 3.20b and 3.21b show that the ground-return inductance increases regularly about a 5% each time ρ is tripled.

3.7 Input Data Preparation

3.7.1 Introduction

Actual cable designs can be different from those assumed by CC routines, which are used to obtain cable parameters. In particular, users can be forced to decide how to represent the core stranding, the inner semiconducting screen, the outer semiconducting screen, or the wire screen (sheath).

CC routines assume that the relative permittivity of each insulating layer is real and frequency independent; therefore, any relaxation phenomenon in the insulation is neglected. The general form of the equations of a cable becomes:

$$Z(\omega) = R(\omega) + j\omega L(\omega) \tag{3.44}$$

$$Y(\omega) = j\omega C(\omega) \tag{3.45}$$

In addition, CC routines take into account the frequency-dependent skin effect in the conductors, but neglect the proximity effect between parallel cables. This means that a cylindrically symmetrical current distribution is assumed in all conductors and the helical winding effect of the wire screen is not taken into account.

For situations with straight sheaths (i.e., no cross-bonding), high frequency transients propagate mainly as uncoupled coaxial waves within each SC cable. The earth characteristics have in this situation only a mild effect on the resulting phase voltages and phase currents.

The following subsection is focused on the representation of a cable within the protective jacket (oversheath).

3.7.2 Conversion Procedure

3.7.2.1 Core

CC routines require the core data to be given by the resistivity ρ_c and the radius r_1. However, the core conductor is often of stranded design, whereas CC routines assume a homogenous (solid) conductor. This makes it necessary to increase the resistivity of the core material to take into account the space between strands:

$$\rho_c = \rho_c' \frac{\pi r_1^2}{A_c} \tag{3.46}$$

where A_c is the efficient (nominal) cross-sectional area of the core. The resistivity ρ_c' to be used for annealed copper and hard drawn aluminum at 20°C is shown in Table 3.1.

If the manufacturer provides the DC resistance of the core, the sought resistivity can alternatively be calculated as follows:

$$\rho_c = R_{DC} \frac{\pi r_1^2}{l} \tag{3.47}$$

3.7.2.2 Insulation and Semiconducting Screens

The semiconducting screens can have a substantial effect on the propagation characteristics of a cable in terms of velocity, surge impedance, and possibly attenuation [33,34]. CC routines do not allow an explicit representation of the semiconducting screens, so an approximate data conversion procedure must be applied:

1. Calculate the inner radius of the sheath, r_2, as the outer radius of the core, r_1, plus the sum of the thickness of the semiconducting screens and the main insulation.
2. Calculate the relative permittivity ε_{r1} as follows:

$$\varepsilon_{r1} = \frac{C \ln(r_2/r_1)}{2\pi\varepsilon_0} \tag{3.48}$$

where
C is the cable capacitance stated by the manufacturer
$\varepsilon_0 = 8.854\text{E-}12$ (F/m)

If C is unknown, ε_{r1} can instead be calculated based on the relative permittivity ε_{rins} of the main insulation [35]:

$$\varepsilon_{r1} = \varepsilon_{rins} \frac{\ln(r_2/r_1)}{\ln(b/a)} \tag{3.49}$$

where a and b are, respectively, the inner and the outer radius of the insulation.
For XLPE ε_{rins} equals 2.3.

The inner and outer semiconducting screens have a relative permittivity of the order of 1000, due to the large amount of carbon they contain. This implies that the capacitance of the screens is much higher than that of the insulation and will tend to act as a short circuit when calculating the shunt admittance between core and sheath. A similar effect is caused by the ohmic conductivity of the semiconducting screens, which is required by norm to be higher than 1.0E-3 S/m. At the same time, the conductivity of the semiconducting screens is much lower than that of the core and the sheath conductors, so the semiconducting screens do not contribute to the longitudinal current conduction. This implies that, when entering the geometrical data in a CC routine, the user should let the XPLE insulation extend to the surface of the core conductor and the sheath conductor, and increase the relative permittivity to leave the capacitance unaltered. Note that this modeling neglects the possible attenuation caused by the semiconducting screens. The attenuation could have a strong impact on very high frequency transients. This is discussed in Example 3.3, Section 3.8.1 (see Figure 3.26).

3.7.2.3 Wire Screen

When the sheath conductor consists of a wire screen, the most practical procedure is to replace the screen with a tubular conductor having a cross-sectional area equal to the total wire area A_s. With an inner sheath radius of r_2, the outer radius r_3 becomes:

$$r_3 = \sqrt{\frac{A_s}{\pi} + r_2^2} \tag{3.50}$$

3.8 Cases Studies

Example 3.3

The procedures outlined above, to prepare input data, will be applied to a 66 kV cable as shown in Figure 3.22. The following data were provided by the manufacturer (see Section 3.7 for notations) [35]:

- Core cross-sectional area: $A_c = 1000\,\text{mm}^2$
- $C = 0.24\,\text{nF/m}$

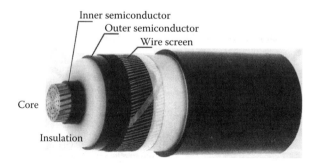

FIGURE 3.22
Example 3.3: SC XLPE test cable.

- R_{DC} = 2.9E-5 Ω/m
- r1 = 19.5 mm
- Thickness of inner insulation screen: 0.8 mm
- Thickness of insulation: 14 mm
- Thickness of outer insulation screen: 0.4 mm
- Wire screen area: A_s = 50 mm^2

This example is aimed at illustrating how to apply the previous recommendations to account for semiconducting screens in CC routines. It also serves to show how to account for variations and uncertainties on manufacturers' data. Finally, it provides a comparison between measurements and simulations of inrush currents. The comparison shows that it is best to obtain capacitance data from a specimen and that a lower accuracy is obtained by neglecting the semiconducting screens.

a. *Data consistency*
Insulation screens can be represented as short circuits when calculating the shunt admittance. This is equivalent to a capacitance between two cylindrical shells deduced from Equation 3.20. In this case a = (19.5 + 0.8) mm = 20.3 mm, and b = a + 14 = 34.3 mm. With a relative permittivity of 2.3 for XLPE, this results in a capacitance of 0.244 nF/m, which is in agreement with the capacitance of 0.24 nF/m provided by the manufacturer.

b. *Data conversion*
- Core
From the manufacturer: r_1 = 19.5 mm.
The resistivity is calculated by means of Equation 3.47: ρ_c = 3.4643E-8 $\Omega \cdot$m.
- Insulation and insulation screens
$r_2 = r_1 + (0.8 + 14 + 0.4) = 34.7$.
From Equation 3.49, ε_{r1} = 2.486.
- Wire screen
The outer radius is calculated using Equation 3.50: r_3 = 34.93 mm.
ρ_s = 1.718E-8 $\Omega \cdot$m (copper).

c. *Manufacturer's data inaccuracy*
Relevant cable standards (e.g., IEC 840, IEC 60502) limit the minimum thickness of each cable layer (in relation to the nominal thickness), but not the maximum thickness. Therefore, the manufacturer is free to use layers thicker than the nominal ones; e.g., to account for dispersity in production and aging effects. This situation is prevalent for the main insulation, the oversheath, and the semiconducting screens.
   After measuring a specimen of the 66 kV cable, it was found that the insulation and, in particular, the semiconducting screens were thicker than stated in the data sheets:

- Thickness of inner insulation screen: 1.5 mm.
- Thickness of insulation: 14.7 mm.
- Thickness of outer insulation screen: 1.1 mm.
- Separation between the outer insulation screen and the center of each conductor in wire screen: 1 mm.

This gives the following modified model:

- r_1 = 19.5 mm
- r_2 = 37.8 mm
- ε_{r2} = 2.856 (from Equation 3.48).

d. *Sensitivity analysis*

At high frequencies, the asymptotic (lossless) propagation velocity and surge impedance are given as

$$v = 1/\sqrt{L_0 C} \tag{3.51}$$

$$Z_c = \sqrt{L_0/C} \tag{3.52}$$

where

$$L_0 = \frac{\mu_0}{2\pi} \ln(r_2/r_1) \tag{3.53}$$

with $\mu_0 = 4\pi E\text{-}7$.

Consider the following three cases, which are used to compare the asymptotic propagation characteristics:

- *Case 1*: The semiconducting screens are neglected; both capacitance and inductance are calculated using Equations 3.20 and 3.53 with $a = r_1$, $b = r_2$.
- *Case 2*: The semiconducting screens are taken into account; capacitance and geometrical data are provided by the manufacturer.
- *Case 3*: The semiconducting screens are taken into account; the capacitance is provided by the manufacturer; geometrical data are deduced from cable specimen.

Table 3.5 shows input data for each case, and the values deduced for the velocity and the characteristic impedance, using the inductance calculated from Equation 3.53. It is obvious from these results that the cable propagation characteristics are highly sensitive to the representation of the core–sheath layers.

e. *Field tests and time-domain simulation*

A field test was carried out on a cable of 6.05 km length, see Figure 3.23. One core conductor was charged up to a 5 kV DC voltage and then shorted to ground. Thus, a negative step voltage was in effect applied to the cable end.

Figure 3.24 shows the measured initial inrush current flowing into the core conductor in p.u. of the DC-voltage. The initial current corresponds to the surge admittance of the cable core–sheath loop, which is the inverse of the surge impedance.

TABLE 3.5

Example 3.3: Sensitivity of Cable Propagation Characteristics

	Case 1	Case 2	Case 3
R_1 (mm)	19.5	19.5	19.5
R_2 (mm)	33.5	34.7	37.8
ε_{r1}	2.300	2.486	2.856
v (m/μs)	197.7	190.1 (−3.8%)	177.4 (−10.3%)
Z_c (Ω)	21.39	21.91 (+2.4%)	23.49 (+9.8%)

Source: Gustavsen, B., *IEEE PES Winter Meeting*, January 28–February 1, 2001. With permission.

The inrush current was simulated using the so-called Universal Line model (ULM) [5]. The CC routine was applied for the three different cases defined above. It is seen that using the cable representation in Case 3 gives a calculated response which is in fairly close agreement with the measured response. The two other representations have a much larger discrepancy. The spike occurring at about 50 µs resulted because of long leads connecting the two cable sections.

f. *Improved modeling of semiconducting screens*
Stone and Boggs [33] suggest modeling the admittance between the core and the sheath by means of the circuit shown in Figure 3.25, in which each semiconducting screen is modeled by a conductance in parallel with a capacitor. With component values obtained from measurements, a good agreement between measured attenuation and calculated attenuation can be obtained in the range of 1–125 MHz. The attenuation effect of the semiconducting screens was strong. Weeks

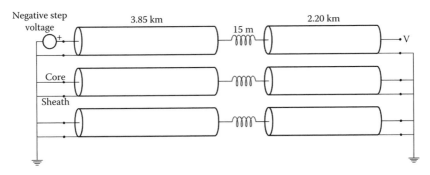

FIGURE 3.23
Example 3.3: Cable test setup.

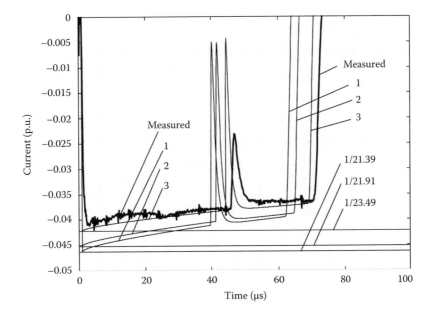

FIGURE 3.24
Example 3.3: Measured and simulated inrush current. (From Gustavsen, B., *IEEE PES Winter Meeting*, January 28–February 1, 2001. With permission.)

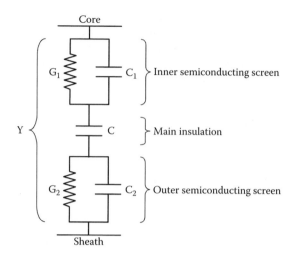

FIGURE 3.25
Example 3.3: Improved model of insulation screens. (From Stone, G.C. and Boggs, S.A., *Proceedings of the Conference on Electrical Insulation and Dielectric Phenomena*, National Academy of Sciences, Washington, DC, pp. 275–280, October 1982. With permission.)

and Min Diao [34] give a systematic investigation of the effects of semiconducting screens on propagation characteristics.

The conductivity and permittivity of the semiconducting screens depend very much on the amount of carbon added, the structure of the carbon, and the type of base polymer. Very high carbon concentrations are used (e.g., 35%). IEC 840 recommends a resistivity below $1000\,\Omega\cdot m$ for the inner screen and below $500\,\Omega\cdot m$ for the outer screen. One manufacturer stated that they use a much lower resistivity, typically 0.1–$10\,\Omega\cdot m$. The relative permittivity is very high, typically of the order of 1000. The permittivity and conductivity can present a strong frequency dependency.

In order to investigate the possible attenuation effects of the insulation screens of the cable considered in this example, a representation as in Figure 3.25 was employed assuming frequency independent conductances and capacitances. The component values were calculated as follows:

$$C = 0.24 \text{ nF/m (from manufacturer)}$$

$$C_1 = 2\pi\varepsilon_0\varepsilon_r/\ln(r_2/a)$$

$$C_2 = 2\pi\varepsilon_0\varepsilon_r/\ln(a/r_1)$$

$$G_1 = 2\pi\sigma/\ln(r_2/b)$$

$$G_2 = 2\pi\sigma/\ln(a/r_1)$$

where
  a is the outer radius of inner semiconducting screen
  b is the inner radius of outer semiconducting screen
  ε_r is the relative permittivity of semiconducting screens
  σ is the conductivity of semiconducting screens

Figure 3.26 shows the attenuation per km, for a few combinations of σ and ε_r. The curves define to which peak value a sinusoidal voltage of 1 p.u. peak value decays over a distance of 1 km. (The signal decays exponentially as function of length.) The model predicts a significant contribution from the semiconducting screens for a low value of both the relative permittivity (10, 100) and the conductivity (0.001). With the high permittivity (1000), the capacitance tends to short out the conductance, and no appreciable increase of the attenuation is seen. The lowest value for the permittivity (10) is probably unrealistic.

Example 3.4

Consider the three 145 kV SC cable system shown in Figure 3.27. The cable design uses a copper core and XLPE insulation, the core radius and insulation thicknesses being those shown in Table 3.6. Semiconductor layers are taken into account by using Equation 3.49.

Using the ULM [5], the voltage caused by a step voltage excitation is calculated at the receiving end of a 5 km cable, see Figure 3.28. All sheaths are treated as continuously grounded. This example has the purpose of showing the effect of sheath resistances and of insulation losses on the transient response of cables. It also shows the effects of the semiconducting screens on the speed of transient waves.

1. *Sensitivity to sheath resistance*
The resulting step voltage is calculated for the following cable sheaths: 1 mm Pb; 2 mm Pb; 3 mm Pb; 0.215 mm Cu (which represents a 50 mm² wire screen).

The receiving end voltages are shown in Figure 3.29, assuming 1 mm semiconducting layers. It can be seen that reducing the thickness of the lead sheath from 2 mm to 1 mm leads to a strong increase in the attenuation, whereas a reduction from 3 mm to 2 mm has little effect. This can be understood by considering that the dominant frequency component of the transient is about

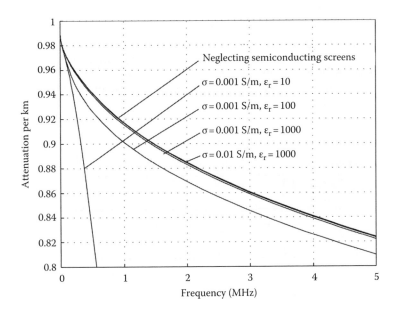

FIGURE 3.26
Example 3.3: Effect of semiconducting screens on attenuation. (From Gustavsen, B., *IEEE PES Winter Meeting*, January 28–February 1, 2001. With permission.)

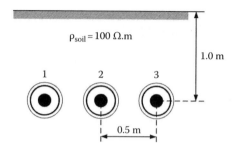

FIGURE 3.27
Example 3.4: Cable configuration.

TABLE 3.6

Example 3.4: Test Cable Parameters

	Radius (m)	Thickness (m)	ρ (Ω·m)	ε_r
Core	20E-3		1.72E-8	
Main insulation		15E-3		2.3
Outer insulation		5E-3		2.3

Source: Gustavsen, B. et al., *IEEE Trans. Power Deliv.*, 20(3), 2045, 2005. With permission.

10 kHz. At this frequency, the penetration depth in lead is 2.6 mm, according to Equation 3.43. Thus, increasing the thickness of the lead sheath beyond 2.6 mm will not lead to a significant change in the response.

2. *Sensitivity to semiconductor thickness*
Assuming a 0.215 mm Cu sheath, the step response is calculated for different thicknesses of the semiconductor layers: 0; 1; 2; 3 mm.

The responses in Figure 3.30 show that the semiconductors lead to a decrease in the propagation speed, as previously explained in Section 3.7.

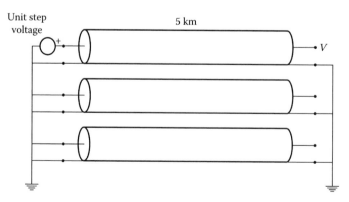

FIGURE 3.28
Example 3.4: Step voltage excitation.

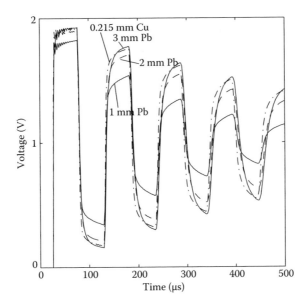

FIGURE 3.29
Example 3.4: Effect of sheath design on overvoltage. (From Gustavsen, B. et al., *IEEE Trans. Power Deliv.*, 20, 2045, 2005. With permission.)

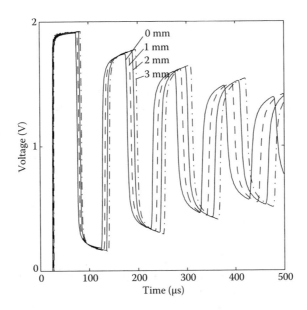

FIGURE 3.30
Example 3.4: Effect of semiconductor thickness on overvoltage. (From Gustavsen, B. et al., *IEEE Trans. on Power Deliv.*, 20, 2045, 2005. With permission.)

3. *Sensitivity to insulation losses*
In this new calculation the XLPE main insulation is replaced by a paper–oil insulation. It is further assumed that the cable has a 2 mm lead sheath and no semiconducting screens. The open end voltage is calculated by applying a 2 μs and a 10 μs width square voltage pulse. The simulation is performed with the following representations of the main insulation:

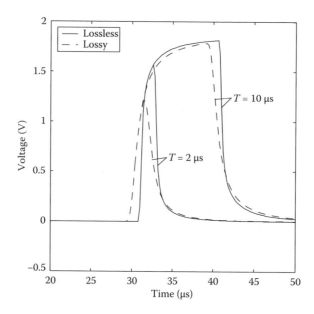

FIGURE 3.31
Example 3.4: Effect of insulation losses on overvoltage. (From Gustavsen, B. et al., *IEEE Trans. Power Deliv.*, 20, 2045, 2005. With permission.)

  a. Lossless insulation, $\varepsilon_r = 3.44$; i.e., DC-value in Equation 3.4.
  b. Lossy insulation by Equation 3.4.

Figure 3.31 shows an expanded view of the initial transient (receiving) end. It can be seen that the lossy insulation gives a much stronger reduction of the peak value for the narrow pulse (2 μs) than that for the lossless insulation. This reduction is an effect of both attenuation and frequency-dependent velocity. It is further seen that the travel time of the lossy insulation is smaller than that of the lossless insulation, which is caused by the reduction in permittivity at high frequencies, according to Equation 3.4.

Example 3.5

In this example an armor of 5 mm steel wires and a 5 mm outer insulation are incorporated into the cable design. It is further assumed XLPE main insulation, a 2 mm lead sheath, and 1 mm semi-conducting screens. Only one cable is considered.

The resulting voltage of the open-circuit step response is calculated for different values of the armor permeability: $\mu_r = 1, 10, 100$, being the cable length 50 km, see Figure 3.32. It is seen that increasing the permeability strongly increases the effective attenuation of the voltage. The reason is that a permeability increase reduces the penetration depth in the armor, thus increasing the resistance of the inner armor surface impedance. For a 5 km cable length, the significance of the armor was found to be small as the magnetic field would not appreciably penetrate the sheath conductor, due to the increased frequency of the transient.

Example 3.6

This example shows the differences that one can obtain when predicting the transient response of an underground cable system as various approximations are used to evaluate ground-return

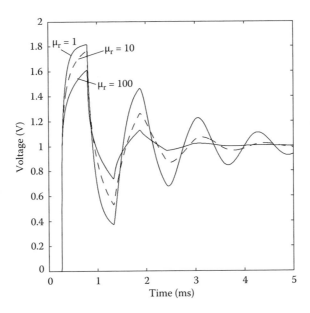

FIGURE 3.32
Example 3.5: Effect of armor permeability on overvoltage. (From Gustavsen, B. et al., *IEEE Trans. Power Deliv.*, 20, 2045, 2005. With permission.)

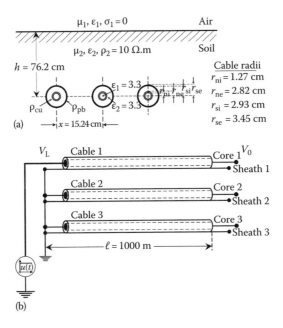

FIGURE 3.33
Example 3.6: Underground transmission system. (a) Transversal configuration; (b) longitudinal configuration.

impedances. Consider the system depicted by Figure 3.33a and b. The first figure provides the transversal geometry, as well as the ground resistivity. All the permitivities and permeabilities are considered equal to those of vacuum. The second figure provides the longitudinal geometry and the system configuration. The cable is energized by injecting a 1 p.u. step at the core of cable 1

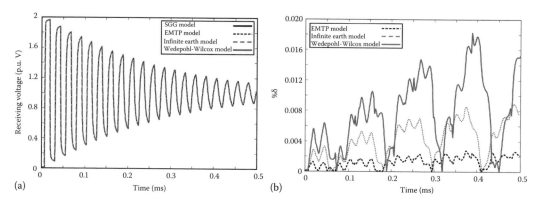

FIGURE 3.34

Example 3.6: Transient response of energized core. (a) Voltage waveforms as obtained with four different methods to estimate Z_g; (b) percent relative differences taking SGG results as reference.

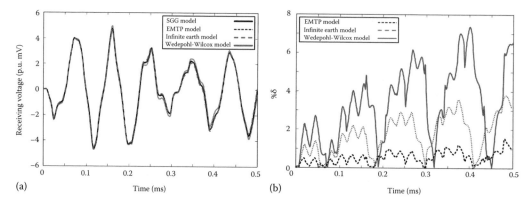

FIGURE 3.35

Example 3.6: Induced transient. (a) Voltage waveforms as obtained with four different methods to estimate Z_g; (b) percent relative differences taking SGG results as reference.

by an ideal voltage source. Note in Figure 3.33b that unenergized cores and sheaths are solidly grounded at the source end and opened at the load-end. Note also in Figure 3.33a that semiconducting screens are not included. The transient response of the system is now obtained by using the numerical Laplace transform (NLT) technique [37], and by applying four different methods for calculating Z_g. These methods are: (1) Carson approximation (Equation 3.24), (2) SGG formula (Equation 3.25), (3) WW formula (Equation 3.26), and (4) Infinite Earth Model formulas (Equations 3.27 and 3.28).

Figure 3.34a shows the four waveforms obtained at the load-end of the energized core. Since the differences among these plots are difficult to detect by eye, Figure 3.34b provides their percent differences taking as reference the one obtained with the SGG formula. Note that the differences are negligible.

Figure 3.35a shows the waveforms of voltage induced at the load-end of core 3 as obtained with the four methods for evaluating Z_g. Now the differences among the four results are more noticeable than those in Figure 3.34a. Figure 3.35b provides the percent differences by taking the SGG results as reference. It can be observed from this last figure that the error at approximating Z_g has a larger effect on the calculation of induced voltages as the distance between the cables is increased. This observation can be of particular relevance when analyzing transient inductions that are caused by underground cables on pipeline systems.

3.9 Conclusions

This chapter has presented conversion procedures aimed at preparing available cable data for application of CC routines and considered cable data for simulating transients on phase conductors of SC cables, three-phase cables, and pipe-type cables. The main conclusions can be summarized as follows:

1. CC routines do not directly apply to SC cables with semiconducting screens, so a conversion procedure is needed before entering the cable data. This chapter describes the needed conversions and also describes the conversions needed for handling the core stranding and wire screens.
2. The nominal thickness of the various insulation and semiconducting cable screens, as stated by manufacturers, can be smaller than those found in actual cables. This can result in a significant error for the propagation characteristics of the cable model.
3. CC routines do not take into account any additional attenuation at very high frequencies resulting from the semiconducting screens.

The chapter has focused also on the importance of a correct modeling of semiconducting screens of SC coaxial-type cables. It is shown that a careless modeling can produce a model with too low surge impedance and a too high propagation velocity. The importance of accurate modeling is strongly dependent on the type of transient study. If the cable is part of a resonant overvoltage phenomenon, an accurate representation of the surge impedance and the propagation velocity is crucial. The conclusions derived from the study on input data can be summarized as follows:

a. It is always necessary to accurately specify the geometry and material properties of the core conductor, the main insulation, and the sheath conductor. It is also important to take the semiconducting layers into account. A simple procedure for achieving the latter is shown in Section 3.4.
b. Lossy effects of paper–oil insulation lead to a strong attenuation and dispersion of narrow pulses. At present, none of the existing CC routines can take this into account.
c. The representation of insulating layers external to the sheath conductors is not very important when the sheaths are grounded at both ends.
d. The representation of metallic conductors external to the sheath conductors is important at low frequencies where the penetration depth exceeds the sheath thickness.
e. Transient voltages can be strongly sensitive to the permeability of any steel armoring when the magnetic field penetrates the sheaths.
f. Ground-return impedance is a more influential parameter in arrays of parallel coaxial cables than in three-phase self-contained cables or in pipe-type cables. For the latter ones, the ground-return currents could even be ignored.
g. Error maps provided in Figures 3.9 and 3.10 can be used on a case by case basis for an assessment of two methods of ground-return impedance estimation.
h. Ground-return impedance presents a low sensitivity to changes in burial depth of cables. Ground-return resistance has a low sensitivity to changes in transversal

distances and in ground resistivity, whereas ground-return inductance presents a higher sensitivity to these two physical parameters. Nevertheless, variations either in cable distances or in ground resistivity produce changes one order of magnitude lower on ground-return inductances.

References

1. L.M. Wedepohl and D.J. Wilcox, Transient analysis of underground power-transmission systems. System-model and wave-propagation characteristics, *Proceedings of the IEEE*, 120(2), 253–260, February 1973.
2. A. Ametani, A general formulation of impedance and admittance of cables, *IEEE Transactions on Power Apparatus and Systems*, 99(3), 902–909, May/June 1980.
3. L. Marti, Simulation of transients in underground cables with frequency-dependent modal transformation matrices, *IEEE Transactions on Power Delivery*, 11(3), 1099–1110, July 1988.
4. T. Noda, N. Nagaoka, and A. Ametani, Phase domain modeling of frequency-dependent transmission line models by means of an ARMA model, *IEEE Transactions on Power Delivery*, 11(1), 401–411, January 1996.
5. A. Morched, B. Gustavsen, and M. Tartibi, A universal model for accurate calculation of electromagnetic transients on overhead lines and underground cables, *IEEE Transactions on Power Delivery*, 14(3), 1032–1038, July 1999.
6. H.W. Dommel, *Electromagnetic Transients Program Reference Manual (EMTP Theory Book)*, Bonneville Power Administration, Portland, August 1986.
7. Y. Yin and H.W. Dommel, Calculation of frequency-dependent impedances of underground power cables with finite element method, *IEEE Transactions on Magnetics*, 25(4), 3025–3027, July 1989.
8. G. Bianchi and G. Luoni, Induced currents and losses in single-core submarine cables, *IEEE Transactions on Power Apparatus and Systems*, 95(1), 49–58, January/February 1976.
9. O. Breien and I. Johansen, Attenuation of traveling waves in single-phase high-voltage cables, *Proceedings of the IEEE*, 118(6), 787–793, June 1971.
10. Private communication with K. Steinbrich and Prof. H. Brakelmann, University of Duisburg-Essen, Duisburg, Germany.
11. A. Ametani, Y. Miyamoto, and N. Nagaoka, Semiconducting layer impedance and its effect on cable wave-propagation and transient characteristics, *IEEE Transactions on Power Delivery*, 19(4), 1523–1531, October 2004.
12. E. Comellini, A. Invernizzi, and G. Manzoni, A computer program for determining electrical resistance and reactance of any transmission line, *IEEE Transactions on Power Apparatus and Systems*, 92(1), 308–314, January/February 1973.
13. R. Lucas and S. Talukdar, Advances in finite element techniques for calculating cable resistances and inductances, *IEEE Transactions on Power Apparatus and Systems*, 97(3), 875–883, May/June 1978.
14. W.T. Weeks, L.L. Wu, M.F. MacAllister, and A. Singh, Resistive and inductive skin effect in rectangular conductors, *IBM Journal of Research and Development*, 23(6), 652–660, November 1979.
15. P. de Arizon and H.W. Dommel, Computation of cable impedances based on subdivision of conductors, *IEEE Transactions on Power Delivery*, 2(1), 21–27, January 1987.
16. D. Zhou and J.R. Marti, Skin effect calculations in pipe-type cables using a linear current subconductor technique, *IEEE Transactions on Power Delivery*, 9(1), 598–604, January 1994.
17. R.A. Rivas and J.R. Marti, Calculation of frequency-dependent parameters of power cables: Matrix partitioning techniques, *IEEE Transactions on Power Delivery*, 17(4), 1085–1092, October 2002.

18. S.A. Schelkunoff, The electromagnetic theory of coaxial transmission lines and cylindrical shields, *Bell System Technical Journal*, 13(4), 532–579, October 1934.

19. F. Pollaczek, On the field produced by an infinitely long wire carrying alternating current, *Elektrische Nachrichtentechnik*, 3, 339–359, 1926.

20. F. Pollaczek, On the induction effects of a single phase ac line, *Elektrische Nachrichtentechnik*, 4, 18–30, 1927.

21. D.A. Tsiamitros, G.K. Papagiannis, and P.S. Dokopoulos, Homogeneous earth approximation of two-layer labridis earth structures: An equivalent resistivity approach, *IEEE Transactions on Power Delivery*, 22(1), 658–666, January 2007.

22. T. Noda, Numerical techniques for accurate evaluation of overhead line and underground cable constants, *Transactions on Electrical and Electronic Engineering*, 3, 549–559, 2008.

23. M. Abramowitz and I.A. Stegun, *Handbook of Mathematical Functions With Formulas, Graphs, and Mathematical Tables*, Dover, New York, 1965.

24. O. Saad, G. Gaba, and M. Giroux, A closed-form approximation for ground return impedance of underground power cables, *IEEE Transactions on Power Delivery*, 11(3), 1536–1545, July 1996.

25. F.A. Uribe, J.L. Naredo, P. Moreno, and L. Guardado, Algorithmic evaluation of underground cable earth impedances, *IEEE Transactions on Power Delivery*, 19(1), 316–322, January 2004.

26. W. Gander and W. Gautschi, Adaptive quadrature—Revisited, *BIT*, 40, 84–101, 2000.

27. X. Legrand, A. Xémard, G. Fleury, P. Auriol, and C.A. Nucci, A quasi-Monte Carlo integration method applied to the computation of the Pollaczek integral, *IEEE Transactions on Power Delivery*, 23(3), 1527–1534, July 2008.

28. G.W. Brown and R.G. Rocamora, Surge propagation in three-phase pipe-type cables, Part I—Unsaturated pipe, *IEEE Transactions on Power Apparatus and Systems.*, 95(1), 89–95, January/February 1976.

29. G.W. Brown and R.G. Rocamora, Surge propagation in three-phase pipe-type cables, Part II—Duplication of field tests including the effects of neutral wires and pipe saturation, *IEEE Transactions on Power Apparatus and Systems*, 96(3), 826–833, May/June 1977.

30. Subcommittee for Power Stations and Substations, Lightning Protection Design Committee of CRIEPI, Guide to Lightning Protection Design of Power Stations, Substations and Underground Transmission Lines, CRIEPI Reports, No. T40, 1995.

31. J.P. Noualy and G. Le Roy, Wave-propagation modes on high-voltage cables, *IEEE Transactions on Power Apparatus and Systems*, 96(1), 158–165, January/February 1977.

32. A. Ametani, Wave propagation characteristics of cables, *IEEE Transactions on Power Apparatus and Systems.*, 99(2), 499–505, March/April 1980.

33. G.C. Stone and S.A. Boggs, Propagation of partial discharge pulses in shielded power cable, *Proceedings of the Conference on Electrical Insulation and Dielectric Phenomena*, National Academy of Sciences, Washington, DC, October 1982, pp. 275–280.

34. W.L. Weeks and Y. Min Diao, Wave propagation in underground power cable, *IEEE Transactions on Power Apparatus and Systems*, 103(10), 2816–2826, October 1984.

35. B. Gustavsen, Panel session on data for modeling system transients insulated cables, *IEEE PES Winter Meeting*, January 28–February 1, 2001.

36. B. Gustavsen, J.A. Martinez, and D. Durbak, Parameter determination for modeling systems transients. Part II: Insulated cables, *IEEE Transactions on Power Delivery*, 20(3), 2045–2050, July 2005.

37. F.A. Uribe, J.L. Naredo, P. Moreno, and L. Guardado, Electromagnetic transients in underground transmission systems through the Numerical Laplace Transform, *International Journal of Electrical Power and Energy Systems*, 24(3), 215–221, March 2002.

4

<hr>

Transformers

<hr>

Francisco de León, Pablo Gómez, Juan A. Martinez-Velasco, and Michel Rioual

CONTENTS

4.1 Introduction 178
4.2 Modeling Guidelines 179
4.3 Standard Tests..181
    4.3.1 Introduction 181
    4.3.2 IEEE Standards..182
    4.3.3 IEC Standards..186
4.4 Parameter Determination for Low-Frequency and Switching Transients.................187
    4.4.1 Introduction 187
    4.4.2 Two-Winding Transformers 188
        4.4.2.1 Single-Phase Transformers...188
        4.4.2.2 Three-Phase Transformers 195
    4.4.3 Three-Winding Transformers..204
        4.4.3.1 Introduction 204
        4.4.3.2 Traditional Equivalent Circuit of a Three-Winding Transformer 205
        4.4.3.3 Saturable Transformer Component...207
        4.4.3.4 Duality Model 207
    4.4.4 Eddy Currents 214
        4.4.4.1 Introduction 214
        4.4.4.2 Eddy Currents in Windings..214
        4.4.4.3 Eddy Currents in Core 215
    4.4.5 Assembling the Models..218
        4.4.5.1 Single-Phase Transformers 218
        4.4.5.2 Three-Phase Transformers 219
4.5 Parameter Determination for Fast and Very Fast Transients.......................220
    4.5.1 Introduction 220
    4.5.2 High-Frequency Transformer Models 221
        4.5.2.1 Lumped-Parameter Model Based on State-Space Equations...........223
        4.5.2.2 Lumped-Parameter Model from Network Analysis 224
        4.5.2.3 Distributed-Parameter Model Based on Single-Phase Transmission Line Theory..225
        4.5.2.4 Distributed-Parameter Model Based on Multiconductor Transmission Line Theory..225
        4.5.2.5 Distributed-Parameter Model Based on Combined STL and MTL Theories..226
        4.5.2.6 Modeling for Transference Analysis 226

          4.5.2.7   Admittance Model from Terminal Measurements (Black Box
                    Model)...228
    4.5.3   Parameter Determination Procedures for High-Frequency Models 230
          4.5.3.1   Capacitance..230
          4.5.3.2   Inductance..233
          4.5.3.3   Conductor Losses..236
          4.5.3.4   Core Losses 237
          4.5.3.5   Loss Component of Capacitance (Turn-to-Turn and
                    Turn-to-Ground)..237
    4.5.4   Examples 238
          4.5.4.1   Capacitance Calculation and Initial Voltage Distribution 238
          4.5.4.2   Behavior of Copper and Core Impedances at High Frequencies..... 240
          4.5.4.3   Sensitivity of Transient Winding Response to Geometrical
                    Inductance Calculation..242
          4.5.4.4   Experimental Setup for Measurement of the Terminal
                    Admittance Matrix of a Distribution Transformer...........................243
References...246

4.1   Introduction

After transmission lines, transformers are the most important components of transmission and distribution systems. The modeling of transmission lines for the study of electromagnetic transients is highly sophisticated. This is so, in part, because all the necessary geometrical data are available to system analysts. For transformers the geometrical data is only available to the manufacturers and it is considered proprietary.

Accurate modeling is very important when the transient is dominated by the transformer behavior. Examples include connection and disconnection of transformers, resonance and ferroresonance, and lightning or switching surges traveling on lines toward a transformer. Many of these transients require proper modeling of the nonlinear behavior of the transformer iron-core caused by magnetization and hysteresis. Others require adequate representation of the frequency dependence of the leakage and/or magnetizing parameters. Very high-frequency transients require the accurate representation of all capacitances: to ground, between winding, intersection and even interturn.

Since not every characteristic of the transformer plays a part in all transients, the commonly used modeling guidelines for transformers given in Table 4.1 are recommended. It is possible to see that there is a clear division between low-frequency and slow-front (switching) transients and fast and very fast transients. In this chapter, after discussing the recommended modeling guidelines in Section 4.2, the methodologies for the estimation of parameters is separated into low-frequency and switching transients and fast and very fast transients.

There exists enough theoretical background for the satisfactory modeling of the transformer subjected to all kind of transients. However, frequently it is not possible to obtain the necessary data to compute the model parameters. To estimate the parameters of a transformer model adequate for the calculation of some electromagnetic transients, one requires knowledge of internal construction details. Some parameters cannot be obtained from terminal measurements only. System analysts frequently cannot determine the parameters for an accurate model and have to carry on by guessing data or select simpler

TABLE 4.1

Modeling Guidelines for Transformers

Parameter/Effect	Low-Frequency Transients	Slow-Front Transients	Fast-Front Transients	Very Fast-Front Transients
Short circuit impedance	Very important[a]	Very important	Important	Negligible
Saturation	Very important[b]	Important	Negligible	Negligible
Iron losses	Important[c]	Negligible	Negligible	Negligible
Eddy currents	Very important	Important	Negligible	Negligible
Capacitive coupling	Negligible[d]	Important	Very important	Very important

[a] Unimportant for ferroresonance. However, it may have an effect on the frequency of the upstream network, especially critical under 300 Hz (harmonics range).

[b] Unimportant for most control interaction cases, harmonic conditions not caused by saturation, and other nonsaturation cases.

[c] Only for resonance phenomena.

[d] Capacitances can be very important for some ferroresonance cases.

models for which parameters exist. This is perhaps the greatest challenge for the proper computation of transients when transformers are involved. Since transformer design information is proprietary, a suggestion to solve the problem is for transformer manufacturers to deliver a set of equivalent circuits with various degrees of sophistication. In this way a manufacturer can provide useful information to the systems analysts without revealing trade secrets.

4.2 Modeling Guidelines

Several factors make transformer modeling a difficult task: transformer's behavior is non-linear and frequency dependent; many topological variations on core and coil construction are possible, see Figures 4.1 and 4.2; there are many physical attributes whose behavior may need to be correctly represented (self- and mutual inductances between coils, leakage fluxes, skin effect and proximity effect in coils, magnetic core saturation, hysteresis and eddy current losses in core, and capacitive effects). Transients during energization are a good example to illustrate difficulties related with transformer modeling [1]: before the flux penetrates the ferromagnetic core, the inductance is basically that of an air core and losses are basically originated in conductors and dielectric; after the flux has penetrated the core completely, the inductance becomes that of an iron core, and losses occur in conductors, core, dielectric, and transformer tank.

Table 4.1 shows the importance of some parameters and effects when modeling a transformer for a specific frequency range. The table is a modified version of that proposed by the CIGRE WG 33-02 [2]. For more details on modeling guidelines for simulation of low-frequency transients see also references [3–7].

Most transformer models available in simulation tools for the analysis of electromagnetic transients are suitable only for low- and mid-frequency transients; that is, phenomena well below the first winding resonance (several kHz). The transients for which adequate

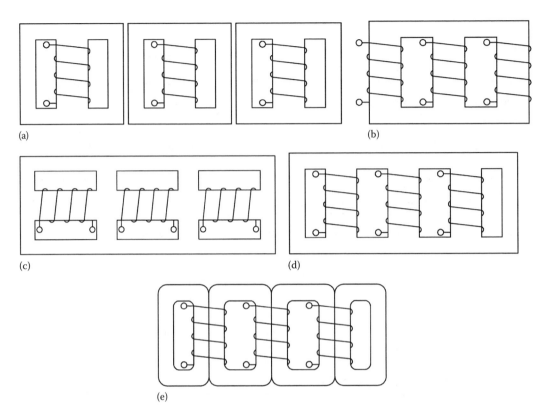

FIGURE 4.1
Three-phase core designs. (a) Triplex core; (b) three-legged stacked-core; (c) shell core; (d) five-legged stacked-core; (e) five-legged wound-core.

models exist include most line switching, capacitor switching, harmonic interactions, ferroresonance, and controller interactions.

Simulation of fast-front transients, for example lightning, should be based on a detailed winding model that is usually not available as a standard model. This type of model could also be developed as a stand-alone model using the transmission line theory (for a strict distributed-parameter representation) or a state-space representation (using a ladder connection of lumped-parameter segments). Assembly of such a model, however, requires a high degree of expertise by the user, as well as substantial knowledge of the transformer construction details. The latter is not typically available except to the transformer manufacturer. System analysts frequently must rely on gray or black box models.

Although much effort has been dedicated to the calculation of transformer parameters from geometric information, they are usually derived from nameplate or test data. The following information is usually available for any transformer: power rating (S), voltage ratings (V_p, V_s), excitation current (I_{exc}), nominal current (I_n), excitation voltage (V_{exc}), nominal voltage (V_n), excitation losses (P_{exc}), core losses (P_{core}), Joule losses (P_J), short circuit current (I_{sh}), short circuit voltage (V_{sh}), short circuit losses (P_{sh}), and in some cases information on the saturation characteristics and capacitances.

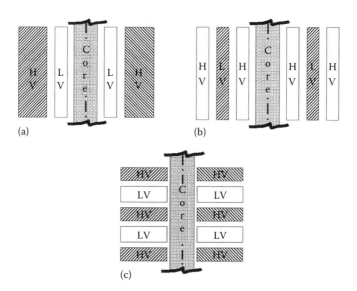

FIGURE 4.2
Winding designs. (a) Concentric design; (b) interleaved design; (c) "pancake" design.

Excitation and short circuit currents, voltages, and losses, must be provided from both positive- and zero-sequence measurements. Standards (e.g., IEC and IEEE) recommend procedures for measuring the above values and provide specifications and requirements for conducting tests [8–17]. However, they neither include expressions for parameter determination nor cover all tests needed to derive the parameters that must be specified in some transformer models. There exists a wide gamut of models and procedures for parameter estimation; however, many of those procedures have been validated by analyzing steady-state conditions only and cannot be extrapolated to transformers of different rating or when the phenomena includes high frequencies; see [5,7] for an overview.

4.3 Standard Tests

4.3.1 Introduction

Parameters to be specified in transformer models can be obtained from laboratory tests. Setups and measurements to be performed in those tests are established in standards. Presently, the most recognized transformer standards are supported by the International Electrotechnical Commission (IEC) and the Institute of Electrical and Electronics Engineers (IEEE). Other transformer standards are recognized at a national level; for example, NEMA and UL in the United States, BS in the United Kingdom, VDE in Germany, AFNOR in France, UNE in Spain, etc.

Unfortunately, present standards do not cover the testing setups and requirements that are needed to obtain parameters for transformer models for higher than power frequency

(50/60 Hz). Basically, IEC and IEEE standards establish tests that are useful for determining parameters of some low-frequency transformer models. Although there is some information on high-frequency transformer models in standards, they are rather exceptional and the provided information is just informative. For instance, Appendix A in IEEE Std C57.138-1998 on distribution transformer voltage distribution depicts an equivalent network for a single-section, layer-wound transformer coil, which is valid for fast-front transients, but no information on how to obtain the parameters of such equivalent network is provided [11].

4.3.2 IEEE Standards

Presently there is a great number of standards that can be applied to transformers. A complete list of IEEE transformer standards classified into ascending order of power ratings has been presented in [12]. In general, power transformers for high-voltage (HV) applications are of two types: dry-type and liquid-immersed transformers.

Tests in standards are classified into three groups [12]:

1. *Routine tests*: Tests made for quality control by the manufacturer on every device or representative samples, or on parts or materials as required, to verify during production that the product meets the design specifications.
2. *Design tests* (type tests in IEC): Tests made to determine the adequacy of the design of a particular type, style, or model of equipment or its component parts to meet its assigned ratings, and to operate satisfactorily under normal service conditions or under special conditions if specified, and to demonstrate compliance with appropriate standards of the industry.
3. *Other tests*: Tests that may be specified by the purchaser in addition to design and routine tests (Examples: impulse, insulation power factor, and audible sound.).

Tables 4.2 and 4.3 show the tests proposed in IEEE standards for liquid-immersed and dry-type transformers, respectively [8,9,13,14]; only the tests related to the determination of transformer models are shown in the tables.

From these tables the following tests and measurements are useful to determine parameters of low-frequency transformer models:

- Resistance measurements
- (Turns) Ratio tests
- No-load loss and excitation (no-load) current tests
- Load loss and impedance voltage tests
- Zero-phase-sequence impedance test

Transformer losses may be classified as no-load loss (excitation loss), load loss (impedance loss), and total loss (the sum of no-load loss and load loss). Load loss is subdivided into I^2R loss and stray loss. Stray loss is determined by subtracting the I^2R loss (calculated from the measured resistance) from the measured load loss (impedance, loss). Stray loss can be defined as the loss due to stray electromagnetic flux in the windings, core, clamps, magnetic shields, enclosure, and tank walls. The stray loss is subdivided into winding stray loss and stray loss in components other than the windings. The winding stray loss includes winding conductor strand eddy-current loss and loss due to circulating currents

TABLE 4.2

Liquid-Immersed Transformer Tests

Tests	Test Classification					
	≤500 kVA			≥501 kVA		
	Routine	Design	Other	Routine	Design	Other
Resistance measurements						
Of all windings on the rated voltage tap, and at tap extremes of the first unit made on a new design		X		X		
Ratio						
Tests on the rated voltage connection and on all tap connections	X			X		
Polarity and phase relation						
Tests on the rated voltage connection	X			X		
Single-phase excitation						
Tests on the rated voltage connection			X			X
No-load losses and excitation current						
100% and 110% of rated voltage and at rated power frequency on the rated voltage tap connection(s) at rated voltage on the rated voltage connection	X		X	X		X
Impedance voltage and load loss						
At rated current and rated frequency on the rated voltage connection and at the tap extremes of the first unit of a new design		X	X	X		
Zero-phase-sequence impedance voltage						X

Source: IEEE Std C57.12.00-2000, IEEE Standard General Requirements for Liquid-Immersed Distribution, Power, and Regulating Transformers. With permission.

between strands or parallel winding circuits, and they may be considered to constitute winding eddy-current loss.

A short description of the purpose of these tests is presented in the following paragraphs. For a definition of the other transformer tests see IEEE Std C57.12.80-2002 [12].

1. Resistance measurement is, in the context of this chapter, of fundamental importance for the calculation of the I^2R component of conductor losses. Since conductor resistance depends on temperature, a reference value is always required and corrections should be included to account for conductor temperature deviation.

2. The (turns) ratio test is aimed at determining the actual ratio of the number of turns in a higher voltage winding to that in a lower voltage winding.

3. No-load losses are those losses that are due to the excitation of the transformer; they include core loss, dielectric loss, conductor loss in the winding due to excitation current, and conductor loss due to circulating current in parallel windings. The main component is the core loss in the transformer core, which is a function of the magnitude, frequency, and waveform of the test voltage. No-load losses change with the excitation voltage, vary with temperature, and are particularly sensitive to waveform distortion. Several other factors affect the no-load losses and current

TABLE 4.3

Dry-Type Transformer Tests

	Test Classification					
	≤500 kVA			≥501 kVA		
Tests	**Routine**	**Design**	**Other**	**Routine**	**Design**	**Other**
Resistance measurements						
Of all windings on the rated voltage tap, and at tap extremes of the first unit made on a new design		X		X		
Ratio						
Tests on the rated voltage connection	X			X		
Polarity and phase relation						
Tests on the rated voltage connection	X			X		
No-load losses and excitation current						
At rated voltage on the rated voltage connection	X[a]			X		
Impedance voltage and load loss						
At rated current and rated frequency on the rated voltage connection and at the tap extremes of the first unit of a new design		X	X	X		
Zero-phase-sequence impedance voltage						X

Source: IEEE Std C57.12.01-1998, IEEE Standard General Requirements for Dry-Type Distribution and Power Transformers Including Those with Solid-Cast and/or Resin-Encapsulated Windings. With permission.

[a] Statistical sampling may be used for this test.

of a transformer. The design-related factors include the type and thickness of core steel, the core configuration, the geometry of core joints, and the core flux density. Factors that cause differences in the no-load losses of transformers of the same design include variability in characteristics of the core steel, mechanical stresses induced in manufacturing, and variation in the joints gaps.

Figure 4.3 shows the connections for no-load loss and excitation current tests using the three-wattmeter method for a three-phase transformer.

4. Excitation (no-load) current is the current that flows in any winding used to excite the transformer when all other windings are open-circuited. The excitation current maintains the rated magnetic flux excitation in the core of the transformer. The excitation current is usually expressed in per unit or in percent of the rated line current of the winding in which it is measured. Excitation current measurements are usually carried out simultaneously with no-load loss tests. For a three-phase transformer, the excitation current is calculated by taking the average of the magnitudes of the three line currents.

5. The load losses of a transformer are those losses incident to a specified load carried by the transformer. Load losses include I^2R loss in the windings due to load current and stray losses due to eddy currents induced by leakage flux in the windings, core clamps, magnetic shields, tank walls, and other conducting parts. Stray losses may also be caused by circulating currents in parallel windings or strands. Load losses are measured by applying a short circuit across either the HV winding or the low-voltage (LV) winding, and applying sufficient voltage across the other winding to cause a specified current to flow in the windings. The power loss

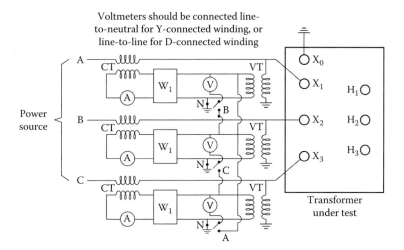

FIGURE 4.3
Connections for no-load loss and excitation current tests (open circuit test) using the three-wattmeter method for a three-phase transformer. (From IEEE Std. C57.12.90-1999, IEEE Standard Test Code for Liquid-Immersed Distribution, Power, and Regulating Transformers; IEEE Std. C57.12.91-2001, IEEE Standard Test Code for Dry-Type Distribution and Power Transformers. With permission.)

within the transformer under these conditions equals the load losses of the transformer at the temperature of test for the specified load current. Figure 4.4 shows the connections for load loss and impedance voltage tests using three-wattmeter method for a three-phase transformer.

6. The impedance voltage of a transformer is the voltage required to circulate rated current through one of two specified windings when the other winding is short-circuited, with the windings connected as for rated-voltage operation. Impedance

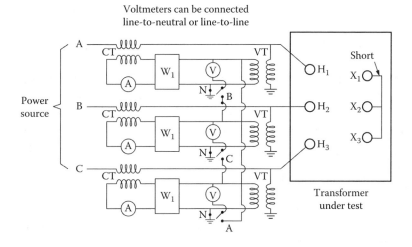

FIGURE 4.4
Connections for load loss and impedance voltage tests (short circuit test) using three-wattmeter method for a three-phase transformer. (From IEEE Std. C57.12.90-1999, IEEE Standard Test Code for Liquid-Immersed Distribution, Power, and Regulating Transformers; IEEE Std. C57.12.91-2001, IEEE Standard Test Code for Dry-Type Distribution and Power Transformers. With permission.)

voltage is usually expressed in per unit, or percent, of the rated voltage of the winding across which the voltage is applied and measured. The impedance voltage is the phasor sum of its two components: the resistive component, called the resistance drop, which is in phase with the current and corresponds to the load losses; the reactive component, called the reactance drop, which is in quadrature with the current and corresponds to the leakage-flux between the windings.

7. The zero-phase-sequence impedance characteristics of three-phase transformers depend on the winding connections and in three-phase transformers on the core construction. Zero-phase-sequence impedance tests apply only to transformers having one or more windings with a physical neutral brought out for external connection. In all tests, one such winding shall be excited at rated frequency between the neutral and the three line terminals connected together. External connection of other windings shall be as described in standards. Transformers with connections other than as described in standards, must be tested as determined by those responsible for design and application.

4.3.3 IEC Standards

The IEC standard for transformers is the Publication 60076, which is divided into several parts, with a combination between technological aspects and electrical ones, see [15–17]. For the determination of transformer impedances, IEC Std 60076-8 [17] describes the tests to be performed.

The tests described in the IEC standards are very similar to those of the IEEE standards. According to IEC Std 60076-1 [16], the tests are also classified into three groups: routine, type, and special tests. Below, all standardized tests are listed, but only those relevant to parameter determination are described with some details.

Routine tests include the following:

1. Measurement of winding resistance. Direct current shall be used for all resistance measurements and care shall be taken that the effects of self-induction are minimized. The resistance should be corrected using the appropriate formula to the reference temperature.

2. Measurement of voltage ratio and check of phase displacement. The voltage ratio shall be measured on each tap. The polarity of single-phase transformers and the connection symbol of three-phase transformers shall be checked.

3. Measurement of short circuit impedance and load loss. The short circuit impedance and load loss for a pair of windings shall be measured at rated frequency with approximately sinusoidal voltage applied to the terminals of one winding, with the terminals of the other winding short-circuited, and with possible other windings open-circuited. The measurements shall be performed quickly so that the temperature rises do not cause significant errors. The difference in temperature between the top oil and the bottom oil shall be small enough to enable the mean temperature to be determined accurately. The short circuit impedance is represented as reactance and ac resistance in series. The impedance is corrected to reference temperature assuming that the reactance is constant.

   On a three-winding transformer, measurements are performed on the three different two-winding combinations. The results are recalculated, allocating impedances and losses to individual windings. Total losses for specified loading cases involving all these windings are determined accordingly.

4. Measurement of no-load loss and current. The no-load loss and the no-load current shall be measured on one of the windings at rated frequency and at a voltage corresponding to rated voltage. The remaining winding or windings shall be left open-circuited and any windings which can be connected in open delta shall have the delta closed. The transformer shall be approximately at factory ambient temperature. For a three-phase transformer the selection of the winding and the connection to the test-power source shall be made to provide, as far as possible, symmetrical and sinusoidal voltages across the three wound limbs.

5. Dielectric routine tests.

6. Tests on under-load tap-changers, where appropriate.

The type tests are

1. Temperature-rise test.

2. Dielectric type tests.

The special tests are

1. Dielectric special tests.

2. Determination of capacitances windings-to-earth, and between windings.

3. Determination of transient voltage transfer characteristics.

4. Measurement of zero-sequence impedance(s) on three-phase transformers. The zero-sequence impedance is measured at rated frequency between the line terminals of a star-connected or zigzag-connected winding connected together and its neutral terminal. In the case of transformers having more than one star-connected winding with neutral terminal, the zero-sequence impedance is dependent upon the connection.

5. Short circuit withstand test.

6. Determination of sound levels.

7. Measurement of the harmonics of the no-load current.

8. Measurement of the power taken by the fan and oil pump motors.

9. Measurement of insulation resistance to earth of the windings, and/or measurement of dissipation factor (tan δ) of the insulation-system capacitances.

4.4 Parameter Determination for Low-Frequency and Switching Transients

4.4.1 Introduction

There are two basic methodologies to determine transformer parameters for low-frequency transients:

1. Experimental determination through laboratory tests.
2. Through topological information and design data, from electromagnetic field simulations; for example, using finite elements.

Electromagnetic field simulations are only possible for the transformer designer. Some utilities also perform these simulations when the internal construction is known. They are useful for obtaining information of the flux distribution in core and air [18], or to compute the electrodynamical forces applied on the coils. Most commonly, however, transformer parameters are estimated from test measurements. The procedures presented in this chapter assume by default that usual nameplate data are available. However, there could be some situations for which a more detailed and accurate model should be used, and some additional information could be needed.

Before model parameters are determined one should clearly know the model to be used. For example, there is no point of implementing a complicated laboratory experiment to determine the hysteresis cycle for a model to be used in a transient where hysteresis plays a negligible role.

Most low-frequency transients are dominated by the behavior of the core (saturation, hysteresis) and its modeling may be a very important issue for those phenomena [2]. Therefore, it is important to properly model the topology of the core according to Figure 4.1. For the study of low-frequency electromagnetic transients where transformers are involved, topology models [7] [19–24] are superior to matrix-based models [25,26]. The most important difference is the way in which the core and winding models are assembled. In the matrix formulation the location of the core model in the equivalent circuit is arbitrary, which may lead to wrong results for some applications; e.g., transformer energization, where the accurate representation of the saturation is mandatory. In topology (duality-based) models it is clear where in the circuit the core should be connected. Additionally, in topology models each inductor representing the core can be related one-to-one with a section of the core, therefore the nonlinear characteristics of the core are properly considered.

4.4.2 Two-Winding Transformers

4.4.2.1 Single-Phase Transformers

4.4.2.1.1 Equivalent Circuit

The two-winding single-phase transformer **T**-equivalent circuit (Figure 4.5) has existed for a very long time. It has been used successfully for many years in steady-state studies and some low-frequency transients. However, there are transient phenomena for which a duality-derived model is better (Figure 4.6). The reason is that the parameters of the

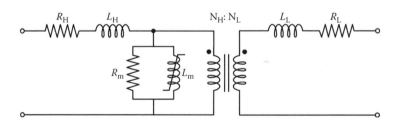

FIGURE 4.5
Traditional equivalent circuit of a single-phase core.

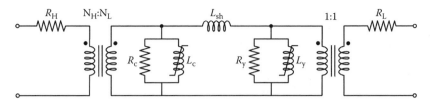

FIGURE 4.6
π-shaped model for a single-phase transformer.

traditional model do not have a direct relationship with the physical components of the transformer. Note, however, that the experimental procedures and required information to determine the parameters of both models is the same [8–10].

Most textbooks, except perhaps only reference [27], use the traditional T-model to describe the behavior of a single-phase transformer. R_H and R_L are the series resistances, which include the Joule losses and eddy current losses in the windings (when data are available). L_H and L_L represent the leakage inductance (or series inductance), which has been divided among the two windings. R_m and L_m, on the shunt branch, describe the core behavior including nonlinearities (saturation and hysteresis) and eddy current phenomena. Note that except for the winding resistances, the other model components cannot be related in a topological sense to specific physical regions of the transformer.

The T-model is relatively easy to implement, but its main disadvantages is the quasi-arbitrary division of the leakage inductance. The leakage inductance, accounting for the voltage drop across a pair of mutually coupled coils, is defined as (with flux in opposite direction) [28]

$$L_{leak} = L_1 + L_2 - 2M \tag{4.1}$$

where
L_1 and L_2 are the self-inductances of the windings
M is the mutual inductance

Therefore, L_{leak} cannot be divided into two (winding) inductances while keeping a physical meaning. The division of the leakage inductance may also result in a wrong location of the core model in the transformer's equivalent circuit.

The Π-model for a single-phase transformer is shown in Figure 4.6. In this model the internal (to the ideal transformers) elements represent physically the magnetic circuit. There is only one leakage inductance in the middle (L_{sh}) and two magnetizing branches, which can have different values. The leakage inductance L_{sh} is linear and frequency dependent while the magnetizing branches are nonlinear and frequency dependent. The parameters of this model can be obtained using the same information as for the T-model, namely, open circuit and short circuit tests. When internal information is not known to physically separate the magnetizing effects into two branches for the Π-model, one can simply double the measured magnetizing parameters to build the model. The relationships between the parameters of the T-model and the Π-model, when no internal information is known, are

$$L_{sh} = L_H + L_L$$

$$R_c = R_y = 2R_m$$

$$L_c = L_y = 2L_m \tag{4.2}$$

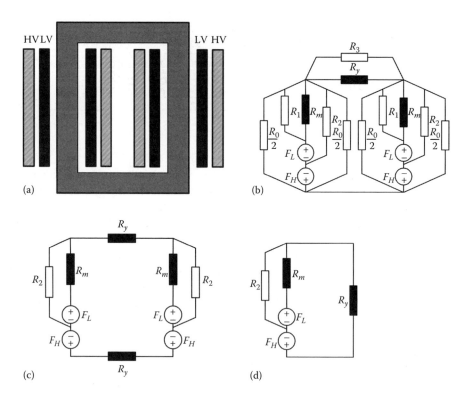

FIGURE 4.7
Duality-derived equivalent circuit of a single-phase type-core transformer. (a) Physical structure; (b) magnetic circuit; (c) simplified magnetic circuit 1; (d) simplified magnetic circuit 2.

The Π-model represents topologically both single-phase transformer configurations: core-type and shell-type as shown next. Figure 4.7 illustrates the process for obtaining the dual electric equivalent circuits from flux paths (or reluctances) for both transformer types. Figure 4.8 shows that the same equivalent circuit is obtained for both single-phase transformer geometries. The circuit is obtained from the core-type transformer circuit by combining the identical parallel branches (series *Leg–Air*). The same circuit is obtained for the shell-type equivalent circuit by merging the series *Air–Air* and *Yoke–Yoke* circuit elements. As with any duality-derived model, the parameters can be accurately estimated when the design dimensions are known.

Example 4.1

This example shows how to determine parameters of this transformer model from manufacturer data and on-site tests. Table 4.4 presents the results of the standardized tests performed on a single-phase transformer at 60 Hz.

The magnetizing parameters are computed from the open circuit test as follows:

$$R_m = \frac{\left(V_1^{open}\right)^2}{P_1^{open}} = \frac{2400^2}{171.1} = 33.665 \ \text{k}\Omega$$

$$L_m = \frac{1}{\omega}\frac{\left(V_1^{open}\right)^2}{Q_1^{open}} = \left(\frac{1}{377}\right)\frac{2400^2}{1148} = 13.3 \ \text{H}$$

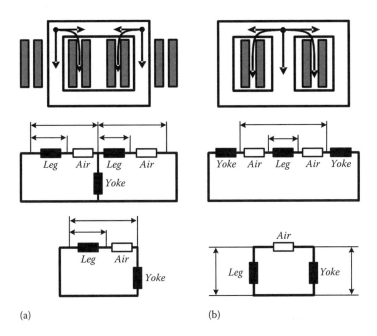

(a)   (b)

FIGURE 4.8
Equivalent circuit for both single-phase transformer types ((a) core and (b) shell) obtained from parallel and series circuit reduction of elements, respectively.

The series (leakage) parameters are computed from the short circuit test as follows:

$$R_s = \frac{P_1^{short}}{\left(I_1^{short}\right)^2} = \frac{642.1}{20.83^2} = 1.48 \ \Omega$$

$$L_{sh} = \frac{1}{\omega}\frac{Q_1^{short}}{\left(I_1^{short}\right)^2} = \left(\frac{1}{377}\right)\frac{846.3}{20.83^2} = 5.284 \ \text{mH}$$

The above parameters can be obtained with better accuracy when more measured information is available. Commonly, the dc resistance of the winding is also measured. Therefore,

TABLE 4.4

Example 4.1: Transformer Parameters Calculated from On-Site Tests Data

Variable	Open Circuit Test	Short Circuit Test
V_1 (V)	2.40	51.87
I_1 (A)	0.48	20.83
P_1 (core loss) (W)	171.10	—
P_1 (winding loss) (W)	—	642.10
V_2 (V)	240.00	0
I_2 (A)	0	208.30
$Q_1 = \sqrt{(V_1 I_1)^2 - P_1^2}$ (VAR)	1148.00	864.30

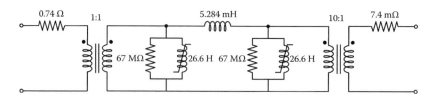

FIGURE 4.9
Example 4.1: Duality-derived circuit equivalent from the measurement of Table 4.4.

the (small) winding-loss component during the open circuit tests can be subtracted from the core loss.

To assemble the Π-model, the computed magnetizing parameters are multiplied by two and the series resistance is divided by two. Then, the resistance of the LV side is referred dividing by the square of the turns ratio (in this example the turns ratio is 10). Figure 4.9 shows the equivalent circuit from duality.

For the traditional **T** circuit the magnetizing parameters are those computed from test, while the leakage parameters are divided by two. One of the branches can be reflected to the other side as shown in Figure 4.10.

4.4.2.1.2 Nonlinear Core Model

The most usual representation is a parallel combination of a nonlinear inductance L_m, representing magnetic core saturation, and a constant resistance R_m, representing core losses. This model is accurate enough for most transformer energization studies when a circuit breaker closes. However, more sophisticated models are mandatory for high-frequency transients, for example the study of magnetizing current chopping. In the latter case, the modeling of hysteresis is a major issue; of particular importance is the determination of the residual flux [29].

Magnetizing inductance: Transformer saturation is an important component of many low-frequency electromagnetic transient phenomena, including ferroresonance, temporary overvoltages during load rejection, and transformer energization leading to inrush currents. In general, saturation needs to be included in transients involving high flux. The total flux in the iron core during an energization is the sum of two fluxes, the residual flux and the forced flux, the latter depends on the closing times of the circuit-breaker poles.

Many modelers are unduly concerned with the details of the transformer saturation curve. For most phenomena, the critical transformer saturation parameters are, as shown in Figure 4.11:

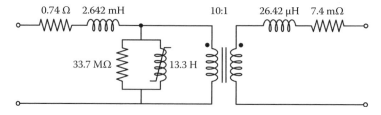

FIGURE 4.10
Example 4.1: Traditional equivalent circuit from the measurement of Table 4.4.

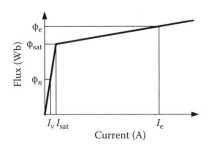

FIGURE 4.11
Description of the saturation curve of a transformer (main parameters).

- The slope (air-core inductance)
- The zero-current intercept of the saturation curve
- The location of the saturation representation in the transformer model topology
- The nominal flux and the corresponding excitation current

Figure 4.12 shows the saturation curve in the case of a 2 MVA stepup transformer for wind-farms. This transformer has a flat saturation region (air-core reactance). This feature may result in high inrush currents (up to 9 p.u. or more) [30].

In most cases, the value of the air-core reactance ranges between 0.1 and 0.8 p.u. Large transformers (600 MVA) have values between 0.7 and 0.8 p.u. while small transformers (100 MVA auxiliary transformer as an example) may have values in the order of 0.2 p.u.

The manufacturers may provide the *V–I* curve for the transformer, up to a voltage value of 1.1 p.u. for HV transformer and up to 1.3 p.u. for lower voltages. Above those values, the voltage waveshape may be distorted, and damage the laminations of the iron core during the test. This explains why the available saturation curves are limited to the knee of the curve.

From Figure 4.13, the hysteretic curve may be built using the following additional parameters:

- The iron core losses, and especially the hysteresis losses, described by the area of the cycle
- The coercive current, which may be given by simple formulas
- The residual flux at the intercept of zero current, equal to 0.8 p.u. in Figure 4.13

Saturation can be incorporated into a power transformer model using test data/manufacturer's curves or estimating the key parameters from transformer geometry. When the first approach is used, some aspects are to be taken into account:

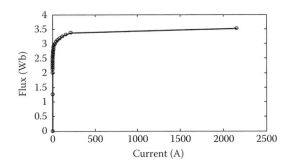

FIGURE 4.12
Saturation curve of a 2 MVA transformer for wind-farm. (From Rioual, M. and Sow, M., *IEEE Workshop on Wind Integration in Power Systems*, May 26–27, 2008. With permission.)

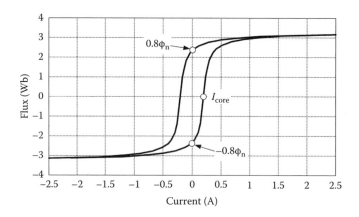

FIGURE 4.13
Hysteresis curve.

- The exciting current includes core loss and magnetizing components.
- Manufacturers usually provide RMS currents, not peak.

In some cases the following can be important:

- Winding capacitance can affect low-current data.
- Hysteresis biases saturation curve; see Figure 4.13.

The following procedure is suggested to obtain the excitation curve:

1. Extract loss component from excitation current for each current point:

$$I_m \approx \sqrt{I_{exc}^2 - (P_{exc}/V_{exc})^2} \tag{4.3}$$

2. Convert the voltage–current RMS curve to an instantaneous flux–current relationship. To perform this conversion the SATURATION supporting routine, available in some Electromagnetic Transients Program (EMTP)-type programs, can be used [26].

3. Compensate for effect of winding capacitance. Winding capacitance can dominate magnetizing reactance causing "cobra" flux–current curves, see Figure 4.14, and cancel much of the magnetizing current [31]. However, these capacitances are seldom given by the manufacturers.

When opening the circuit breaker linked to the transformer, the initial value of the residual flux is the value at which the curve intercepts the zero value of the current

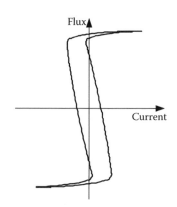

FIGURE 4.14
Hysteresis characteristic with capacitive effect included.

(e.g., 0.8 p.u. on the preceding curve). From that initial value, the residual flux may decrease, caused generally by the oscillation of the circuit constituted by the inductance of the transformer and capacitances (stray capacitances of the circuit breaker). The final value of the residual flux will depend on the following parameters:

- Initial value on the hysteretic curve when intercepting the zero current
- Chopped current of the circuit breaker

The saturation characteristic can be modeled by a piece-wise linear inductance with two slopes, since increasing the number of slopes does not significantly improve the accuracy. However, there are some cases (e.g., ferroresonance), for which a more detailed representation of the saturation characteristic may be required.

Except for very specific applications, a very accurate hysteresis model is not required. In addition, a good knowledge of parameters to be specified in some models is usually lacking. Reference [32] presents a summary of the different methodologies available for representing hysteresis in transformers. Promising results including the complex physics of hysteresis when required can be found in [33].

Core losses: They can also be critically important in some low-current phenomena involving saturation, such as ferroresonance. These losses are a very small part of the power rating and include magnetization losses, dielectric losses, and winding losses. As a first approach, core losses can be represented as a linear resistance obtained from test measurements at rated voltage as follows:

$$R_m = \frac{V_{exc}^2}{P_{exc} - R\,I_{exc}^2} \qquad (4.4)$$

where
V_{exc} is the rated voltage
R is the ac resistance of the winding used during the excitation test

4.4.2.2 Three-Phase Transformers

Guidelines on parameter estimation for three-phase core configurations are presented below. They have been separated into two subsections, assuming that the transformer model to be used in transient simulations can be linear or must incorporate nonlinear behavior, respectively [34].

4.4.2.2.1 Matrix Representation

This representation is usually derived from tests performed by the manufacturers, "acceptance tests," mainly short circuit tests and no-load tests showing that the transformers meet the specifications before being installed in the network. In order to get model parameters directly from those tests some supporting routines (e.g., BCTRAN) have been implemented in some EMTP-type tools, which can be applied to both single- and multiphase multiwinding transformers.

A detailed description of the procedure implemented in this capability is out of the scope of this book. Only a summary of its main principles is described below.

Readers are referred to [25,26] for more details. It should be noted that this approach, based on measurements made at steady state (50/60 Hz), may only give accurate information for low-frequency transients. This method may not be applicable to higher frequencies.

The elements of the impedance matrix [Z] can be derived from the open circuit and short circuit tests. By ignoring the excitation losses, the imaginary parts of submatrix diagonal elements can be derived from the positive- and zero-sequence excitation tests:

$$X_{s-ii} = (X_{0-ii} + 2X_{1-ii})/3 \qquad (4.5a)$$

$$X_{m-ii} = (X_{0-ii} - X_{1-ii})/3 \qquad (4.5b)$$

X_{1-ii} and X_{0-ii} are the positive- and zero-sequence reactances measured by tests
X_{s-ii} represents the self-reactance of winding i (diagonal elements)
X_{m-ii} is the mutual reactance (off-diagonal elements of the winding submatrix) in p.u. values

Excitation losses are not included in [Z], but they can be added as shunt resistances across one or more windings. Winding resistances are obtained from the load losses computed during the short circuit tests at power frequency.

The off-diagonal elements of the submatrix are calculated from the short circuit impedances. First, the positive- and zero-sequence values are computed using the following expression:

$$Z_{ik} = Z_{ki} = \sqrt{(Z_{ii} - Z_{ik(sh)})/Z_{kk}} \qquad (4.6)$$

Then, these values are converted using Equations 4.5a and 4.5b. Their p.u. real parts are the p.u. load losses, while the imaginary parts are calculated as follows:

$$X_{ik(sh)} = \sqrt{|Z_{ik(sh)}|^2 - (R_i + R_k)^2} \qquad (4.7)$$

where $(R_i + R_k)$ are the p.u. winding resistances, which can be calculated from the p.u. losses measured in short circuit tests.

There could be some accuracy problems with the above calculations if exciting currents are low or ignored because the matrix becomes singular. To those problems an admittance matrix representation should be used. For a discussion of these problems and more details of the alternative representation see [25,26].

This approach has a very simple usage, as it is based on nameplate data. However, as mentioned before, these models are linear and theoretically valid only for the frequency at which the nameplate data was obtained. Nonlinear behavior can be incorporated by externally attaching the core representation at the model terminals in the form of nonlinear elements. Attention must be paid as the real part of the elements of the matrix may be negative in certain cases. Other aspects discussed in the original reference are the modifications to be used with delta-connected transformers.

4.4.2.2.2 Topology-Based Models

Several difficulties arise when modeling and simulating three-phase core configurations: Three-phase transformers have magnetic coupling between phases; see Figure 4.1. Four- and five-legged transformers, as well as shell-type transformers, have low reluctance for zero-sequence fluxes, as they can circulate directly through the core. Three-legged core-type transformers have a high reluctance path for zero-sequence fluxes, which close through the air and the transformer tank.

Different approaches have been proposed to date for representation of the same core configurations. Some differences exist between models for representing the same core configuration, even when the same principle (e.g., duality) is used. As a consequence, different procedures for parameter estimation have been proposed for models of the same core configuration by the model' authors. Tests to obtain some characteristics values to be specified in several models are not covered by present standards. Only parameter-estimation procedures for stacked-core transformers which have been derived from the principle of duality are analyzed in this chapter.

Three-legged transformers: Figure 4.15a shows the core configuration of a three-legged stacked-core transformer. The magnetic circuit is depicted in Figure 4.15b [34]. $\mathfrak{I}_H$ and $\mathfrak{I}_L$ represent the magnetomotive forces (MMFs) at the HV and the LV side, respectively. Reluctances due to paths through iron are shown as solid rectangles: $\mathfrak{R}_m$ for the legs, $\mathfrak{R}_y$ for the yokes. Flux leakage paths are shown as open rectangles: $\mathfrak{R}_1$ for paths between the legs and the innermost windings, $\mathfrak{R}_2$ for paths between the legs and the space between the two windings, $\mathfrak{R}_0$ for paths between the legs and outside the windings, and $\mathfrak{R}_3$ for leakage path through the air in parallel with the yokes. Reluctances $\mathfrak{R}_0$, also known as zero-sequence paths, have been halved [34].

After using duality, reluctances due to paths through iron will transform into nonlinear inductances, while flux paths in air will transform into linear inductances. Figure 4.15c shows the equivalent circuit, where the effect of reluctances $\mathfrak{R}_1$ and $\mathfrak{R}_3$ has been neglected. In the figure, R_H and R_L represent the winding resistances of the HV and LV windings; the linear inductances L_{sh} represent the flux leakages between both windings; zero-sequence flux paths are represented by the linear inductances L_0; eddy current losses resulting from zero-sequence fluxes are represented by resistances R_0; the parallel combination of resistances R_m and saturable inductances L_m represent the flux paths through legs; each parallel combination of R_y and L_y represents the yoke section between a pair of phases [34].

Capacitances between terminals and ground (core and tank), between windings and between phases could be added to the equivalent circuit. The elements L_0 and R_0 represent leakage fluxes and associated losses that circulate outside the concentric coil pairs on each phase. Dividing the flux paths into two parts results in a more symmetric and convenient connection for the core equivalent in the resulting electrical equivalent circuit.

In the case of unbalanced excitation that includes zero sequence, the total fluxes linked by the three sets of coils on the phase legs will not add to zero, and the zero-sequence flux will circulate through the surrounding oil and air and through fittings and tank walls. Following the assumptions made in the development of the equivalent circuit, this flux is distributed across the top of the coils on the three phase legs. Alternatively, the zero-sequence effect might have been concentrated all at the center phase, or divided into two parts at each of the other phases. In any case, the total effect is the same. In the equivalent electrical circuit, it is seen that zero-sequence operation results in an additional current

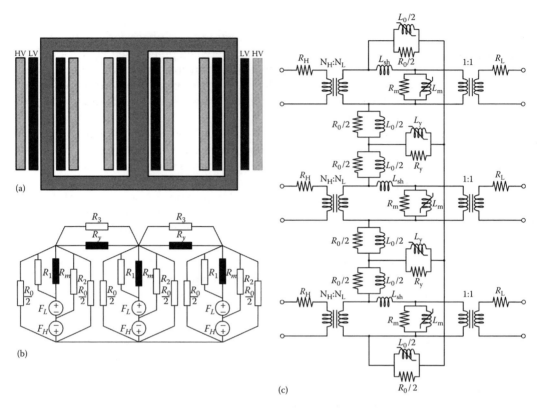

FIGURE 4.15
Duality derived equivalent circuit of a three-legged stacked-core transformer. (From Martínez, J.A. et al., *IEEE Trans. Power Delivery*, 20, 2051, July 2005. With permission.)

(it can be seen in terms of superposition) which circulates through the R_0 and L_0 elements. Regardless of how the R_0 and L_0 effect is distributed, the total impedance of the zero-sequence path must be the same, as shown in Table 4.5, where the zero-sequence test is described.

The portion of the equivalent circuit inside the ideal transformers is valid regardless of winding connections; delta or wye connections are made up at the outside. However, procedures for parameter determination will depend on transformer connections since the same zero-sequence tests cannot be applied to all transformers: for wye-wye connected transformers, data can be derived from excitation and short circuit tests at both positive and zero sequences; for delta-wye connected transformers, zero-sequence tests from delta side cannot be performed and excitation and short circuit tests at zero sequence provide the same results; for delta-delta connected transformers, zero-sequence tests are meaningless.

As for single-core units, parameters of the model depicted in Figure 4.15c can be split into two groups: winding parameters (R_H, R_L, L_{sh}, R_0, and L_0) and core parameters (R_m, L_m, R_y, and L_y).

Tables 4.5 through 4.7 show the diagram of the tests proposed for parameter estimation. Note that the tests used for winding parameters are the standard positive-sequence

TABLE 4.5

Tests for the Determination of Winding Parameters of Three-Legged Stacked-Core Wye-Wye Transformers

Test Diagram	Equivalent Circuit

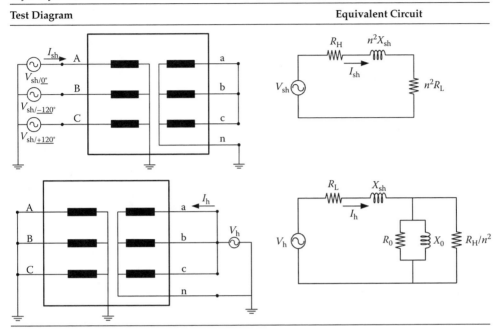

Source: Martínez, J.A. et al., *IEEE Trans. Power Delivery*, 20, 2051, July 2005. With permission.

TABLE 4.6

Tests for the Determination of Winding Parameters of Three-Legged Stacked-Core Delta-Wye Transformers

Test Diagram	Equivalent Circuit

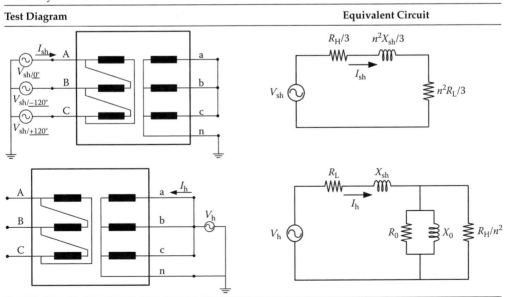

Source: Martínez, J.A. et al., *IEEE Trans. Power Delivery*, 20, 2051, July 2005. With permission.

TABLE 4.7

Tests for the Determination of Core Parameters of Three-Legged Stacked-Core Transformers

Test Diagram	Equivalent Circuit

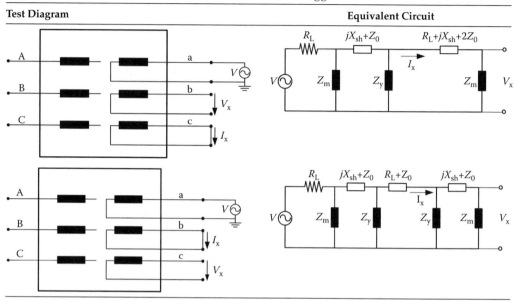

Source: Stuehm, D.L., Three phase transformer core modeling, Bonneville Power Administration, Award No. DE-BI79-92BP26700, February 1993.

excitation test and the zero-sequence tests, respectively. Tests for core parameter estimation were first proposed in [21], and they are not supported by any standard. The procedures to obtain all parameters are detailed in the following paragraphs [34].

Winding parameters: Winding resistances can be obtained by standard tests; if these values are not available, short circuit test data should be used, following the same procedure that was used with single-phase transformers.

Data from the short circuit positive-sequence test are used to obtain R_H, R_L, and X_{sh}, as follows [34]:

$$R_H = k \cdot \alpha \cdot \frac{P_{sh}/3}{I_{sh}^2} \tag{4.8a}$$

$$R_L = k \cdot \frac{1-\alpha}{n^2} \cdot \frac{P_{sh}/3}{I_{sh}^2} \tag{4.8b}$$

$$X_{sh} = \frac{k}{n^2} \cdot \sqrt{\left(\frac{V_{sh}}{I_{sh}}\right)^2 - \left(\frac{(P_{sh}/3)}{I_{sh}^2}\right)^2} \tag{4.8c}$$

where
P_{sh} is the three-phase active power
n is the turn ratio
k is a factor that depends on the transformer connection at the source side: for wye connection $k = 1$, for delta connection $k = 3$

As for single-phase units, α is the percentage of the resistance to be placed at the HV side (in the previous section 0.5 for 50% was used).

Data from zero-sequence tests can be used to obtain R_0 and X_0 according to the following sequence of calculations [34]:

$$V_0 = V_h - (R_L + jX_{sh})I_h \qquad (4.9a)$$

$$P_0 = \frac{P_h}{3} - R_L I_h^2 - \frac{V_0^2}{R_H/n^2} \qquad (4.9b)$$

$$Q_0 = \sqrt{(V_h I_h)^2 - (P_h/3)^2} - X_{sh} I_h^2 \qquad (4.9c)$$

$$R_0 = \frac{V_0^2}{P_0} \qquad (4.9d)$$

$$X_0 = \frac{V_0^2}{Q_0} \qquad (4.9e)$$

where

V_h and I_h are the RMS values of voltage and current, respectively
P_h is the three-phase active power, measured in the zero-sequence test

Core parameters: Diagram of the two tests required to obtain the parameters of the core are shown in Table 4.7. In both tests, the delta-connected HV side is left open. The first one is performed by shorting one outer leg and exciting the other one; it will be used to determine leg characteristic; i.e., R_m, L_m. The second test is performed by shorting the center leg and exciting an outer leg; after obtaining the combined leg and yoke characteristic, the leg characteristic, obtained in the previous test, is subtracted.

The tests are also used to obtain core losses and therefore the full representation of legs and yokes. As for single-phase units, some care is needed to account for capacitances effects during excitation (open circuit) tests. Parameters R_0 and X_0 can be also derived from a zero-sequence open circuit test for both transformer connections. As for the tests shown in Tables 4.5 through 4.7, the delta side of a delta-wye connected transformer should also be open. In such case, the equivalent circuit would be that shown in Figure 4.16. A modified version of Equations 4.9a through 4.9e must be used. In fact only Equation 4.9b has to be modified

$$P_0 = \frac{P_h}{3} - R_L I_h^2 \qquad (4.10)$$

Discussion: Tests shown in Table 4.7 cannot be always performed, since opening the delta side or even separating the three windings of the wye side is only possible for very particular transformers. Therefore, an alternative procedure should be considered. An efficient approach can be used if transformer geometry is known. As illustrated

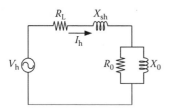

FIGURE 4.16
Equivalent circuit of a zero-sequence open circuit test (delta side must be open in a delta-wye connection).

in several works (see for instance [20,35]) an accurate enough estimation of the different core reluctances, and therefore of the saturable inductances, can be performed by using some simple expressions presented in textbooks [36].

Other tests to obtain transformer parameters have been proposed by some authors. Those tests fit the model they are using for representing a three-legged transformer. For instance, Figure 4.17 shows the diagram of the measurements proposed in [37] for estimation of the nonlinear characteristics of the three legs and the zero-sequence parameters. However, one can also note that both types of measurements are based on open circuit tests. Therefore, the same concern for delta-connected transformers still remains.

Some researchers suggest a core loss representation using nonlinear resistors in parallel with each nonlinear inductance, with the nonlinear resistance values determined from test data at different excitation levels. However, as discussed earlier for single-phase transformers, there are significant limitations of nonlinear resistance representation because the hysteresis losses depend on the maximum flux level and not the maximum voltage level. A fit of nonlinear resistances to loss data which is accurate for one excitation waveform may not provide a match for excitation of a different frequency or waveform.

Five-legged transformers: Figure 4.18 shows the construction of a 600 MVA, five-legged transformer.

The derivation of the equivalent circuit is similar to that used with the three-legged transformer. Figure 4.19 shows the equivalent circuit of a five-legged stacked-core transformer. In this case, the outer legs provide a closed iron path for zero-sequence fluxes; these legs and their yokes are represented as two parallel combinations of R_l and L_l. Therefore, the circuit is obtained from that of a three-legged transformer after removing the linear inductances L_0 representing zero-sequence flux paths and their associated resistances R_0, and adding the parallel combinations R_l–L_l that represent the outer legs.

Parameter estimation can be based on measurements similar to those proposed for a three-legged transformer. Data from the short circuit positive sequence test is used to obtain R_H, R_L, and L_{sh}, see Equations 4.8a through 4.8c. Core parameters can be estimated

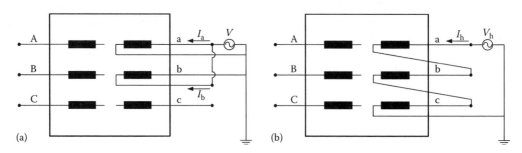

FIGURE 4.17
Measurement of characteristics of a three-legged stacked-core transformer. (a) Setup for measurement of the nonlinear leg characteristics. (b) Setup for measurement of the zero-sequence characteristic. (From Fuchs, E.F. and You, Y., *IEEE Trans. Power Deliv.*, 17(4), 983, October 2002. With permission.)

FIGURE 4.18
Five-legged transformer (Alstom 600 MVA transformer).

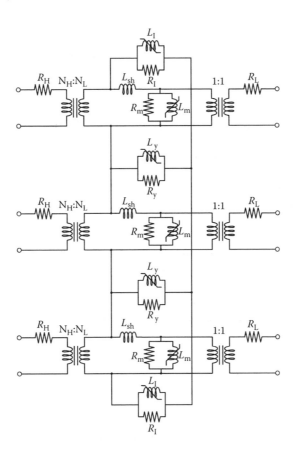

FIGURE 4.19
Duality-derived equivalent circuit of a five-legged stacked-core transformer. (From Stuehm, D.L., Three phase transformer core modeling, Bonneville Power Administration, Award No. DE-BI79-92BP26700, February 1993.)

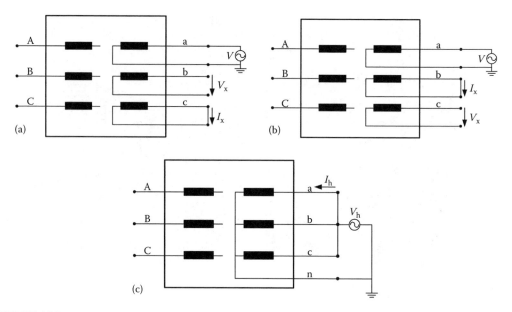

FIGURE 4.20
Circuits for measurement of characteristics of a five-legged stacked-core transformer. (a) Open circuit test of center leg; (b) open circuit test of outer phase; (c) zero-sequence open circuit test. (From Stuehm, D.L., Three phase transformer core modeling, Bonneville Power Administration, Award No. DE-BI79-92BP26700, February 1993.)

from special test measurements. Figure 4.20 shows the scheme of tests proposed for a five-legged stacked-core transformer [21], where R_m, L_m can be estimated from measurements obtained with the open circuit test of the center leg; the combined characteristic of leg and yoke (R_m, L_m, R_y, and L_y) can be obtained from the open circuit test of the outer phase; R_l, L_l can be derived from measurements obtained with the zero-sequence open circuit test. For delta-connected transformers, open-delta tests are required, but the same concerns mentioned above still remain.

Other comments for the three-legged transformer apply to five-legged transformers. That is, the performance of the equivalent circuit can be improved by representing core losses as a nonlinear resistance, compensation of capacitive effects should be considered to obtain a correct saturation curve, and capacitive effects should not be always neglected. For equivalent circuit derivation and parameter estimation of a five-legged wound-core transformer, see [21,24]. In [37] the parameters of a hybrid model for five-legged transformers are estimated.

4.4.3 Three-Winding Transformers

4.4.3.1 Introduction

The traditional equivalent circuit for a three-winding transformer was obtained by Boyajian in 1924 [39], see Figure 4.21. The circuit often contains a negative inductance. Notwithstanding that the negative inductance is not realizable, it has not presented problems with frequency-domain studies using phasors [39–42]. The equivalent has been successfully used for many years for the study of power flow, short circuit, transient stability, etc. However, when computing electromagnetic transients (time-domain modeling), the negative inductance has been identified as the source of spurious oscillations [26,43–46].

To correct the inconsistency, a circuit derived from the principle of duality is used in [47]. The model can be constructed using mutually coupled inductances readily available in any time-domain simulation program, such as the EMTP and the like. The parameters of the equivalent circuit can be obtained in two ways: (1) from the design data and (2) from terminal leakage inductance measurements of two windings at a time.

4.4.3.2 Traditional Equivalent Circuit of a Three-Winding Transformer

The conventional model for the leakage inductance of a three-winding transformer is a star-connected circuit [1–4], see Figure 4.21. L_1, L_2, and L_3 are commonly referred as the windings' leakage inductances. However, L_1, L_2, and L_3 do not correspond to leakage flux paths as the components of duality-derived models.

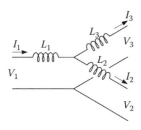

FIGURE 4.21
Traditional equivalent circuit for the leakage inductance of three-winding transformers.

4.4.3.2.1 Model Derived from Terminal Measurements

The parameters of the circuit can be obtained by matching the inductive network (Figure 4.21) to leakage inductances measured at the terminals. The measurements are performed by taking two windings at a time. One winding is energized with a second one short-circuited while keeping the third winding open. Three inductances Ls_{12}, Ls_{23}, and Ls_{13} can be obtained from terminal measurements as follows:

Ls_{12} is obtained when energizing winding number 1 while winding 2 is short-circuited and winding 3 is open.

Ls_{23} is obtained by energizing winding 2 and short-circuiting winding 3 with winding 1 left open.

Ls_{13} is obtained when winding 1 is energized and winding 3 is in short circuit with winding 2 open.

Table 4.8 describes the testing setup. For convenience, here all inductances are referred to a common number of turns (N). The inductances of the network of Figure 4.21 are computed by solving a set of equations such that the terminal measurements are matched. In the calculation no consideration is given to the physical meaning of the inductances.

By inspection of the circuit of Figure 4.21 the following relations are obtained:

TABLE 4.8

Leakage Inductance Tests for a Three-Winding Transformer

Parameter	Test
Ls_{12}	W1 — W2, W3
Ls_{23}	W1 — W2, W3
Ls_{13}	W1 — W2, W3

$$L_1 + L_2 = Ls_{12}$$

$$L_2 + L_3 = Ls_{23}$$

$$L_1 + L_3 = Ls_{13} \qquad (4.11)$$

L_1, L_2, and L_3 are computed to match leakage measurements and therefore one should not assign a physical meaning to them. Solving Equations 4.11 we get

$$L_1 = \frac{1}{2}\left(Ls_{12} - Ls_{23} + Ls_{13}\right)$$

$$L_2 = \frac{1}{2}\left(Ls_{12} + Ls_{23} - Ls_{13}\right)$$

$$L_3 = \frac{1}{2}\left(-Ls_{12} + Ls_{23} + Ls_{13}\right) \tag{4.12}$$

For a standard transformer design $Ls_{12} + Ls_{23} < Ls_{13}$ and therefore L_2 becomes negative. The negative inductance appears in the traditional model (Figure 4.21) because it does not consider the mutual couplings that take place in the region of the middle winding.

4.4.3.2.2 Model Derived from Design Information

The leakage inductances for a pair of windings can be computed from the design parameters assuming a trapezoidal flux distribution [19,48], see Figure 4.22. Given the flux distribution of Figure 4.22 we obtain the following expressions [19]:

$$Ls_{12} = \frac{\mu_0 N^2 l}{ls}\left[\frac{a_1}{3} + d_{12} + \frac{a_2}{3}\right]$$

$$Ls_{23} = \frac{\mu_0 N^2 l}{ls}\left[\frac{a_2}{3} + d_{23} + \frac{a_3}{3}\right]$$

$$Ls_{13} = \frac{\mu_0 N^2 l}{ls}\left[\frac{a_1}{3} + d_{12} + a_2 + d_{23} + \frac{a_3}{3}\right] \tag{4.13}$$

where
　N is the common (or base) number of turns
　l is the mean length of the winding turn

Substituting Equation 4.13 into Equation 4.12 we obtain

$$L_1 = \frac{\mu_0 N^2 l}{ls}\left[\frac{a_1}{3} + d_{12} + \frac{a_2}{2}\right]$$

$$L_2 = \frac{\mu_0 N^2 l}{ls}\left[-\frac{a_2}{6}\right]$$

$$L_3 = \frac{\mu_0 N^2 l}{ls}\left[\frac{a_2}{2} + d_{23} + \frac{a_3}{3}\right] \tag{4.14}$$

Clearly the value of L_2 is always negative since the thickness of the middle winding (a_2) is always positive. It is interesting to note that all inductances (Ls_{12}, Ls_{13}, Ls_{23}, L_1, L_2, and L_3) are functions of the thickness of the winding in the center.

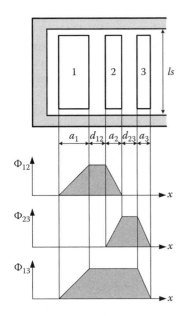

FIGURE 4.22
Geometrical arrangement of windings in a transformer window. (From de León, F. and Martinez, J.A., *IEEE Trans. Power Delivery*, 24, 160, January 2009. With permission.)

4.4.3.3 Saturable Transformer Component

A single-phase multiwinding transformer model can be based on a star circuit representation, see Figure 4.23 [26]. This model is of limited application, even for single-phase units. An important aspect is the node to which the magnetizing inductance is connected. It is unimportant if the inductance is not saturated, because of its high value; but it can be important if the inductance is saturated. The star point is not the correct topological connecting point. That is, the traditional model for three-winding transformers has two serious drawbacks: (1) the existence of the negative inductance that yields numerical unstable transient simulations; (2) the arbitrary location of the magnetizing branch that is arbitrarily set in the middle.

4.4.3.4 Duality Model

Duality models, where an electrical equivalent is used to describe the behavior of the magnetic core, can very accurately represent any type of core design in low-frequency transients. The application of the principle of duality results in models that include the effects of saturation in each individual leg of the core, interphase magnetic coupling, and leakage effects [7,19–24]. In the equivalent magnetic circuit, windings appear as MMF sources, leakage paths appear as

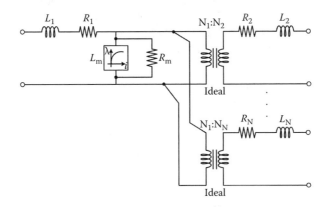

FIGURE 4.23
Star circuit representation of single-phase N-winding transformers.

linear reluctances, and magnetic cores appear as saturable reluctances. The mesh and node equations of the magnetic circuit are duals of the electrical equivalent node and mesh equations, respectively. Winding resistances, core losses, and capacitive coupling effects are not obtained directly from the transformation, but can be added to the equivalent circuit.

Duality models are obtained from the geometrical arrangement of windings in the transformer window. No attention is paid to the terminal leakage measurements and therefore there is usually an inconsistency between duality models and terminal leakage measurements. The easiest way to build a duality model is to establish the flux paths in the transformer window and to assign an inductance to each one. The process is illustrated in Figure 4.24.

The magnetizing flux is represented by the nonlinear inductances (L_{m1}, L_{m2}, and L_{m3}) and the leakage flux by the two linear inductances (L_{12} and L_{23}). The values of these two inductances match the measured leakage inductances Ls_{12} and Ls_{23}. However, there is no match for the leakage inductance Ls_{13}. In the duality model, the following relation holds (neglecting magnetizing): $Ls_{13} = Ls_{12} + Ls_{23}$. One can obtain from Equation 4.13

$$Ls_{12} + Ls_{23} = \frac{\mu_0 N^2 l}{ls} \left[\frac{a_1}{3} + d_{12} + \frac{2 a_2}{3} + d_{23} + \frac{a_3}{3} \right]$$

$$Ls_{13} = \frac{\mu_0 N^2 l}{ls} \left[\frac{a_1}{3} + d_{12} + a_2 + d_{23} + \frac{a_3}{3} \right] \qquad (4.15)$$

Accordingly, the duality model wrongly accounts for the leakage inductance between the internal and the external windings, and Ls_{13} is short by

$$\frac{\mu_0 N^2 l}{ls} \left[\frac{a_2}{3} \right] \qquad (4.16)$$

4.4.3.4.1 Duality Model Matching Terminal Measurements

The leakage inductance circuit is derived from the duality model by adding a mutual coupling M between L_{12} and L_{23} as shown in Figure 4.25. This allows compensating for the missing factor (Equation 4.16). The dot marks have been selected in such a way that the total inductance increases for the test of Ls_{13}.

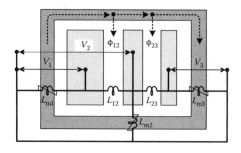

FIGURE 4.24
Duality-derived model for a three-winding transformer. (From de León, F. and Martinez, J.A., *IEEE Trans. Power Deliv.*, 24, 160, January 2009. With permission.)

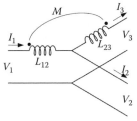

FIGURE 4.25
Duality-derived model for a three-winding transformer. (From de León, F. and Martinez, J.A., *IEEE Trans. Power Deliv.*, 24, 160, January 2009. With permission.)

Applying the three tests depicted in Table 4.8 to the circuit of Figure 4.25 we get

$$L_{12} = Ls_{12}$$

$$L_{23} = Ls_{23}$$

$$L_{12} + L_{23} + 2M = Ls_{13} \qquad (4.17)$$

M can be determined from Equation 4.17 in a straightforward manner yielding

$$M = \frac{1}{2}(Ls_{13} - Ls_{12} - Ls_{23}) \qquad (4.18)$$

Equation 4.18 describes the computation of the compensating mutual inductance directly from the leakage inductance tests. M is positive in most cases because $Ls_{13} > Ls_{12} + Ls_{23}$ for standard designs. By substituting Equation 4.15 into Equation 4.18 one can obtain an expression for M as a function of the design parameters as

$$M = \frac{\mu_0 N^2 l}{ls}\left[\frac{a_2}{6}\right] \qquad (4.19)$$

Note that, as expected, M is half the factor (Equation 4.16) because M enters twice in the total series-inductance calculation. Additionally, note that $M = -L_2$.

The complete dual equivalent circuit, including the magnetizing branches and the winding resistances, is shown in Figure 4.26. This circuit has the magnetic and electric elements separated by three ideal transformers. Three magnetizing branches represent the leg, the yokes, and the flux return (dually) connected at the terminals of the ideal transformers.

Magnetizing parameters: The inductances representing the leakage flux are computed either from tests, Equations 4.17 and 4.18, or from the design parameters, Equations

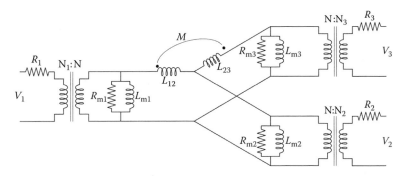

FIGURE 4.26
Duality-derived model for a three-winding transformer including magnetizing branches and winding resistances. (From de León, F. and Martinez, J.A., *IEEE Trans. Power Deliv.*, 24, 160, January 2009. With permission.)

4.13 and 4.19. To compute the magnetizing inductances L_{m1}, L_{m2}, and L_{m3} one must know the construction and dimensions of the core. However, this is rarely available. The equivalent circuit of Figures 4.25 and 4.26 apply to both constructions: shell-type and core-type.

Calculation of parameter for single-phase transformers: The magnetizing inductance and its associated shunt resistance (used to represent hysteresis and eddy current losses) of a transformer are obtained from the open circuit test. Only one magnetizing inductance L_m and one resistance R_m are determined from the measurements. From Figure 4.26 one can see that during an open circuit test the three magnetizing branches are in parallel since the voltage drop in the leakage inductances L_{12}, L_{23}, and M is negligible. Then the relationship between the measured magnetizing values (R_m and L_m) and the model values is given by

$$R_m = \frac{1}{\dfrac{1}{R_{m1}} + \dfrac{1}{R_{m2}} + \dfrac{1}{R_{m3}}}; \quad L_m = \frac{1}{\dfrac{1}{L_{m1}} + \dfrac{1}{L_{m2}} + \dfrac{1}{L_{m3}}} \quad (4.20)$$

A sensible approximation for L_{m1}, L_{m2}, L_{m3}, R_{m1}, R_{m2}, and R_{m3} can be obtained from the values of L_m and R_m when the geometrical information is not known by assuming that the window approximates a square. Thus, the length of the yokes is the same as the length of the legs ($l_0 = l_{yoke} = l_{leg}$). A reasonable way to divide the yoke is in 3 sections as shown in Figure 4.27. The flux length path for L_{m1} and L_{m3} (and R_{m1} and R_{m3}) becomes $^5/_3 l_0$ while the flux length path for L_{m2} (and R_{m2}) is $^2/_3 l_0$ (there are two yokes represented by the pair L_{m2}, R_{m2}).

Resistances are directly proportional to the path length while inductances are inversely proportional to length. The resistance and inductance associated with a leg length are

$$R_{leg} = \frac{\rho}{Area} l_0; \quad L_{leg} = \frac{\mu N^2 Area}{l_0} \quad (4.21)$$

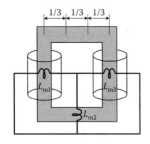

FIGURE 4.27
Duality model for the iron-core of a core-type transformer. (From de León, F. and Martinez, J.A., *IEEE Trans. Power Deliv.*, 24, 160, January 2009. With permission.)

Then we have that

$$R_{m1} = R_{m3} = \frac{5}{3} R_0; \quad R_{m2} = \frac{2}{3} R_0$$

$$L_{m1} = L_{m3} = \frac{3}{5} L_0; \quad L_{m2} = \frac{3}{2} L_0 \quad (4.22)$$

By substituting Equation 4.22 in Equation 4.20 and after some algebra we obtain

$$R_0 = \frac{27}{10} R_m; \quad L_0 = 4 L_m \quad (4.23)$$

Substituting Equation 4.23 in Equation 4.22 we get

$$R_{m1} = R_{m3} = \frac{9}{2} R_m; \quad R_{m2} = \frac{9}{5} R_m$$

$$L_{m1} = L_{m3} = \frac{12}{5} L_m; \quad L_{m2} = 6 L_m \tag{4.24}$$

When the transformer is tall and slim, with a small leakage inductance, then $l_{leg} > l_{yoke}$. As an extreme case it can be assumed that $l_{leg} = 2 l_{yoke}$. Consequently, the flux length path for R_{m1}, R_{m2}, L_{m1}, and L_{m3} is $^4/_3 l_0$ while the flux length path for R_{m2} and L_{m2} is $^1/_3 l_0$. When the transformer is short and wide, with a large leakage inductance, then $l_{yoke} > l_{leg}$. As the other extreme consider that $l_{yoke} = 2 l_{leg}$. Now the length of the flux path for R_{m1}, R_{m2}, L_{m1}, and L_{m3} is $^7/_3 l_0$ while it is $^4/_3 l_0$ for R_{m2} and L_{m2}. Table 4.9 summarizes the standard, maximum and minimum values for the magnetizing inductances L_{mi} and resistances R_{mi} as function of L_m and R_m.

Figure 4.28a and b show the models for three-phase three-winding transformers (core-type and shell-type respectively). The magnetizing losses, the ideal transformers, and the winding's resistance can be added in a similar fashion as for single-phase transformers. Note that there are similarities between the core-type and shell-type models, but there are four more magnetizing branches in the shell-type.

The magnetizing parameters are much more difficult to obtain for a three-phase transformer than for a single-phase transformer. There are no standardized tests that would allow an accurate determination of the parameters. One needs to find a way to energize one of the windings in every leg with all other coils in the transformer opened as previously discussed. Although the tests can always be performed at the factory, it may not be possible to test when only the transformer terminals (after connections) are available in the field. Cooperation from transformer manufacturers will be most probably needed to properly determine the magnetizing parameters of a dual equivalent circuit. Leakage parameters can be computed using the procedures for single-phase transformers described above.

Added complications are the facts that all legs are mutually coupled and the iron-core is highly nonlinear. Thus during tests the different components of the core could be excited at a different flux density than during normal operation, therefore rendering the tests meaningless.

Example 4.2

As an illustration example consider the transformer data presented in [45]. The rated transformer data are given in Table 4.10 (note that winding numbers 2 and 3 are switched with respect to reference [45]). The values of the magnetizing pair, resistance and inductance, derived from an open circuit test and referred to the LV side (13.8 kV) are $R_m = 254\ \Omega$ and $L_m = 0.337\ H$. The short circuit tests have given the following per unit leakage reactances: $Ls_{12} = 0.10$, $Ls_{23} = 0.84$, and $Ls_{13} = 0.96$.

TABLE 4.9

Relationships between the Model Leakage Inductances and Resistances (Figure 4.26) to Those Obtained from Terminal Tests

	Case	$L_{m1} = L_{m3}$	L_{m2}	$R_{m1} = R_{m3}$	R_{m2}
Standard	$l_{yoke} = l_{leg}$	$(12/5)\ L_m$	$6.0\ L_m$	$(9/2)\ R_m$	$(9/5)\ R_m$
Tall	$l_{leg} = 2\ l_{yoke}$	$(9/4)\ L_m$	$9.0\ L_m$	$6.0\ R_m$	$(3/2)\ R_m$
Short	$l_{yoke} = 2\ l_{leg}$	$(18/7)\ L_m$	$4.5\ L_m$	$(15/4)\ R_m$	$(15/7)\ R_m$

Source: de León, F. and Martinez, J.A., *IEEE Trans. Power Deliv.*, 24, 160, January 2009. With permission.

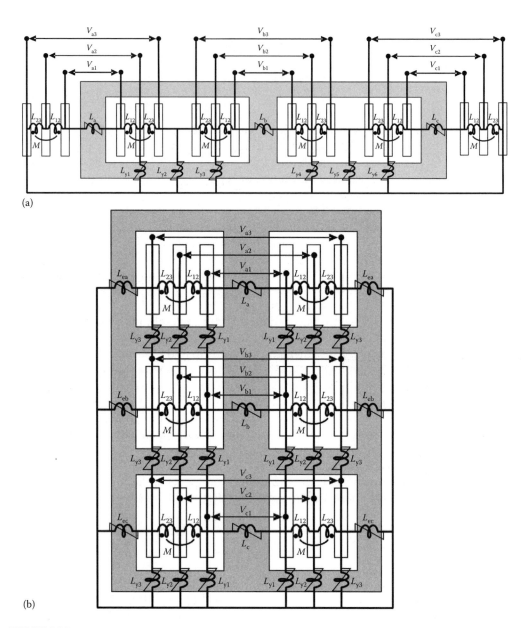

(a)

(b)

FIGURE 4.28
Model derived from the principle of duality for three-winding transformers matching terminal leakage measurements: (a) Core-type; (b) shell-type. (From de León, F. and Martinez, J.A., *IEEE Trans. Power Deliv.*, 24, 160, January 2009. With permission.)

The model leakage inductances are computed in per unit from Equation 4.17 and Equation 4.18 as follows:

$$L_{12} = 0.10; \quad L_{23} = 0.84$$

$$M = \frac{1}{2}(0.96 - 0.10 - 0.84) = 0.01$$

TABLE 4.10

Example 4.2: Test Transformer Data

Winding	Power (MVA)	Voltage (kV)	Resistance (mΩ)
1	300	13.8	1.25
2	300	199.2	260
3	50	19.92	2.60

The leakage inductance values can be computed for the LV side using the impedance and inductance base: $Z_{base} = 0.635\ \Omega$, $L_{base} = 1.684\ mH$. Thus, we have $L_{12} = 0.168\ mH$, $L_{23} = 1.414\ mH$, and $M = 0.017\ mH$. The model is shown in Figure 4.29. The test consists in energizing the HV winding with a cosinusoidal function at $t = 0$ while keeping the other two windings open. Figure 4.30a shows that the voltage on the LV terminal is stable, while Figure 4.30b taken from [45] shows numerical instability for the same case.

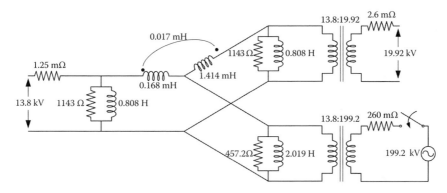

FIGURE 4.29
Example 4.2: Equivalent circuit of three-winding transformer without negative inductance. (From de León, F. and Martinez, J.A., *IEEE Trans. Power Deliv.*, 24, 160, January 2009. With permission.)

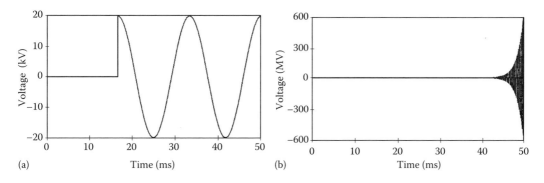

FIGURE 4.30
Example 4.2: Results comparison between dual model and traditional model. (a) Stable simulation with the dual circuit of Figure 4.29. The steady-state voltage at the LV terminals is 13.8 kV. (From de León, F. and Martinez, J.A., *IEEE Trans. Power Deliv.*, 24, 160, January 2009. With permission.) (b) Unstable simulation with the traditional model. (From Chen, X., *IEEE Trans. Power Deliv.*, 15, 1199, October 2000. With permission.)

4.4.4 Eddy Currents

4.4.4.1 Introduction

An important aspect of transformer modeling for low-frequency and switching transients is the proper estimation of the frequency dependence of the model parameters, or eddy-current effects. Eddy currents are induced in conductors (windings and core) when the magnetic flux changes with time.

In steady state the eddy currents are undesirable because they produce losses. However, when a transformer is subjected to a transient the induced eddy currents are beneficial because they add damping to the transients and the proper modeling of this damping could be important.

4.4.4.2 Eddy Currents in Windings

Series winding losses are frequency dependent. Figure 4.31 shows the frequency dependence of a winding resistance R as a function of the factor X/R, where X is the leakage reactance. Winding resistances must incorporate an ac component due to eddy currents in windings, skin effect, and stray losses. The ac resistance is commonly approximated by the following expression [49]:

$$R = R_{dc} + \Delta R_{ac} \cdot \left(f/f_0\right)^m \tag{4.25}$$

where
   f_0 is the power frequency
   m is a factor between 1.2 and 2

Figure 4.32 shows a series–parallel R–L circuit (Foster equivalent) that can be used for an accurate representation of the winding resistance and the leakage inductance in low-frequency and switching transients. To obtain parameters of such circuit, a frequency response test must be performed and a fitting procedure applied [50,51]. If such information is not available, then effective resistance approximations presented in the literature could be considered, see for instance [49]. A correction of the resistance value to account for temperature effect should be also considered.

The series Foster equivalent circuit of Figure 4.32 is only a terminal model. The elements of the circuit are not related to a section or region of the winding in a dual way.

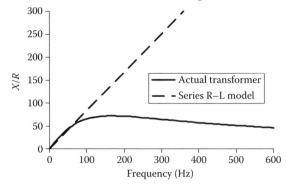

FIGURE 4.31
Frequency dependency of the factor X/R. (From Martínez, J.A. et al., *IEEE Trans. Power Deliv.*, 20, 2051, July 2005. With permission.)

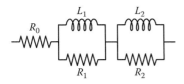

FIGURE 4.32
Simplified frequency-dependent representation of winding.

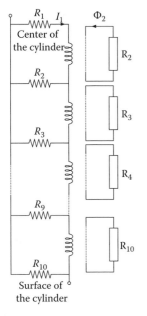

FIGURE 4.33
Dual representation of eddy currents in a transformer winding.

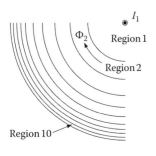

FIGURE 4.34
Physical interpretation of the equivalent circuit of Figure 4.33.

The elements of a Cauer equivalent circuit to represent the eddy-current effects proposed in [52] can be related to currents and magnetic fluxes in the duality sense. Figure 4.33 shows the Cauer model to represent a winding (or a conductor). The resistive (series) elements represent the actual paths of the eddy currents while the (shunt) inductive elements represent magnetic flux paths. The circuit can be interpreted as a subdivision of the conductor in layers as shown in Figure 4.34.

4.4.4.3 Eddy Currents in Core

The magnetization curves presented above are only valid for slowly varying phenomena as it has been assumed that the magnetic field can penetrate the core completely. For high frequency, this is not true. A change in the magnetic field induces eddy currents in the iron. As a consequence of this, the flux density will be lower than that given by the normal magnetization curve. For high frequencies, the flux will be confined to a thin layer close to the lamination surface, whose thickness decreases as the frequency increases. This indicates that inductances representing iron path magnetization and resistances representing eddy-current losses are frequency dependent [53].

The circulation of these eddy currents introduces additional losses. To limit their influence, a transformer core is built up from a large number of parallel laminations. Eddy-current models intended for simulation of the frequency dependence of the magnetizing inductance as well as losses can be classified into two categories obtained, respectively, by the realization of the analytical expression for the magnetizing impedance as a function of frequency, and by subdivision of the lamination into sublaminations and the generation of their electrical equivalents [50–53].

Computationally efficient models have been derived by synthesizing a Foster or a dual Cauer equivalent circuit to match the equivalent impedance of either a single lamination or a coil wound around a laminated iron core leg [49,50]. The accuracy of the standard Cauer representation over a defined frequency range depends on the number of terms retained in a partial fraction expansion and, therefore, on the number of sections. To represent the frequency range up to 200 kHz with an error of less than 5%, only four terms are required [49]. The first section governs its characteristics at frequencies up to a few kilohertz; each subsequent section comes into play as the frequency increases.

Another form of the Cauer model would have shunt resistances and series inductances; see Figure 4.35. The inductances represent (using duality) the flux paths, and the resistances are in the path of eddy currents. The high-frequency response is defined by the blocks near the terminals. The blocks of this model can be thought as being a discretization of the lamination. The parameters of this circuit can be calculated by means of an iterative method that could be seen as an optimization of the discretizing distances for the

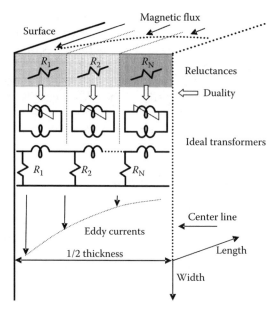

FIGURE 4.35
Illustrating the physical significance of the dual Cauer circuit for the representation of eddy currents in the transformer iron-core.

used fitting frequencies. Below 200 kHz, small errors of less than 1% can be achieved with a model order of 4. A model order of 3 or less will suffice for transients below 10 kHz. Inductive components of the models representing the magnetizing reactances have to be nonlinear to account for the hysteresis and saturation effects. The inductances and resistances in this model represent a physical part of the iron-core lamination, therefore the incorporation of these effects is straightforward. Note, however, since the high-frequency components do not contribute appreciably to the flux in the transformer core, it can be assumed that only low-frequency components are responsible for driving the core into saturation. It may, therefore, be justifiable to represent as nonlinear only the first section of the model.

The effect of eddy currents due to flux penetration in the core can be described by means of the following expression [50–53]:

$$Z_{core} = \frac{4N^2 A \rho_{core}}{l d^2} x \tanh x \qquad (4.26)$$

where

$$x = \frac{d}{2} \sqrt{\frac{j\omega\mu_{core}}{\rho_{core}}} \qquad (4.27)$$

and

l is the length of the represented magnetic element (leg or yoke)
d is the thickness of the lamination
μ_{core} is the permeability of the core material
ρ_{core} is the resistivity of the core material
N is the number of turns
A is the total cross-sectional area of all laminations

The parameters of this circuit are calculated with the iterative method described in [50]. We start by evaluating Equation 4.26 at logarithmically spaced N frequencies and assigning an initial guess to every component. Then we compute the next L_k and R_k in the iterative process from the following equations:

$$b\left(\omega_k L_k + \omega_k L_{kR}\right)^2 - \left(\omega_k L_k + \omega_k L_{kR}\right) + b R_{kR}^2 = 0 \quad \text{for} \quad k = 1, 2, \ldots, N-1 \qquad (4.28a)$$

$$R_{kR} = \left(a - G_k\right)\left(R_{kR}^2 + \left(\omega_k L_k + \omega_k L_{kR}\right)^2\right) \quad \text{for} \quad k = 1, 2, \ldots, N-1 \qquad (4.28b)$$

where

$$a = \text{Re}\left(\frac{1}{Z_{kL}}\right); \quad b = \text{Im}\left(\frac{1}{Z_{kL}}\right) \qquad (4.29)$$

Z_{kR} and Z_{kL} are the complex impedances to the right (sub R) and to the left (sub L) of the elements being computed, which can be easily evaluated from the ladder circuit. The innermost resistance and inductance of the ladder can be computed as

$$G_N = \left(\frac{\sum_{i=1}^{N-1} L_i}{L_N}\right)\left(G_{dc} - \sum_{k=1}^{N-1} \frac{\sum_{i=k}^{N-1} L_i}{\sum_{i=1}^{N-1} L_i} G_k\right) \qquad L_N = L_{dc} - \sum_{k=1}^{N-1} L_k \qquad (4.30)$$

where G_{dc} and L_{dc} are the dc conductance and inductance of the core, respectively.

The calculation of losses may also be done by electromagnetic 3D simulations. This requires a detailed representation of the iron core and also the windings around the core, as shown in Figure 4.36 [18].

Electromagnetic field simulations allow to accurately compute the flux around the core. Figure 4.37 shows how the legs are not saturated in the same way when energizing a transformer [18].

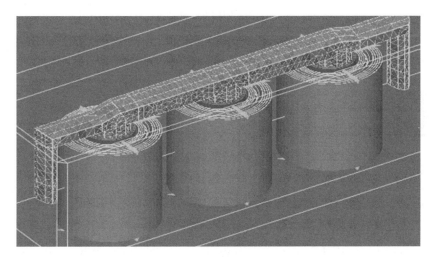

FIGURE 4.36
Modeling of a 600 MVA transformer with an electromagnetic field simulation program (Flux 3D). (From Rioual, M. et al., *IPST 2007*, Lyon, June 4–7, 2007.)

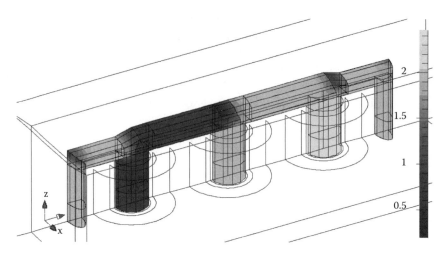

FIGURE 4.37
Description of the flux flowing in the five-legged core (electromagnetic 3D program), when energizing a transformer. (From Rioual, M. et al., *IPST 2007*, Lyon, June 4–7, 2007.)

Some electromagnetic field simulation programs allow for the connection of input and output signals with an electromagnetic transients program. Thus, the effects of the network can be introduced in details. These elaborated simulations may be useful to determine the losses dissipated in a transformer and the electrodynamical forces. The simulations are time consuming and need the knowledge of the electrical and geometrical data of the transformer.

4.4.5 Assembling the Models

4.4.5.1 Single-Phase Transformers

The recommended models for the calculation of slow and switching transients are those derived from the principle of duality. The complete model will include a dual derived structure, saturation, hysteresis, eddy currents in both the windings and the core, and capacitances. Figure 4.38 shows the equivalent for a single-phase two-winding transformer including all the mentioned phenomena.

The circuit of Figure 4.38 has the dc resistance of the windings and the capacitances outside the ideal transformers. This duality-derived circuit for a single-phase, two-winding transformer uses Cauer circuits to represent eddy-current effects in both the windings and the core. This circuit can be obtained from the equivalent of Figure 4.4 by exchanging the leakage and magnetizing branches with frequency-dependent ones.

The circuit of Figure 4.38 is the most physical representation of a transformer yet available for slow and switching transient studies. All inductances representing the air are linear while those representing the core are nonlinear. However, apart from the first inductance, that should be nonlinear to represent saturation and hysteresis, the other inductances for the core can be linear since they represent flux paths for high frequencies and thus it is unlikely that they will saturate. Figure 4.38 shows only the first inductance as nonlinear.

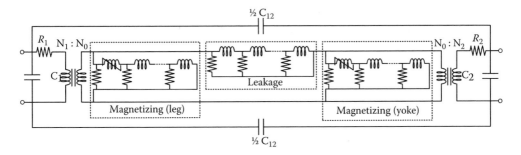

FIGURE 4.38
Dual circuit model for a single-phase two-winding transformer including saturation, hysteresis, eddy currents in the core and eddy currents in the windings, and capacitances.

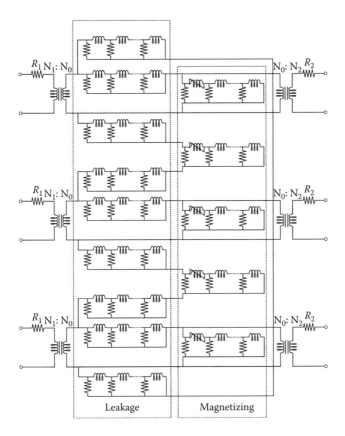

FIGURE 4.39
Dual circuit model for a three-phase three-legged two-winding transformer, including saturation, hysteresis, and eddy currents in the core and eddy currents in the windings. Capacitances are not shown, but they may be needed.

4.4.5.2 Three-Phase Transformers

The three-phase three-leg equivalent model considering eddy currents in core and windings is shown in Figure 4.39. This circuit is obtained from the one in Figure 4.18 by exchanging the linear and nonlinear Cauer models for the windings and core, respectively. The model has nine linear Cauer equivalent circuits representing the leakage and five nonlinear Cauer equivalents to represent the magnetizing phenomena of the core.

4.5 Parameter Determination for Fast and Very Fast Transients

4.5.1 Introduction

A transformer connected to a power network is subjected to various types of transient voltages whose magnitudes and waveshapes must be identified in order to define the stresses undergone by the insulation.

Voltage distribution along the transformer windings depends greatly on the waveshape of the voltage applied to the windings. It can be noticed that at low frequencies the distribution is linear along the windings. In the case of fast transients, a larger portion of the voltage applied is distributed on the first few turns of the winding. Transformers are designed to withstand such stresses and the performance is checked by lightning and switching impulse laboratory tests. High frequency oscillations may generate internal resonances that can damage the insulation. These resonances can be better understood when considering the internal structure of a transformer.

Computation of parameters for high-frequency transformer models is based on similar approaches from those applied for low- and midfrequency models. Three basic methodologies can be distinguished:

1. Application of formulae:
   a. From simplified geometrical configuration
   b. Empirical equations from measurements
2. Experimental determination through laboratory tests
3. Electromagnetic field simulations

Regardless of the model employed for the simulation of the transformer transient, inductive, capacitive, and loss components of the model are in general required to accurately describe the behavior at high frequencies.

The flux penetration into the core is usually neglected for very fast transients, such as those related to switching operations in gas insulated substations (GIS), considering that the core acts as a flux barrier at these high frequencies. However, the flux-penetration dynamics in the core should be taken into account for fast transients due to switching or lightning, particularly when information about the internal behavior of the transformer is required. The core inductance is considered to behave as a completely linear element since high frequencies yield reduced magnetic flux density.

Frequency dependence of both series and shunt elements of the windings at high frequencies should be considered. Skin and proximity effects produce frequency dependence of winding and core impedances because of the reduced flux penetration. At very high frequency, the conductance representing the capacitive loss in the winding's dielectric also depends on frequency. This happens when the polarization time constant of the dielectric is of the same order of magnitude (or larger) than the frequency of the applied electric field.

An important issue in the parameter determination for high-frequency transformer models is that it requires very detailed information of the transformer geometrical configuration which is only available to manufacturers. However, if overvoltages generated within the windings are not required, parameters can be obtained from terminal measurements for the desired frequency range. Note that this is highly dependent on the measurement setup and related instrumentation, and therefore an accurate derivation of parameters can be a complicated task.

4.5.2 High-Frequency Transformer Models

Electromagnetic transients in transformers due to high-frequency waves (also known as fast-front waves) are commonly studied using internal (gray box) models, which consider the propagation and distribution of the incident impulse along the transformer windings. A typical internal representation for a differential segment Δz of a single transformer winding is shown in Figure 4.40 [54,55]. Parameters per unit length are defined as follows:

- L is the series inductance of the winding
- R is the series resistance of the winding (i.e., the loss component of L)
- C_s is the series capacitance of the winding (turn-to-turn)
- R_s is the loss component of C_s
- C_g is the capacitance to ground of the winding (turn-to-earth)
- R_g is the loss component of C_g

Behavior of the impulse propagation at different periods of time can be described by means of voltage-distribution curves. Three significant periods can be distinguished [1,56,57]:

a. Initial voltage distribution: At the beginning of the phenomenon (wave front), the capacitances of the circuit are the predominant elements, producing a nonuniform voltage distribution.

b. Final voltage distribution: At the end of the phenomenon (wave tail), the resistances govern the response of the circuit, resulting in a uniform potential distribution.

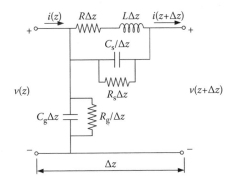

FIGURE 4.40
Equivalent circuit per unit length of a transformer winding.

c. Transient voltage distribution: Between the periods of time described in (a) and (b), interaction of the electrical and magnetic energies stored in the capacitances and inductances of the circuit is developed, giving rise to an oscillatory period in which overvoltages can be obtained at different points along the winding.

Figure 4.41 shows typical voltage-distribution curves for the 3 periods of time aforementioned. Any point along the winding may be stressed at the transitional period, given that voltages to earth may develop in the main body of the winding that considerably exceed the magnitude of the incident impulse, which can also give rise to risk of failure of the insulation between turns of the winding.

The model in Figure 4.40 can be described either by a distributed-parameter representation, using the transmission line theory, or as a ladder connection of lumped-parameter segments. The latter models can be entered into a transients simulation program (e.g., an EMTP-type tool) or by solving the corresponding state-space equations.

Proper choice of the segment length for lumped-parameter modeling is fundamental. Analysis of fast-front transients (in the order of hundreds of kilohertz) using one segment per coil of the winding can be sufficient, whereas very fast front transients (in the order of megahertz) might require considering one segment per turn. Therefore, the resulting circuit can be very large and computationally expensive.

Wave propagation phenomena along the winding can only be reproduced with a distributed-parameter model. However, the need to consider the turn-to-turn inductance can be a serious shortcoming. A model based on the multiconductor transmission line theory, initially proposed by Rabins [58] and developed for electrical machines by Guardado and Cornick [59], has been successfully used to account for the mutual inductance between turns in a distributed-parameter model for transformer windings [60–65]. The model is based on a zigzag connection of the different conductors, each of them representing a turn of the complete winding.

In reference [66], results from lumped- and distributed-parameter models of windings are compared with frequency-domain measurements. It is concluded that the lumped-parameter model can give adequate results for fast transients (up to 1 MHz). However, for very fast transients (above 1 MHz) a distributed-parameter model provides better results.

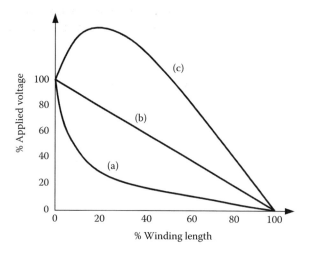

FIGURE 4.41
Impulse voltage distribution: (a) initial; (b) final; (c) transient (time of maximum).

Surges affecting one of the transformer windings can give rise to overvoltages in the other windings. The analysis of this transference phenomenon can also be of importance at the design stage of winding insulation. Lumped- and distributed-parameter models have been used for these studies [67–70]. Additionally, an admittance matrix model (black box model) based on terminal measurements has been presented in [71,72]. This model can be implemented in time-domain simulation programs by means of a rational approximation procedure.

4.5.2.1 Lumped-Parameter Model Based on State-Space Equations

The network equations for the circuit shown in Figure 4.40, considering a cascaded connection (ladder network) of n equal segments, are described in nodal form as follows [73]:

$$\hat{\mathbf{C}}\frac{\mathrm{d}^2\hat{\mathbf{v}}(t)}{\mathrm{d}t^2}+\hat{\mathbf{G}}\frac{\mathrm{d}\hat{\mathbf{v}}(t)}{\mathrm{d}t}+\hat{\mathbf{\Gamma}}\hat{\mathbf{v}}(t)=\mathbf{0} \tag{4.31}$$

In Equation 4.31 the loss component of the series branch has been neglected. $\hat{\mathbf{C}}$, $\hat{\mathbf{G}}$, and $\hat{\mathbf{\Gamma}}$ are the nodal matrices of capacitances, conductances (obtained from the inverse of elements R_g), and inverse inductances, respectively, including the input node, while $\hat{\mathbf{v}}(t)$ is the output vector of node voltages including also the input node. The number of equations in Equation 4.31 is reduced by extracting the input node k of known voltage $u(t)$:

$$\mathbf{C}\frac{\mathrm{d}^2\mathbf{v}(t)}{\mathrm{d}t^2}+\mathbf{G}\frac{\mathrm{d}\mathbf{v}(t)}{\mathrm{d}t}+\mathbf{\Gamma}\mathbf{v}(t)=-\mathbf{C}_k\frac{\mathrm{d}^2u(t)}{\mathrm{d}t^2}-\mathbf{G}_k\frac{\mathrm{d}u(t)}{\mathrm{d}t}-\mathbf{\Gamma}_k u(t) \tag{4.32}$$

where
   $\mathbf{C}$, $\mathbf{G}$, and $\mathbf{\Gamma}$ are nodal matrices of capacitances, conductances, and inverse inductances, respectively, with the kth row and column removed
   $\mathbf{v}(t)$ is the output vector of the $k-1$ node voltages that remain unknown
   $\mathbf{C}_k$, $\mathbf{G}_k$, and $\mathbf{\Gamma}_k$ are the kth columns of $\hat{\mathbf{C}}$, $\hat{\mathbf{G}}$, and $\hat{\mathbf{\Gamma}}$ without the kth row

The state variables are chosen as follows:

$$\mathbf{x}_1(t)=\mathbf{v}(t)+\mathbf{C}^{-1}\mathbf{C}_k u(t) \tag{4.33a}$$

$$\mathbf{x}_2(t)=\frac{\mathrm{d}\mathbf{x}_1(t)}{\mathrm{d}t}+\mathbf{C}^{-1}\mathbf{G}\mathbf{x}_1(t)-\mathbf{C}^{-1}\left(\mathbf{G}\mathbf{C}^{-1}\mathbf{C}_k-\mathbf{G}_k\right)u(t) \tag{4.33b}$$

After manipulation of Equations 4.32 and 4.33, the desired state-space system of equations is obtained [73]:

$$\frac{\mathrm{d}\mathbf{x}(t)}{\mathrm{d}t}=\mathbf{A}\mathbf{x}(t)+\mathbf{B}u(t) \tag{4.34a}$$

$$\mathbf{v}(t)=\mathbf{F}\mathbf{x}(t)+\mathbf{D}u(t) \tag{4.34b}$$

where

$$\mathbf{x}(t) = \begin{bmatrix} \mathbf{x}_1(t) \\ \mathbf{x}_2(t) \end{bmatrix} \tag{4.35a}$$

$$\mathbf{A} = \begin{bmatrix} -\mathbf{C}^{-1}\mathbf{G} & \mathbf{U} \\ -\mathbf{C}^{-1}\mathbf{\Gamma} & \mathbf{0} \end{bmatrix} \tag{4.35b}$$

$$\mathbf{B} = \begin{bmatrix} \mathbf{C}^{-1}\left(\mathbf{GC}^{-1}\mathbf{C}_k - \mathbf{G}_k\right) \\ \mathbf{C}^{-1}\left(\mathbf{\Gamma}\mathbf{C}^{-1}\mathbf{C}_k - \mathbf{\Gamma}_k\right) \end{bmatrix} \tag{4.35c}$$

$$\mathbf{F} = \begin{bmatrix} \mathbf{U} & \mathbf{0} \end{bmatrix} \tag{4.35d}$$

$$\mathbf{D} = -\mathbf{C}^{-1}\mathbf{C}_k \tag{4.35e}$$

The set of equations given by Equation 4.34 can be solved by numerical integration, or by other numerical evaluation techniques for the solution of the space transition matrix.

4.5.2.2 Lumped-Parameter Model from Network Analysis

Equation 4.32 expressed in the Laplace domain becomes

$$s\mathbf{C}\mathbf{V}(s) + \mathbf{G}\mathbf{V}(s) + \frac{\mathbf{\Gamma}}{s}\mathbf{V}(s) = -s\mathbf{C}_k U(s) - \mathbf{G}_k U(s) - \frac{\mathbf{\Gamma}_k}{s} U(s) \tag{4.36}$$

which can be rearranged as follows:

$$\mathbf{I}(s) = \mathbf{Y}(s)\mathbf{V}(s) \tag{4.37}$$

where
   $\mathbf{Y}(s)$ is the nodal-admittance matrix of the circuit
   $\mathbf{I}(s)$ is the nodal current vector, given by

$$\mathbf{Y}(s) = s\mathbf{C} + \mathbf{G} + \frac{\mathbf{\Gamma}}{s} \tag{4.38a}$$

$$\mathbf{I}(s) = -s\mathbf{C}_k U(s) - \mathbf{G}_k U(s) - \frac{\mathbf{\Gamma}_k}{s} U(s) \tag{4.38b}$$

Voltage propagation along the winding can be computed by solving Equation 4.37 for $\mathbf{V}(s)$. Then, the time response of the circuit can be obtained by either an algorithm of numerical frequency–time transformation [74,75], or a rational-approximation procedure to describe the admittance matrix [76].

4.5.2.3 Distributed-Parameter Model Based on Single-Phase Transmission Line Theory

From the elements included in the circuit presented in Figure 4.40, a series impedance Z and a shunt admittance Y per unit length can be defined as

$$Z = \frac{R+sL}{1+(R+sL)(sC_s+1/R_s)} \tag{4.39a}$$

$$Y = sC_g + 1/R_g \tag{4.39b}$$

Modeling can be reduced to the well-known Telegrapher equations of a single-phase transmission line (STL), which are defined in the Laplace domain as follows:

$$\frac{dV(z,s)}{dz} = -Z(s)I(z,s) \tag{4.40a}$$

$$\frac{dI(z,s)}{dz} = -Y(s)V(z,s) \tag{4.40b}$$

where $V(z,s)$ and $I(z,s)$ are the voltage and current at point z of the winding. Numerical solutions to Equation 4.40 have been widely studied for several decades, and include time-domain techniques such as those based on the Bergeron method [77], finite-differences methods [78,79], as well as the direct solution in the frequency domain, and subsequent numerical frequency–time transformation [74,75].

4.5.2.4 Distributed-Parameter Model Based on Multiconductor Transmission Line Theory

The model described in the previous section cannot take into account the mutual inductance between turns of a winding coil. These mutual inductances can be a parameter of importance when computing the interturn voltages due to a fast-front incident impulse. To overcome this problem, a model based on representing the winding by a multiconductor transmission line (MTL) can be applied, as shown in Figure 4.42. In this model, each conductor represents a winding section (disc or turn). To preserve continuity, the end of each conductor is topologically connected to the beginning of the next conductor, resulting in a zigzag connection [60–65]. This is defined simply as

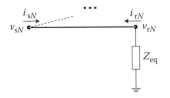

$$v_{ri} = v_{s(i+1)} \qquad i_{ri} = -i_{s(i+1)}, \qquad i = 1 \dots n-1 \tag{4.41}$$

Taking Equation 4.41 into account, the winding model is obtained from the telegrapher equations of an MTL, which are defined in the Laplace domain as

$$\frac{d\mathbf{V}(z,s)}{dz} = -\mathbf{Z}(s)\mathbf{I}(z,s) \tag{4.42a}$$

$$\frac{d\mathbf{I}(z,s)}{dz} = -\mathbf{Y}(s)\mathbf{V}(z,s) \tag{4.42b}$$

FIGURE 4.42
Multiconductor transmission line model for the transformer winding.

$\mathbf{Z}$ and $\mathbf{Y}$ are the $N \times N$ matrices of series impedances and shunt conductances per unit length, where N is the number of conductors (discs or turns)

$\mathbf{V}(z,s)$ and $\mathbf{I}(z,s)$ are the vectors of voltages and currents at point z of the winding

In Figure 4.42, the equivalent impedance Z_{eq} connected at the end of the N-th element can be used to represent the remaining part of the winding, when only a section of the winding needs to be modeled in detail. Z_{eq} can also represent the neutral impedance.

The numerical solution of Equations 4.42 can be obtained using the frequency- and time-domain techniques described above.

4.5.2.5 Distributed-Parameter Model Based on Combined STL and MTL Theories

For very fast transients (e.g., transient generated by switching operations in gas-insulated substations of up to 30 MHz), all turns and coils of the winding might need to be considered in the study. In this case, an MTL-based model would result in very large matrix operations and, as a consequence, a significant computational effort. This problem can be addressed by combining the STL- and MTL-based models described in Sections 4.5.2.3 and 4.5.2.4. In [61–63], the problem is solved in two steps: First, each coil is represented by an STL model and voltages at the coil's ends are obtained. Then, each coil is represented by an MTL model to compute the distribution of the interturn overvoltages independently from the other coils, using the voltages computed in the previous step as inputs. This is illustrated in Figure 4.43. Since the first coils are usually exposed to the highest stress, the MTL model can be considered only for these coils, so that the voltage distribution can be studied in detail.

4.5.2.6 Modeling for Transference Analysis

Due to capacitive and inductive coupling, an incident impulse propagating along a transformer winding will be transferred to the other windings. A typical example would be the case of a lightning surge propagating along the HV/primary side of a transformer, and partially transferred to its LV/secondary side. This phenomenon can be analyzed starting from a representation similar to that of Figure 4.40. Considering the case of a single-phase, 2-winding transformer, the equivalent circuit for a differential segment Δz is shown in Figure 4.44 [55]. Similar to the single winding representation, this model can be described by means of lumped or distributed parameters.

If a distributed-parameter modeling is chosen, a representation of the two-winding transformer similar to a two-phase transmission line can be used. The capacitive and inductive couplings of the line correspond to the coupling between windings. After some circuit analysis, propagation equations are given in [55] as

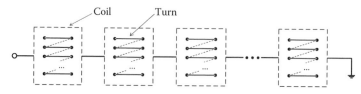

FIGURE 4.43
Combined winding model.

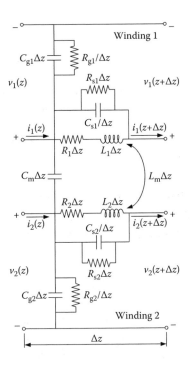

FIGURE 4.44
Equivalent circuit per unit length of a single-phase two-winding transformer.

$$
\begin{bmatrix} \dfrac{dV_1(z,s)}{dz} \\ \dfrac{dV_2(z,s)}{dz} \end{bmatrix} = \dfrac{1}{D(s)} \begin{bmatrix} Z_1 + Z_1 Y_2 Z_2 - Z_m^2 Y_2 & Z_m \\ Z_m & Z_2 + Z_2 Y_1 Z_1 - Z_m^2 Y_1 \end{bmatrix} \begin{bmatrix} I_1(z,s) \\ I_2(z,s) \end{bmatrix}
\tag{4.43a}
$$

$$
\begin{bmatrix} \dfrac{dI_1(z,s)}{dz} \\ \dfrac{dI_2(z,s)}{dz} \end{bmatrix} = \begin{bmatrix} Y_{g1} + Y_m & -Y_m \\ -Y_m & Y_{g2} + Y_m \end{bmatrix} \begin{bmatrix} V_1(z,s) \\ V_2(z,s) \end{bmatrix}
\tag{4.43b}
$$

where

$$
D(s) = 1 + Z_1 Y_1 + Z_2 Y_2 + Z_1 Z_2 Y_1 Y_2 - Z_m^2 Y_1 Y_2
\tag{4.44a}
$$

$$
Z_i = R_i + s L_i \quad i = 1, 2
\tag{4.44b}
$$

$$
Y_i = s C_{si} + 1/R_{si} \quad i = 1, 2
\tag{4.44c}
$$

$$
Y_{gi} = s C_{gi} + 1/R_{gi} \quad i = 1, 2
\tag{4.44d}
$$

$$
Z_m = s L_m
\tag{4.44e}
$$

$$
Y_m = s C_m
\tag{4.44f}
$$

Since this model is based on the STL theory, it cannot take into account the mutual inductance between turns. An alternative, which in contrast cannot reproduce the wave propagation along the winding, is a lumped-parameter representation based also in Figure 4.44. A different approach could be the extension of the MTL theory for a multi-winding case, but this has not been reported in the literature to date in a final stage.

As for the lumped-parameter model, both network analysis and state-space approaches can be applied. The resulting models for each case are similar to those described in Sections 4.5.2.1 to 4.5.2.5 considering also the mutual elements between windings in the capacitive and inductive matrices.

4.5.2.7 Admittance Model from Terminal Measurements (Black Box Model)

If the computation of internal stresses along the windings is not required, a transformer terminal model can be used. This model is described in the frequency domain in terms of its admittance matrix as [72]:

$$\mathbf{I}_t(s) = \mathbf{Y}_t(s)\mathbf{V}_t(s) \tag{4.45}$$

Considering a transformer with n terminals, $\mathbf{Y}_t$ is the admittance matrix of size $n \times n$ which relates terminal currents and voltages given by vectors $\mathbf{I}_t$ and $\mathbf{V}_t$, respectively, both of length n. If a 1-p.u. voltage is applied at node j of the transformer while the remaining terminals are short-circuited, the jth column of $\mathbf{Y}_t$ will be equivalent to the currents measured from ground to each terminal. Applying this procedure, direct measurement of all elements of $\mathbf{Y}_t$ can be completed, as presented in Figure 4.45.

If the measured admittance matrix is used with different terminal conditions (e.g., when some terminals are open), measurement errors may become magnified. In order to validate the model for different terminal conditions, the situation that the transformer terminals are divided into two groups, denoted by A and B, respectively, is considered. Equation 4.45 can be partitioned as follows:

$$\begin{bmatrix} \mathbf{I}_A \\ \mathbf{I}_B \end{bmatrix} = \begin{bmatrix} \mathbf{Y}_{AA} & \mathbf{Y}_{AB} \\ \mathbf{Y}_{BA} & \mathbf{Y}_{BB} \end{bmatrix} \begin{bmatrix} \mathbf{V}_A \\ \mathbf{V}_B \end{bmatrix} \tag{4.46}$$

If the terminals of set A are open-circuited, so that $\mathbf{I}_A = \mathbf{0}$, the following voltage ratio is obtained:

$$\mathbf{V}_{AB} = -\mathbf{Y}_{AA}^{-1}\mathbf{Y}_{AB} \tag{4.47}$$

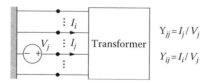

FIGURE 4.45
Measurement procedure for the jth column of the admittance matrix. (From Gustavsen, B., *IEEE Trans. Power Deliv.*, 19, 414, January 2004. With permission.)

For a 3-phase 2-winding transformer, considering the case that all terminals of one winding are open-circuited, Equation 4.47 can be used to obtain the corresponding voltage ratios:

$$\mathbf{V}_{HL} = -\mathbf{Y}_{HH}^{-1}\mathbf{Y}_{HL} \tag{4.48a}$$

$$\mathbf{V}_{LH} = -\mathbf{Y}_{LL}^{-1}\mathbf{Y}_{LH} \tag{4.48b}$$

where H and L denote the HV and LV windings. $\mathbf{V}_{HL}$ and $\mathbf{V}_{LH}$ are matrices of size 3×3. Validation of the resulting $\mathbf{Y}_t$ can be attained by comparing measured values of the voltage ratios with those computed from Equation 4.48. Moreover, the accuracy of the transformer model in open circuit conditions can be increased by introducing the measured voltage ratios into the model. From Equation 4.48, a modified admittance matrix is defined:

$$\mathbf{Y}'_{LL} = \mathbf{Y}_{LL} \tag{4.49a}$$

$$\mathbf{Y}'_{LH} = -\mathbf{Y}'_{LL}\mathbf{V}_{LH} \tag{4.49b}$$

$$\mathbf{Y}'_{HL} = \mathbf{Y}'_{LH} \tag{4.49c}$$

$$\mathbf{Y}'_{LL} = \mathbf{Y}'_{HL}\mathbf{V}_{HL} \tag{4.49d}$$

To obtain the time-domain response of the system, the modified admittance matrix is approximated with rational functions [76]. The output of this approximation will be matrices **A**, **B**, **C**, **D**, and **E**, which define a space equation realization (SER) of the form

$$\mathbf{I}_t(s) \cong \mathbf{Y}_{fit}(s)\mathbf{V}_t(s) = \left[\mathbf{C}(s\mathbf{U} - \mathbf{A})^{-1}\mathbf{B} + \mathbf{D} + s\mathbf{E}\right]\mathbf{V}_t(s) \tag{4.50}$$

where $\mathbf{Y}_{fit}$ represents the rational approximation of the modified admittance matrix. If a set of N poles is used, with $n = 6$ for the 3-phase 2-winding transformer, the matrix dimensions are **A**: $N \times N$, **B**: $N \times 6$, **C** $= 6 \times N$, **D**: 6×6, **E**: 6×6.

Next, $\mathbf{Y}_{fit}$ is represented in the form of an electrical network, whose branches are calculated as follows:

$$y_i = \sum_{j=1}^{n} Y_{fit,ij}, \quad y_{ij} = -Y_{fit,ij} \tag{4.51}$$

where y_i and y_{ij} represent admittance branches between node i and ground and between nodes i and j, respectively. Each branch in Equation 4.51 is described as a rational function:

$$y(s) = \sum_{m=1}^{N} \frac{c_m}{s - a_m} + d + se \tag{4.52}$$

Finally, each branch can be represented by an electrical network as shown in Figure 4.46. R_0 and C_0 are computed as

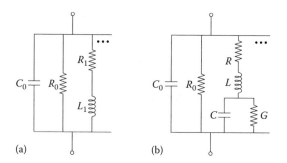

FIGURE 4.46
Synthesization of electrical network from rational approximation: (a) real poles; (b) complex conjugate pairs. (From Gustavsen, B., *IEEE Trans. on Power Deliv.*, 17, 1093, October 2002. With permission.)

$$C_0 = e, \quad R_0 = 1/d \qquad (4.53)$$

Real poles result in R–L branches:

$$R_1 = -a/c, \quad L_1 = 1/c \qquad (4.54)$$

while complex conjugate pairs of the form

$$\frac{c' + jc''}{s - (a' + ja'')} + \frac{c' - jc''}{s - (a' - ja'')}, \qquad (4.55)$$

result in RLC branches:

$$L = 1/(2c') \qquad (4.56a)$$

$$R = \left[-2a' + 2(c'a' + c''a'')L \right] L \qquad (4.56b)$$

$$1/C = \left[a'^2 + a''^2 + 2(c'a' + c''a'')R \right] L \qquad (4.56c)$$

$$G = -2(c'a' + c''a'')CL \qquad (4.56d)$$

Alternatively, a recursive convolution procedure can be used to include the rational approximation in the form of a companion network in EMTP-type programs.

Table 4.11 summarizes the different models discussed in this section. Their characteristics, application, and limitations are listed.

4.5.3 Parameter Determination Procedures for High-Frequency Models

4.5.3.1 Capacitance

The most common and straightforward approach to compute the winding capacitances is based on the well-known formula for parallel plates. STL-based winding model requires

TABLE 4.11

Summary of the Characteristics of High-Frequency Transformer Models

| Model | | Characteristics | |
		Limitations	Application
Lumped-parameter	State-space analysis	Large matrices, computer time, difficult inclusion of frequency dependence	Fast transients
	Network analysis	Large matrices, computer time	
	Simulation program[a]	Size of circuit, inclusion of inductive coupling between turns, difficult inclusion of frequency dependence	
Distributed-parameter	STL theory	Inclusion of inductive coupling between turns	Fast and very fast transients
	MTL theory	Complexity of the solution	
	Combined STL and MTL	Complexity of the solution	

[a] It refers to direct model construction with the different EMTP-type programs, not relying on user-defined programming and modeling.

computing only the series (turn-to-turn) capacitance, C_s, and the capacitance to ground (turn-to-earth), C_g, both per unit length. For the series capacitance, the following equation can be used:

$$C_s = \frac{\varepsilon_0 \varepsilon_r h}{d_s} \qquad (4.57)$$

where
  ε_0 is the free space permittivity
  ε_r is the relative permittivity of the dielectric material between turns
  h is the rectangular conductor's height
  d_s is the distance between turns

The relative permittivity can be distinguished according to the insulation materials of corresponding local space.

The capacitance to ground can be computed in a similar manner:

$$C_g = \frac{\varepsilon_0 \varepsilon_r w}{d_g} \qquad (4.58)$$

where
  w is the rectangular conductor's width
  d_g is the distance between turn and ground plane

For the MTL-based winding model, a capacitance matrix is formed starting from the representation of Figure 4.47 [59], where the 2 first discs of the HV side are considered. In the figure,

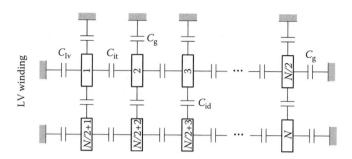

FIGURE 4.47
Representation of two discs of a transformer winding.

- C_{lv} is the capacitance between the HV and LV sides.
- C_{it} is the capacitance between adjacent turns.
- C_g is the capacitance between turn and ground.
- C_{id} is the capacitance between adjacent discs.

Computation of these parameters can be completed similarly to Equations 4.57 and 4.58, by means of simple parallel-plate formulations, considering adequate values for dielectric permittivity, distance between elements, and transversal area for each element. Capacitances between nonadjacent turns can also be included, although values for distant turns are considered negligible.

However, Equations 4.57 and 4.58 do not take into account the fringe effects and related stray capacitances. Expressions for the turn-to-turn and disk-to-disk (axial) capacitances based also on geometrical considerations which take into account these effects have been obtained in [57].

$$C_{it} = \frac{\varepsilon_0 \varepsilon_{it} \left(w + d_{it} \right)}{d_{it}} \tag{4.59a}$$

$$C_{id} = \varepsilon_0 \left(\frac{k}{d_{it}/\varepsilon_{it} + d_{id}/\varepsilon_{oil}} + \frac{1-k}{d_{it}/\varepsilon_{it} + d_{id}/\varepsilon_{id}} \right) (R + d_{id}) \tag{4.59b}$$

where
ε_{it} and ε_{id} are the relative permittivities of the insulation between turns and between discs
ε_{oil} is the relative permittivity of the oil insulation
d_{it} and d_{id} are the distances between turns and between discs
k is a fraction of circumferential space occupied by oil
R is the winding radial depth

The terms d_{it} and d_{id}, added respectively to w and R in Equations 4.59a and 4.59b, are used to account for the fringe effects.

A more general and accurate capacitance calculation can be obtained from electrostatic field simulation. A technique based on the finite element method (FEM) to evaluate the elements of the winding capacitance matrix from rather basic electrostatic analysis has been described in [64]. Considering N winding turns, the jth segment potential is set to a value $U = 1\,V$ while the potential of all other turns is set to zero. The electrostatic energy stored in j is defined as

$$W_j = \frac{1}{2}\sum_{i=1}^{N} C_{ij}\Delta U_{ij}^2 \tag{4.60}$$

where
C_{ij} is the capacitance between turns i and j (not a matrix element)
ΔU_{ij} is the potential difference between such turns
W_j is obtained using an FEM package

If the geometrical arrangement of Figure 4.47 is considered, four different types of capacitive elements need to be computed. Using Equation 4.60, the following expression can be obtained:

$$\frac{1}{2}\begin{bmatrix} 1 & 1 & 1 & 1 \\ 0 & 1 & 2 & 1 \\ 0 & 2 & 1 & 1 \\ 0 & 1 & 1 & 1 \end{bmatrix}\begin{bmatrix} C_{lv} \\ C_g \\ C_{it} \\ C_{id} \end{bmatrix} = \begin{bmatrix} W_1 \\ W_2 \\ W_{N/2} \\ W_N \end{bmatrix} \tag{4.61}$$

Equation 4.61 is solved for the capacitances. Finally, the capacitance matrix is formed similar to a nodal admittance matrix: C_{ii} is given by the addition of all capacitances converging to node (turn) i, while C_{ij} is given by the mutual capacitance between nodes (turns) i and j.

4.5.3.2 Inductance

An exact expression for the mutual inductance between the two thin-wire coaxial loops a and b shown in Figure 4.48, with radii r_a and r_b, and spaced a distance d apart was defined by Maxwell as [1,80]

$$L_{ab} = \frac{2\mu_0}{k}\sqrt{r_a r_b}\left[\left(1 - \frac{k^2}{2}\right)K(k) - E(k)\right] \tag{4.62}$$

where
μ_0 is the permeability of free space
$K(k)$ and $E(k)$ are complete elliptic integrals of the first and second kind, respectively

$$k = \sqrt{\frac{4 r_a r_b}{(r_a + r_b)^2 + d^2}} \tag{4.63}$$

FIGURE 4.48
Two thin-wire coaxial loops.

For rectangular cross-section coils, Lyle's method in conjunction with Equation 4.62 can be used for a more accurate determination

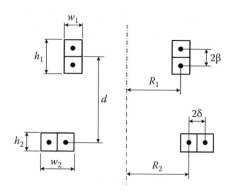

FIGURE 4.49
Lyle's method for rectangular cross-section coils.

of the mutual inductance [1,81]. The method consists of replacing each coil of rectangular cross section by two equivalent thin-wire loops. The corresponding dimensions are shown in Figure 4.49. For $h > w$, the coil is replaced by the loops 1–1′ and 2–2′ with the radii given by

$$r_1 = R_1\left(1 + \frac{w_1^2}{24R_1^2}\right)$$ (4.64)

The loops are spaced on each side of the median plane of the coil by a distance β given by

$$\beta = \sqrt{\frac{h_1^2 - w_1^2}{12}}$$ (4.65)

If $w > h$, the coil is replaced by the loops 3–3′ and 4–4′ lying in the median plane of the coil, with radii $(r_2 + \delta)$ and $(r_2 - \delta)$, respectively, where

$$r_2 = R_2\left(1 + \frac{h_2^2}{24R_2^2}\right)$$ (4.66)

and

$$\delta = \sqrt{\frac{w_2^2 - h_2^2}{12}}$$ (4.67)

Since the two coils of rectangular cross section are replaced by four fictitious thin-wire loops, four combinations of mutual inductances are computed from Equation 4.62, and the mutual inductance between the coils is obtained as an average of those values:

$$L_{ab} = \frac{L_{13} + L_{14} + L_{23} + L_{24}}{4}$$ (4.68)

On the other hand, the self-inductance of a single-turn circular coil of square cross section with an average radius a and square side length c has been defined by Grover as [82]

$$L_s = \mu_0 a \left[\frac{1}{2}\left(1 + 1/6\left(\frac{c}{2a}\right)^2\right)\ln\left(\frac{8}{(c/2a)^2}\right) - 0.84834 + 0.2041\left(\frac{c}{2a}\right)^2 \right] \quad (4.69)$$

Equation 4.69 applies for a small cross section ($c/2a > 0.2$). If the cross section is not square, it can be subdivided into a number of squares and Equation 4.62 together with Equation 4.69 can be used to compute the self-inductance more accurately. Gray obtained the following alternative expression for a coil of rectangular cross section $w \times h$ [83]:

$$L_s = \mu_0 a \ln\left(\frac{8a}{\text{GMD}} - 2\right) \quad (4.70)$$

where

$$\ln \text{GMD} = \frac{1}{2}\ln\left(h^2 + w^2\right) + \frac{2w}{3h}\tan^{-1}\frac{h}{w} + \frac{2h}{3w}\tan^{-1}\frac{w}{h}$$

$$- \frac{w^2}{12h^2}\ln\left(1 + \frac{h^2}{w^2}\right) - \frac{h^2}{12w^2}\ln\left(1 + \frac{w^2}{h^2}\right) - \frac{25}{12} \quad (4.71)$$

Finally, the required series and mutual inductances per unit length are obtained as

$$L = L_s/l_t \quad (4.72a)$$

$$L_m = L_{ab}/l_t \quad (4.72b)$$

where l_t is the turn length in meters.

Another approach for winding inductance calculation, which is based on the MTL theory and therefore is more suitable for the MTL-based model, is by defining an inductance matrix per unit length, divided in a geometrical inductance $\mathbf{L}_g$ matrix and a conductor inductance matrix $\mathbf{L}_c$, such that

$$\mathbf{L} = \mathbf{L}_g + \mathbf{L}_c \quad (4.73)$$

The easiest way to obtain the geometrical inductance matrix $\mathbf{L}_g$ is directly from the capacitance matrix:

$$\mathbf{L}_g = \frac{\varepsilon_r}{c^2}\mathbf{C}^{-1} \quad (4.74)$$

where
ε_r is the relative permittivity of the dielectric material
c is the velocity of light in free space

$\mathbf{C}$ is the capacitance matrix calculated as described in Section 4.5.3.1

The conductor inductance matrix is computed as

$$\mathbf{L}_c = \frac{\mathrm{Im}(Z_c)}{\omega}\mathbf{U} \tag{4.75}$$

where
  Z_c is the conductor impedance due to skin effect, which is defined in Section 4.5.3.3
  ω is the angular frequency in rad/s

Similar to the capacitance matrix, the inductance matrix can be computed directly from the FEM analysis using the energy method to obtain more accurate results for realistic arrangements [84,85]. Self-inductance L_{ii} can be computed from the magnetic energy obtained when applying a current I_i to the ith section (turn or group of turns) of the winding:

$$W_{\mathrm{mag},i} = \frac{1}{2}L_{ii}\,I_i^2 \tag{4.76}$$

Mutual inductance L_{ij} is computed from the magnetic energy obtained when applying current at both elements i and j:

$$W_{\mathrm{mag},ij} = L_{ij}\,I_iI_j + \frac{1}{2}\left(L_{ii}I_i^2 + L_{jj}I_j^2\right) \tag{4.77}$$

The matrix diagonal elements must be calculated first from Equation 4.76, in order to obtain the off-diagonal elements from Equation 4.77.

4.5.3.3 Conductor Losses

Taking into account the skin effect at high frequencies, and considering again a rectangular cross section, conductor impedance per unit length is computed from the dc resistance R_{dc} and the impedance at high frequencies Z_{hf} [86]:

$$Z_c = \sqrt{R_{dc}^2 + Z_{hf}^2} \tag{4.78}$$

where

$$R_{dc} = \frac{\rho_c}{wh} \tag{4.79a}$$

$$Z_{hf} = \frac{\rho_c}{2p(w+h)} \tag{4.79b}$$

ρ_c is the resistivity of the conductor material
w and h are the conductor cross-sectional dimensions
p is the complex penetration depth due to skin effect, defined as follows:

$$p = \sqrt{\frac{\rho_c}{j\omega\mu_c}} \tag{4.80}$$

where μ_c is the permeability of the conductor material.

4.5.3.4 Core Losses

The effect of eddy currents due to flux penetration in the core can be described by means of the following expression, see [50,51,53]:

$$Z_{\text{core}} = \frac{4N^2 A \rho_{\text{core}}}{l d^2} x \tanh x \tag{4.81}$$

where

$$x = \frac{d}{2} \sqrt{\frac{j\omega\mu_{\text{core}}}{\rho_{\text{core}}}} \tag{4.82}$$

and
   l is the length of the magnetic path
   d is the thickness of the lamination
   μ_{core} is the permeability of the core material
   ρ_{core} is the resistivity of the core material
   N is the number of turns per coil
   A is the total cross-sectional area of all laminations

Expression 4.81 represents the frequency-dependent impedance of a coil wound around a laminated iron core; it was derived by solving Maxwell's equations assuming that the electromagnetic field distribution is identical in all laminations.

4.5.3.5 Loss Component of Capacitance (Turn-to-Turn and Turn-to-Ground)

The capacitive loss in the insulation material can be computed directly from the capacitance matrix making use of the loss factor $\tan\delta$, and is defined in terms of a conductance matrix [86]:

$$\mathbf{G} = \omega \tan\delta\, \mathbf{C} \tag{4.83}$$

From Figure 4.40, G_{ii} corresponds to the addition of elements $1/R_s$ and $1/R_g$ converging at node i, while G_{ij} is given by the element $1/R_s$ connected between nodes i and j with a minus sign. However, it can be noticed from Equation 4.83 that this element is a linear function of frequency; moreover, it has been observed that the loss factor $\tan\delta$ in oil-treated cellulose papers is also frequency dependent [66,87]. Figure 4.50 illustrates this

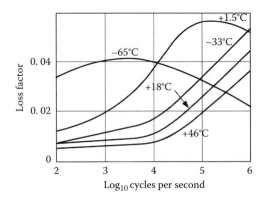

FIGURE 4.50
Loss factor of oil-impregnated cellulose versus frequency. (From Hosseini, S.M.H. et al., *IEEE Trans. Power Deliv.*, 23, 733, April 2008. With permission.)

variation up to 1 MHz. For example, at 46°C, the loss factor can be estimated as 0.005 in the frequency band of 0–40 kHz and then it increases linearly with frequency and at 1 MHz it is equal to 0.036.

4.5.4 Examples

4.5.4.1 Capacitance Calculation and Initial Voltage Distribution

The geometrical configuration presented in Figure 4.51 is considered [57]. It consists of two discs with six turns per disk. Relative permittivities are taken as $\varepsilon_{it} = \varepsilon_{id} = 3.5$ and $\varepsilon_{oil} = 2.2$. Capacitances are computed from analytical expressions and from an FEM technique, see Section 4.5.3.1. Table 4.12 shows a comparison of the results.

Parallel-plate formulation with no fringe effects considered (see Equations 4.57 and 4.58) is in agreement with FEM results only for the turn-to-turn capacitance, C_{it}. However, for the other capacitances, whose fictitious parallel plates have a larger separation, the values obtained with both methods are quite different. Figure 4.52 shows the simulation of

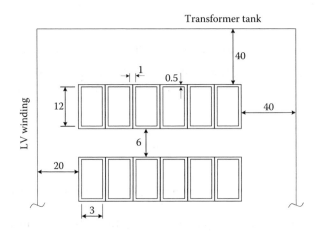

FIGURE 4.51
Geometrical arrangement of winding for capacitance calculation (units in millimeters). (From Kulkarni, S.V. and Khaparde, S.A., *Transformer Engineering: Design and Practice*, CRC Press, New York, 2004. With permission.)

TABLE 4.12

Comparison of Capacitance Calculation
with Analytical Expressions and FEM

	Analytical (pF/m)	FEM (pF/m)
C_{it}	371.360/402.310[a]	383.410
C_{id}	46.42/88.266[a]	80.540
C_{lv}	11.670	4.255
C_g	1.458	0.323

[a] Including fringe effects (Equation 4.57).

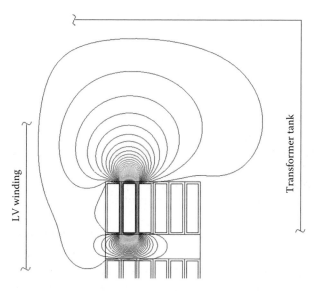

FIGURE 4.52
Contour of electric potential obtained by FEM.

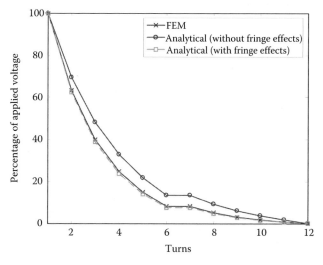

FIGURE 4.53
Initial voltage distribution.

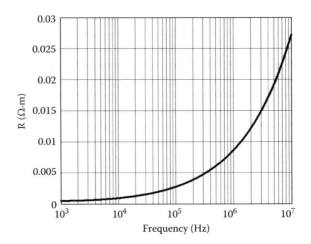

FIGURE 4.54
Series resistance versus frequency.

electric potential when turn 2 is excited and all other turn voltages are set to zero. Complex fringe effects are readily seen.

From the capacitive values obtained by both methods, the capacitance matrix is formed and the voltage distribution along the disc pair is computed, as shown in Figure 4.53. A 1-p.u. voltage is applied to turn 1, while turn 12 is grounded. Turns 6 and 7 are connected by means of a low impedance to preserve continuity.

4.5.4.2 Behavior of Copper and Core Impedances at High Frequencies

Frequency dependence of copper and core impedances within a range of 10 kHz to 10 MHz is analyzed, considering the transformer data shown in Table 4.13. Equation 4.78 is used to obtain the series (copper) resistance and inductance, respectively, as presented in Figures 4.54 and 4.55. Equation 4.81 is applied to compute the core resistance and inductance, which are plotted in Figures 4.56 and 4.57. Relative permeability of the core material is varied from 5,000 to 50,000. Only one turn of the coil is considered.

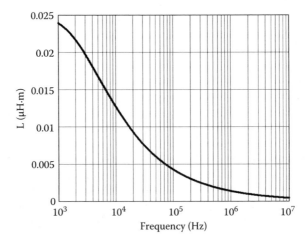

FIGURE 4.55
Series inductance versus frequency.

TABLE 4.13

Core and Winding Parameters

	Core		Winding
l	$1\,\text{m}$	w	$0.003\,\text{m}$
d	$0.003\,\text{m}$	h	$0.012\,\text{m}$
ρ_{core}	$9.58\times10^{-8}\,\Omega/\text{m}$	μ_c	$1.0\mu_0\,\text{H/m}$
A	$0.54\,\text{m}^2$	ρ_c	$1.68\times10^{-8}\,\Omega/\text{m}$

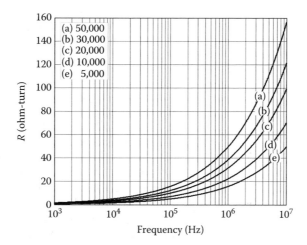

FIGURE 4.56
Core resistance versus frequency, with different relative permeabilities.

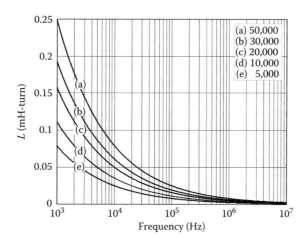

FIGURE 4.57
Core inductance versus frequency, with different relative permeabilities.

From Figures 4.54 and 4.56, it can be noticed that both resistive components (conductor and core) exhibit a similar behavior, although core resistance is considerably higher. On the other hand, Figures 4.55 and 4.57 show that very small inductive values are obtained at high frequencies, and therefore the geometrical inductance will generally be the dominant factor. Besides, deviation of core resistance values obtained with different relative permeabilities becomes larger as the frequency is increased, while the trend is different for core inductance values.

4.5.4.3 Sensitivity of Transient Winding Response to Geometrical Inductance Calculation

This example deals with the simple winding configuration shown in Figure 4.58, extracted from [88], which consists of 16 disks wound onto one of the outer legs of a commercial three-phase 25 kVA core. The two ends of each disk were brought out to terminals on terminal board for connection as a single winding (eight disks of 12 turns).

Capacitances obtained for the winding configuration are rather low, resulting in much higher natural resonant frequencies than would occur in real-size transformers. To obtain conditions similar to those of large transformers, external capacitances of higher values than the natural capacitances were included in the experimental setup of [88].

The winding inductance matrix was computed with FEM from the energy method, considering frequency-dependent effects, and from analytical formulae, both described in Section 4.5.3.2. Self- and mutual inductances were obtained considering as basic element one disc of the winding. An important issue to consider in the computation of these values from FEM is that since the coil is wound to an external leg of the core, the inductance matrix will be different in the coil sections corresponding to regions inside and outside the core window. If the simulation is not performed using a 3D arrangement, the inductance matrix can still be approximated by considering separated simulations in the different regions and computing an equivalent inductance (for this particular case, 25% inside the window and 75% outside the window). Values obtained for the self-inductance L_{11} and the mutual inductance L_{12} at different frequencies between 100 Hz and 10 MHz are shown in Figure 4.59. It can be noticed that these values are very similar due to the strong influence of the iron core for the complete frequency range.

The unit-step response of the winding (with its far end grounded) was computed and compared with the experimental result reported in [88] (Figure 4.60). The MTL winding model was applied to obtain the transient.

Only the geometrical inductance for air core and the internal inductance due to skin effect were considered when computing the inductance from analytical formulae assuming, as done in several references [60–64], that the flux does not penetrate the core at high frequencies, acting as a flux barrier. However, it is readily seen that the result from the energy method (i.e., FEM) is in much better agreement with the experimental measurement than with the result from the analytical formulae. This is due to the fact that, given the low equivalent conductivity of the core material for this example (97.09 S-m), the magnetic flux still penetrates into the core at the frequencies considered in the calculation.

Figure 4.61 shows the resulting transient when only one of the regions is taken into account, as well as the result from averaging the inductance matrix from the two regions. Comparing the waveforms with the measurement presented in Figure 4.60, one can notice the importance of taking into account the different geometrical regions of the core in the parameters computation.

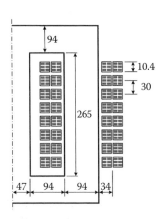

FIGURE 4.58
Geometrical arrangement.

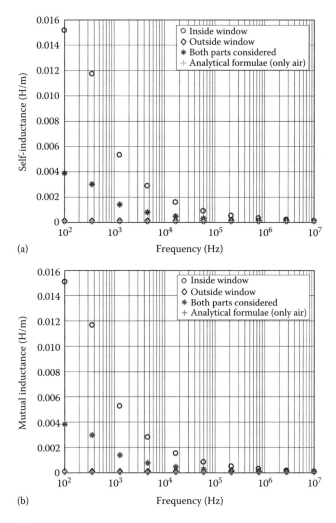

FIGURE 4.59
Values obtained from FEM and analytical formulae for (a) self-inductance and (b) mutual inductance.

4.5.4.4 Experimental Setup for Measurement of the Terminal Admittance Matrix of a Distribution Transformer

A setup designed in [72] for measuring the elements of the admittance matrix $\mathbf{Y}_t$ for the black box model presented in Section 4.5.2.6 is described in this section. A 30 kVA, 10 kV/400 V, 3-phase distribution transformer is considered. Both HV and LV windings are connected as a grounded star.

The setup uses a network analyzer, a connection board with a built-in current sensor and shielded cables. The cable shields are grounded at both ends; at the transformer and at the connection board. All connections for measuring different matrix elements are done using jumper connections. This is done so to produce repeatable results. The cables between the network analyzer and the connection board are made as short as possible. The frequency dependency of the current sensor, voltage probes, and attenuator are taken into account with the analyzer built-in functionality. Figure 4.62a shows the connection board for the

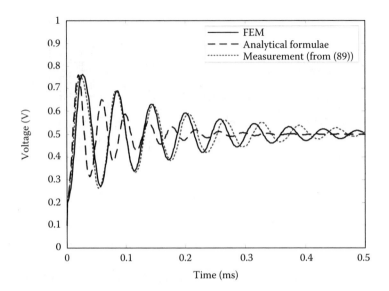

FIGURE 4.60
Unit-step response of the winding.

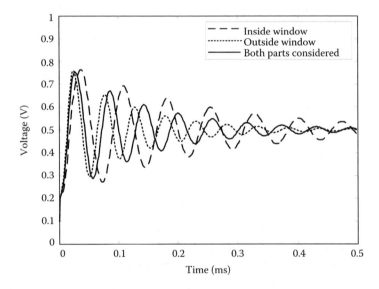

FIGURE 4.61
Unit-step response of the winding, with the inductance matrix computed at different regions of the core.

measurement of the diagonal element (Y_{11}). Figure 4.62b shows the configuration used for measuring the off-diagonal element (Y_{13}). The following components were used [72]: (1) network analyzer: HP3577A; (2) attenuator: HP8491A, 30 dB. This attenuator has low-pass characteristic when connecting the source to the LV winding; (3) current sensor: Pearson model 2100, 1 V/1A.

The measurements were carried out using the following settings for the network analyzer: logarithmically distributed samples between 50 Hz and 2 MHz, 100 s sweep time, and 10 Hz resolution bandwidth.

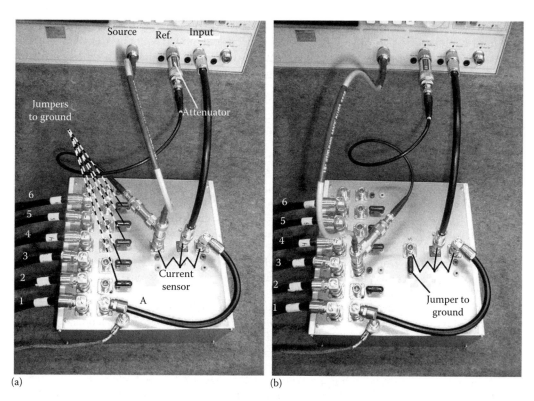

FIGURE 4.62
Measuring setup: (a) diagonal element Y_{11}; (b) off-diagonal element Y_{13}. (From Gustavsen, B., *IEEE Trans. Power Deliv.*, 19, 414, January 2004. With permission.)

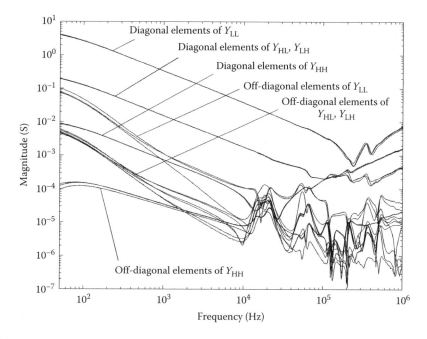

FIGURE 4.63
Measured elements of $\mathbf{Y}_t$. (From Gustavsen, B., *IEEE Trans. Power Deliv.*, 19, 414, January 2004. With permission.)

Figure 4.63 shows the magnitude of the 36 measured elements of $\mathbf{Y_t}$ obtained with measurement cables of 5 m length. All elements exhibit a strong frequency dependency and some elements are several orders of magnitude smaller than others. These large differences are mainly the result of the large voltage ratio of the transformer (25:1).

References

1. A. Greenwood, *Electrical Transients in Power Systems*, 2nd edn., Wiley, New York, 1991.
2. CIGRE Working Group 02 (SC 33), Guidelines for Representation of Network Elements when Calculating Transients, CIGRE Technical Brochure no. 39, 1990.
3. IEEE TF on Slow Transients, Modeling and analysis guidelines for slow transients. Part I: Torsional oscillations; transient torques; turbine blade vibrations; fast bus transfer, *IEEE Transactions on Power Delivery*, 10(4), 1950–1955, October 1995.
4. IEEE TF on Slow Transients, Modeling and analysis guidelines for slow transients. Part II: Controller interactions; harmonic interactions, *IEEE Transactions on Power Delivery*, 11(3), 1672–1677, July 1996.
5. IEEE TF on Slow Transients, Modeling and analysis guidelines for slow transients. Part III: The study of ferroresonance, *IEEE Transactions on Power Delivery*, 15(1), 255–265, January 2000.
6. A.M. Gole, J.A. Martinez-Velasco, and A.J.F. Keri (Eds.), Modeling and Analysis of System Transients Using Digital Programs, IEEE PES Special Publication, TP-133-0, 1999.
7. J.A. Martinez and B. Mork, Transformer modeling for low- and mid-frequency transients: A review, *IEEE Transactions on Power Delivery*, 20(2), 1625–1632, April 2005.
8. IEEE Std. C57.12.90-1999, IEEE Standard Test Code for Liquid-Immersed Distribution, Power, and Regulating Transformers.
9. IEEE Std. C57.12.91-2001, IEEE Standard Test Code for Dry-Type Distribution and Power Transformers.
10. IEEE Std. C57.123-2002, IEEE Guide for Transformer Loss Measurement.
11. IEEE Std C57.138-1998, IEEE Recommended Practice for Routine Impulse Test for Distribution Transformers.
12. IEEE Std C57.12.80-2002, IEEE Standard Terminology for Power and Distribution Transformers.
13. IEEE Std C57.12.00-2000, IEEE Standard General Requirements for Liquid-Immersed Distribution, Power, and Regulating Transformers.
14. IEEE Std C57.12.01-1998, IEEE Standard General Requirements for Dry-Type Distribution and Power Transformers Including Those with Solid-Cast and/or Resin-Encapsulated Windings.
15. P.J. Hopkinson, United States power transformer equipment standards and processes, Chapter 24 of *Electric Power Transformer Engineering*, J.H. Harlow (Ed.), 2nd edn., CRC Press, Boca Raton, FL, 2007.
16. IEC Standard 60076-1, Power transformers—Part 1: General, 2000.
17. IEC Standard 60076-8, Power transformers—Part 8: Application guide, 1997.
18. M. Rioual, S. Dennetière, and Y. Guillot, A link between EMTP-RV and FLUX3D: Application to transformer energization, *IPST 2007*, Lyon, June 4–7, 2007.
19. G.R. Slemon, Equivalent circuits for transformers and machines including non-linear effects, *Proceedings of IEE*, 100, Part IV, 129–143, 1953.
20. C.M. Arturi, Transient simulation and analysis of a three-phase five-limb step-up transformer following an out-of-phase synchronization, *IEEE Transactions on Power Delivery*, 6(1), 196–207, January 1991.
21. D.L. Stuehm, Three phase transformer core modeling, Bonneville Power Administration, Award No. DE-BI79-92BP26700, February 1993.

22. F. de León and A. Semlyen, Complete transformer model for electromagnetic transients, *IEEE Transactions on Power Delivery*, 9(1), 231–239, January 1994.
23. A. Narang and R.H. Brierley, Topology based magnetic model for steady-state and transient studies for three-phase core type transformers, *IEEE Transactions on Power Systems*, 9(3), 1337–1349, August 1994.
24. B.A. Mork, Five-legged wound-core transformer model: Derivation, parameters, implementation, and evaluation, *IEEE Transactions on Power Delivery*, 14(4), 1519–1526, October 1999.
25. V. Brandwajn, H.W. Dommel, and I.I. Dommel, Matrix representation of three-phase n-winding transformers for steady-state and transient studies, *IEEE Transactions on Power Apparatus and Systems*, 101(6), 1369–1378, June 1982.
26. H.W. Dommel, *Electromagnetic Transients Program Reference Manual* (*EMTP Theory Book*), Bonneville Power Administration, Portland, August 1986.
27. G.R. Slemon, *Electric Machines and Drives*, Addison-Wesley Publishing Company, Reading, MA, 1992.
28. M.A. Plonus, *Applied Electromagnetics*, McGraw-Hill, New York, 1978.
29. M. Rioual and Y. Husianicia, Determination of the residual fluxes when de-energizing a power transformer & comparison with on site tests, *IEEE PES General Meeting*, San Francisco, CA, June 2006.
30. M. Rioual and M. Sow, Study of the sympathetic interactions when energizing transformers for wind-farms: Description of the phenomena involved and determination of the stresses during their energization, *IEEE Workshop on Wind Integration in Power Systems*, France, May 26–27, 2008.
31. R.A. Walling, K.D. Barker, T.M. Compton, and I.E. Zimmerman, Ferroresonant overvoltages in grounded padmount transformers with low-loss silicon-steel cores, *IEEE Transactions on Power Delivery*, 8(3), 1647–1660, July 1993.
32. F. de Leon and A. Semlyen, A simple representation of dynamic hysteresis losses in power transformers, *IEEE Transactions on Power Delivery*, 10, 315–321, January 1995.
33. W. Chandrasena, P.G. McLaren, U.D. Annakkage, and R.P. Jayasinghe, An improved low-frequency transformer model for use in GIC studies, *IEEE Transactions on Power Delivery*, 19(2), 643–651, April 2004.
34. J.A. Martínez, R. Walling, B. Mork, J. Martin-Arnedo, and D. Durbak, Parameter determination for modeling systems transients. Part III: Transformers, *IEEE Transactions on Power Delivery*, 20(3), 2051–2062, July 2005.
35. N.D. Hatziargyriou, J.M. Prousalidis, and B.C. Papadias, Generalised transformer model based on the analysis of its magnetic core circuit, *IEE Proceedings C*, 140(4), 269–278, July 1993.
36. V. Del Toro, *Electric Machines and Power Systems*, Prentice-Hall, Englewood Cliffs, NJ, 1985.
37. E.F. Fuchs and Y. You, Measurement of v-i characteristics of asymmetric three-phase transformers and their applications, *IEEE Transactions on Power Delivery*, 17(4), 983–990, October 2002.
38. B.A. Mork, D. Ishchenko, F. Gonzalez, and S.D. Cho, Parameter estimation methods for five-limb magnetic core model, *IEEE Transactions on Power Delivery*, 23(4), 2025–2032, October 2008.
39. A. Boyajian, Theory of three-circuit transformers, *AIEE Transactions*, 508–528, February 1924.
40. F. Starr, Equivalent circuits-I, *AIEE Transactions*, 57, 287–298, June 1932.
41. L.F. Blume, A. Boyajian, G. Camilli, T.C. Lennox, S. Minneci, and V.M. Montsinger, *Transformer Engineering*, Wiley, New York, 1951 (Chapter V by A. Boyajian).
42. Electric Systems Technology Institute, *Electrical Transmission and Distribution Reference Book*, ABB, Raleigh, NC, 1997. (Westinghouse T&D Book).
43. W.S. Meyer and T.-H. Liu, Unstable saturable transformer, *Can/Am EMTP News*, 93(2), 15–16, April 1993.
44. P.S. Holenarsipur, N. Mohan, V.D. Albertson, and J. Cristofersen, Avoiding the use of negative inductances and resistances in modeling three-winding transformers for computer simulations, *1999 IEEE/PES Winter Meeting*, New York, January 1999.

45. X. Chen, Negative inductance and numerical instability of the saturable transformer component in EMTP, *IEEE Transactions on Power Delivery*, 15(4), 1199–1204, October 2000.

46. T. Henriksen, How to avoid unstable time domain responses caused by transformer models, *IEEE Transactions on Power Delivery*, 17(2), 516–522, April 2002.

47. F. de León and J.A. Martinez, Dual three-winding transformer equivalent circuit matching leakage measurements, *IEEE Transactions on Power Delivery*, 24(1), 160–168, January 2009.

48. K. Karsai, D. Kerenyi, and L. Kiss, *Large Power Transformers*, Elsevier, New York, 1987.

49. E.E. Fuchs, D. Yildirim, and W.M. Grady, Measurement of eddy-current loss coefficient PEC-R, derating of single-phase transformers, and comparison with K-factor approach, *IEEE Transactions on Power Delivery*, 15(1), 148–154, January 2000.

50. F. de León and A. Semlyen, Time domain modeling of eddy current effects for transformer transients, *IEEE Transactions on Power Delivery*, 8(1), 271–280, January 1993.

51. E.J. Tarasiewicz, A.S. Morched, A. Narang, and E.P. Dick, Frequency dependent eddy current models for nonlinear iron cores, *IEEE Transactions on Power Systems*, 8(2), 588–597, May 1993.

52. P. Holmberg, M. Leijon, and T. Wass, A wideband lumped circuit model of eddy current losses in a coil with a coaxial insulation system and a stranded conductor, *IEEE Transactions on Power Delivery*, 18(1), 50–60, January 2003.

53. J. Avila-Rosales and F.L. Alvarado, Nonlinear frequency dependent transformer model for electromagnetic transient studies in power systems, *IEEE Transactions on Power Apparatus and Systems*, 101(11), 4281–4288, November 1982.

54. W.J. Mc Nutt, T.J. Blalock, and R.A. Hinton, Response of transformer windings to system transient voltages, *IEEE Transactions on Power Apparatus and Systems*, 93(2), 457–466, March/April 1974.

55. A.S. AlFuhaid, Frequency characteristics of single phase two winding transformer using distributed parameter modeling, *IEEE Transactions on Power Delivery*, 16(4), 637–642, October 2001.

56. M. Heathcote, *J P Transformer Book*, 13th edn., Newnes, 2007.

57. S.V. Kulkarni and S.A. Khaparde, *Transformer Engineering: Design and Practice*, CRC Press, Boca Raton, FL, 2004.

58. L. Rabins, A New approach to the analysis of impulse voltages and gradients in transformer windings, *AIEE Transactions*, 79(4), 1784–1791, February 1960.

59. J.L. Guardado and K.J. Cornick, A computer model for calculating steep-fronted surge distribution in machine windings, *IEEE Transactions on Energy Conversion*, 4(1), 95–101, March 1989.

60. K.J. Cornick, B. Filliat, C. Kieny, and W. Muller, Distribution of very fast transient overvoltages in transformer windings, *CIGRE Session*, Report 12–204, 1992.

61. Y. Shibuya, S. Fujita, and N. Hosokawa, Analysis of very fast transient overvoltage in transformer winding, *IEE Proceedings C, Generation, Transmission and Distribution*, 144(5), 461–468, September 1997.

62. Y. Shibuya, S. Fujita, and E. Tamaki, Analysis of very fast transients in transformer, *IEE Proceedings C, Generation, Transmission and Distribution*, 148(5), 377–383, September 2001.

63. M. Popov, L.V. Sluis, and G.C. Paap, Computation of very fast transient overvoltages in transformer windings, *IEEE Transactions on Power Delivery*, 18(4), 1268–1274, October 2003.

64. G. Liang, H. Sun, X. Zhang, and X. Cui, Modeling of transformer windings under very fast transient overvoltages, *IEEE Transactions on Electromagnetic Compatibility*, 48(4), November 2006.

65. M. Popov, L. van der Sluis, R.P.P. Smeets, and J. Lopez Roldan, Analysis of very fast transients in layer-type transformer windings, *IEEE Transactions on Power Delivery*, 22(1), 238–247, January 2007.

66. S.M.H. Hosseini, M. Vakilian, and G.B. Gharehpetian, Comparison of transformer detailed models for fast and very fast transient studies, *IEEE Transactions on Power Delivery*, 23(2), 733–741, April 2008.

67. A. Miki, T. Hosoya, and K. Okuyama, A calculation method for impulse voltage distribution and transferred voltage in transformer windings, *IEEE Transactions on Power Apparatus and Systems*, 97(3), 930–939, May/June 1978.

68. D.J. Wilcox and T.P. McHale, Modified theory of modal analysis for the modeling of multiwinding transformers, *IEE Proceedings C, Generation, Transmission and Distribution*, 139(6), 505–512, November 1992.

69. K. Ragavan and L. Satish, An efficient method to compute transfer function of a transformer from its equivalent circuit, *IEEE Transactions on Power Delivery*, 20(2), 780–788, April 2005.

70. M.H. Nazemi and G.B. Gharehpetian, Influence of mutual inductance between HV and LV windings on transferred overvoltages, *XIVth International Symposium on High Voltage Engineering*, Beijing, China, August 25–29, 2005.

71. B. Gustavsen and A. Semlyen, Application of vector fitting to the state equation representation of transformers for simulation of electromagnetic transients, *IEEE Transactions on Power Delivery*, 13(3), 834–842, July 1998.

72. B. Gustavsen, Wide band modeling of power transformers, *IEEE Transactions on Power Delivery*, 19(1), 414–422, January 2004.

73. P.I. Fergestad and T. Henriksen, Transient oscillations in multiwinding transformers, *IEEE Transactions on Power Apparatus and Systems*, 93(2), 500–509, March/April 1974.

74. J. Wilcox, Numerical Laplace transformation and inversion, *International Journal of Electrical Engineering Education*, 15, 247–265, 1978.

75. P. Moreno and A. Ramirez, Implementation of the numerical Laplace transform: A review, *IEEE Transactions on Power Delivery*, 23(4), 2599–2609, October 2008.

76. B. Gustavsen, Computer code for rational approximation of frequency dependent admittance matrices, *IEEE Transactions on Power Delivery*, 17(4), 1093–1098, October 2002.

77. L. Bergeron, *Water Hammer in Hydraulics and Wave Surges in Electricity*, Wiley, New York, 1961.

78. J.L. Naredo, A.C. Soudack, and J.R. Marti, Simulation of transients on transmission lines with corona via the method of characteristics, *IEE Proceedings C, Generation, Transmission and Distribution*, 142(1), 81–87, 1995.

79. C.R. Paul, Incorporation of terminal constraints in the FDTD analysis of transmission lines, *IEEE Transactions on Electromagnetic Compatibility*, 36(2), 85–91, May 1994.

80. J.C. Maxwell, *A Treatise on Electricity and Magnetism*, The Clarendon Press, Oxford, 1904.

81. T.R. Lyle, Magnetic shell equivalent of circular coils, *Philosophical Magazine*, 3, 310–329, 1902.

82. F.W. Grover, *Inductance Calculations*, Dover Publications Inc., New York, 1973.

83. A. Gray, *Absolute Measurements in Electricity and Magnetism*, Macmillan, London, 1921.

84. Z. Azzouz, A. Foggia, L. Pierrat, and G. Meunier, 3D finite element computation of the high frequency parameters of power transformer windings, *IEEE Transactions on Magnetics*, 29(2), 1407–1410, March 1993.

85. E. Bjerkan and H.K. Høidalen, High frequency FEM-based power transformer modeling: Investigation of internal stresses due to network-initiated overvoltages, *IPST 2005*, Montreal, Canada, June 19–23, 2005.

86. C. Paul, *Analysis of Multiconductor Transmission Lines*, Wiley, New York, 1994.

87. F.M. Clark, *Insulating Materials for Design and Engineering Practice*, Wiley, New York, 1962.

88. D.J. Wilcox, W.G. Hurley, T.P. McHale, and M. Conlon, Application of modified modal theory in the modelling of practical transformers, *IEE Proceedings C, Generation, Transmission and Distribution*, 139(6), 513–520, November 1992.

5

Synchronous Machines

Ulas Karaagac, Jean Mahseredjian, and Juan A. Martinez-Velasco

CONTENTS

5.1 Introduction 252
5.2 Synchronous Machine Equations 253
    5.2.1 Synchronous Machine Circuit Equations 253
    5.2.2 Flux Linkage Equations 254
    5.2.3 Voltage Equations 260
    5.2.4 Electrical Torque Equation 261
5.3 Park Transformation 261
    5.3.1 Introduction 261
    5.3.2 Flux Linkage Equations in $dq0$ Components 263
    5.3.3 Voltage Equations in $dq0$ Components 264
    5.3.4 Electrical Torque Equation in $dq0$ Components 266
    5.3.5 Choice of k_d, k_q, and k_0 Constants in Park Transformation 267
5.4 Per Unit Representation 268
5.5 Equivalent Circuits 271
5.6 Synchronous Machine Parameters 274
    5.6.1 Three-Phase Short Circuit at the Terminals of a Synchronous Machine 275
    5.6.2 Subtransient and Transient Inductances and Time Constants 276
    5.6.3 Armature Time Constant 281
    5.6.4 Operational Parameters 282
5.7 Data Conversion Procedures 286
    5.7.1 Determination of Fundamental Parameters from Operational
        Parameters 287
    5.7.2 Procedure for Complete Second-Order d-Axis Model 287
    5.7.3 Procedure for Simplified Second-Order d-Axis Model 289
    5.7.4 Procedure for Complete Second-Order d-Axis Model with Canay's
        Characteristic Inductance 292
    5.7.5 Procedure for Second-Order q-Axis Model 295
    5.7.6 Determination of Fundamental Parameters from Transient and
        Subtransient Inductances and Time Constants 297
5.8 Magnetic Saturation 298
    5.8.1 Open Circuit and Short Circuit Characteristics 298
    5.8.2 Representation of the Magnetic Saturation 300
5.9 Test Procedures for Parameter Determination 303
    5.9.1 Introduction 303
    5.9.2 Steady-State Tests 304
        5.9.2.1 Open Circuit Characteristic Curve 304

              5.9.2.2   Short Circuit Characteristic Curve 305
              5.9.2.3   Determination of *d*-Axis Synchronous Impedance from
                        OCC and SCC 305
              5.9.2.4   Slip Test and Measuring *q*-Axis Synchronous Impedances 306
       5.9.3 Short Circuit Tests 307
              5.9.3.1   Sudden Short Circuit with the Machine Operating
                        Open-Circuited at Rated Speed 307
              5.9.3.2   Sudden Short Circuit of Armature and Field with the
                        Machine Operating Open-Circuited at Rated Speed 313
              5.9.3.3   Sudden Short Circuit with Machine Operating on Load at
                        Low Voltage 314
       5.9.4 Decrement Tests 315
              5.9.4.1   Voltage Recovery Test 315
              5.9.4.2   Disconnecting the Armature with the Machine Running
                        Asynchronously on Load 319
              5.9.4.3   Disconnecting the Armature at a Very Low Slip 320
       5.9.5 Standstill Frequency Response Tests 320
              5.9.5.1   Introduction 320
              5.9.5.2   SSFR Test Measurements 321
              5.9.5.3   Parameter Identification from SSFR Test Results 322
       5.9.6 Online Tests 325
              5.9.6.1   Online Frequency Response Test 326
              5.9.6.2   Load Rejection Test 326
              5.9.6.3   Time-Domain Small Disturbance Test 326
              5.9.6.4   Time-Domain Large Disturbance Test 326
5.10 Models for High-Frequency Transient Simulations 326
       5.10.1 Winding Models 326
       5.10.2 Calculation of Electrical Parameters 329
5.11 Synchronous Machine Mechanical System Equations 331
       5.11.1 Lumped Multimass Model 331
       5.11.2 Mechanical System Model in Terms of Modal Parameters 334
       5.11.3 Determination of Mechanical Parameters 338
       5.11.4 Determination of Damping Factors of the
              Mass–Spring–Damper Model 339
Appendix A. Third-Order Equivalent Circuits 340
       5.A.1 Algorithm for Parameter Calculation from $G(s)$ and $L_d(s)$ 340
       5.A.2 Algorithm for Parameter Calculation from $L_d(s)$ and L_C 342
       5.A.3 Algorithm for Parameter Calculation from $L_d(s)$ 344
       5.A.4 Algorithm for Parameter Calculation from $L_q(s)$ 345
References 346

5.1   Introduction

A synchronous machine is a complex component whose behavior is the result of the inter-action of electrical and mechanical systems. The detail with which a synchronous machine must be represented depends on the frequency range of the transients to be analyzed.

TABLE 5.1

Modeling Guidelines for Synchronous Machines

TOPIC	Low-Frequency Transients	Slow-Front Transients	Fast-Front Transients	Very-Fast-Front Transients
Representation	A detailed representation of both mechanical and electrical parts, including saturation effects	A simplified representation of the electrical part: an ideal ac source behind the frequency-dependent transient impedance	A linear per phase circuit which matches the frequency response of the machine	A capacitance-to-ground per phase
Voltage control	Very important	Negligible	Negligible	Negligible
Speed control	Important	Negligible	Negligible	Negligible
Capacitance	Negligible	Important	Important	Very important
Frequency-dependent parameters	Important	Important	Negligible	Negligible

Source: CIGRE WG 33.02, Guidelines for Representation of Network Elements when Calculating Transients, CIGRE Brochure no. 39, 1990. With permission.

Table 5.1 presents a summary of the guidelines proposed by the CIGRE WG 33.02 for representing synchronous machines [1]. Note that the representation of mechanical and control systems is crucial in low-frequency and slow-front frequency transient studies, but it can be neglected for higher-frequency transients.

This chapter deals with the determination of electrical and mechanical parameters needed to represent three-phase synchronous machines in low- and midfrequency transient studies, such as transient stability, subsynchronous resonance, load rejection, generator tripping, generator synchronization, and inadverted energization. The different sections of the chapter can be therefore grouped into two parts dedicated respectively to modeling and parameter determination of the electrical and the mechanical systems of a synchronous machine.

A significant effort has been made to derive the electrical parameters needed to represent synchronous machines during system transients. This effort is reflected in standards, which are regularly updated. Test setups and testing conditions, as well as procedures for the determination of characteristic electrical parameters, are described and justified in IEEE and IEC standards, see [2,3]. This chapter presents some conversion procedures that can be used to pass from test or manufacturer's data to basic or fundamental electrical parameters, those needed by digital programs to construct internal circuits and equations.

A short section has been included to introduce the models that can be required for representing a machine in high-frequency (fast and very fast front) transients.

Readers are referred to the specialized literature for more details, see Refs. [4–8].

5.2 Synchronous Machine Equations

5.2.1 Synchronous Machine Circuit Equations

Figure 5.1 illustrates the circuits involved in the analysis of the machine. The stator circuits are composed of three-phase armature windings and the rotor circuits are composed of the field and the damper windings. Although a large number of circuits are

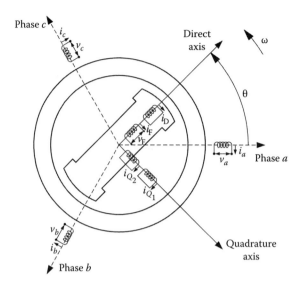

FIGURE 5.1
Stator and rotor circuits of a synchronous machine. *a,b,c*: Stator phase windings, F: field winding, D: *d*-axis damper winding, Q_1: first *q*-axis damper winding, Q_2: second *q*-axis damper winding, θ: angle by which *d*-axis leads the magnetic axis of phase *a* winding, in electrical rad, ω: rotor angular velocity, in electrical rad/s.

used to represent damper effects in machine analysis, a limited number of circuits may be used in power system analysis depending on the type of rotor construction and the frequency range of interest. Usually the damper effects are represented with three damper windings: one located at *d*-axis, and other two located at *q*-axis. The rotor circuits have therefore three damper windings; the machine equations will be derived based on this assumption, see Figure 5.1.

In the derivation of basic equations of the machine, the following assumptions are made:

- The magnetomotive force (mmf) in the air-gap is sinusoidally distributed and the space harmonics are neglected.
- The effect of stator slots on the rotor inductances is neglected; i.e., saliency is restricted to the rotor.
- Magnetic hysteresis is neglected.
- Magnetic saturation effects are neglected.

The omission of magnetic saturation effects is made to deal with linear coupled circuits and make superposition applicable. However, saturation effects can be significant and the methods of accounting their effects will be discussed in Section 5.8.

5.2.2 Flux Linkage Equations

The stator and rotor flux linkages can be written as

$$\begin{bmatrix} \psi_s \\ \psi_r \end{bmatrix} = \begin{bmatrix} L_{ss} & L_{sr} \\ L_{rs} & L_{rr} \end{bmatrix} \begin{bmatrix} i_s \\ i_r \end{bmatrix} \tag{5.1}$$

where

$i_s^t = \begin{bmatrix} i_a & i_b & i_c \end{bmatrix}$: instantaneous stator phase currents

$i_r^t = \begin{bmatrix} i_F & i_D & i_{Q_1} & i_{Q_2} \end{bmatrix}$: field and damper circuit currents

$\psi_s^t = \begin{bmatrix} \psi_a & \psi_b & \psi_c \end{bmatrix}$: flux linkage in stator windings

$\psi_r^t = \begin{bmatrix} \psi_F & \psi_D & \psi_{Q_1} & \psi_{Q_2} \end{bmatrix}$: flux linkage in rotor windings

L_{ss}: stator–stator inductances

L_{sr}, L_{rs}: stator–rotor and rotor–stator inductances

L_{rr}: rotor–rotor inductances

Stator–Stator Inductances: The flux produced by a stator winding follows a path through the stator iron, across the air-gap, through the rotor iron, and back across the air-gap. As the permeance of the magnetic flux path varies with the rotor position due to nonuniform air-gap, both the self- and mutual inductances of the stator circuits vary with the rotor position. This is pronounced for not only the salient-pole machines but also for the round rotor machines in which the permeances in the two axis differ mainly due to the large number of slots associated with field winding. The variations in permeance of this path can be approximated as follows:

$$P = P_0 + P_2 \cos 2\alpha \tag{5.2}$$

where α is the angular distance from the *d*-axis measured along the periphery, as illustrated in Figure 5.2.

The double frequency variation in Equation 5.2 is due to the equal permeances of the north and south poles.

The self-inductance of phase *a* winding, l_{aa}, is equal to the sum of the self-inductance l_{gaa} of phase *a* due to air-gap flux and the leakage inductance L_{al} which represents the leakage flux not crossing the air-gap.

$$l_{aa} = L_{al} + l_{gaa} \tag{5.3}$$

l_{gaa} is directly proportional to the permeance of the flux path described above and has a second harmonic variation. It can be found by evaluating the air-gap flux linking phase *a* winding, when only phase *a* is excited.

Let MMF$_a$ be the mmf of phase *a*, having a sinusoidal distribution in space with a peak amplitude $N_a i_a$ centered on the phase *a* axis where N_a is the effective turns per phase. MMF$_a$ can be decomposed into two sinusoidally distributed mmfs, one centered on *d*-axis and the other centered on *q*-axis. This decomposition is illustrated in Figure 5.3.

The peak values of these two component waves are

$$\text{peak MMF}_{ad} = N_a i_a \cos\theta \tag{5.4a}$$

$$\text{peak MMF}_{aq} = N_a i_a \sin\theta \tag{5.4b}$$

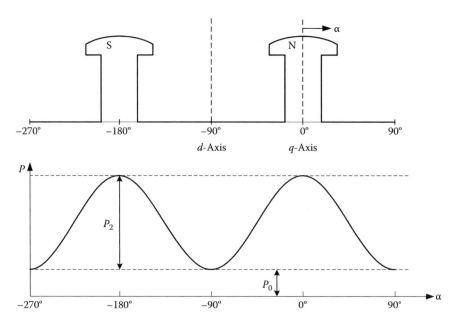

FIGURE 5.2
Variation of permeance with rotor position.

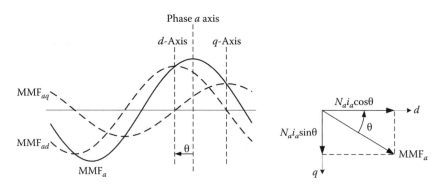

FIGURE 5.3
Decomposition of phase *a* mmf.

The air-gap fluxes per pole along these two axes are

$$\Phi_{gad} = \left(N_a i_a \cos\theta\right)P_d \tag{5.5a}$$

$$\Phi_{gaq} = \left(N_a i_a \sin\theta\right)P_q \tag{5.5b}$$

where P_d and P_q are the permeance coefficients of the *d*- and *q*-axes including the required factors to relate flux per pole with the peak value of mmf wave.

The total air-gap flux linking phase a is

$$\Phi_{gaa} = \Phi_{gad} \cos\theta + \Phi_{gaq} \sin\theta$$

$$= N_a i_a (P_d \cos^2\theta + P_q \sin^2\theta)$$

$$= N_a i_a \left(\frac{P_d + P_q}{2} + \frac{P_d - P_q}{2} \cos 2\theta \right) \tag{5.6}$$

From this result, the self-inductance l_{gaa} of phase a due to air-gap flux can be expressed as follows:

$$l_{gaa} = \frac{N_a \Phi_{gaa}}{i_a}$$

$$= N_a^2 \left(\frac{P_d + P_q}{2} + \frac{P_d - P_q}{2} \cos 2\theta \right)$$

$$= L_{g0} + L_{g2} \cos 2\theta \tag{5.7}$$

The self-inductance of phase a winding, l_{aa}, can be found by substituting Equation 5.7 into Equation 5.3

$$l_{aa} = L_{al} + L_{g0} + L_{g2} \cos 2\theta$$

$$= L_{aa0} + L_{aa2} \cos 2\theta \tag{5.8a}$$

Since the windings of phase b and c are identical to the winding of phase a, and displaced from it by 120° and 240°, respectively, the self-inductances of phase b and c windings are

$$l_{bb} = L_{aa0} + L_{aa2} \cos 2\left(\theta - \frac{2\pi}{3} \right)$$

$$= L_{aa0} + L_{aa2} \cos\left(2\theta + \frac{2\pi}{3} \right) \tag{5.8b}$$

$$l_{cc} = L_{aa0} + L_{aa2} \cos 2\left(\theta + \frac{2\pi}{3} \right)$$

$$= L_{aa0} + L_{aa2} \cos\left(2\theta - \frac{2\pi}{3} \right) \tag{5.8c}$$

Similar to the self-inductances, the mutual inductance between any two windings also has a second harmonic variation due to the rotor shape. It is always negative and reaches its maximum absolute value when the north and south poles are equidistant from the centers of the considered two windings.

The mutual inductance between phases a and phase b windings can be found by evaluating the air-gap flux linking phase b winding, when only phase a is excited or vice versa. As phase b winding is displaced from phase a winding by 120°, the flux linking of phase b due to mmf of phase a can be found by replacing θ in Equation 5.6 with $\theta - \dfrac{2\pi}{3}$

$$\Phi_{gba} = \Phi_{gad} \cos\left(\theta - \frac{2\pi}{3}\right) + \Phi_{gaq} \sin\left(\theta - \frac{2\pi}{3}\right)$$

$$= N_a i_a \left(P_d \cos\theta \cos\left(\theta - \frac{2\pi}{3}\right) + P_q \sin\theta \sin\left(\theta - \frac{2\pi}{3}\right) \right)$$

$$= N_a i_a \left(-\frac{P_d + P_q}{4} + \frac{P_d - P_q}{2} \cos\left(2\theta - \frac{2\pi}{3}\right) \right) \qquad (5.9)$$

From this result, the mutual inductance l_{gba} between phases a and b due to air-gap flux is as follows:

$$l_{gba} = \frac{N_a \Phi_{gba}}{i_a}$$

$$= -\frac{1}{2} L_{g0} + L_{g2} \cos\left(2\theta - \frac{2\pi}{3}\right) \qquad (5.10)$$

It should be noted that there is a small amount of mutual flux that does not cross the air-gap. By adding the inductance which represents this flux, the mutual inductance between phases a and b can be written as follows:

$$l_{ab} = l_{ba} = L_{ab0} + L_{ab2} \cos\left(2\theta - \frac{2\pi}{3}\right) \qquad (5.11)$$

Similarly,

$$l_{bc} = l_{cb} = L_{ab0} + L_{ab2} \cos 2\theta \qquad (5.12a)$$

$$l_{ca} = l_{ac} = L_{ab0} + L_{ab2} \cos\left(2\theta + \frac{2\pi}{3}\right) \qquad (5.12b)$$

From Equations 5.8a and 5.10, it can be seen that $L_{ab2} = L_{aa2}$. This is expected due to the fact that the variation of the permeance of magnetic flux path with the rotor position produces the second harmonic terms of both self- and mutual inductances.

The stator–stator inductance matrix can be then written as

$$[L_{ss}] = \begin{bmatrix} L_{aa0} & L_{ab0} & L_{ab0} \\ L_{ab0} & L_{aa0} & L_{ab0} \\ L_{ab0} & L_{ab0} & L_{aa0} \end{bmatrix} + L_{aa2} \begin{bmatrix} \cos 2\theta & \cos\left(2\theta - \frac{2\pi}{3}\right) & \cos\left(2\theta + \frac{2\pi}{3}\right) \\ \cos\left(2\theta - \frac{2\pi}{3}\right) & \cos\left(2\theta + \frac{2\pi}{3}\right) & \cos 2\theta \\ \cos\left(2\theta + \frac{2\pi}{3}\right) & \cos 2\theta & \cos\left(2\theta - \frac{2\pi}{3}\right) \end{bmatrix} \qquad (5.13)$$

Stator–Rotor Inductances: Since the effects of the stator slots are neglected, the rotor circuits see a constant permeance. Therefore, the variation of the mutual inductance is only due to the relative motions of the windings.

With a sinusoidal distribution of mmf and flux waves, the mutual inductance between *d*-axis rotor windings and phase *a* winding varies as $\cos\theta$; thus

$$l_{aF} = L_{aF} \cos\theta \tag{5.14a}$$

$$l_{aD} = L_{aD} \cos\theta \tag{5.14b}$$

$$l_{aQ_1} = L_{aQ_1} \cos\left(\theta - \frac{\pi}{2}\right)$$
$$= L_{aQ_1} \sin\theta \tag{5.14c}$$

$$l_{aQ_2} = L_{aQ_2} \cos\left(\theta - \frac{\pi}{2}\right)$$
$$= L_{aQ_2} \sin\theta \tag{5.14d}$$

In the expressions for the mutual inductance between phase *b* winding and the rotor circuits, θ is replaced by $\theta - \frac{2\pi}{3}$; for the mutual inductance between phase *c* winding and the rotor circuits, θ is replaced by $\theta + \frac{2\pi}{3}$. Then, the stator–rotor inductance matrix can be written as

$$[L_{sr}] = \begin{bmatrix} L_{aF} \cos\theta & L_{aD} \cos\theta & L_{aQ_1} \sin\theta & L_{aQ_2} \sin\theta \\ L_{aF} \cos\left(\theta - \frac{2\pi}{3}\right) & L_{aD} \cos\left(\theta - \frac{2\pi}{3}\right) & L_{aQ_1} \sin\left(\theta - \frac{2\pi}{3}\right) & L_{aQ_2} \sin\left(\theta - \frac{2\pi}{3}\right) \\ L_{aF} \cos\left(\theta + \frac{2\pi}{3}\right) & L_{aD} \cos\left(\theta + \frac{2\pi}{3}\right) & L_{aQ_1} \sin\left(\theta + \frac{2\pi}{3}\right) & L_{aQ_2} \sin\left(\theta + \frac{2\pi}{3}\right) \end{bmatrix} \tag{5.15}$$

Since $l_{aF} = l_{Fa}$, $l_{aD} = l_{Da}$, $l_{aQ1} = l_{Q1}a$ and $l_{aQ2} = l_{Q2a}$, the rotor–stator inductance matrix can be written as

$$[L_{rs}] = [L_{sr}]^t \tag{5.16}$$

Rotor–Rotor Inductances: Since the effects of the stator slots are neglected, the rotor circuits see a constant permeance. Therefore, the self-inductances of the rotor circuit and the mutual inductances between each other do not change with the rotor position; i.e., they are constant. The constant rotor–rotor inductance matrix can be written as

$$[L_{rr}] = \begin{bmatrix} L_{FF} & L_{FD} & 0 & 0 \\ L_{FD} & L_{DD} & 0 & 0 \\ 0 & 0 & L_{Q_1Q_1} & L_{Q_1Q_2} \\ 0 & 0 & L_{Q_1Q_2} & L_{Q_2Q_2} \end{bmatrix} \tag{5.17}$$

5.2.3 Voltage Equations

The voltage equations for the stator and rotor windings are

$$-\frac{d\psi_s}{dt} - [R_s]i_s = v_s \tag{5.18a}$$

$$-\frac{d\psi_r}{dt} - [R_r]i_r = v_r \tag{5.18b}$$

where

$v_s^t = \begin{bmatrix} v_a & v_b & v_c \end{bmatrix}$ is the vector of instantaneous stator phase voltages

$v_r^t = \begin{bmatrix} -v_F & 0 & 0 & 0 \end{bmatrix}$ is the vector of field and damper winding voltages

$[R_s] = \begin{bmatrix} R_a & 0 & 0 \\ 0 & R_a & 0 \\ 0 & 0 & R_a \end{bmatrix}$ is the matrix of stator resistances

$[R_r] = \begin{bmatrix} R_F & 0 & 0 & 0 \\ 0 & R_D & 0 & 0 \\ 0 & 0 & R_{Q_1} & 0 \\ 0 & 0 & 0 & R_{Q_2} \end{bmatrix}$ is the matrix of rotor resistances

It should be noted that generator convention is used while expressing the voltage equations; that is, the currents are assumed to be leaving the winding at the terminals and the terminal voltages are assumed to be the voltage drops in the direction of currents as illustrated in Figure 5.4.

The state space form of the voltage equations can be written based on either the flux linkages or the currents as state variables:

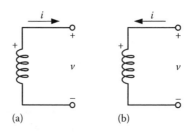

(a)   (b)

FIGURE 5.4
Illustration of generator and motor convention. (a) Generator, (b) motor.

$$\begin{bmatrix} \dfrac{d\psi_s}{dt} \\[2mm] \dfrac{d\psi_r}{dt} \end{bmatrix} = - \begin{bmatrix} R_s & 0 \\ 0 & R_r \end{bmatrix} \begin{bmatrix} L_{ss} & L_{sr} \\ L_{rs} & L_{rr} \end{bmatrix}^{-1} \begin{bmatrix} \psi_s \\ \psi_r \end{bmatrix} - \begin{bmatrix} v_s \\ v_r \end{bmatrix} \tag{5.19a}$$

$$\begin{bmatrix} \dfrac{di_s}{dt} \\[2mm] \dfrac{di_r}{dt} \end{bmatrix} = - \begin{bmatrix} L_{ss} & L_{sr} \\ L_{rs} & L_{rr} \end{bmatrix}^{-1} \left(\begin{bmatrix} R_s & 0 \\ 0 & R_r \end{bmatrix} + \frac{d\theta}{dt} \begin{bmatrix} \dfrac{\partial L_{ss}}{\partial \theta} & \dfrac{\partial L_{sr}}{\partial \theta} \\[2mm] \dfrac{\partial L_{rs}}{\partial \theta} & \dfrac{\partial L_{rr}}{\partial \theta} \end{bmatrix} \right) \begin{bmatrix} i_s \\ i_r \end{bmatrix} - \begin{bmatrix} v_s \\ v_r \end{bmatrix} \tag{5.19b}$$

5.2.4 Electrical Torque Equation

The electromagnetic torque expression can be found from the coenergy function, W', of the magnetic field in the air-gap

$$T_e = -\frac{\partial W'}{\partial \theta_m} \tag{5.20}$$

where the coenergy function is expressed as

$$W' = \frac{1}{2} \begin{bmatrix} i_s^t & i_r^t \end{bmatrix} \begin{bmatrix} L_{ss} & L_{sr} \\ L_{rs} & L_{rr} \end{bmatrix} \begin{bmatrix} i_s \\ i_r \end{bmatrix} \tag{5.21}$$

For a two-pole machine, $\theta_m = \theta$. However, for a p-pole machine, Equation 5.20 becomes

$$T_e = -\frac{p}{2} \frac{\partial W'}{\partial \theta} \tag{5.22}$$

By substituting Equation 5.21 into Equation 5.22, and considering the fact that $[L_{rr}]$ is constant, the electromagnetic torque expression can be written as follows:

$$T_e = -\frac{p}{2} \left(i_s^t \left[\frac{\partial L_{ss}}{\partial \theta} \right] i_s + 2 i_s^t \left[\frac{\partial L_{sr}}{\partial \theta} \right] i_r \right) \tag{5.23}$$

5.3 Park Transformation

5.3.1 Introduction

The electrical performance of the synchronous machine can be completely described by using either Equation 5.19a or Equation 5.19b with flux linkage equations given in Equation 5.1. Although these equations can be solved numerically, it is almost impossible to obtain an analytical solution due to the time-varying inductances.

The time-varying machine equations can be transformed to a time invariant set with a proper transformation of variables. This transformation provides a set of fictitious currents, voltages, and flux linkages, as well as new equations which will be solved for the new variables as a function of time. From the results based on new variables, the actual electrical quantities can be found as a function of time. It should be noted that the new variables may be purely mathematical and may not give any physical interpretation.

The time invariant set of machine equations can be obtained by utilizing Park transformation [9,10]. The new quantities are obtained from the projection of the actual stator variables along three axes, which are the direct axis of the rotor winding (*d*-axis), the neutral axis of field winding or quadrature axis (*q*-axis), and the stationary axis. In other words, all the stator quantities are transformed into new variables in which the reference frame rotates with the rotor. Thus, by definition

$$f_s = [P] f_{dq0} \tag{5.24}$$

where

f_s are the stator phase quantities that can be either voltages, currents, or flux linkages of the stator windings

f_{dq0} are the new fictitious quantities

$$[P] = \begin{bmatrix} k_d \cos\theta & k_q \sin\theta & k_0 \\ k_d \cos\left(\theta - \dfrac{2\pi}{3}\right) & k_q \sin\left(\theta - \dfrac{2\pi}{3}\right) & k_0 \\ k_d \cos\left(\theta + \dfrac{2\pi}{3}\right) & k_q \sin\left(\theta + \dfrac{2\pi}{3}\right) & k_0 \end{bmatrix} \tag{5.25}$$

The constants k_d, k_q, and k_0 are arbitrary and their values may be chosen to simplify the numerical coefficients in performance equations.

The inverse transformation is

$$f_{dq0} = [P]^{-1} f_s \tag{5.26}$$

where

$$[P]^{-1} = \begin{bmatrix} k_1 \cos\theta & k_1 \cos\left(\theta - \dfrac{2\pi}{3}\right) & k_1 \cos\left(\theta + \dfrac{2\pi}{3}\right) \\ k_2 \sin\theta & k_2 \sin\left(\theta - \dfrac{2\pi}{3}\right) & k_2 \sin\left(\theta + \dfrac{2\pi}{3}\right) \\ k_3 & k_3 & k_3 \end{bmatrix} \tag{5.27a}$$

and

$$k_1 = \frac{2}{3k_d}, \quad k_2 = \frac{2}{3k_q}, \quad k_3 = \frac{1}{3k_0} \tag{5.27b}$$

5.3.2 Flux Linkage Equations in *dq*0 Components

The flux linkage in *dq*0 components can be found by applying Park transformation to the basic equations for the phase flux linkage equations given in Equation 5.1.

$$
\begin{bmatrix} \Psi_{dq0} \\ \Psi_r \end{bmatrix} = \begin{bmatrix} P^{-1} & 0 \\ 0 & I_4 \end{bmatrix} \begin{bmatrix} L_{ss} & L_{sr} \\ L_{rs} & L_{rr} \end{bmatrix} \begin{bmatrix} P & 0 \\ 0 & I_4 \end{bmatrix} \begin{bmatrix} i_{dq0} \\ i_r \end{bmatrix}
$$

$$
= \begin{bmatrix} P^{-1}L_{ss}P & P^{-1}L_{sr} \\ L_{rs}P & L_{rr} \end{bmatrix} \begin{bmatrix} i_{dq0} \\ i_r \end{bmatrix}
$$

$$
= \begin{bmatrix} L'_{ss} & L'_{sr} \\ L'_{rs} & L_{rr} \end{bmatrix} \begin{bmatrix} i_{dq0} \\ i_r \end{bmatrix} \tag{5.28}
$$

where

$$
[L'_{ss}] = \begin{bmatrix} L_d & 0 & 0 \\ 0 & L_q & 0 \\ 0 & 0 & L_0 \end{bmatrix} \tag{5.29a}
$$

$$
L_d = L_{aa0} - L_{ab0} + \frac{3}{2}L_{aa2}
$$

$$
L_q = L_{aa0} - L_{ab0} - \frac{3}{2}L_{aa2}
$$

$$
L_0 = L_{aa0} + 2L_{ab0} \tag{5.29b}
$$

$$
[L'_{sr}] = \begin{bmatrix} \dfrac{L_{aF}}{k_d} & \dfrac{L_{aD}}{k_d} & 0 & 0 \\ 0 & 0 & \dfrac{L_{aQ_1}}{k_q} & \dfrac{L_{aQ_2}}{k_q} \\ 0 & 0 & 0 & 0 \end{bmatrix} \tag{5.29c}
$$

$$
[L'_{rs}] = \begin{bmatrix} \dfrac{3}{2}L_{aF}k_d & 0 & 0 \\ \dfrac{3}{2}L_{aD}k_d & 0 & 0 \\ 0 & \dfrac{3}{2}L_{aQ_1}k_q & 0 \\ 0 & \dfrac{3}{2}L_{aQ_2}k_q & 0 \end{bmatrix} \tag{5.29d}
$$

Equation 5.28 shows the transformation of the stator phase windings to fictitious *dq*0 windings. It can be seen from the above equations that the fictitious 0 winding, in which the

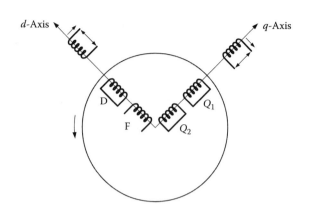

FIGURE 5.5
Synchronous machine with rotating armature windings.

zero-sequence currents flow, has no coupling with the rotor windings and can be neglected in balanced conditions. Moreover, the fictitious d winding is aligned with the d-axis and the fictitious q winding is aligned with the q-axis, as illustrated in Figure 5.5. Therefore, there is no coupling between the fictitious $d(q)$ winding and the rotor windings on $q(d)$-axis.

It can be seen from Equations 5.29 that the stator-to-rotor and rotor-to-stator mutual inductances will not be equal unless the following condition satisfies

$$k_d^2 = \frac{2}{3}, \quad k_q^2 = \frac{2}{3} \tag{5.30}$$

It should be also noted that, the rotor self-inductances remain unchanged, since there is no transformation of rotor quantities.

5.3.3 Voltage Equations in *dq*0 Components

Applying Park transformation, Equation 5.18 can be rewritten as follows:

$$-\frac{d}{dt}\left([P]\psi_{dq0}\right)-[R_s][P]i_{dq0} = [P]v_{dq0} \tag{5.31}$$

The first term on the left-hand side in Equation 5.31 can be expressed as

$$-\frac{d}{dt}\left([P]\psi_{dq0}\right)=-\frac{\partial[P]}{\partial\theta}\frac{d\theta}{dt}\psi_{dq0}-[P]\frac{d\psi_{dq0}}{dt} \tag{5.32}$$

where

$$\frac{\partial[P]}{\partial\theta}=\begin{bmatrix} -k_d\sin\theta & k_q\cos\theta & 0 \\ -k_d\sin\left(\theta-\frac{2\pi}{3}\right) & k_q\cos\left(\theta-\frac{2\pi}{3}\right) & 0 \\ -k_d\sin\left(\theta+\frac{2\pi}{3}\right) & k_q\cos\left(\theta+\frac{2\pi}{3}\right) & 0 \end{bmatrix}=[P][X] \tag{5.33}$$

and

$$[X] = \begin{bmatrix} 0 & \dfrac{k_q}{k_d} & 0 \\[2ex] -\dfrac{k_q}{k_d} & 0 & 0 \\[2ex] 0 & 0 & 0 \end{bmatrix} \tag{5.34}$$

Upon substitution of Equation 5.32 into Equation 5.31, and after some manipulations, the stator voltage equations in $dq0$ components become

$$-\frac{d\psi_{dq0}}{dt} - \omega[X]\psi_{dq0} - R_a i_{dq0} = v_{dq0} \tag{5.35}$$

where $\omega = \dfrac{d\theta}{dt}$.

Equation 5.35 can be expanded as follows:

$$-\frac{d\psi_d}{dt} - \omega\frac{k_q}{k_d}\psi_q - R_a i_d = v_d$$

$$-\frac{d\psi_q}{dt} + \omega\frac{k_q}{k_d}\psi_d - R_a i_q = v_q$$

$$-\frac{d\psi_0}{dt} - R_a i_0 = v_0 \tag{5.36}$$

The above equations have a form similar to those of a static coil, except for the $\omega\dfrac{k_q}{k_d}\psi_q$ and $\omega\dfrac{k_q}{k_d}\psi_d$ terms, which are called speed voltages. These terms result from the transformation of the reference frame from stationary to rotating, and represent the fact that the flux wave rotating in synchronism with the rotor creates voltages in the stationary armature coil.

The rotor voltage equations remain unchanged and can be written in the expanded form as

$$\frac{d\psi_F}{dt} + R_F i_F = v_F$$

$$\frac{d\psi_D}{dt} + R_D i_D = 0$$

$$\frac{d\psi_{Q_1}}{dt} + R_{Q_1} i_{Q_1} = 0$$

$$\frac{d\psi_{Q_2}}{dt} + R_{Q_2} i_{Q_2} = 0 \tag{5.37}$$

5.3.4 Electrical Torque Equation in *dq*0 Components

Applying Park transformation, Equation 5.23 can be rewritten as follows:

$$T_e = -\frac{p}{2}\left(i_{dq0}^t \left[P\right]^t \left[\frac{\partial L_{ss}}{\partial \theta}\right]\left[P\right]i_{dq0} + 2i_{dq0}^t \left[P\right]^t \left[\frac{\partial L_{sr}}{\partial \theta}\right]i_r \right) \tag{5.38}$$

where

$$\left[\frac{\partial L_{ss}}{\partial \theta}\right] = -2L_{aa2}\begin{bmatrix} \sin 2\theta & \sin\left(2\theta - \frac{2\pi}{3}\right) & \sin\left(2\theta + \frac{2\pi}{3}\right) \\ \sin\left(2\theta - \frac{2\pi}{3}\right) & \sin\left(2\theta + \frac{2\pi}{3}\right) & \sin 2\theta \\ \sin\left(2\theta + \frac{2\pi}{3}\right) & \sin 2\theta & \sin\left(2\theta - \frac{2\pi}{3}\right) \end{bmatrix} \tag{5.39a}$$

$$\left[\frac{\partial L_{ss}}{\partial \theta}\right]\left[P\right] = -3L_{aa2}\left[P\right]\begin{bmatrix} 0 & \frac{k_q}{k_d} & 0 \\ \frac{k_q}{k_d} & 0 & 0 \\ 0 & 0 & 0 \end{bmatrix} \tag{5.39b}$$

$$\left[\frac{\partial L_{sr}}{\partial \theta}\right] = \begin{bmatrix} -L_{aF}\sin\theta & -L_{aD}\sin\theta & L_{aQ_1}\cos\theta & L_{aQ_2}\cos\theta \\ -L_{aF}\sin\left(\theta - \frac{2\pi}{3}\right) & -L_{aD}\sin\left(\theta - \frac{2\pi}{3}\right) & L_{aQ_1}\cos\left(\theta - \frac{2\pi}{3}\right) & L_{aQ_2}\cos\left(\theta - \frac{2\pi}{3}\right) \\ -L_{aF}\sin\left(\theta + \frac{2\pi}{3}\right) & -L_{aD}\sin\left(\theta + \frac{2\pi}{3}\right) & L_{aQ_1}\cos\left(\theta + \frac{2\pi}{3}\right) & L_{aQ_2}\cos\left(\theta + \frac{2\pi}{3}\right) \end{bmatrix} \tag{5.39c}$$

$$\left[P\right]^t\left[\frac{\partial L_{sr}}{\partial \theta}\right] = \begin{bmatrix} 0 & 0 & \frac{2}{3}k_d L_{aQ_1} & \frac{2}{3}k_d L_{aQ_2} \\ -\frac{2}{3}k_q L_{aF} & -\frac{2}{3}k_q L_{aD} & 0 & 0 \\ 0 & 0 & 0 & 0 \end{bmatrix} \tag{5.39d}$$

After some manipulations, the electrical torque expression reduces to the following one:

$$T_e = \frac{p}{2}\frac{3}{2}k_d k_q\left(i_q\left(\frac{L_{aF}}{k_d}i_F + \frac{L_{aD}}{k_d}i_D + \frac{3}{2}L_{aa2}i_d\right) - i_d\left(\frac{L_{aQ_1}}{k_q}i_{Q_1} + \frac{L_{aQ_2}}{k_q}i_{Q_2} - \frac{3}{2}L_{aa2}i_q\right) \right) \tag{5.40}$$

Since

$$\psi_d = L_d i_d + \frac{L_{aF}}{k_d} i_F + \frac{L_{aD}}{k_d} i_D \qquad (5.41a)$$

$$\psi_q = L_q i_q + \frac{L_{aQ_1}}{k_q} i_{Q_1} + \frac{L_{aQ_2}}{k_q} i_{Q_2} \qquad (5.41b)$$

and

$$L_d - \frac{3}{2} L_{aa2} = L_q + \frac{3}{2} L_{aa2} = L_{aa0} - L_{ab0} \qquad (5.42)$$

the electrical torque expression in $dq0$ components becomes

$$T_e = \frac{p}{2}\frac{3}{2} k_d k_q \left(i_q \left(\psi_d - \left(L_d - \frac{3}{2} L_{aa2} \right) i_d \right) - i_d \left(\psi_q - \left(L_q + \frac{3}{2} L_{aa2} \right) i_q \right) \right)$$

$$= \frac{p}{2}\frac{3}{2} k_d k_q \left(i_q \psi_d - i_d \psi_q \right) \qquad (5.43)$$

5.3.5 Choice of k_d, k_q, and k_0 Constants in Park Transformation

The constants k_d, k_q, and k_0, defined by Equation 5.25 in the Park transformation, are arbitrary, and their values may be chosen to simplify the numerical coefficients in performance equations. In the original Park transformation, $k_d = 1$, $k_q = -1$, and $k_0 = 1$, where the q-axis was assumed to lead d-axis. When the original Park transformation is used for balanced conditions, the peak values of i_d and i_q are equal to the peak value of the stator currents. However, it should be noted that the original Park transformation is not orthogonal, and hence it is not power invariant as illustrated below. Using the relationship $P_e = v_s^t i_s$ and taking into account the transformation defined in Equation 5.25

$$v_s^t i_s = v_{dq0}^t \left[P \right]^t \left[P \right] i_{dq0}$$

$$= v_{dq0}^t \begin{bmatrix} \frac{3}{2} k_d^2 & 0 & 0 \\ 0 & \frac{3}{2} k_q^2 & 0 \\ 0 & 0 & 3k_0^2 \end{bmatrix} i_{dq0} \qquad (5.44)$$

With $k_d = 1$, $k_q = -1$, and $k_0 = 1$, the electrical power expression in $dq0$ components becomes

$$P_e = \frac{3}{2} \left(v_d i_d + v_q i_q + 2 v_0 i_0 \right) \qquad (5.45)$$

In order to obtain a power invariant transformation (i.e., $[P]^t [P] = I_3$), the choice of k_d, k_q, and k_0 should be as follows:

$$k_d = \pm\sqrt{\frac{2}{3}}, \quad k_q = \pm\sqrt{\frac{2}{3}}, \quad k_0 = \pm\sqrt{\frac{1}{3}} \tag{5.46}$$

By taking the positive values of constants, the power invariant Park transformation can be written as

$$[P] = \frac{1}{\sqrt{3}} \begin{bmatrix} \sqrt{2}\cos\theta & \sqrt{2}\sin\theta & 1 \\ \sqrt{2}\cos\left(\theta - \frac{2\pi}{3}\right) & \sqrt{2}\sin\left(\theta - \frac{2\pi}{3}\right) & 1 \\ \sqrt{2}\cos\left(\theta + \frac{2\pi}{3}\right) & \sqrt{2}\sin\left(\theta + \frac{2\pi}{3}\right) & 1 \end{bmatrix} \tag{5.47}$$

The positive value of k_q indicates that the q-axis is lagging d-axis (see Figure 5.1). In the original Park transformation, q-axis was assumed to be leading d-axis, and k_q had a negative value.

The major advantage of this transformation is that, all the transformed mutual inductances are reciprocal; i.e., $[L'_{sr}] = [L'_{rs}]$.

5.4 Per Unit Representation

In power system analysis, it is common to express the system variables in per unit by choosing appropriate base quantities. A properly chosen per unit system can increase computational efficiency, simplify evaluation, and facilitate understanding of system characteristics. In addition to this, machine impedances lie in a reasonably narrow numerical range when expressed in a per unit system related to their ratings. And this enables rapid check of their correctness.

Per Unit System for the Stator Quantities: The common practice is to use the machine ratings as base quantities for stator quantities. The base quantities for the stator windings can be chosen as follows:

Base power, S_B = Three-phase rated power, VA
Base voltage, V_B = Rated line-to-line RMS voltage, V (same for phase voltages)
Base current, $I_B = \dfrac{S_B}{V_B}$, A (for phase currents), $I_{abcB} = \dfrac{S_B}{\sqrt{3}V_B}$, A
Base impedance, $Z_B = \dfrac{V_B^2}{S_B}$, Ω
Base frequency, f_B = Rated-frequency, Hz
Base electrical angular frequency, $\omega_B = 2\pi f_B$, rad/s
Base mechanical angular frequency, $\omega_{mB} = \omega_B \dfrac{2}{p}$, rad/s

Base inductance, $L_B = \dfrac{Z_B}{\omega_B}$, H

Base flux linkage, $\Psi_B = \dfrac{V_B}{\omega_B}$, Wb·turns

Note that different base current values are used for phase and for d-q axis currents.

Per Unit System for the Rotor Quantities: The choice of base current is based on equal mutual flux linkages in which the base field current or the d-axis damper current will produce the same space fundamental air-gap flux as produced by the base stator current acting in the fictitious d-axis winding. The mutual flux linkages in the d-axis are as follows:

$$\psi_{ad} = \left(L_d - L_{al}\right)I_B = L_{ad}I_B = L_{dF}I_{FB} = L_{dD}I_{DB} \tag{5.48}$$

where

$$L_{dF} = \frac{L_{aF}}{k_d} = \sqrt{\frac{3}{2}}L_{aF}, \quad L_{dD} = \frac{L_{aD}}{k_d} = \sqrt{\frac{3}{2}}L_{aD} \tag{5.49}$$

and L_{al} is the stator leakage inductance.

I_{FB} and I_{DB} are the base currents of the field and the d-axis damper windings, and can be found from Equation 5.48 as follows:

$$I_{FB} = \frac{L_{ad}}{L_{dF}}I_B, \quad I_{DB} = \frac{L_{ad}}{L_{dD}}I_B \tag{5.50}$$

The base flux linkages for the rotor circuits are chosen as follows:

$$\psi_{FB}I_{FB} = \psi_{DB}I_{DB} = \psi_B I_B \tag{5.51}$$

The per unit quantities are chosen in such a way that the per unit mutual inductances are equal; i.e., $\bar{L}_{dF} = \bar{L}_{dD} = \bar{L}_d - \bar{L}_{al}$, in which the bar over the parameters indicates that they are in per unit quantities.

Similarly, the base quantities for the currents and the flux linkages of the q-axis damper windings can be chosen as

$$I_{Q_1B} = \frac{L_{aq}}{L_{qQ_1}}I_B, \quad I_{Q_2B} = \frac{L_{aq}}{L_{qQ_2}}I_B \tag{5.52a}$$

$$\psi_{Q_1B} = \frac{\psi_B I_B}{I_{Q_1B}}, \quad \psi_{Q_2B} = \frac{\psi_B I_B}{I_{Q_2B}} \tag{5.52b}$$

where

$$L_{aq} = L_q - L_{al}, \quad L_{qQ_1} = \frac{L_{aQ_1}}{k_q} = \sqrt{\frac{3}{2}}L_{aQ_1}, \quad L_{qQ_2} = \frac{L_{aQ_2}}{k_q} = \sqrt{\frac{3}{2}}L_{aQ_2} \tag{5.53}$$

Let

$$N_F = \frac{L_{dF}}{L_{ad}}, \quad N_D = \frac{L_{dD}}{L_{ad}}, \quad N_{Q_1} = \frac{L_{qQ_1}}{L_{aq}}, \quad N_{Q_2} = \frac{L_{qQ_2}}{L_{aq}} \tag{5.54}$$

The rotor base quantities will be as follows:

Field winding base current, $I_{FB} = \dfrac{I_B}{N_F}$, A

d-Axis damper winding base current, $I_{DB} = \dfrac{I_B}{N_D}$, A

First q-axis damper winding base current, $I_{Q_1B} = \dfrac{I_B}{N_{Q_1}}$, A

Second q-axis damper winding base current, $I_{Q_2B} = \dfrac{I_B}{N_{Q_2}}$, A

Field winding base flux linkage, $\psi_{FB} = N_F \psi_B$, Wb · turns

d-Axis damper winding base flux linkage, $\psi_{DB} = N_D \psi_B$, Wb · turns

First q-axis damper winding base flux linkage, $\psi_{Q1B} = N_{Q1} \psi_B$, Wb · turns

Second q-axis damper winding base flux linkage, $\psi_{Q2B} = N_{Q2} \psi_B$, Wb · turns

Field winding base voltage, $V_{FB} = N_F V_B$, V

d-Axis damper winding base voltage, $V_{DB} = N_D V_B$, V

First q-axis damper winding base voltage, $V_{Q1B} = N_{Q1} V_B$, V

Second q-axis damper winding base voltage, $V_{Q2B} = N_{Q2} V_B$, V

Field winding base impedance, $Z_{FB} = N_F^2 Z_B$, Ω

d-Axis damper winding base impedance, $Z_{DB} = N_D^2 Z_B$, Ω

First q-axis damper winding base impedance, $Z_{Q1B} = N_{Q1}^2 Z_B$, Ω

Second q-axis damper winding base impedance, $Z_{Q2B} = N_{Q2}^2 Z_B$, Ω

Field winding base inductance, $L_{FB} = N_F^2 L_B$, H

d-Axis damper winding base inductance, $L_{DB} = N_D^2 L_B$, H

First q-axis damper winding base inductance, $L_{Q1B} = N_{Q1}^2 L_B$, H

Second q-axis damper winding base inductance, $L_{Q2B} = N_{Q2}^2 L_B$, H

It should be also noted that the selection of base quantities for rotor circuits can be eliminated by referring the rotor quantities to stator using N_F, N_D, N_{Q1}, and N_{Q2} as the related turn ratios.

The complete set of electrical equations can be written as follows:

- *Per unit stator voltage equations*

$$-\frac{1}{\omega_B} \frac{d\bar{\psi}_d}{dt} - \bar{\omega}\bar{\psi}_q - \bar{R}_a \bar{i}_d = \bar{v}_d$$

$$-\frac{1}{\omega_B} \frac{d\bar{\psi}_q}{dt} + \bar{\omega}\bar{\psi}_d - \bar{R}_a \bar{i}_q = \bar{v}_q$$

$$-\frac{1}{\omega_B} \frac{d\bar{\psi}_0}{dt} - \bar{R}_a \bar{i}_0 = \bar{v}_0 \tag{5.55}$$

- *Per unit stator flux linkage equations*

$$\overline{\psi}_d = \overline{L}_d \overline{i}_d + \overline{L}_{ad} \overline{i}_F + \overline{L}_{ad} \overline{i}_D$$

$$\overline{\psi}_q = \overline{L}_q \overline{i}_q + \overline{L}_{aq} \overline{i}_{Q_1} + \overline{L}_{aq} \overline{i}_{Q_2}$$

$$\overline{\psi}_0 = \overline{L}_0 \overline{i}_0 \qquad (5.56)$$

- *Per unit rotor voltage equations*

$$\frac{1}{\omega_B} \frac{d\overline{\psi}_F}{dt} + \overline{R}_F \overline{i}_F = \overline{v}_F$$

$$\frac{1}{\omega_B} \frac{d\overline{\psi}_D}{dt} + \overline{R}_D \overline{i}_D = 0$$

$$\frac{1}{\omega_B} \frac{d\overline{\psi}_{Q_1}}{dt} + \overline{R}_{Q_1} \overline{i}_{Q_1} = 0$$

$$\frac{d\overline{\psi}_{Q_2}}{dt} + \overline{R}_{Q_2} \overline{i}_{Q_2} = 0 \qquad (5.57)$$

- *Per unit rotor flux linkage equations*

$$\overline{\psi}_F = \overline{L}_{ad} \overline{i}_d + \overline{L}_{FF} \overline{i}_F + \overline{L}_{FD} \overline{i}_D$$

$$\overline{\psi}_D = \overline{L}_{ad} \overline{i}_d + \overline{L}_{FD} \overline{i}_F + \overline{L}_{DD} \overline{i}_D$$

$$\overline{\psi}_{Q_1} = \overline{L}_{aq} \overline{i}_q + \overline{L}_{Q_1Q_1} \overline{i}_{Q_1} + \overline{L}_{Q_1Q_2} \overline{i}_{Q_2}$$

$$\overline{\psi}_{Q_2} = \overline{L}_{aq} \overline{i}_q + \overline{L}_{Q_1Q_2} \overline{i}_{Q_1} + \overline{L}_{Q_2Q_2} \overline{i}_{Q_2} \qquad (5.58)$$

- *Per unit air-gap torque*

$$T_e = \overline{\psi}_d \overline{i}_q - \overline{\psi}_q \overline{i}_d \qquad (5.59)$$

The above equations are in a suitable form for time-domain solution, where time is expressed in seconds. The $\dfrac{1}{\omega_B}$ that accompanies the time-derivative terms can be eliminated by expressing the time in per unit by choosing $t_B = \dfrac{1}{\omega_B}$.

5.5 Equivalent Circuits

The equivalent circuits of the synchronous machine represent its complete electrical characteristics, and are developed from Equations 5.55 to 5.58. Since all quantities will be in per unit, including time, the superbar notation is dropped in this part.

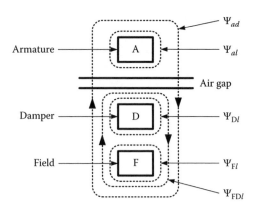

FIGURE 5.6
d-Axis flux paths.

Let L_{Fl}, L_{Dl} be the leakage inductances of the F and D windings, and L_{Fdl} be the leakage inductance that represents the flux linking both the field and the damper but not the *d*-axis winding as illustrated in Figure 5.6. The *d*-axis flux linkages can be rewritten as

$$\psi_d = \left(L_{ad} + L_{al}\right)i_d + L_{ad}i_F + L_{ad}i_D$$

$$\psi_F = L_{ad}i_d + \left(L_{ad} + L_{Fdl} + L_{Fl}\right)i_F + \left(L_{ad} + L_{Fdl}\right)i_D$$

$$\psi_D = L_{ad}i_d + \left(L_{ad} + L_{Fdl}\right)i_F + \left(L_{ad} + L_{Fdl} + L_{Dl}\right)i_D \tag{5.60}$$

where

$$L_d = \left(L_{ad} + L_{al}\right)$$

$$L_{FF} = \left(L_{FD} + L_{Fl}\right)$$

$$L_{DD} = \left(L_{FD} + L_{Dl}\right)$$

$$L_{FD} = \left(L_{ad} + L_{Fdl}\right) \tag{5.61}$$

Following a procedure similar to that used to develop the equivalent circuit of a transformer, the equivalent circuit that represents the Equation 5.60 can be as depicted in Figure 5.7.

The inductance L_{Fdl} is usually omitted. This may be reasonable due to the fact that the damper windings are near the air-gap and the flux linking damper circuit is nearly equal to that linking the armature. However, this approach is not valid especially for the short-pitched damper circuits and solid rotor iron paths [11].

In case of *q*-axis, there is no field winding and the solid rotor offers paths for eddy currents having equivalent effect of damper winding currents. Therefore, it is reasonable to assume that the armature and damper circuits all link a single ideal mutual flux represented by L_{aq} [7]. Similar to *d*-axis, *q*-axis flux linkages can be rewritten as follows:

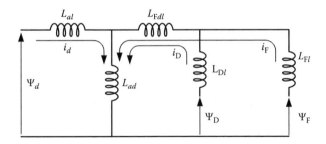

FIGURE 5.7
d-Axis equivalent circuit illustrating ψ–*i* relationship.

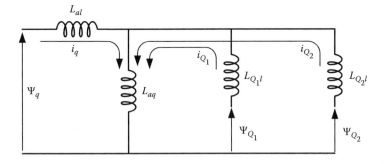

FIGURE 5.8
q-Axis equivalent circuit illustrating ψ–*i* relationship.

$$\psi_q = \left(L_{aq} + L_{al}\right)i_d + L_{aq}i_{Q_1} + L_{aq}i_{Q_2}$$

$$\psi_{Q_1} = L_{aq}i_q + \left(L_{aq} + L_{Q_1l}\right)i_{Q_1} + L_{aq}i_{Q_2}$$

$$\psi_{Q_2} = L_{aq}i_q + L_{aq}i_{Q_1} + \left(L_{aq} + L_{Q_2l}\right)i_{Q_2} \tag{5.62}$$

where $L_{Q_1}l$, $L_{Q_2}l$ are the leakage inductances of the first and second damper windings.

Following the procedure of developing equivalent circuit of transformer, the equivalent circuit that represents the Equation 5.62 can be obtained as illustrated in Figure 5.8.

The *d*- and *q*-axes equivalent circuits that include the voltage equations are illustrated in Figures 5.9 and 5.10, respectively.

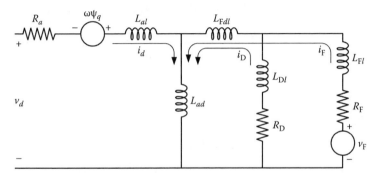

FIGURE 5.9
d-Axis equivalent circuit representing the complete characteristic.

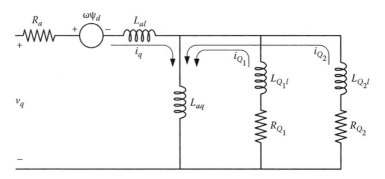

FIGURE 5.10
q-Axis equivalent circuit representing the complete characteristic.

TABLE 5.2

Synchronous Machine Models for Transient Studies

	q-Axis			
d-Axis	No Damper Circuit	One Damper Circuit	Two Damper Circuits	Three Damper Circuits
Field circuit only	MODEL 1.0	MODEL 1.1	Not used	Not used
Field circuit + one damper circuit	Not used	MODEL 2.1	MODEL 2.2	MODEL 2.3
Field circuit + two damper circuit	Not used	Not used	Not used	MODEL 3.3

Source: ANSI/IEEE Std 1110-1991, IEEE Guide for Synchronous Generator Modeling Practices in Stability Analysis. With permission.

These equivalent circuits have been widely used in many transient studies. However, equivalent circuits with different degrees of complexity have been applied. Table 5.2 shows the matrix of equivalent circuits with those model structures proposed in IEEE Std. 1110 [12]. Note that up to 12 combinations are possible, but only 7 are considered. The selection of a model is usually based on the type of machine, the study to be performed, the user's experience, and the available information. Depending on the characteristic parameter source, the most complex models very often cannot be used due to lack of data. As mentioned in the introduction, these representations are suitable for simulation of low-frequency transients, such as transient stability studies, subsynchronous resonance, load rejection, generator tripping, generator synchronization, and inadverted energization.

The conversion procedures that have been proposed for determination of the parameters for these models are discussed in the subsequent sections.

5.6 Synchronous Machine Parameters

Equations 5.55 through 5.58, which represent the complete electrical characteristics of the machine, require the inductances and resistances of the stator and rotor circuits as parameters. These parameters are called fundamental or basic parameters, and are identified as the elements of the equivalent circuits shown in Figures 5.9 and 5.10. They cannot be directly determined from test/field measurements. The parameters usually provided by the manufacturer are obtained by means of standardized procedures and they need to be converted to fundamental parameters. Before introducing the machine data expressed in terms of test measurements, it may be useful to investigate the synchronous machine behavior to a sudden three-phase short circuit at the armature terminals.

5.6.1 Three-Phase Short Circuit at the Terminals of a Synchronous Machine

Let the machine armature be open-circuited and the excitation system have no voltage regulator. When a sudden three-phase fault is suddenly applied to the terminals of the machine, the three-phase armature currents will be as illustrated in Figure 5.11.

Before the short circuit, the field current is constant and the corresponding flux linkage of each armature winding varies sinusoidally in time having a 120° phase difference between adjacent windings. The short circuit generates currents in the armature in such a way that the armature winding flux linkages are maintained at the instant of the short circuit. In general the fault current in each phase has two distinct components: a fundamental frequency ac component corresponding to the armature current opposing to the time-varying flux produced by the rotor windings and a dc component corresponding to the initial flux linkage at the instant of the short circuit.

A similar situation occurs on the rotor windings. Following the short circuit, the rotor windings are rotating in a stationary trapped armature flux wave and respond with a fundamental frequency ac current component to oppose the tendency of this trapped armature flux to change the flux linkages of the rotor windings. In addition, a dc component is also induced on the rotor windings to oppose the synchronously rotating component of the flux generated by the ac components of the armature current which tends to demagnetize the flux generated by the rotor windings.

The dc component of each phase current is determined from the initial flux in the related winding at the short circuit instant. As the armature currents are balanced, similar to ac components, the sum of the dc components is also zero at any instant. The ac components of the rotor windings are due to the dc components of the armature currents, and they decay together with a rate determined by the armature resistance.

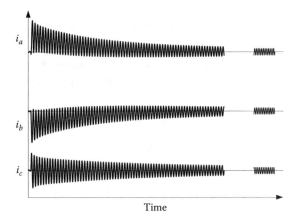

FIGURE 5.11
Three-phase armature currents following a stator short circuit.

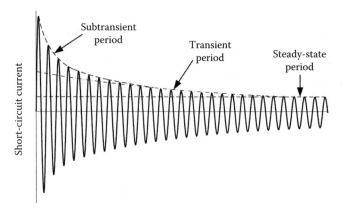

FIGURE 5.12
Fundamental frequency ac component of the armature current.

The waveform of the fundamental frequency ac component of the armature current following a three-phase short circuit illustrated in Figure 5.12 can be divided into three distinct time periods:

- The subtransient period, which includes the first few cycles and during which the current amplitude decays rapidly.
- The transient period, which spans a longer time and during which the current amplitude decays slowly.
- The steady-state period, during which the current amplitude remains constant.

The amplitude of the fundamental frequency ac components of the armature current is a function of rotor flux linkages and therefore it is not constant. The initial rapid decay is due to the rapid decay of the flux that links the subtransient circuit; i.e., the damper windings D and Q_2 in Figures 5.9 and 5.10. The slowly decaying part is due to the relatively slow decay of flux linking the transient circuits; i.e., the field and damper winding Q_1 in Figures 5.9 and 5.10.

5.6.2   Subtransient and Transient Inductances and Time Constants

According to superposition principle, application of a sudden three-phase short circuit to armature terminals produces the same armature currents produced by the application of sudden balanced three-phase voltages to the armature terminals with the field winding in short circuit as illustrated in Figure 5.13.

Let the voltages suddenly applied to the stator be given by

$$
v_s = \begin{bmatrix} v_a \\ v_b \\ v_c \end{bmatrix} = \begin{bmatrix} \cos\theta \\ \cos\left(\theta - \dfrac{3\pi}{2}\right) \\ \cos\left(\theta + \dfrac{3\pi}{2}\right) \end{bmatrix} \sqrt{2}Vu(t) \tag{5.63}
$$

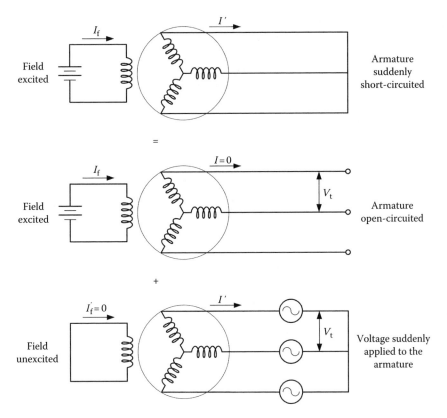

FIGURE 5.13
Application of the superposition principle.

where $u(t)$ is the unit step function. The transformed voltages are as follows:

$$v_{dq0} = \begin{bmatrix} v_d \\ v_q \\ v_0 \end{bmatrix} = \begin{bmatrix} \sqrt{3}Vu(t) \\ 0 \\ 0 \end{bmatrix} \tag{5.64}$$

Initially, the transient currents are strong enough to keep the flux linkage ψ_F and ψ_D constant at zero. Hence, at $t = 0^+$,

$$\psi_F = L_{ad}i_d + L_{FF}i_F + L_{FD}i_D = 0$$

$$\psi_D = L_{ad}i_d + L_{FD}i_F + L_{DD}i_D = 0 \tag{5.65}$$

From these results, the following field and d-axis damper currents are obtained

$$i_F = -\frac{L_{ad}L_{DD} - L_{ad}L_{FD}}{L_{FF}L_{DD} - L_{FD}^2} i_d, \quad i_D = -\frac{L_{ad}L_{FF} - L_{ad}L_{FD}}{L_{FF}L_{DD} - L_{FD}^2} i_d \tag{5.66}$$

The initial flux of the fictitious d-axis winding can be calculated in terms of the current through this winding. Upon substitution of Equation 5.66 into Equation 5.56 it becomes

$$\psi_d = \left(L_d - \frac{L_{ad}^2 L_{DD} + L_{ad}^2 L_{FF} - 2L_{ad}^2 L_{FD}}{L_{FF}L_{DD} - L_{FD}^2} \right) i_d \tag{5.67}$$

The subtransient inductance is defined as the initial stator flux linkage per unit of stator current with all rotor circuits short-circuited. Thus,

$$\psi_d = L_d'' i_d \tag{5.68}$$

where L'' is the d-axis subtransient inductance and it can be found from Equations 5.67 and 5.61 as follows:

$$L_d'' = L_{al} + \frac{L_{ad}\left(L_{Dl}L_{Fl} + L_{Dl}L_{Fdl} + L_{Fl}L_{Fdl} \right)}{L_{ad}L_{Fl} + L_{ad}L_{Dl} + L_{Dl}L_{Fl} + L_{Dl}L_{Fdl} + L_{Fl}L_{Fdl}} \tag{5.69}$$

When the balance voltages described by Equation 5.63 are suddenly applied to a machine having no damper winding, the same procedure gives

$$\psi_d = L_d' i_d \tag{5.70}$$

where L_d' is the d-axis transient reactance

$$L_d' = L_{al} + \frac{L_{ad}\left(L_{Fl} + L_{Fdl} \right)}{L_{ad} + L_{Fl} + L_{Fdl}} \tag{5.71}$$

The decrement of the damper winding current is very rapid when compared to the decrement of field winding current, but it becomes negligible after a few cycles, when the effective armature reactance has increased from the subtransient to the transient value.

The d-axis short circuit subtransient and transient time constants, T_d'' and T_d', determine the decrement of both the amplitude of the fundamental frequency ac components of the armature currents and the amplitude of the dc component of the rotor currents. Their values can be found as follows. From Equations 5.56 to 5.58, the d-axis rotor voltage equations can be written as

$$\left(L_{FF} - \frac{L_{ad}^2}{L_d} \right)\frac{di_F}{dt} + \left(L_{FD} - \frac{L_{ad}^2}{L_d} \right)\frac{di_D}{dt} + \frac{L_{ad}}{L_d}\frac{d\psi_d}{dt} + R_F i_F = 0$$

$$\left(L_{FD} - \frac{L_{ad}^2}{L_d} \right)\frac{di_F}{dt} + \left(L_{DD} - \frac{L_{ad}^2}{L_d} \right)\frac{di_D}{dt} + \frac{L_{ad}}{L_d}\frac{d\psi_d}{dt} + R_D i_D = 0 \tag{5.72}$$

Let

$$L_{sF} = \left(L_{FF} - \frac{L_{ad}^2}{L_d} \right), \quad L_{sD} = \left(L_{DD} - \frac{L_{ad}^2}{L_d} \right), \quad L_{MD} = \left(L_{FD} - \frac{L_{ad}^2}{L_d} \right) \tag{5.73}$$

From these results, and after some manipulations

$$
\begin{bmatrix} \dfrac{di_F}{dt} \\[2mm] \dfrac{di_D}{dt} \end{bmatrix} = \frac{1}{L_{sF}L_{sD} - L_{MD}^2} \begin{bmatrix} L_{sD} & -L_{MD} \\ -L_{MD} & L_{sF} \end{bmatrix} \left(\begin{bmatrix} -R_F & 0 \\ 0 & -R_D \end{bmatrix} \begin{bmatrix} i_F \\ i_D \end{bmatrix} + \begin{bmatrix} -\dfrac{L_{ad}}{L_d}\dfrac{d\psi_d}{dt} \\[2mm] -\dfrac{L_{ad}}{L_d}\dfrac{d\psi_d}{dt} \end{bmatrix} \right)
$$

$$
= \frac{1}{L_{sF}L_{sD} - L_{MD}^2} \begin{bmatrix} -L_{sD}R_F & L_{MD}R_D \\ L_{MD}R_F & -L_{sF}R_D \end{bmatrix} \begin{bmatrix} i_F \\ i_D \end{bmatrix} + \frac{1}{L_{sF}L_{sD} - L_{MD}^2} \begin{bmatrix} -\dfrac{(L_{sD} - L_{MD})L_{ad}}{L_d}\dfrac{d\psi_d}{dt} \\[2mm] -\dfrac{(L_{sF} - L_{MD})L_{ad}}{L_d}\dfrac{d\psi_d}{dt} \end{bmatrix}
$$

$$(5.74)$$

The time constants of transient d-axis rotor currents can be found by calculating the opposite of the two eigenvalues of the matrix that relates the currents to their derivative when the excitation voltage is zero. Thus, the time constants satisfy the following equation:

$$
\left(\frac{-R_F L_{sD}}{L_{sF}L_{sD} - L_{MD}^2} + \frac{1}{T} \right)\left(\frac{-R_D L_{sF}}{L_{sF}L_{sD} - L_{MD}^2} + \frac{1}{T} \right) - \frac{R_F R_D L_{MD}^2}{\left(L_{sF}L_{sD} - L_{MD}^2 \right)^2} = 0 \qquad (5.75)
$$

Hence the d-axis short circuit subtransient time constant, T_d'', and the d-axis short circuit transient time constant, T_d', satisfy the following equations:

$$
T_d'' + T_d' = \frac{L_{sF}}{R_F} + \frac{L_{sd}}{R_D}
$$

$$
= \frac{1}{R_F}\left(L_{FF} - \frac{L_{ad}^2}{L_d} \right) + \frac{1}{R_D}\left(L_{DD} - \frac{L_{ad}^2}{L_d} \right) \qquad (5.76a)
$$

$$
T_d'' T_d' = \frac{L_{sF}L_{sD} - L_{MD}^2}{R_F R_D}
$$

$$
= \frac{1}{R_F R_D}\left[\left(L_{FF} - \frac{L_{ad}^2}{L_d} \right)\left(L_{DD} - \frac{L_{ad}^2}{L_d} \right) - \left(L_{FD} - \frac{L_{ad}^2}{L_d} \right)^2 \right]
$$

$$
= \frac{1}{R_F R_D}\left(L_{FF}L_{DD} - L_{FD}^2 - \frac{L_{ad}^2\left(L_{Fl} + L_{Dl} \right)}{L_d} \right) \qquad (5.76b)
$$

Repeating the same procedure for q-axis, the q-axis subtransient and transient inductances (L_q'' and L_q') and the q-axis short circuit subtransient and transient time constants (T_q'' and T_q') become

$$
L_q'' = L_{al} + \frac{L_{aq}L_{Q_1l}L_{Q_2l}}{L_{aq}L_{Q_1l} + L_{aq}L_{Q_2l} + L_{Q_1l}L_{Q_2l}} \qquad (5.77a)
$$

$$L_q' = L_{al} + \frac{L_{aq}L_{Q_1l}}{L_{aq} + L_{Q_1l}} \tag{5.77b}$$

$$T_q'' + T_q' = \frac{1}{R_{Q_1}}\left(L_{Q_1Q_1} - \frac{L_{aq}^2}{L_q}\right) + \frac{1}{R_{Q_2}}\left(L_{Q_2Q_2} - \frac{L_{aq}^2}{L_q}\right) \tag{5.77c}$$

$$T_q''T_q' = \frac{1}{R_{Q_1}R_{Q_2}}\left(L_{Q_1Q_1}L_{Q_2Q_2} - L_{aq}^2 - \frac{L_{aq}^2\left(L_{Q_1l} + L_{Q_2l}\right)}{L_q}\right) \tag{5.77d}$$

where the voltages suddenly applied to the stator are

$$v_s = \begin{bmatrix} v_a \\ v_b \\ v_c \end{bmatrix} = \begin{bmatrix} \sin\theta \\ \sin\left(\theta - \dfrac{3\pi}{2}\right) \\ \sin\left(\theta + \dfrac{3\pi}{2}\right) \end{bmatrix} \sqrt{2}Vu(t) \tag{5.78}$$

It should be noted that this machine model has two rotor circuits in each axis and in general is not applicable for a laminated salient-pole machine having one damper winding on the q-axis, unless the second rotor circuit denoted by subscript Q_2 is ignored. Therefore, the first rotor circuit denoted by subscript Q_1 is considered to represent the rapidly decaying subtransient effects and no distinction is made between transient and steady-state conditions. As a result, the expressions for the q-axis inductances and short circuit time constants of the salient-pole machine become

$$L_q'' = L_{al} + \frac{L_{aq}L_{Q_1l}}{L_{aq} + L_{Q_1l}} \tag{5.79a}$$

$$T_q'' = \frac{1}{R_{Q1}}\left[L_{Q_1Q_1} - \frac{L_{aq}^2}{L_q}\right] \tag{5.79b}$$

If a step change is applied to the field voltage v_F, the amplitude of the dc component of the d-axis rotor currents decay with the time constants T_d'' and T_d' when the stator is short-circuited. T_{d0}'' and T_{d0}' are the d-axis open circuit subtransient and transient time constants which control the decrement of the d-axis rotor currents following a step change in the applied field voltage when the stator is open-circuited.

Let $v_F = u(t)$ be the voltage applied to the field winding when the stator is open-circuited, $u(t)$ being the unit step function. From Equations 5.57 and 5.58, the following equation is obtained

$$\frac{d}{dt}\begin{bmatrix} i_F \\ i_D \end{bmatrix} = \frac{1}{L_{FF}L_{DD} - L_{FD}^2}\begin{bmatrix} L_{DD} & -L_{FD} \\ -L_{FD} & L_{FF} \end{bmatrix}\left(\begin{bmatrix} -R_F & 0 \\ 0 & -R_D \end{bmatrix}\begin{bmatrix} i_F \\ i_D \end{bmatrix} + \begin{bmatrix} v_F \\ 0 \end{bmatrix}\right)$$

$$= \frac{1}{L_{FF}L_{DD} - L_{FD}^2}\begin{bmatrix} -L_{DD}R_F & L_{FD}R_D \\ L_{FD}R_F & -L_{FF}R_D \end{bmatrix}\begin{bmatrix} i_F \\ i_D \end{bmatrix} + \frac{1}{L_{FF}L_{DD} - L_{FD}^2}\begin{bmatrix} L_{DD}v_F \\ -L_{FD}v_F \end{bmatrix} \tag{5.80}$$

The time constants of transient d-axis rotor currents can be found by calculating the opposite of two eigenvalues of the matrix that relates the currents to their derivative when the excitation voltage is zero. Thus, the time constants satisfy the following equation:

$$\left(\frac{-R_F L_{DD}}{L_{FF} L_{DD} - L_{FD}^2} + \frac{1}{T} \right) \left(\frac{-R_D L_{FF}}{L_{FF} L_{DD} - L_{FD}^2} + \frac{1}{T} \right) - \frac{R_F R_D L_{FD}^2}{\left(L_{FF} L_{DD} - L_{FD}^2 \right)^2} = 0 \tag{5.81}$$

Hence the d-axis open circuit subtransient time constant, T_{d0}'' and the d-axis open circuit transient time constant, T_{d0}' satisfy the following equations:

$$T_{d0}'' + T_{d0}' = \frac{L_{FF}}{R_F} + \frac{L_{DD}}{R_D} \tag{5.82a}$$

$$T_{d0}'' T_{d0}' = \frac{L_{FF} L_{DD} - L_{FD}^2}{R_F R_D} \tag{5.82b}$$

After similar analysis for the q-axis, the equations for q-axis open circuit subtransient and transient time constants, T_{q0}'' and T_{q0}', are as follows:

$$T_{q0}'' + T_{q0}' = \frac{L_{Q_1 Q_1}}{R_{Q_1}} + \frac{L_{Q_2 Q_2}}{R_{Q_2}} \tag{5.83a}$$

$$T_{q0}'' T_{q0}' = \frac{L_{Q_1 Q_1} L_{Q_2 Q_2} - L_{aq}^2}{R_{Q_1} R_{Q_2}} \tag{5.83b}$$

For a laminated salient-pole machine having one damper winding on q-axis, T_{q0}' is not defined and T_{q0}'' is obtained as

$$T_{q0}'' = \frac{L_{Q_1 Q_1}}{R_{Q_1}} \tag{5.84}$$

5.6.3 Armature Time Constant

As discussed in Section 5.6.1, the dc component of the short circuit phase currents induce a fundamental ac frequency component of the currents in the rotor circuits, keeping the flux linking with these circuits constant. The flux path due to the dc component of the armature current is similar to those corresponding to subtransient inductances L_d'' and L_q''. As the rotor rotates at synchronous speed with respect to the stationary mmf wave produced by the dc current component of the short circuit phase currents, the effective inductance seen by these currents lies between L_d'' and L_q'', and is equal to the inductance L_2 [8], seen by the negative sequence currents applied to the armature windings. Thus,

$$T_a = \frac{L_2}{R_a} \tag{5.85}$$

where L_2 is the negative sequence inductance, which is given by

$$L_2 = \frac{L_d'' + L_q''}{2} \tag{5.86}$$

5.6.4 Operational Parameters

The relationship between the armature and field terminal quantities can be expressed in the operational form as follows (see Figure 5.14):

$$\psi_d(s) = G(s)v_f(s) + L_d(s)i_d(s)$$

$$\psi_q(s) = L_q(s)i_q(s) \tag{5.87}$$

where

 $G(s)$ is the armature-to-field transfer function. It is the Laplace ratio of the d-axis armature flux linkages to the field voltage, with the armature open-circuited.
 $L_d(s)$ is the d-axis operational inductance. It is the Laplace ratio of the d-axis armature flux linkages to the d-axis current, with the field voltage short-circuited.
 $L_q(s)$ is the q-axis operational inductance. It is the Laplace ratio of the q-axis armature flux linkages to the q-axis current.

The Laplace transforms of d-axis flux linkages can be found from Equations 5.56 and 5.58 as

$$\psi_d(s) = L_d i_d(s) + L_{ad} i_F(s) + L_{ad} i_D(s) = 0$$

$$\psi_F(s) = L_{ad} i_d(s) + L_{FF} i_F(s) + L_{FD} i_D(s) = 0$$

$$\psi_D(s) = L_{ad} i_d(s) + L_{FD} i_F(s) + L_{DD} i_D(s) = 0 \tag{5.88}$$

The Laplace transforms of d-axis rotor voltages can be found from Equation 5.57 as

$$v_F(s) = s\psi_F(s) - \psi_F(0) + R_F i_F(s) = 0$$

$$0 = s\psi_D(s) - \psi_D(0) + R_D i_D(s) \tag{5.89}$$

where $\psi_F(0)$ and $\psi_D(0)$ denote the initial values of flux linkages. In order to eliminate initial values, incremental values around initial operating condition can be used. Upon

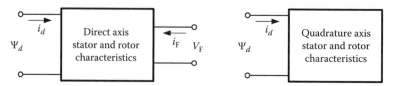

FIGURE 5.14
d- and q-axes representations based on Equation 5.87.

substitution of Equation 5.88 into Equation 5.89, the voltage equations in incremental form become

$$\Delta v_F(s) = sL_{ad}\Delta i_d(s) + (R_F + sL_{FF})\Delta i_F(s) + sL_{FD}i_D(s) = 0$$

$$0 = sL_{ad}\Delta i_d(s) + sL_{FD}\Delta i_F(s) + (R_D + sL_{DD})i_D(s) \tag{5.90}$$

From these results, the rotor currents in terms of the terminal quantities v_F and i_d become

$$\Delta i_F(s) = \frac{1}{D(s)}\left((R_D + sL_{DD})\Delta v_f(s) - sL_{ad}(R_D + sL_{DI})\Delta i_d(s)\right) \tag{5.91a}$$

$$\Delta i_D(s) = \frac{1}{D(s)}\left(-sL_{FD}\Delta v_f(s) - sL_{ad}(R_F + sL_{FI})\Delta i_d(s)\right) \tag{5.91b}$$

where

$$D(s) = R_D R_F + s(R_D L_{FF} + R_F L_{DD}) + s^2(L_{FF}L_{DD} - L_{FD}^2) \tag{5.92}$$

After substituting Equations 5.91a and 5.91b into the incremental form of the fictitious d-axis winding flux linkage expression, derived from Equation 5.88, the relationship between the d-axis quantities can be expressed in the desired form

$$\Delta \psi_d(s) = G(s)\Delta v_f(s) + L_d(s)\Delta i_d(s)$$

The expressions for the d-axis parameters are as follows:

$$G(s) = \frac{L_{ad}}{R_F} \frac{(1 + sT_{kd})}{1 + (T'_{do} + T''_{do})s + (T'_{do}T''_{do})s^2} \tag{5.93a}$$

$$L_d(s) = L_d \frac{1 + (T'_d + T''_d)s + (T'_d T''_d)s^2}{1 + (T'_{do} + T''_{do})s + (T'_{do}T''_{do})s^2} \tag{5.93b}$$

where $T_{kd} = \dfrac{L_{DI}}{R_D}$, and $(T'_d + T''_d), (T'_d T''_d), (T'_{do} + T''_{do}), (T'_{do}T''_{do})$ are from Equations 5.76 and 5.82.

Under steady-state conditions (i.e., $s = 0$), Equation 5.93b becomes

$$L_d(0) = L_d \tag{5.94}$$

During fast transients (i.e., $s \to \infty$), the limiting value of $L_d(s)$ is

$$L_d(s) = L_d \frac{(T'_d T''_d)}{(T'_{do}T''_{do})} \tag{5.95}$$

From Equations 5.76b and 5.82b

$$L_d \frac{(T_d' T_d'')}{(T_{d0}' T_{d0}'')} = L_d \frac{\dfrac{1}{R_F R_D}\left(\left(L_{FF} - \dfrac{L_{ad}^2}{L_d}\right)\left(L_{DD} - \dfrac{L_{ad}^2}{L_d}\right) - \left(L_{FD} - \dfrac{L_{ad}^2}{L_d}\right)^2\right)}{\dfrac{L_{FF}L_{DD} - L_{FD}^2}{R_F R_D}}$$

$$= \frac{\dfrac{L_{FF}L_{DD} - L_{FD}^2}{R_F R_D}\left(L_d - \dfrac{L_{ad}^2\left(L_{DD} + L_{FF} - 2L_{FD}\right)}{L_{FF}L_{DD} - L_{FD}^2}\right)}{\dfrac{L_{FF}L_{DD} - L_{FD}^2}{R_F R_D}}$$

$$= L_{al} + \frac{L_{ad}\left(L_{Dl}L_{Fl} + L_{Dl}L_{Fdl} + L_{Fl}L_{Fdl}\right)}{L_{ad}L_{Fl} + L_{ad}L_{Dl} + L_{Dl}L_{Fl} + L_{Dl}L_{Fdl} + L_{Fl}L_{Fdl}} \tag{5.96}$$

This result is equal to the *d*-axis subtransient inductance given in Equation 5.69, as expected.

The basic definition of subtransient and transient inductances (L_d'' and L_d') in ANSI/IEEE Std 115–1995 is [2]:

$$\frac{1}{L_d(s)} = \frac{1}{L_d} + \left(\frac{1}{L_d'} - \frac{1}{L_d}\right)\frac{sT_d'}{1 + sT_d'} + \left(\frac{1}{L_d''} - \frac{1}{L_d'}\right)\frac{sT_d''}{1 + sT_d''} \tag{5.97}$$

Under steady-state conditions (i.e., $s = 0$), this equation becomes

$$\frac{1}{L_d(0)} = \frac{1}{L_d} \tag{5.98}$$

During fast transients (i.e., when $s \to \infty$), the limiting value of $\dfrac{1}{L_d(s)}$ is

$$\frac{1}{L_d(s)} = \frac{1}{L_d} + \left(\frac{1}{L_d'} - \frac{1}{L_d}\right) + \left(\frac{1}{L_d''} - \frac{1}{L_d'}\right) = \frac{1}{L_d''} \tag{5.99}$$

During slow transients, $\dfrac{1}{L_d(s)}$ is

$$\frac{1}{L_d(s)} = \frac{1}{L_d} + \left(\frac{1}{L_d'} - \frac{1}{L_d}\right) = \frac{1}{L_d'} \tag{5.100}$$

From Equation 5.96, the first equation that describes the relation between the open circuit and the short circuit time constants in terms of subtransient and transient inductances can be written as

$$T'_{d0}T''_{d0} = \frac{L''_d}{L_d} T'_d T''_d \tag{5.101}$$

The second equation describing the relation between the open circuit and the short circuit time constants in terms of subtransient and transient inductances can be found by using $1/L_d(s)$, obtained from Equation 5.93b. Expanding it into partial fractions

$$
\begin{aligned}
\frac{1}{L_d(s)} &= \frac{1}{L_d}\frac{1+\left(T'_{d0}+T''_{d0}\right)s+\left(T'_{d0}T''_{d0}\right)s^2}{1+\left(T'_d+T''_d\right)s+\left(T'_d T''_d\right)s^2} \\[2mm]
&= \frac{1}{L_d}\frac{\left(1+sT'_{d0}\right)\left(1+sT''_{d0}\right)}{\left(1+sT'_d\right)\left(1+sT''_d\right)} \\[2mm]
&= \frac{1}{L_d}-\frac{1}{L_d}\frac{\left(T'_d-T'_{d0}\right)\left(T'_d-T''_{d0}\right)}{T'_d\left(T'_d-T''_d\right)}\frac{sT'_d}{1+sT'_d}-\frac{1}{L_d}\frac{\left(T''_d-T'_{d0}\right)\left(T''_d-T''_{d0}\right)}{T''_d\left(T''_d-T'_d\right)}\frac{sT''_d}{1+sT''_d}
\end{aligned}
\tag{5.102}
$$

Taking into account 5.101 and equating the coefficient of the second term in both Equation 5.97 and Equation 5.102, the following result is obtained

$$\left(T'_{d0}+T''_{d0}\right)= T'_d\frac{L_d}{L'_d}+T''_d\left(1-\frac{L_d}{L'_d}+\frac{L_d}{L''_d}\right) \tag{5.103}$$

Similarly, for q-axis, the equations that describe the relation between the open circuit and the short circuit time constants in terms of subtransient and transient inductances can be written as

$$T'_{q0}T''_{q0} = \frac{L''_q}{L_q} T'_q T''_q \tag{5.104a}$$

$$\left(T'_{q0}+T''_{q0}\right)= T'_q\frac{L_q}{L'_q}+T''_q\left(1-\frac{L_q}{L'_q}+\frac{L_q}{L''_q}\right) \tag{5.104b}$$

The magnitude of the *d*-axis operational inductance as a function of frequency is plotted with asymptotic approximation in Figure 5.15 for a typical synchronous machine. The effective inductance is equal to the synchronous inductance at the frequencies below 0.02 Hz, the transient inductance in the range 0.2–2 Hz, and the subtransient inductance beyond 10 Hz.

The magnitude of the armature-to-field transfer function as a function of frequency with normalized dc gain (i.e., $G(0) = 1$) is plotted with asymptotic approximation in Figure 5.16 for a typical synchronous machine. From the plot, it can be seen that the effective gain drops off at high frequencies, indicating that high-frequency variations in field voltage are not transferred to stator flux linkages.

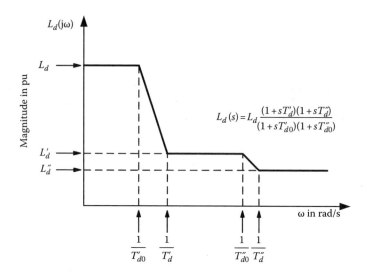

FIGURE 5.15
Variation of magnitude of $L_d(s)$ with frequency.

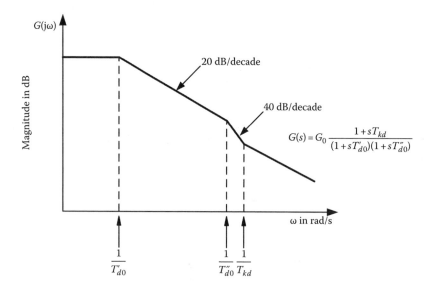

FIGURE 5.16
Variation of magnitude of $G(s)$ with frequency.

5.7 Data Conversion Procedures

A wide range of test methods has been proposed and used in order to obtain derived parameters that characterize the synchronous machine. Therefore, depending on the performed tests, each machine might have different type of data sets in terms of the derived parameters. This part details the techniques which are used to obtain fundamental parameters from the different types of data sets.

5.7.1 Determination of Fundamental Parameters from Operational Parameters

If a synchronous machine has been subjected to a parameter-determination study based on frequency response tests, the operational parameters of the synchronous machine, $G(s)$, $L_d(s)$, and $L_q(s)$, are expected to be available. In this part the necessary procedures are presented in order to obtain second and lower order models of d- and q-axes. The procedures for third-order models of both direct and quadrature axes are presented in the Appendix without derivation.

5.7.2 Procedure for Complete Second-Order d-Axis Model

The parameters of the equivalent circuit shown in Figure 5.9 can be obtained when $G(s)$ and $L_d(s)$ are provided together with the stator leakage impedance L_{al}. Let

$$DS = \left(T_{d0}'' + T_{d0}'\right) - \left(T_d'' + T_d'\right) \tag{5.105a}$$

$$DP = \left(T_{d0}'' T_{d0}'\right) - \left(T_d'' T_d'\right) \tag{5.105b}$$

By substituting Equations 5.76a and 5.82a into Equation 5.105a, and Equations 5.76 and 5.82b into Equation 5.105b, the following equations are deduced

$$\left(\frac{1}{R_F} + \frac{1}{R_D}\right) = DS\frac{L_d}{L_{ad}^2} \tag{5.106a}$$

$$\left(\frac{T_1}{R_D} + \frac{T_2}{R_F}\right) = DP\frac{L_d}{L_{ad}^2} \tag{5.106b}$$

where

$$T_1 = \frac{L_{Fl}}{R_F}$$

$$T_2 = \frac{L_{Dl}}{R_D}$$

By substituting Equation 5.106a into Equation 5.82a, and Equation 5.106b into Equation 5.82b,

$$T_1 + T_2 + L_{FD}\left(DS\frac{L_d}{L_{ad}^2}\right) = T_{d0}'' + T_{d0}' \tag{5.107a}$$

$$T_1 T_2 + L_{FD}\left(DP\frac{L_d}{L_{ad}^2}\right) = T_{d0}'' T_{d0}' \tag{5.107b}$$

in which $T_2 = T_{kd}$, T_1, and L_{FD} are the unknowns.

Taking into account that L_{FD} ($=L_{ad} + L_{Fdl}$), after some manipulations of Equations 5.107 it can be found

$$L_{FD} = \frac{T''_{d0}T'_{d0} - T_2\left(T''_{d0} + T'_{d0}\right) + T_2^2}{DS\dfrac{L_d}{L_{ad}^2}\left(\dfrac{DP}{DS} - T_2\right)} \tag{5.108}$$

The value of T_1 can be obtained after substituting this result into either Equation 5.107a or Equation 5.107b. By using Equation 5.107b

$$T_1 = \frac{1}{T_2}\left[T''_{d0}\,T'_{d0} - L_{FD}\left(DP\,\frac{L_d}{L_{ad}^2}\right)\right] \tag{5.109}$$

The values of R_F and R_D can be derived from Equations 5.106. After that L_{Fl} and L_{Dl} can be obtained from $T_1 = \dfrac{L_{Fl}}{R_F}$ and $T_2 = \dfrac{L_{Dl}}{R_D}$.

Example 5.1

The operational parameters for a 60 Hz generator are given as below with $L_{al} = 0.15$ p.u.

$$L_d(s) = 1.65\frac{(1+0.6564s)(1+0.4615s)}{(1+7.910s)(1+0.463s)}$$

$$G(s) = 7.958\frac{(1+0.0460s)}{(1+7.910s)(1+0.463s)}$$

After some manipulations on the equations used for calculating fundamental parameters of the complete second-order *d*-axis model, the following algorithm for calculation from $G(s)$ and $L_d(s)$ can be obtained [13]:

- Step 1

$$L_{ad} = L_d - L_{al}$$

- Step 2: Calculate intermediate parameters

$$R_p = \frac{L_{ad}^2}{L_d\left[\left(T''_{d0} + T'_{d0}\right) - \left(T''_d + T'_d\right)\right]}$$

$$a = \frac{L_d\left(T''_d + T'_d\right) - L_{al}\left(T''_{d0} + T'_{d0}\right)}{L_{ad}}$$

$$b = \frac{L_d T''_d\, T'_d - L_{al}T''_{d0}T'_{d0}}{L_{ad}}$$

$$c = \frac{T''_{d0}T'_{d0} - T''_d T'_d}{\left(T''_{d0} + T'_{d0}\right) - \left(T''_d + T'_d\right)}$$

- Step 3: Calculate fundamental parameters

$$L_{Fdl} = R_p \frac{b - a T_{kd} + T^2_{kd}}{c - T_{kd}}$$

$$R_F = R_p \frac{L_{Fdl} + R_p \left(2T_{kd} - a\right)}{L_{Fdl} + R_p \left(T_{kd} + c - a\right)}$$

$$R_D = \frac{R_F R_p}{R_F - R_p}$$

$$L_{Fl} = R_F \left(a - T_{kd} \frac{L_{Fdl}}{R_p}\right)$$

$$L_{Dl} = R_D T_{kd}$$

The MATLAB® script shown in Figure 5.17 was developed to perform this procedure.

The fundamental parameters calculated by using the above procedure are shown in Table 5.3. It should be noted that the time constants in the above equations are in per unit. Therefore the time constants in the operational parameters are in seconds and need to be converted into per unit values.

5.7.3 Procedure for Simplified Second-Order *d*-Axis Model

When $G(s)$ is not available, it is not possible to extract the parameters of the equivalent circuit shown in Figure 5.9, unless the so-called Canay's characteristic inductance is provided [14,15]. Therefore, only the parameters of the simplified equivalent circuit, in which L_{Fdl} is omitted, can be obtained. The omission of L_{Fdl} brings the assumption that all *d*-axis rotor circuits link a single ideal mutual flux represented by L_{ad}. It should be noted that most stability software packages use the simplified second-order *d*-axis model, while some Electromagnetic Transients Program (EMTP)-type programs provide the complete second-order *d*-axis model option.

Since L_{FD} is assumed to be equal to L_{ad}, Equations 5.107 become

$$T_1 + T_2 + DS \frac{L_d}{L_{ad}} = T''_{d0} + T'_{d0} \tag{5.110a}$$

$$T_1 T_2 + DP \frac{L_d}{L_{ad}} = T''_{d0} T'_{d0} \tag{5.110b}$$

T_1 and T_2 are the unknowns of these equations, and also the roots of the following equation:

```
% ----------------------------------------------------------------------
Wb = 120*pi;     % Base angular speed
Tb = 1/Wb;       % Base time
% -----mmm-----I----mmm----I-------------          T1 = LF1/RF
%       Lal    I    LFdl    m          m           T2 = LDl/RD
%              m            m LF1      m LD1        Ld  = Lad + Lal
%         m Lad           I          I            LFF = LFD + LF1
%              m            <          <            LDD = LFD + LD1
%         I            < RF       < RD             LFD = Lad + LFdl
%------------ --I-----------I----------- I
Lal = 0.15;                                         % Leakage inductance
% ----------Ld(s)-------
Ld = 1.65;
Td0p = 7.91/Tb;          Tdp = 0.6564/Tb;           % Time constants
Td0pp = 0.0463/Tb;       Tdpp = 0.04615/Tb;         % are punched in pu
% ----------G(s)--------
Tkd = 0.046/Tb;
% ----------------------
Lad = Ld-Lal;
DS = (Td0p+Td0pp)-(Tdp+Tdpp);                       % Equation (5.105a)
DP = (Td0p*Td0pp)-(Tdp*Tdpp);                       % Equation (5.105b)
T2 = Tkd;
LFD = ((Td0p*Td0pp)-T2*(Td0p+Td0pp)+T2^2)/(DP-T2*DS)/(Ld/Lad^2);   % Equation (5.108)
T1 = ((Td0p*Td0pp)- LFD*(DP*Ld/Lad^2))/T2;          % Equation (5.109)
inv_RD= (DP*Ld/Lad^2-T2*DS*Ld/Lad^2)/(T1-T2);    % 1/RD from Equations (5.106a) and (5.106b)
inv _ RF = DS*Ld/Lad^2-inv _ RD;                 % 1/RF from Equation (5.106a) with
                                                    known 1/RD

LFdl = LFD - Lad;
RF = 1/inv _ RF;
RD = 1/inv _ RD;
LF1 = T1 * RF;
LD1 = T2 * RD;
% ----------------------------------------------------------------------
```

FIGURE 5.17
Example 5.1: MATLAB script for fundamental parameter calculation of the complete second-order d-axis model from $G(s)$ and $L_d(s)$.

$$T^2 - \left(T_{d0}'' + T_{d0}' - DS \frac{L_d}{L_{ad}} \right) T + \left(T_{d0}'' T_{d0}' - DP \frac{L_d}{L_{ad}} \right) = 0 \qquad (5.111)$$

Let r_1 and r_2 be the roots of Equation 5.111. It is not possible to know whether r_1 is equal to T_1 or T_2. For that reason Equations 5.107 can be rewritten by considering two rotor windings without assigning them to the field or the damper windings. Assume R_1 and L_1 are the resistance and inductance values of the first rotor winding, and R_2 and L_2 are the resistance and inductance values of the second rotor winding:

$$\left(\frac{1}{R_1} + \frac{1}{R_2} \right) = DS \frac{L_d}{L_{ad}^2} \qquad (5.112a)$$

$$\left(\frac{r_1}{R_2} + \frac{r_2}{R_1} \right) = DP \frac{L_d}{L_{ad}^2} \qquad (5.112b)$$

TABLE 5.3

Example 5.1: Calculated Fundamental Parameters in Per Unit

L_d	1.6500
L_{al}	0.1500
L_{ad}	1.5000
L_{Fdl}	−0.0813
L_{Fl}	0.0684
R_F	4.9915E−4
L_{Dl}	8.7873
R_D	0.5067

where

$$r_1 = \frac{L_1}{R_1}$$

$$r_2 = \frac{L_2}{R_2}$$

R_1, L_1, R_2, and L_2 can be obtained by using Equations 5.112. Since the excitation winding has a larger time constant, the resistance, and the inductance values of the winding having larger time constant can be assigned to excitation winding; i.e., if

$$T_{w1} = \frac{L_{ad} + L_1}{R_1} > T_{w2} = \frac{L_{ad} + L_2}{R_2}$$

then $R_F = R_1$ and $L_{Fl} = L_1$, else $R_F = R_2$ and $L_{Fl} = L_2$

Example 5.2

Consider the same machine as that in Example 5.1. Since $G(s)$ is not available for that machine, only the fundamental parameters of the simplified second-order d-axis model can be calculated.

After some manipulations on the equations used for calculating fundamental parameters of the simplified second-order d-axis model, the following algorithm for calculation from $L_d(s)$ can be obtained [13]:

- Step 1

$$L_{ad} = L_d - L_{al}$$

- Step 2: Calculate intermediate parameters

$$R_p = \frac{L_{ad}^2}{L_d \left[(T_{d0}'' + T_{d0}') - (T_d'' + T_d') \right]}$$

$$a = \frac{L_d (T_d'' + T_d') - L_{al} (T_{d0}'' + T_{d0}')}{L_{ad}}$$

$$b = \frac{L_d T_d'' T_d' - L_{al} T_{d0}'' T_{d0}'}{L_{ad}}$$

$$c = \frac{T_{d0}'' T_{d0}' - T_d'' T_d'}{(T_{d0}'' + T_{d0}') - (T_d'' + T_d')}$$

- Step 3: Calculate fundamental parameters of the rotor windings

$$R_{w1} = \frac{2R_p \sqrt{a^2 - 4b}}{a - 2c + \sqrt{a^2 - 4b}}$$

$$L_{w1} = \frac{R_{w1} \left(a + \sqrt{a^2 - 4b} \right)}{2}$$

$$R_{w2} = \frac{2R_p\sqrt{a^2 - 4b}}{2c - a + \sqrt{a^2 - 4b}}$$

$$L_{w2} = \frac{R_{w2}\left(a - \sqrt{a^2 - 4b}\right)}{2}$$

- Step 4: Identify field and damper winding parameters:

If $\left(T_{w1} = \dfrac{L_{ad} + L_1}{R_1} > T_{w2} = \dfrac{L_{ad} + L_2}{R_2}\right)$

then $R_F = R_{w1}$, $L_{FI} = L_{w1}$, $R_D = R_{w2}$, $L_{DI} = L_{w2}$

else $R_F = R_{w2}$, $L_{FI} = L_{w2}$, $R_D = R_{w1}$, $L_{DI} = L_{w1}$

The MATLAB script developed for calculations is shown in Figure 5.18, and the fundamental parameters calculated by using the above equations are given in Table 5.4.

5.7.4   Procedure for Complete Second-Order *d*-Axis Model with Canay's Characteristic Inductance

In this book, the leakage inductance L_{al} is chosen in such a way that it represents the leakage flux not crossing the air-gap. However, its choice is arbitrary and can be chosen equal to the Canay's characteristic inductance denoted by L_C, which transforms the equivalent circuit given in Figure 5.9 to the form depicted in Figure 5.19 [14,15]. When the Canay's characteristic inductance is available, in addition to $L_d(s)$, it is possible to find the parameters of the complete second-order *d*-axis model.

The new leakage inductance, L_C, modifies the flux common to the fictitious *d*-axis winding, the field winding, and the *d*-axis damper winding. Therefore, the ratio of the number of turns changes, and results in a change of real electrical quantities of the rotor referred to the stator.

The relationship between the Canay's characteristic inductance and the inductance L_{Fdl}, which represents the flux linking both the field and the damper but not the *d*-axis winding, is as follows:

$$\frac{1}{L_C - L_{al}} = \frac{1}{L_{ad}} + \frac{1}{L_{Fdl}} \tag{5.113}$$

Since L_{Fdl} can be found from Equation 5.113, T_1 and T_2 become the unknowns of the Equation 5.107, and they are the roots of the following equation:

$$T^2 - \left(T_{d0}'' + T_{d0}' - L_{FD}DS\frac{L_d}{L_{ad}^2}\right)T + \left(T_{d0}''T_{d0}' - L_{FD}DP\frac{L_d}{L_{ad}^2}\right) = 0 \tag{5.114}$$

Let r_1 and r_2 be the roots of the Equation 5.114. As for the previous case, it is not possible to know whether r_1 is equal to T_1 or T_2. Therefore, the procedure for parameter calculation used with the simplified second-order *d*-axis model will be followed; i.e., Equations 5.112 with the modified rotor winding time constant formulations. The resulting expressions are as follows:

```
% -------------------------------------------------------------------------
Wb = 120*pi;    % Base angular speed
Tb = 1/Wb;      % Base time
%
% -----mmm-----I---------- -I----------- - - -          T1 = LF1/RF
%      Lal       I          C             C            T2 = LD1/RD;
%                I          C LF1         C LD1        Ld  = Lad + Lal
%                C Lad      I             I            LFF = LFD + LF1
%                C          <             <            LDD = LFD + LD1
%                I          < RF          < RD         LFD = Lad
%------------ - - -I--------- - - -I---------- - - -I
%
Lal = 0.15;                                           % Leakage Inductance
% -----------Ld(s)-------
Ld = 1.65;
Td0p = 7.91/Tb;           Tdp = 0.6564/Tb;            % Time constants
Td0pp = 0.0463/Tb;        Tdpp = 0.04615/Tb;          % are punched in pu
% -----------G(s)--------
Tkd = 0.046/Tb;
% ----------------------
Lad = Ld-Lal;
DS = (Td0p+Td0pp)-(Tdp+Tdpp);                         % Equation (5.105a)
DP = (Td0p*Td0pp)-(Tdp*Tdpp);                         % Equation (5.105b)
T1 _ plus _ T2 = (Td0p+Td0pp)-DS*Ld/Lad;
T1 _ prod _ T2 = (Td0p*Td0pp)-DP*Ld/Lad;
[r] = roots([1 -T1 _ plus _ T2 T1 _ prod _ T2]);     % Roots of Equation (5.111)
inv _ R1 = (DP*Ld/Lad^2-r(1)*DS*Ld/Lad^2)/(r(2)-r(1));  % 1/R1 from Equation (5.112a) &
                                                        (5.112b)
inv _ R2 = DS*Ld/Lad^2-inv _ R1;                     % 1/R2 from Equation (5.112a)
R1 = 1/inv _ R1;
R2 = 1/inv _ R2;
L1 = r(1) * R1;
L2 = r(2) * R2;
Tw1 = (Lad+L1)/R1; Tw2 = (Lad+L2)/R2;                % Calculate time constant of windings
if Tw1>Tw2
    RF = R1; LF1 = L1; RD = R2; LD1 = L2;            % Assign values of the winding
else                                                 % having larger time constant
    RF = R2; LF1 = L2; RD = R1; LD1 = L1;            % to the field
end
% -------------------------------------------------------------------------
```

FIGURE 5.18

Example 5.2: MATLAB script for fundamental parameter calculation of the simplified second-order *d*-axis model from $L_d(s)$.

TABLE 5.4

Example 5.2: Calculated Fundamental Parameters in Per Unit

L_d	1.6500
L_{al}	0.1500
L_{ad}	1.5000
L_{Fl}	−0.0133
R_F	5.0232E−4
L_{Dl}	1.2178
R_D	0.0685

$$\left(T_{w1} = \frac{L_{ad} + L_{Fdl} + L_1}{R_1}, \quad T_{w2} = \frac{L_{ad} + L_{Fdl} + L_2}{R_2} \right) \quad (5.115)$$

Example 5.3

Consider the same machine that in Example 5.2, where $L_C = 0.064$ p.u. Since Canay's characteristic inductance is available, fundamental parameters of the complete second-order *d*-axis model can be calculated.

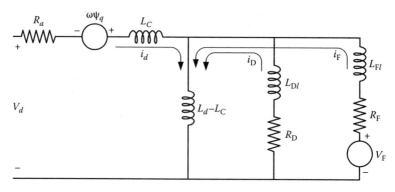

FIGURE 5.19
d-Axis equivalent circuit after transformation.

After some manipulations on the equations utilized for calculating fundamental parameters, the following algorithm for calculation from $L_d(s)$ and L_C can be obtained:

- Step 1

$$L_{ad} = L_d - L_{al}$$

- Step 2

$$L_{FDl} = \frac{L_{ad}\left(L_C - L_{al}\right)}{L_{ad} - L_C + L_{al}}$$

- Step 3: Calculate intermediate parameters

$$R_p = \frac{L_{ad}^2}{L_d\left[\left(T''_{d0} + T'_{d0}\right) - \left(T''_d + T'_d\right)\right]}$$

$$a = \frac{L_d\left(T''_d + T'_d\right) - L_{al}\left(T''_{d0} + T'_{d0}\right)}{L_{ad}}$$

$$b = \frac{L_d T''_d T'_d - L_{al} T''_{d0} T'_{d0}}{L_{ad}}$$

$$c = \frac{T''_{d0} T'_{d0} - T''_d T'_d}{\left(T''_{d0} + T'_{d0}\right) - \left(T''_d + T'_d\right)}$$

$$A = a - \frac{L_{Fdl}}{R_p}$$

$$B = b - L_{Fdl}\frac{c}{R_p}$$

- Step 4: Calculate fundamental parameters of the rotor windings

$$R_{w1} = \frac{2R_p\sqrt{A^2 - 4B}}{A - 2c + \sqrt{A^2 - 4B}}$$

$$L_{w1} = \frac{R_{w1}\left(A + \sqrt{A^2 - 4B}\right)}{2}$$

$$R_{w2} = \frac{2R_p\sqrt{A^2 - 4B}}{2c - A + \sqrt{A^2 - 4B}}$$

$$L_{w2} = \frac{R_{w2}\left(A - \sqrt{A^2 - 4B}\right)}{2}$$

- Step 5: Identify field and damper winding parameters:

If $\left(T_{w1} = \dfrac{L_{ad} + L_{Fdl} + L_1}{R_1} > T_{w2} = \dfrac{L_{ad} + L_{Fdl} + L_2}{R_2}\right)$

then $R_F = R_{w1}, \quad L_{Fl} = L_{w1}, \quad R_D = R_{w2}, \quad L_{Dl} = L_{w2}$

else $R_F = R_{w2}, \quad L_{Fl} = L_{w2}, \quad R_D = R_{w1}, \quad L_{Dl} = L_{w1}$

The related MATLAB script developed for calculation is given in Figure 5.20, and the fundamental parameters calculated by using the above equations are given in Table 5.5.

5.7.5 Procedure for Second-Order *q*-Axis Model

All armature and damper circuits in the *q*-axis link a single ideal mutual flux represented by L_{aq}. Since, the second-order *q*-axis model shown in Figure 5.10 is similar to the simplified second-order *d*-axis model, the procedure used for the simplified second-order *d*-axis model can be also applied to calculate the fundamental parameters of the second-order *q*-axis model, except field winding identification. That is, the algorithm of fundamental parameter calculation from $L_q(s)$ may be as follows:

- Step 1

$$L_{aq} = L_q - L_{al}$$

- Step 2: Calculate intermediate parameters

$$R_p = \frac{L_{aq}^2}{L_q\left[\left(T_{q0}'' + T_{q0}'\right) - \left(T_q'' + T_q'\right)\right]}$$

$$a = \frac{L_q\left(T_q'' + T_q'\right) - L_{al}\left(T_{q0}'' + T_{q0}'\right)}{L_{aq}}$$

```
% --------------------------------------------------------------------
Wb = 120*pi;    % Base angular speed
Tb = 1/Wb;      % Base time
%
% -----mmm-----I----mmm----I-----------                   T1 = LF1/RF
%       Lal    I   LFdl  C               C                 T2 = LD1/RD;
%              I            C LF1     C LD1                 Ld  = Lad + Lal
%              C Lad     I           I                      LFF = LFD + LF1
%              C         <           <                      LDD = LFD + LD1
%              I       < RF        < RD                     LFD = Lad + LFdl
%------------- I--------- - I-----------I
%
Lal = 0.15;     % Leakage inductance
LC  = 0.0640;   % Canay's characteristic inductance
% ----------Ld(s)-------
Ld = 1.65;
Td0p = 7.91/Tb;          Tdp = 0.6564/Tb;        % Time constants
Td0pp = 0.0463/Tb;       Tdpp = 0.04615/Tb;      % are punched in pu
% ----------------------
Lad = Ld-Lal;
LFdl = Lad*(LC-Lal)/(Lad-LC+Lal);                % From Equation (5.113)
LFD = Lad + LFdl;
DS = (Td0p+Td0pp)-(Tdp+Tdpp);                    % Equation (5.105a)
DP = (Td0p*Td0pp)-(Tdp*Tdpp);                    % Equation (5.105b)
T1 _ plus _ T2 = (Td0p+Td0pp)-DS*LFD*Ld/Lad^2;
T1 _ prod _ T2 = (Td0p*Td0pp)-DP*LFD*Ld/Lad^2;
[r] = roots([1 -T1 _ plus _ T2 T1 _ prod _ T2]);    % Roots of Equation (5.114)
inv _ R1 = (DP*Ld/Lad^2-r(1)*DS*Ld/Lad^2)/(r(2)-r(1));  % 1/R1 from Equations (5.112a)
                                                        &  (5.112b)
inv _ R2 = DS*Ld/Lad^2-inv _ R1;                        % 1/R2 from Equation (5.112a)
R1 = 1/inv _ R1;
R2 = 1/inv _ R2;
L1 = r(1) * R1;
L2 = r(2) * R2;
Tw1 = (Lad+L1+LFdl)/R1; Tw2 = (Lad+L2+LFdl)/R2;  % Calculate time constants of
                                                 windings
if Tw1>Tw2
    RF = R1; LF1 = L1; RD = R2; LD1 = L2;        % Assign values of the winding
else                                             % having larger time constant
    RF = R2; LF1 = L2; RD = R1; LD1 = L1;        % to the field
end
% --------------------------------------------------------------------
```

FIGURE 5.20
Example 5.3: MATLAB script for fundamental parameter calculation of the complete second-order d-axis model from $L_d(s)$ and L_C.

$$b = \frac{L_q T_q'' T_q' - L_{al} T_{q0}'' T_{q0}'}{L_{aq}}$$

$$c = \frac{T_{q0}'' T_{q0}' - T_q'' T_q'}{\left(T_{q0}'' + T_{q0}'\right) - \left(T_q'' + T_q'\right)}$$

- Step 3: Calculate fundamental parameters of the rotor windings

$$R_{Q_1} = \frac{2 R_p \sqrt{a^2 - 4b}}{a - 2c + \sqrt{a^2 - 4b}}$$

TABLE 5.5

Example 5.3: Calculated
Fundamental Parameters in
Per Unit

L_d	1.6500
L_{al}	0.1500
L_C	0.064
L_{ad}	1.5000
L_{Fdl}	−0.0813
L_{Fl}	0.0685
R_F	4.9915E−4
L_{Dl}	8.7968
R_D	0.5073

TABLE 5.6

Transient and Subtransient
Inductances and Time Constants

d-Axis Parameters	*q*-Axis Parameters
Transient and subtransient inductances	
L_d'	L_q'
L_d''	L_q''
Transient and subtransient short circuit time constants	
T_d'	T_q'
T_d''	T_q''
Transient and subtransient open circuit time constants	
T_{d0}'	T_{q0}'
T_{d0}''	T_{q0}''

TABLE 5.7

Data Sets to Obtain the Simplified
Second-Order *d*-Axis Model

First data set	$T_{d0}'', T_{d0}', T_d', T_d''$
Second data set	$L_d', L_d'', T_{d0}'', T_{d0}'$
Third data set	L_d', L_d'', T_d', T_d''

$$L_{Ql_1} = \frac{R_{Q_1}\left(a + \sqrt{a^2 - 4b}\right)}{2}$$

$$R_{Q_2} = \frac{2R_p\sqrt{a^2 - 4b}}{2c - a + \sqrt{a^2 - 4b}}$$

$$L_{Ql_2} = \frac{R_{Q_2}\left(a - \sqrt{a^2 - 4b}\right)}{2}$$

5.7.6 Determination of Fundamental Parameters from Transient and Subtransient Inductances and Time Constants

If the synchronous machine has been subjected to parameter determination studies other than the tests based on frequency response, data set of the machine regarding the stability studies may contain the parameters listed in Table 5.6 in addition to the synchronous and leakage inductances based on the standard tests.

The simplified second-order *d*-axis model can be found when the *d*-axis operational inductance $L_d(s)$ is known; i.e., both the *d*-axis open circuit and short circuit time constants, T_d', T_d'', T_{d0}', T_{d0}'', are provided. However, if the *d*-axis subtransient and transient inductances of the machine, L_d', L_d'', are known, one can deduce from Equations 5.101 and 5.103 that the availability of either *d*-axis open circuit time constants T_{d0}', T_{d0}'', or short circuit time constants T_d', T_d'' is sufficient to obtain the second-order *d*-axis model. In other words, the data sets illustrated in Table 5.7 are equivalent and sufficient to obtain the simplified second-order *d*-axis model.

The procedure for the simplified second-order *d*-axis model given in Section 5.7.3 can be directly used if the first data set (i.e., T_{d0}'', T_{d0}', T_d', and T_d'') is available. When either the second or the third data set is available, the first data set should be obtained with the following equations:

Second Data Set → First Data Set

$$\frac{L_d^2}{L_d' L_d''}T^2 - \frac{L_d}{L_d''}\left(T_{d0}' + T_{d0}''\right)T + \left(1 - \frac{L_d}{L_d'} + \frac{L_d}{L_d''}\right)T_{d0}'T_{d0}'' = 0 \qquad (5.116)$$

T_d' and T_d'' are the roots of this equation and the greater root is T_d'.

Third Data Set → First Data Set

TABLE 5.8

Data Sets to Obtain the Simplified Second-Order d-Axis Model

$$T^2 - \left(\frac{L_d}{L_d'}\,T_d' + \left(1 - \frac{L_d}{L_d'} + \frac{L_d}{L_d''}\right)T_d''\right)T + \frac{L_d}{L_d''}\,T_d'T_d'' = 0 \quad (5.117)$$

First data set	$T_{q0}'',\ T_{q0}',\ T_q',\ T_q''$
Second data set	$L_q',\ L_q'',\ T_{q0}'',\ T_{q0}'$
Third data set	$L_q',\ L_q'',\ T_q',\ T_q''$

T_{d0}' and T_{d0}'' are the roots of this equation and the greater root is T_{d0}'.

Since the d-axis operational inductance $L_d(s)$ is available when one of the data sets illustrated in Table 5.7 is available, in case of availability of the Canay's characteristic inductance, the complete second-order d-axis model can be obtained by applying the procedure presented in Section 5.7.4.

Similar to d-axis, the data sets illustrated in Table 5.8 are equivalent and sufficient to obtain the simplified second-order q-axis model.

Second Data Set → First Data Set

$$\frac{L_q^2}{L_q'L_q''}\,T^2 - \frac{L_q}{L_q''}\,(T_{q0}' + T_{q0}'')T + \left(1 - \frac{L_q}{L_q'} + \frac{L_q}{L_q''}\right)T_{q0}'T_{q0}'' = 0 \quad (5.118)$$

T_q' and T_q'' are the roots of this equation and the greater root is T_q'.

Third Data Set → First Data Set

$$T^2 - \left(\frac{L_q}{L_q'}\,T_q' + \left(1 - \frac{L_q}{L_q'} + \frac{L_q}{L_q''}\right)T_q''\right)T + \frac{L_q}{L_q''}\,T_q'T_q'' = 0 \quad (5.119)$$

T_{d0}' and T_{d0}'' are the roots of this equation and the greater root is T_{d0}'.

5.8 Magnetic Saturation

In the derivation of the basic equations of the synchronous machine, magnetic saturation effects are neglected in order to deal with linear coupled circuits and make superposition applicable. However, saturation effects are significant and their effects should be taken into account in power system analysis. Before discussing the methods of representing saturation effects, it may be useful to review the characteristics of a synchronous machine while the stator terminals are open- and short-circuited.

5.8.1 Open Circuit and Short Circuit Characteristics

The open circuit characteristic (OCC) of the machine is the essential data for the representation of the saturation. Since stator terminals are open-circuited while the rotor is rotating at synchronous speed, $i_d = i_q = \psi_q = v_d = 0$ and $v_t = v_q = \psi_d = L_{ad}i_f$, where v_t is the terminal voltage. Thus, the OCCs relating v_t and i_f gives the d-axis saturation characteristics. A typical OCC is illustrated in Figure 5.21.

The straight-line tangent to the lower part of the OCC curve shows the field current that is required to overcome the reluctance of the air-gap, and it is known as air-gap line. Hence, the degree of saturation is the departure of the OCC from the air-gap line.

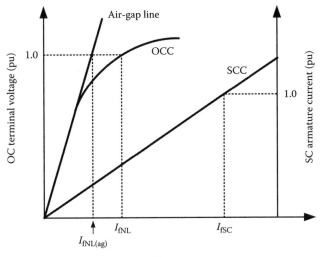

FIGURE 5.21
Open circuit and short circuit characteristics.

The short circuit characteristic (SCC) relates the armature current to the field current in steady state while the rotor is rotating at synchronous speed and the stator terminals are short-circuited. The SCC is linear up to the rated armature current due to very little or no saturation in rotor and stator iron.

The unsaturated d-axis synchronous impedance, X_{du}, can be derived from OCC and SCC. If the effect of the armature resistance is neglected, this synchronous impedance, in per unit, is equal to the ratio of the field current at base armature current from the short circuit test, I_{fSC}, to the field current at base voltage on the air-gap line, $I_{fNL(ag)}$,

$$X_{du} = \frac{I_{fSC}}{I_{fNL(ag)}}$$

(5.120)

The saturated value of d-axis synchronous impedance, X_{ds}, corresponding to the rated voltage is

$$X_{ds} = \frac{I_{fSC}}{I_{fNL}}$$

(5.121)

where I_{fNL} is the field current that produces the rated voltage at no load.

The short circuit ratio (SCR) is defined as the ratio of the field current that produces the rated voltage at no load to the field current that produces the rated armature voltage under a steady three-phase short circuit condition. In other words, it is the reciprocal value of the saturated d-axis synchronous impedance

$$SCR = \frac{I_{fNL}}{I_{fSC}} = \frac{1}{X_{ds}}$$

(5.122)

The SCR would be equal to the reciprocal of the unsaturated d-axis synchronous impedance if there was no saturation. The SCR reflects the degree of saturation, and hence it
gives some idea about both performance and cost of the machine. If a machine has a lower
SCR, it requires a larger change in field current to keep the terminal voltage constant for a
load change. Hence, a machine with a lower SCR requires an excitation system which can
provide large changes in field current to maintain system stability. On the other hand, in a
machine with a lower SCR, both the size and the cost are lower [7].

5.8.2 Representation of the Magnetic Saturation

In the representation of the magnetic saturation in power system analysis, the following
assumptions are usually made:

- The leakage fluxes pass partly through the iron and they are not significantly
  affected by the saturation of the iron portion. Therefore, they are considered to be
  independent of saturation; i.e., only mutual inductances L_{ad} and L_{aq} are saturable.
- The leakage fluxes are usually small and their paths coincide with the main flux
  only for a small part of its path. Therefore, the contribution of the leakage fluxes
  on the iron saturation is neglected; i.e., saturation is determined by the air-gap flux
  linkage.
- The sinusoidal distribution of the magnetic field over the face of the pole is considered to be unaffected by saturation, and d- and q-axes remain uncoupled.

For solid rotor machines, in addition to the assumptions given above, the saturation relationship between the resultant air-gap flux and the mmf under loaded conditions is considered to be the same as at no-load conditions. With this assumption, a single saturation
curve (i.e., open circuit saturation curve) is sufficient to characterize this phenomenon. Let
the saturation factor, K_s, be defined as follows:

$$L_{ad} = K_s L_{adu}, \qquad L_{aq} = K_s L_{aqu} \tag{5.123}$$

where L_{adu} and L_{aqu} are the unsaturated values of the mutual inductances L_{ad} and L_{aq}.

The saturation factor is determined from the OCC as illustrated in Figure 5.22. For an
operating point "A," the saturation factor is

$$K_s = \frac{\psi_{at}}{\psi_{at0}} = \frac{I_0}{I} \tag{5.124}$$

where

$$I = \sqrt{\left(i_d + i_F + i_D\right)^2 + \left(i_q + i_{Q_1} + i_{Q_2}\right)^2} \tag{5.125a}$$

$$\psi_{at} = \sqrt{\psi_{ad}^2 + \psi_{aq}^2} \tag{5.125b}$$

being ψ_{at0} the air-gap flux produced by current I, and I_0 the current required to produce
air-gap flux ψ_{at}, both measured on the air-gap line, see Figure 5.22.

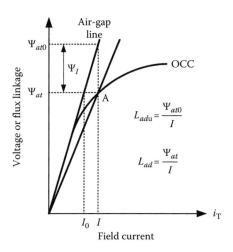

FIGURE 5.22
OCC illustrating effects of saturation.

Upon definition of

$$\psi_I = \psi_{at0} - \psi_{at} \qquad (5.126)$$

the expression of the saturation factor becomes

$$K_s = \frac{\psi_{at}}{\psi_{at} + \psi_I} \qquad (5.127)$$

In Ref. [7], the saturation curve is divided into three segments: unsaturated segment, non-linear segment, and fully saturated linear segment as illustrated in Figure 5.23, in which ψ_{T1} and ψ_{T2} define the boundaries of the segments.

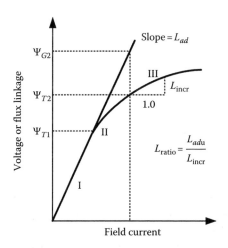

FIGURE 5.23
Representation of saturation characteristic.

Segment-I ($\psi_{at} \le \psi_{T1}$)

$$\psi_I = 0 \tag{5.128}$$

where ψ_{T1} is the boundary of the segments I and II.

Segment-II ($\psi_{T1} < \psi_{at} \le \psi_{T2}$)

$$\psi_I = A_{sat} \exp\left(B_{sat}(\psi_{at} - \psi_{T1})\right) \tag{5.129}$$

where
  A_{sat} and B_{sat} are the constants describing saturation characteristic on segment II
  ψ_{T2} is the boundary of the segments II and III

Segment-III ($\psi_{at} > \psi_{T2}$)

$$\psi_I = \psi_{G2} + L_{ratio}\left(\psi_{at} - \psi_{T2}\right) - \psi_{at} \tag{5.130}$$

where L_{ratio} is the ratio of the slope of the air-gap line to the incremental slope of segment III of the OCC as illustrated in Figure 5.23.

When this representation method is applied, ψ_{T1}, ψ_{T2}, ψ_{G2}, A_{sat}, B_{sat}, and L_{ratio} specify the saturation characteristic of the machine.

It should be noted that in this representation there is a discontinuity at the junction of segments I and II ($\psi_{at} = \psi_{T1}$). However, A_{sat} is normally very small and this discontinuity is inconsequential [7].

In case of salient-pole machines, the magnetic structure differ between the direct and the quadrature axes, and the cross flux is usually neglected in saturation representation [16]. As the path for q-axis flux is largely in air, L_{aq} does not vary significantly due to the saturation of the iron path, so it is usually necessary to adjust only ψ_{ad}; thus

$$L_{aq} = L_{aqu}, \quad L_{ad} = K_s L_{adu} \tag{5.131}$$

where K_s is defined as in Equation 5.124, with Equation 5.125 modified as follows:

$$I = \sqrt{\left(i_d + i_f + i_D\right)^2} \tag{5.132a}$$

$$\psi_{at} = \psi_{ad} \tag{5.132b}$$

Example 5.4

The open circuit curve of a 733.5 MVA generator is shown in Figure 5.24. In order to represent the saturation effect with the presented method, ψ_{T1}, ψ_{T2}, ψ_{G2}, A_{sat}, B_{sat}, and L_{ratio} should be specified from the given open circuit saturation curve.

From Figure 5.24, the boundaries of nonlinear-unsaturated segments and nonlinear-fully satu-rated segments can be found as $\psi_{T1} = 0.675$ pu and $\psi_{T2} = 1.25$ pu, respectively. Hence, for the unsaturated segment ($\psi_{at} \le \psi_{T1} = 0.675$ pu) $\psi_I = 0$, while for the nonlinear segment ($\psi_{T1} = 0.675$ pu <

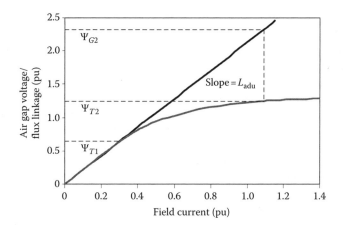

FIGURE 5.24
Example 5.4: Open circuit saturation curve.

$\psi_{at} \leq \psi_{T2} = 1.25$ pu) the saturation characteristics of the machine can be described by the Equation 5.129. A_{sat} and B_{sat} values can be calculated by considering two points on the saturation curve corresponding to $\psi_{at} = 1$ pu and $\psi_{at} = 1.2$ pu.

With $\psi_{at} = 1$ pu, $\psi_I = 0.2$ pu, and with $\psi_{at} = 1.2$ pu, $\psi_I = 0.8$ pu. Hence,

$$0.2 = A_{sat} \exp\left(B_{sat}(1-0.675)\right)$$

and

$$0.8 = A_{sat} \exp\left(B_{sat}(1.2-0.675)\right)$$

Solving the above equations gives $A_{sat} = 0.021$ and $B_{sat} = 6.9315$.

The saturation characteristics of the machine at the fully saturated segment is described with ψ_{G2} and L_{ratio}. These values can be obtained from Figure 5.24 as $\psi_{G2} = 2.35$ pu and $L_{ratio} = 0.1$ pu.

In this example the reciprocal per unit system, initially advocated in 1945 by Rankin [17], is applied. With this system, the per unit field current required to produce 1.0 per unit terminal voltage on the air-gap line, with the generator open-circuited, is equal to $\dfrac{1}{X_{adu}}$, where X_{adu} is in per unit.

On the other hand, most computer programs use a nonreciprocal per unit system for the field current. In the nonreciprocal per unit system, 1.0 per unit field current is required to produce 1.0 per unit generator terminal voltage on the air-gap line. Numerically, the nonreciprocal per unit system is more convenient to use and visualize. Readers are referred to [2,12] for details regarding the utilization of reciprocal and nonreciprocal per unit systems.

5.9 Test Procedures for Parameter Determination

5.9.1 Introduction

Standard testing methods can be classified into three groups [2–4]: (1) acceptance tests; (2) performance tests; and (3) parameter estimation tests. The last group is the most important for the purposes of this chapter, but some steady-state performance tests may provide useful information for synchronous machine studies (e.g., losses, saturation curve, SCC).

It is important to distinguish between: (1) test procedures (e.g., short circuit, open circuit, standstill frequency response (SSFR), load rejection, and partial load rejection tests), (2) determination of characteristic parameters (e.g., open/short circuit reactances and time constants), and (3) determination of parameters to be specified in equivalent models. Test setups and testing conditions, as well as procedures for determination of characteristic electrical parameters, are described and justified in standards [2,3]. However, data conversion procedures for the determination of equivalent circuit parameters are not established in standards. Conversion procedures for passing from manufacturer's data to model parameters were presented in Section 5.7, a review of the most common off-line tests is presented in this section.

Steady-state tests are performed in order to obtain X_d, X_q, negative sequence reactance and resistance (X_2, R_2), zero-sequence reactance and resistance (X_0, R_0), as well as OCC and SCC of the machine. Methods for conducting a wide range of steady-state tests can be found in ANSI/IEEE Std 115–1995 [2].

Short circuit tests are performed to demonstrate that the mechanical design of the machine is adequate to withstand the mechanical stresses arising from short circuit currents. These tests also facilitate the determination of various synchronous machine characteristics quantities used to predict the machine behavior under transient conditions. These test procedures provide X_d', X_d'', T_d', T_d'', T_{d0}', and T_{d0}''. However, they do not provide q-axis transient and subtransient characteristic quantities. In addition, accurate identification of field circuit is not possible with these procedures.

To derive q-axis quantities, special procedures are presented in IEC 34-4 [3]. The methodology of these procedures involves short-circuiting or open-circuiting the synchronous machine operating at low voltage and low power, and connected either synchronously or asynchronously to a power network. Appendix A of IEC 34-4 describes such tests and is titled "Unconfirmed." These q-axis tests are also summarized in Appendix 11A of ANSI/IEEE Std 115-1995 and titled "Informative."

SSFR testing has been developed within the last few years as an alternative to traditional tests. These testing procedures are described in Section 12 of ANSI/IEEE Std 115-1995 [2]. They basically involve exciting the stator or the field of the machine when the machine is off-line and at standstill. The operational parameters of the machine that are required to derive the complete model can be obtained from SSFR test results. In addition, these tests enable to obtain equivalent circuits of order higher than Model 2.2 detailed in the previous sections.

To avoid some of the drawbacks and limitations of off-line tests, the so-called online tests have been developed. These tests can be either time-domain tests, such as the load rejection test, or frequency-domain tests, such as the online frequency response test. A short summary of online test is provided at the end of this section.

5.9.2 Steady-State Tests

5.9.2.1 Open Circuit Characteristic Curve

The OCC curve is obtained when the machine is running at rated speed with the armature terminals open-circuited and the field winding excited. The terminal voltage, field current, and terminal frequency or shaft speed are recorded; the field current is varied. In order to obtain useful data for generator model derivation the records should be distributed as follows [2]:

- Six records below 60% of rated voltage (First reading is at zero excitation).
- A minimum of 10 points are recorded between 60% and 110% of rated voltage. Due to the importance of this area, it is recommended to obtain as many points as the excitation control resolution allows.
- At least two points are recorded above 110% of rated voltage, including one point at about 110% of the rated no-load field current (or at the maximum value recommended by the manufacturer).

Following the correction for the speed, the OCC curve can be plotted as in Figure 5.25. The voltage of a single phase or the average of the three-phase voltages at each excitation value may be used. A typical OCC curve is a straight-line for low voltages; the air-gap line can be obtained by extending that straight-line. As already mentioned, the degree of saturation is the departure of the OCC from the air-gap line.

5.9.2.2 Short Circuit Characteristic Curve

The SCC curve is obtained when the machine is running at rated speed with the armature terminals short-circuited and the field winding excited. The armature current and field current are recorded as the field current is varied from zero to the value that will induce 125% of rated armature current. Since the armature is short-circuited, the machine is operating at a lower level of flux density when compared to the case of OCC. Therefore, the SCC curve is a straight-line, as illustrated in Figure 5.25.

5.9.2.3 Determination of d-Axis Synchronous Impedance from OCC and SCC

The unsaturated d-axis synchronous impedance, X_{du}, can be derived from the results of the OCC and SCC given in Figure 5.25. This synchronous impedance in per unit is equal to the ratio of the field current at base armature current, from the short circuit test, to the field

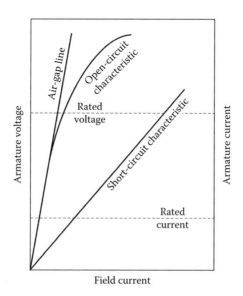

FIGURE 5.25
OCC and SCC curves.

current at base voltage on the air-gap line if the effect of armature resistance is neglected, see Equation 5.120.

Saturated values of the synchronous reactance X_{ds} depend on the machine operating conditions. As explained in Section 5.8.2, X_d is assumed to be composed of unsaturable leakage reactance, X_l, and saturable stator-to-rotor mutual reactance, X_{ad}. Thus,

$$X_{du} = X_l + X_{adu} \tag{5.133}$$

When X_{ad} is saturated to any degree (X_{ads}), then [2]

$$X_{ds} = X_l + X_{ads} \tag{5.134}$$

5.9.2.4   *Slip Test and Measuring q-Axis Synchronous Impedances*

The slip test is conducted by driving the rotor at a speed very close to the synchronous speed with the field open-circuited and the armature energized by a three-phase, rated-frequency, positive-sequence power source at a voltage below the point on the open circuit saturation curve, where the curve deviates from the air-gap line. The armature current, the armature voltage, and the voltage across the open circuit field winding are recorded. Figure 5.26 illustrates the method, although the slip shown to illustrate the relationships is higher than should be used in practice.

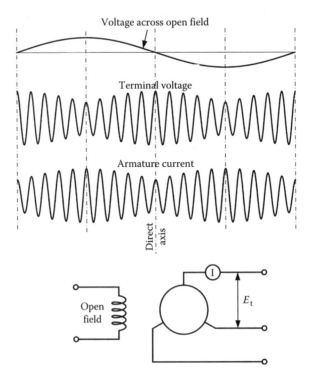

FIGURE 5.26
Slip method of measuring *q*-axis synchronous reactance.

The minimum and maximum ratios of the armature voltage to the armature current are obtained when the slip is very small. The approximate values of d- and q-axes synchronous impedances are obtained as follows:

$$X_{ds} = \frac{E_{max}}{I_{min}} \text{ pu (a certain saturated value)} \tag{5.135a}$$

$$X_{qs} = \frac{E_{min}}{I_{max}} \text{ pu (a certain saturated value)} \tag{5.135b}$$

The most accurate method is to determine the d-axis synchronous reactance (X_{du}) from OCC and SCC curves of the machine, and then to obtain the q-axis synchronous reactance by means of the following equation [2]:

$$X_{qu} = X_{du} \left(\frac{E_{min}}{E_{max}} \right) \left(\frac{I_{min}}{I_{max}} \right) \text{pu} \tag{5.136}$$

5.9.3 Short Circuit Tests

Three short circuit test procedures are presented in this part. The sudden short circuit test at no load, used in the first two procedures, provide the information required for a second-order d-axis model in terms of X'_d, X''_d, T'_d, and T''_d. In the last procedure the sudden short circuit is applied when the machine is on load at low voltage and it provides information for a second-order q-axis model in terms of X'_q, X''_q, T'_q, and T''_q. Readers are referred to Refs. [2,3] for more details.

5.9.3.1 Sudden Short Circuit with the Machine Operating Open-Circuited at Rated Speed

The sudden three-phase short circuit is applied while the machine is operating open-circuited at rated speed. Since this operation imposes severe mechanical stress on the machine, the number of these tests is limited and they are performed at reduced voltage to eliminate the risk of damage. Therefore, the parameters are obtained for lower voltage levels (up to 65% of the rated voltage) and then extrapolated to the rated voltage to obtain the saturated values [18].

During the tests, the field voltage should be kept constant. Hence, these tests cannot be performed when the excitation cannot be supplied from a constant voltage low-impedance source.

The error resulting from minor speed changes is negligible provided the machine is operating at rated speed at the instant the voltages are measured, just before the short circuit. If the initial speed deviates slightly from rated speed, correction may be made by multiplying the voltage E by the ratio of rated speed to actual speed.

When a three-phase short circuit is applied to the open-circuited machine terminals operating at rated speed there will be no q-axis component of flux. Hence, the synchronous reactance (X_d), the transient reactance (X'_d), and the subtransient reactance (X''_d), as well as the transient time constant (T'_d) and the subtransient time constant (T''_d), are required to describe the machine behavior. By neglecting armature resistances and assuming constant exciter voltage, the ac rms components of the current following a three-phase short circuit at no load can be expressed as follows:

$$I = \frac{E}{X_d} + \left(\frac{E}{X_d'} - \frac{E}{X_d} \right) e^{\frac{-t}{T_d'}} + \left(\frac{E}{X_d''} - \frac{E}{X_d'} \right) e^{\frac{-t}{T_d''}}$$

(5.137)

where
  I is the ac rms short circuit current
  E is the ac rms voltage before the short circuit
  t is the time measured from the instant of short circuit

This expression is based on the second-order d-axis model. Hence, it is assumed that the current is composed of a constant term and two decaying exponential terms. By subtracting the first (constant) term and plotting the remainder on semilog paper as a function of time, the curve would appear as a straight-line after the rapidly decaying term decreases to zero. The rapidly decaying portion of the curve is the subtransient portion, while the straight-line is the transient portion, see Figures 5.11 and 5.12.

As the objective is to find X_d, X_d', X_d'', T_d', and T_d'' values in Equation 5.137, Figure 5.12 is analyzed after subtracting out the dc components from each phase. The per unit values of the dc and the time-variant ac rms components of the three-phase short circuit currents can be calculated from the upper and lower envelopes of the currents as follows:

$$I = \frac{\left(I_{\text{upp}} - I_{\text{low}} \right)}{2\sqrt{2} I_{\text{abcB}}}$$

(5.138a)

$$I_{dc} = \frac{\left(I_{\text{upp}} + I_{\text{low}} \right)}{2\sqrt{2} I_{\text{abcB}}}$$

(5.138b)

where I_B is the base current (see Section 5.4).
  Equation 5.137 may be rewritten as follows:

$$I = I_{ss} + I' + I''$$

(5.139)

where
  I_{ss} is the steady state
  I' is the transient
  I' is the subtransient components of I; i.e.,

$$I_{ss} = \frac{E}{X_d}$$

(5.140a)

$$I' = \left(\frac{E}{X_d'} - \frac{E}{X_d} \right) e^{\frac{-t}{T_d'}}$$

(5.140b)

$$I'' = \left(\frac{E}{X_d''} - \frac{E}{X_d'} \right) e^{\frac{-t}{T_d''}}$$

(5.140c)

T''_d is typically small when compared to T'_d, and the subtransient term becomes negligible after a few cycles. Therefore, by subtracting the steady-state term $(I - I_{ss})$ and plotting the remainder with semilog scale, the curve would appear as a straight-line after the rapidly decaying term decreases to zero as illustrated with Curve B in Figure 5.27. The transient component can be approximated by a straight-line (Line C in Figure 5.27) and X'_d can be determined by the zero-time intercept, while T'_d is the slope of this line. T'_d is the time, in seconds, required for the transient alternating component of the short circuit current (line C in Figure 5.27), to decrease to $1/e$ or 0.368 times its initial value.

The subtransient component can be approximated as the difference between Curve B and Line C as shown in Figure 5.27 (Curve A). The subtransient component can be approximated by a straight-line (Line D in Figure 5.27); X''_d is determined from the zero-time intercept, and T''_d is the slope of Line D. T''_d is the time, in seconds, required for the subtransient ac component of the short circuit current (line D in Figure 5.27) to decrease to $1/e$ or 0.368 times its initial value.

Although Figure 5.27 shows an analysis based on only one of the three phases, normally the average of the phase currents is used.

It should be noted that the two exponential functions only approximate the true current behavior and the actual short circuit current may not follow the above form of variation due to several factors, such as saturation and eddy current effects. Therefore, the transient portion of the semilog plot may actually be slightly curved. Any relatively short portion of this curved line can be well approximated by a straight-line, and depending on the choice of the short portion, the slope of this line varies. Since the transient reactance, X'_d, is determined from the zero-time intercept of this line, its value is somewhat arbitrary. In order to establish a test procedure that will produce a definite transient reactance and hence definite transient and subtransient time constants, the range of time needed to make the semilog plot is established as the first second following the short circuit.

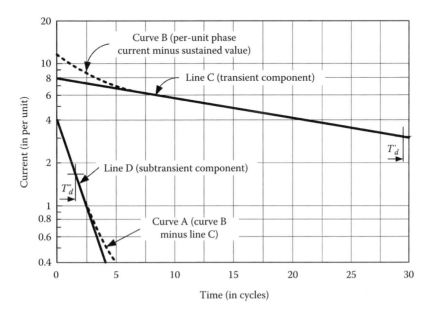

FIGURE 5.27
Analysis of ac components of the short circuit test. (From ANSI/IEEE Std 115-1995, IEEE Guide: Test Procedures for Synchronous Machines. With permission.)

The *d*-axis short circuit transient time constant can be corrected to a specified temperature by using the following formula [2]:

$$T'_d = T'_{dt} \left(\frac{k + t_t}{k + t_s} \right)$$ (5.141)

where

T'_d is the *d*-axis transient short circuit time constant at specified temperature
T'_{dt} is the *d*-axis transient short circuit time constant at test temperature
t_t is the temperature of the field winding before test, in °C
t_s is the specified temperature, in °C
k is 234.5 for pure copper, in °C
k is 225 for aluminum based on a volume conductivity of 62% of pure copper, in °C

No correction for temperature is included for the *d*-axis subtransient short circuit time constant because of the uncertain nature of the correction.

The short circuit armature time constant can be found by analyzing the dc components of the armature currents. The dc component in per unit is calculated for each time value using the following equation [2]:

$$I_{dc} = \sqrt{\frac{4}{27}\left(a^2 + b^2 - ab\right)} + \sqrt{\frac{4}{27}\left(a^2 + c^2 - ac\right)} + \sqrt{\frac{4}{27}\left(b^2 + c^2 - bc\right)}$$ (5.142)

where

a is the largest value of the dc components of the three-phase currents at the selected time, in p.u.
b is the second largest value of the dc components, in p.u.
c is the smallest value, in p.u.

If the values of the currents deduced from Equation 5.142 are plotted as a function of time in a semilog plot, the initial current is obtained by extrapolating the curve back to the moment of the short circuit. The armature time constant is then determined from the time where current reduces to $1/e$ of its initial value.

The armature time constant is usually specified at temperature 75°C. Therefore, it should be corrected using the above formula:

$$T_a = T_{at} \left(\frac{k + t_t}{k + t_s} \right)$$ (5.143)

where T_{at} is the armature time constant at test temperature.

Example 5.5

In order to explain the method, a computer simulation of a sudden three-phase short circuit test is performed by using the EMTP-RV. The line-to-line voltage before the short circuit is 5.75 kV. The parameters of the machine are those given in Table 5.9. The simulated short circuit currents are depicted in Figure 5.28.

TABLE 5.9

Example 5.5: Machine Parameters

S (MVA)	125
V (kV, rms line-to-line)	10.5
X_d (p.u.)	2.21
X'_d (p.u.)	0.225
X''_d (p.u.)	0.176
T'_d (s)	0.77
T''_d (s)	0.0133

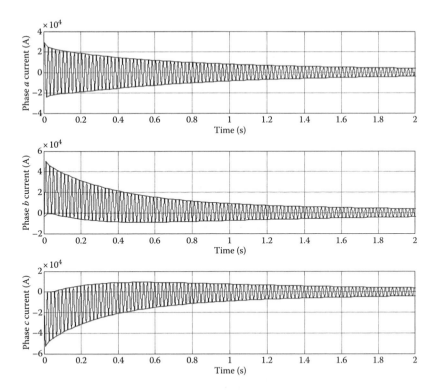

FIGURE 5.28
Example 5.5: Three-phase short circuit currents.

The ac rms short circuit current given in Figure 5.29 is determined from the average of the envelope of the fundamental-frequency ac components which are obtained after subtracting the dc components of the currents.

By subtracting the steady-state term from the ac rms short circuit current $(I - I_{steady\text{-}state})$ and plotting the remainder as a function of time with a semilog scale, the curve appears as a straight-line after the rapid decaying. The transient component can be approximated by this straight-line as illustrated in Figure 5.30. The zero-time intercept of this line is approximately 2.186, and X'_d can be found as follows:

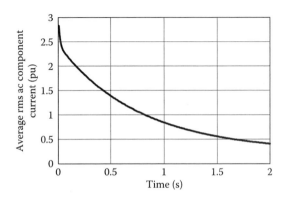

FIGURE 5.29
Example 5.5: ac rms short circuit current.

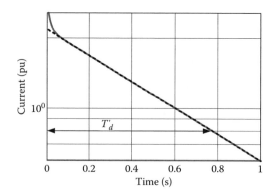

FIGURE 5.30
Example 5.5: ac short circuit current depicted in Figure 5.29 minus steady-state value.

$$X_d' = \left(\frac{E}{I' + \dfrac{E}{X_d}} \right) = \frac{0.5476}{2.186 + \dfrac{0.5476}{2.21}} = 0.225 \, \text{pu}$$

According to Figure 5.30, the transient component decreases to 1/e of its initial value (zero-time intercept of the line) after $T_d' = 0.77$ s.

The difference between the current illustrated in Figure 5.30 and the straight-line approximation of the transient component is plotted with a semilog scale and the subtransient component can be approximated with the straight-line illustrated in Figure 5.31.

The zero-time intercept of the straight-line approximation of the subtransient component is 0.6776, and X_d'' can be found as follows:

$$X_d'' = \left(\frac{E}{I'' + \dfrac{E}{X_d'}} \right) = \frac{0.5476}{0.6776 + \dfrac{0.5476}{0.225}} = 0.176 \, \text{pu}$$

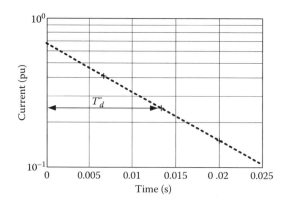

FIGURE 5.31
Example 5.5: Current depicted in Figure 5.30—straight-line approximation.

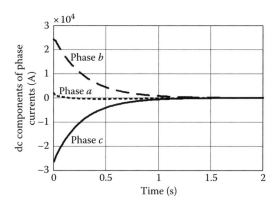

FIGURE 5.32
Example 5.5: dc components of phase currents.

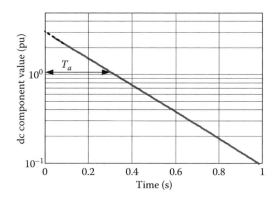

FIGURE 5.33
Example 5.5: dc component value.

According to Figure 5.31, the subtransient component decreases to 1/e of its initial point (zero-time intercept of the line) after $T''_d = 0.0133$ s.

The armature time constant can be found by analyzing the dc components of the armature currents depicted in Figure 5.32. The values resulting from Equation 5.142 are given in Figure 5.33. The armature time constant can be determined from the time where the current decreases to 1/e of its initial value; in this case $T_{at} = 0.28$ s.

5.9.3.2 Sudden Short Circuit of Armature and Field with the Machine Operating Open-Circuited at Rated Speed

In this method, a three-phase short circuit is suddenly applied to the armature of the machine simultaneously with a short circuit to the field winding while the machine is operating at rated speed, the armature being open-circuited and the field excited with current corresponding to the desired voltage. This method may be used when the excitation cannot be supplied from a constant voltage low-impedance source. This may result from the necessity of using a remote exciter, or if the effect of the heavy transient exciter currents causes it to change the operating point on its hysteresis curve, resulting in a significantly different field current after steady state is reached from that which existed prior to the short circuit.

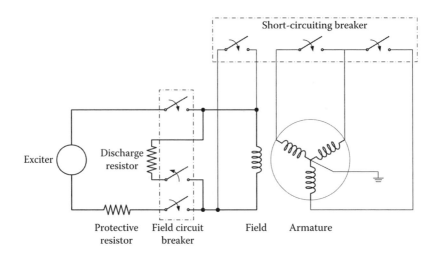

FIGURE 5.34

Connection diagram for sudden short circuit of armature and field test. (From ANSI/IEEE Std 115-1995, IEEE Guide: Test Procedures for Synchronous Machines. With permission.)

Figure 5.34 shows the connection diagram that may be used for this test. One pole of the short-circuiting breaker may be used to short-circuit the field winding. A protective resistor is used to prevent short-circuiting the exciter. An exciter circuit breaker is controlled to open shortly after the short-circuiting breaker is closed.

Equation 5.137 applies to the current in this test if the two terms of the form E/X_d are eliminated. The steady-state armature current is zero; hence the portion of the current which decays according to the transient time constant is larger than in the previous method. The test is made and analyzed in a manner similar to previous method.

5.9.3.3 Sudden Short Circuit with Machine Operating on Load at Low Voltage

Since the sudden short circuit is applied in the previous procedures at no-load conditions, there will be no initial q-axis component of flux. Therefore, they can be used for determining d-axis transient and subtransient characteristic quantities. In order to determine q-axis transient and subtransient characteristic quantities, the sudden short circuit must be applied when the machine is running on load with the armature winding connected to a rated-frequency symmetrical three-phase low voltage supply (at about 10% of its nominal value) and the excitation winding is short-circuited.

The voltage should be chosen properly in order to load the machine up to 90 ± 20 electrical degrees and to prevent damage during short-circuiting. Following the short circuit, the voltage supply is disconnected.

The d- and q-axes components of voltage and current are determined by means of positioning signals on the shaft. The positioning signals are set to correspond to the quadrature (magnetic axis) by means of a line-to-line voltage at no load. The transmitter producing the positioning signal should be displaced round the shaft until its signal coincides with the instant when the line-to-line voltage between phases 1 and 2, V_{12} passes through zero. At this position the signal will coincide with the maximum of the third phase voltage, V_3, which, at no load, corresponds to a q-axis voltage. When the machine is loaded, the instantaneous values of the phase voltage V_3 and of the phase current I_3, coincide with the q-axis components of this voltage and current.

The test is performed by short-circuiting the machine after it has reached an angle approaching 90 electrical degrees ($t = 0$). It may also slip very slowly. The q-axis current is obtained by using the measured phase current I_3, while the d-axis voltage is obtained from the measured line-to-line voltage V_{12} at the time of the q-axis signal. The values of $I_q(0)$ and $E_d(0)$, and following the short circuit, I_q values (i.e., I_3 values at the time of the q-axis signal), are measured on an oscillograph. The plot of current value against time on a natural time scale is to be obtained from the oscillogram; the initial values $\Delta I_q'(0)$ and $\Delta I_q''(0)$, as well as variations of transient $\Delta I_q'$ and subtransient $\Delta I_q''$ current components with respect to time, are determined and drawn on a semilog plot.

The q-axis transient and subtransient inductances can be obtained from the test results as follows:

$$X_q' = \frac{E_d(0)}{\left(I_q(0) + I_q'(0)\right)} \tag{5.144a}$$

$$X_q'' = \frac{E_d(0)}{\left(I_q(0) + \Delta I_q'(0) + \Delta I_q''(0)\right)} \tag{5.144b}$$

where
$\quad I_q(0)$ and $E_d(0)$ are the q-axis current and d-axis voltage at the time of short circuit
$\quad \Delta I_q'(0)$ and $\Delta I_q''(0)$ are the transient and subtransient changes in the q-axis current at the time of short circuit

The q-axis transient time constant (T_q') is the time required for the transient component (i.e., the slowly changing component of q-axis short circuit current, $\Delta I_q'$) to decrease to $1/e$ or 0.368 times its initial value. The time required for the subtransient component (i.e., the rapid changing component of q-axis short circuit current, $\Delta I_q''$) to decrease to $1/e$ or 0.368 times its initial value is the q-axis subtransient time constant, T_q''.

5.9.4 Decrement Tests

Three decrement test procedures are briefly presented in this part. The first procedure is based on the removal of a short circuit and can be used to determine the parameters of a second-order d-axis model, in terms of X_d', X_d'', T_{d0}', and T_{d0}''. The second and third procedures are based on the disconnection of the machine from a low-voltage supply; both of them can be used to obtain the parameters of a second-order q-axis model, in terms of X_q', X_q'', T_{q0}', and T_{q0}''. Readers are referred to Ref. [2] for more details.

5.9.4.1 Voltage Recovery Test

This test is performed by the sudden removal of a steady-state three-phase short circuit at stator terminals when the machine is running at rated speed with a selected excitation value. During the test, the circuit breaker should open all three phases as simultaneously as possible.

By neglecting armature resistances and assuming constant exciter voltage, the rms voltage following the sudden removal of a steady-state three-phase short circuit of the armature is as follows:

$$E = I_0 X_d + I_0 \left(X_d' - X_d\right)e^{\frac{-t}{T_{d0}'}} + I_0 \left(X_d'' - X_d'\right)e^{\frac{-t}{T_{d0}''}} \tag{5.145}$$

where
   I_0 is the rms short circuit current, in p.u.
   E is the root-mean-square voltage after short circuit removal, in p.u.
   t is the time measured from the instant of short circuit removal, in s

Equation 5.145 may be rewritten as

$$E = E_{ss} + E' + E'' \tag{5.146}$$

where E_{ss}, E', and E'' are respectively the steady state, the transient, and the subtransient components of E; i.e.

$$E_{ss} = I_0 X_d \tag{5.147a}$$

$$E' = I_0 \left(X'_d - X_d \right) e^{\frac{-t}{T'_{d0}}} \tag{5.147b}$$

$$E'' = I_0 \left(X''_d - X'_d \right) e^{\frac{-t}{T''_{d0}}} \tag{5.147c}$$

Defining the differential voltage, E_Δ, as the difference between the average of the three rms voltages obtained from the oscillogram and the average of the three rms steady-state voltages; i.e.

$$E_\Delta = E_{ss} - E = -(E' + E'') \tag{5.148}$$

and plotting it with a semilog scale, the curve would appear as a straight-line except during the first few cycles of rapid change, as illustrated in Figure 5.35 (Curve B). The transient component of the differential voltage is the slowly varying portion; by neglecting the first few cycles of rapid change, it can be approximated with a straight-line, Line C in Figure 5.35.

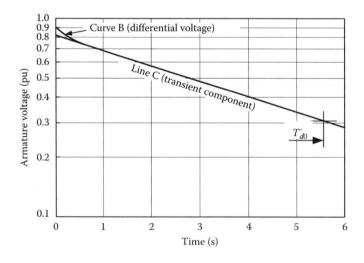

FIGURE 5.35
Analysis of the voltage recovery test for calculation X'_d and T'_{d0}. (From ANSI/IEEE Std 115-1995, IEEE Guide: Test Procedures for Synchronous Machines. With permission.)

Let $E'_{\Delta 0}$ be the zero-time intercept of Line C, then X'_d can be obtained as

$$X'_d = \frac{E_{ss} - E'_{\Delta 0}}{I_0} \tag{5.149}$$

T'_{d0} is the time, in seconds, required for the transient component of the differential voltage (line C in Figure 5.35) to decrease to $1/e$ or 0.368 times its initial value.

If the speed of the machine differs from the rated speed or varies during the test, it is necessary, before calculating the differential voltage, to correct voltages as follows:

$$E_C = E \frac{n_R}{n_t} \tag{5.150a}$$

$$E_{ssC} = E_{ss} \frac{n_R}{n_{ss}} \tag{5.150b}$$

where
E is the measured voltage as a function of time
E_C is the corrected value of E
E_{ss} is the steady-state voltage
E_{ssC} is the corrected steady-state voltage
n_R is the rated speed
n_t is the speed at the time E is measured (it can be obtained approximately by linear interpolation from initial speed to the steady-state speed)
n_{ss} is the speed corresponding to the steady-state voltage

The subtransient voltage (Curve A in Figure 5.36) can be obtained by subtracting the transient component of the differential voltage (Line C) from the differential voltage (Curve B) and approximated with a straight-line (line D in Figure 5.36). Let $E''_{\Delta 0}$ be the zero-time intercept of Line D, then X''_d can be obtained as follows:

$$X''_d = \frac{E_{ss} - E'_{\Delta 0} - E''_{\Delta 0}}{I_0} \tag{5.151}$$

T''_{d0} is the time, in seconds, required for the transient component of differential voltage (line D in Figure 5.36), to decrease to $1/e$ or 0.368 times its initial value.

Example 5.6

To illustrate the method, a computer simulation of a voltage recovery test for the machine having the parameters given in Example 5.5 is performed by using EMTP-RV. In the simulations, the excitation voltage is 0.1 p.u. and the machine speed is kept constant.

The simulated phase a current (i_a) and the line-to-line voltage (v_{ab}) during the first second period are depicted in Figure 5.37. To analyze the test results for calculation of X'_d and T'_{d0}, the average of the rms differential voltage ($E_\Delta = E_{ss} - E$) is plotted with a semilog scale in Figure 5.38 and zoomed in Figure 5.39. X'_d can be then obtained as follows:

$$X'_d = \frac{E_{ss} - E'_{\Delta 0}}{I_0} = \frac{10^{-1} - 10^{-1.0466}}{0.0452} = 0.225 \text{ pu}$$

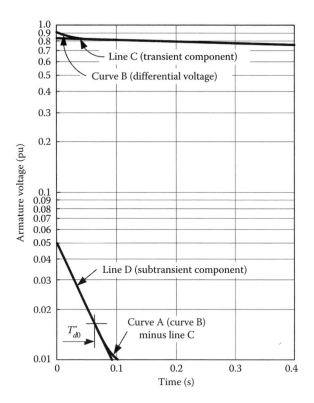

FIGURE 5.36
Analysis of the voltage recovery test for calculation of X_d'' and T_{d0}''. (From ANSI/IEEE Std 115-1995, IEEE Guide: Test Procedures for Synchronous Machines. With permission.)

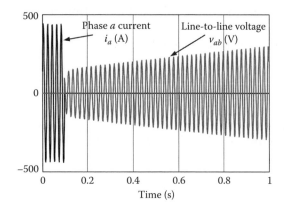

FIGURE 5.37
Example 5.6: EMTP-RV simulation of the voltage recovery test—phase a current (i_a) and line-to-line voltage (v_{ab}) during the first second period.

From Figure 5.38, it can be seen that the transient component decreases to $1/e$ times its initial value (zero-time intercept of the line) after $T_{d0}' = 7.56\,\text{s}$.

The subtransient component of the rms differential voltage can be obtained by subtracting the transient component of the differential voltage (Line B in Figure 5.39) from the differential voltage (Curve A in Figure 5.39) as illustrated in Figure 5.40. X_d' can be then obtained as follows:

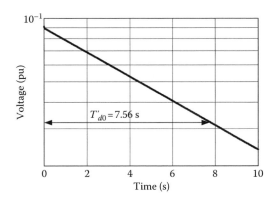

FIGURE 5.38
Example 5.6: Average rms differential voltage.

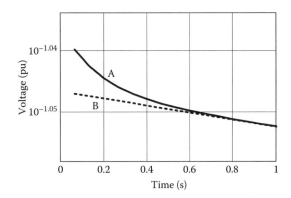

FIGURE 5.39
Example 5.6: Zoomed version of Figure 5.38—Curve A: average rms differential voltage; Line B: straight-line approximation of the transient component.

$$X_d'' = \frac{E_{ss} - E_{\Delta 0}' - E_{\Delta 0}''}{I_0} = \frac{10^{-1} - 10^{-1.0466} - 10^{-2.6542}}{0.0452} = 0.176 \text{ pu}$$

From Figure 5.40, it can be seen that the subtransient component decreases to $1/e$ times of its initial value (zero-time intercept of the line) after $T_{d0}'' = 0.017\,\text{s}$.

The ac rms short circuit current given in Figure 5.29 is determined from the average of the envelope of the fundamental frequency ac components which are obtained after subtracting the dc components of the currents.

5.9.4.2 Disconnecting the Armature with the Machine Running Asynchronously on Load

In this method, the rated-frequency symmetrical three-phase low-voltage supply is suddenly disconnected when the machine is running asynchronously on load and the excitation winding is short-circuited to avoid any saturation effect.

The positioning signals on the shaft are adjusted in a similar way to that described in Section 5.9.3.3. The test is performed by disconnecting the machine after it has reached an internal angle approaching 90 ± 20 electrical degrees. The q-axis current is obtained from

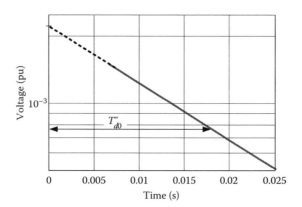

FIGURE 5.40
Example 5.6: Subtransient component of the average rms differential voltage.

the measured phase current I_3, and the d-axis voltage is obtained from the measured line-to-line voltage V_{12} at the time of the q-axis signal. The values of $I_q(0)$, $E_d(0)$, and, following the disconnection of the machine, E_d (i.e., the V_{12} value at the time of the q-axis signal) are measured. The voltage vs. time plot on a natural time scale is to be obtained from the oscillogram; the initial values $\Delta E_q'(0)$ and $\Delta E_q''(0)$, as well as the variations of transient $\Delta E_q'$ and subtransient $\Delta E_q''$ voltage components, with respect to time, are determined and drawn on a semilog scale.

5.9.4.3 Disconnecting the Armature at a Very Low Slip

In this method, the rated-frequency symmetrical three-phase low-voltage supply (5%–10% of normal voltage) is suddenly disconnected when the machine is running at a slip considerably less than 0.01 p.u., being the excitation circuit open. The sudden disconnection should be performed when the rotor is magnetized in the q-axis. The rotor position is checked by measuring the internal angles between the armature voltage and the quadrature magnetic rotor axis. Armature current and voltage, as well as rotor position indication, are measured and recorded.

At the instant of switching off the machine from the low-voltage supply, the armature (primary) winding voltage suddenly drops to a particular value and then gradually decays. This initial voltage drop is independent of the residual voltage. In determining time constants, the residual voltage must be less than 0.2 of the applied voltage, and accordingly its value need not be taken into account for determining time constants along the q-axis with the required accuracy.

In order to determine quantities in d-axis, excitation circuit should be short-circuited and the sudden disconnection should be performed when the rotor is magnetized in the d-axis.

5.9.5 Standstill Frequency Response Tests

5.9.5.1 Introduction

Short circuit and decrement test presented in Sections 5.9.3 and 5.9.4 provides the second-order transfer functions of d- and q-axes operational inductances ($L_d(s)$, $L_q(s)$). However, armature to field transfer function, $G(s)$, should be taken into account when defining machine model and structure for precise modeling of the field circuit and excitation effects. Moreover, depending on the rotor construction, equivalent circuits of order higher than Model 2.2 (see Table 5.2) may be more appropriate in some cases.

In order to perform an accurate identification of synchronous machine parameters, SSFR testing has been developed. The actual procedures for testing are described in Section 12 of ANSI/IEEE Std 115-1995 [2]. These tests are becoming a widely used alternative to short circuit tests due to the advantages listed below:

- SSFR testing can be performed either in the factory or in site at a relatively low cost.
- Equivalent circuits of order higher than Model 2.2 (see Table 5.2) can be derived.
- Identification of field responses is possible.

SSFR tests as a mean to estimate synchronous machine parameters were introduced to avoid onerous short circuit tests, to enable q-axis parameters to be measured, and parameters of high-order models to be determined. Since this technique was first proposed in 1956 [19], SSFR tests have played an important role in the determination of synchronous machine parameters. Refs. [19–49] are just a sample of the effort dedicated to this subject during the last 30 years. The list should be also complemented by those works motivated by the improvements suggested for a better modeling of synchronous machines. Ref. [21] provides a short story of the work performed since 1956.

5.9.5.2 SSFR Test Measurements

Before SSFR testing, the machine shall be shut down, disconnected from its turning gear, and electrically isolated. Since the SSFR tests are performed independently for direct and quadrature axes, it is also necessary to align the rotor to two particular positions with respect to the stator.

Table 5.10 taken from Ref. [50] shows the test setups for each of the measurable parameters and the main relationships derived from each test. Figure 5.41 depicts typical responses for $L_d(s)$, $L_q(s)$, and $sG(s)$.

- $Z_d(s)$ and $Z_q(s)$ are the direct and quadrature operational impedances as viewed from the armature terminals. The operational inductances can be computed by subtracting the armature resistance from these impedances:

$$L_d(s) = \frac{Z_d(s) - R_a}{s} \tag{5.152a}$$

$$L_q(s) = \frac{Z_q(s) - R_a}{s} \tag{5.152b}$$

where R_a is the dc resistance of one armature phase and $s = j\omega$.
- The function $sG(s)$ rather than $G(s)$ is used since it can be measured at the same time as $Z_d(s)$.
- The parameters of d-axis equivalent circuit can be obtained using the transfer functions $Z_d(s)$ and $sG(s)$.

TABLE 5.10

Standard SSFR Tests

Test Diagram	Measurement	Relationships
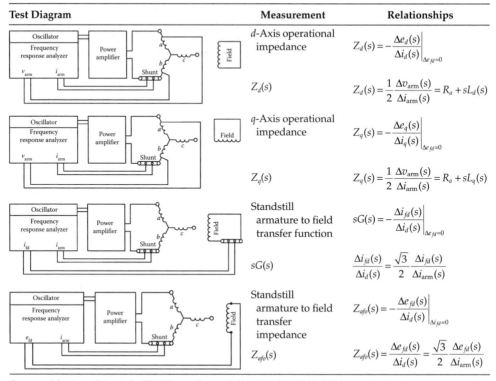	*d*-Axis operational impedance	$Z_d(s) = -\dfrac{\Delta e_d(s)}{\Delta i_d(s)}\bigg\|_{\Delta e_{fd}=0}$
	$Z_d(s)$	$Z_d(s) = \dfrac{1}{2}\dfrac{\Delta v_{\text{arm}}(s)}{\Delta i_{\text{arm}}(s)} = R_a + sL_d(s)$
	q-Axis operational impedance	$Z_q(s) = -\dfrac{\Delta e_q(s)}{\Delta i_q(s)}\bigg\|_{\Delta e_{fd}=0}$
	$Z_q(s)$	$Z_q(s) = \dfrac{1}{2}\dfrac{\Delta v_{\text{arm}}(s)}{\Delta i_{\text{arm}}(s)} = R_a + sL_q(s)$
	Standstill armature to field transfer function	$sG(s) = -\dfrac{\Delta i_{fd}(s)}{\Delta i_d(s)}\bigg\|_{\Delta e_{fd}=0}$
	$sG(s)$	$\dfrac{\Delta i_{fd}(s)}{\Delta i_d(s)} = \dfrac{\sqrt{3}}{2}\dfrac{\Delta i_{fd}(s)}{\Delta i_{\text{arm}}(s)}$
	Standstill armature to field transfer impedance	$Z_{afo}(s) = -\dfrac{\Delta e_{fd}(s)}{\Delta i_d(s)}\bigg\|_{\Delta i_{fd}=0}$
	$Z_{afo}(s)$	$Z_{afo}(s) = \dfrac{\Delta e_{fd}(s)}{\Delta i_d(s)} = \dfrac{\sqrt{3}}{2}\dfrac{\Delta e_{fd}(s)}{\Delta i_{\text{arm}}(s)}$

Source: Martinez, J.A. et al., *IEEE Trans. Power Deliv.*, 20, 2063, 2005. With permission.

- *q*-Axis equivalent circuit parameters can be obtained using the transfer function $Z_q(s)$.
- The armature to field transfer impedance, $Z_{afo}(s)$, is used for determining the effective stator-to-rotor turns ratio.

5.9.5.3 Parameter Identification from SSFR Test Results

The procedure for identification of *d*-axis parameters from SSFR can be summarized as follows [2,50]:

1. The best available estimate for the armature leakage inductance L_ℓ is the value supplied by the manufacturer.

2. Obtain $L_d(0)$, the low-frequency limit of $L_d(s)$, then determine

$$L_{ad}(0) = L_d(0) - L_\ell \tag{5.153}$$

3. As the $L_{ad}(0)$ is determined, by using $Z_{afo}(s)$, the field-to-armature turns ratio can be found as

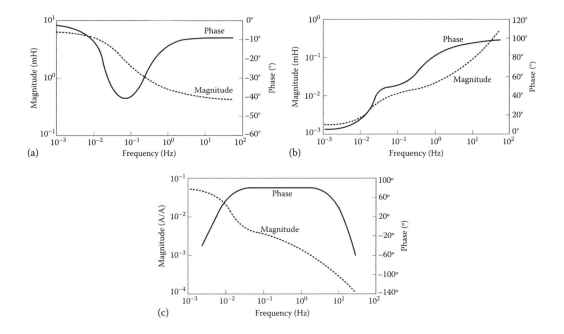

FIGURE 5.41
Typical responses of SSFR tests. (a) $L_d(j\omega)$ response; (b) $L_q(j\omega)$ response; (c) $j\omega G(j\omega)$ response.

$$N_{af}(0) = \frac{1}{sL_{ad}(0)} \lim_{s \to 0} \left[\frac{\Delta e_{fd}(s)}{\Delta i_d(s)} \right] \qquad (5.154)$$

For a discussion on the factors to be used for adjusting this value, see Ref. [2].

4. Calculate the field resistance referred to the armature winding

$$R_F = \frac{sL_{ad}(0)}{\lim\limits_{s \to 0} \left[\dfrac{\Delta i_{fd}(s)}{\Delta i_d(s)} \right] \dfrac{2}{3} N_{af}(0)} \qquad (5.155)$$

5. Choose an equivalent circuit structure for the d-axis.
6. Use the available parameters and a fitting technique to find values for the unknown parameters that produce the best fit for $L_d(s)$ and $sG(s)$.
7. Adjust L_{ad} to its unsaturated value L_{adu}.

SSFR tests are performed using very low currents (typically 40 A), compared to rated armature current. Therefore, the low-level iron nonlinearity cannot be ignored. That is, the values of iron-dependent inductance measured during SSFR tests will be lower than unsaturated values on the air-gap line. Therefore, L_{ad} and L_{aq} in the equivalent circuits derived to match standstill test data need to be adjusted upward to achieve an unsaturated model for the machine. The unsaturated value for L_{ad} can be calculated from the rated-speed open circuit curve [2]:

$$L_{adu} = \frac{3}{2}\left(\frac{1}{N_{af}(0)}\right)\left(\frac{V_t}{\omega I_{fd}}\right)$$ (5.156)

where

V_t and I_{fd} define a point on the air-gap line

ω is the electrical speed in electrical radians per second

V_t is the line-to-neutral peak voltage and I_{fd} is in dc A. L_{adu} is substituted for L_{ad} in d-axis equivalent circuit calculated from SSFR data. L_{aq} can be adjusted to its unsaturated value by multiplying by $\dfrac{L_{adu}}{L_{ad}}$; i.e., the same factor used in the d-axis.

8. Measure the field winding resistance, convert it to the desired operating temperature, and refer it to the stator

$$R_F \text{ at } t_s = \left[\frac{k+t_s}{k+t_t}\right][R_{Ft}]\left[\frac{3}{2}\frac{1}{N_{af}(0)}\right]^2$$ (5.157)

where

R_{Ft} is the field resistance measured at field terminals

t_t is average field temperature during measurement (°C)

t_s is the specified temperature (°C)

k is defined in Equation 5.141

Note that L_{adu} and R_F substitute the parameters used to find unknown values during the fitting procedure. Once the parameters have been determined they can be normalized to per unit values.

The procedure for identification of q-axis parameters from SSFR can be summarized as follows [2]:

1. Use the armature leakage inductance L_ℓ supplied by the manufacturer.
2. Obtain $L_q(0)$, the low-frequency limit of $L_q(s)$, then determine

$$L_{aq}(0) = L_q(0) - L_\ell$$ (5.158)

3. Choose an equivalent circuit structure for the d-axis.
4. Use the available parameters and a fitting technique to find values for the unknown parameters that produce the best fit for $L_q(s)$.
5. Adjust L_{aq} to its unsaturated value L_{aqu}.

As with the d-axis, these equivalent circuit parameters can be normalized to per unit values.

The most complex step in the detailed procedure is the application of a curve-fitting technique to derive the value of the equivalent circuit parameters that match SSFR curves. Many techniques have been proposed for this purpose [2]. One of the first techniques was the nonlinear least squares Marquadt–Levenberg algorithm [22]. Although this or similar algorithms are still used, the gained experience has fired different approaches to cope with some of the inconveniences related to SSFR tests.

Other techniques presented for parameters estimation from frequency response during the last years are based on a noniterative parameter-identification procedure [43] or pattern search methods [2].

Noise is always present in field tests and it can significantly impact on parameter estimation from SSFR data [34]. To solve this aspect a maximum likelihood estimation technique was proposed [33]: transfer function models are first determined using a least square error method; they are then transformed into time-domain responses; finally, a maximum likelihood technique is used to estimate parameters as a constrained minimization problem.

Not much work has been made on the influence of saturation on the parameter estimation from SSFR data [45], although it is recognized that it can be important [38,39].

A permanent subject of discussion has been the order of the equivalent circuits. One of the advantages of manipulating frequency response data is that the complexity of the equivalent circuits is not fixed. Several works have been dedicated to analyze the improvement in accuracy that higher-order equivalent circuits can introduce; see, for instance [38–40,44].

Salient-pole machines have several features distinct from round–rotor machines [46–48]: the ratio L_q/L_d is about 0.6, as opposed to nearly unity for round rotor machines; a salient-pole armature winding is usually designed with a fractional number of slots per pole; most salient-pole machines have damper windings in the pole faces; the field winding in round rotor machines is distributed while in salient-pole machines it is concentrated. The positioning of the rotor for salient-pole machines has been a challenge when using SSFR measurements for parameter estimation, since a small error in the mechanical alignment could result in a large error in the electrical degree. A methodology for estimation of salient-pole machine parameters was presented [49]; it avoids the requirement of a precise rotor mechanical alignment to obtain accurate parameter values from standstill measurements. The experience gained with SSFR testing and analysis of salient-pole machines was summarized in Ref. [48]; see also Refs. [46,47].

SSFR tests have disadvantages also [20,39,50]:

- Eddy current losses on the armature resistance during the SSFR are not accounted for when the operating reactances are deduced using the following expression

$$sL_d(s) = Z_d(s) - R_a \qquad (5.159)$$

  where R_a is the dc armature resistance.
- The test equipment requires very linear, very high-power amplifiers.
- Standstill measurements are made at low currents; however, $sL_d(s)$ and $sL_q(s)$ can vary up to 20% in the range from no load to rated current.
- Tests are conducted at unsaturated conditions.
- Centrifugal forces on damper windings are not accounted for, it being difficult to assess the error introduced by them.
- The resistance in the contact points of damper windings can be higher at standstill than it is during running.

5.9.6 Online Tests

Online tests have been developed in order to avoid some of the drawbacks and limitations of off-line tests. These tests can be either time-domain or frequency-domain tests. Some of these tests are briefly described in the following paragraphs [50].

5.9.6.1   Online Frequency Response Test

This test is carried out with the machine running at rated speed and near rated or at reduced load. The frequency response is obtained by modulating the excitation system with sinusoidal signals and measuring the steady-state changes in field voltage and current, rotor speed, terminal voltage, and active and reactive power outputs [20,51].

5.9.6.2   Load Rejection Test

This test is carried out with the machine running at rated speed, with power injected to the system as near to zero as possible and the excitation system on manual control [20]. The generator circuit breaker is opened and terminal voltage, field voltage, and field current are recorded. This test is performed for both under- and overexcited conditions in order to obtain unsaturated and saturated values.

5.9.6.3   Time-Domain Small Disturbance Test

The linear parameters of the machine are identified from lightly loaded, underexcited conditions, while saturation characteristics are identified from a wide range of operating conditions [52].

5.9.6.4   Time-Domain Large Disturbance Test

A sudden large disturbance is introduced in the excitation reference voltage while the machine is operating under normal conditions [53]. The recorded variables are terminal and field voltages, armature and field currents, as well as rotor speed.

Other online methods are based on a multivariable linear transfer function [54], or the use of an observer for damper winding currents [55–57].

5.10   Models for High-Frequency Transient Simulations

5.10.1   Winding Models

Steep-fronted transient voltages on multiturn coils of rotating machines can be caused by lightning or by switching operations; fast front or very fast front transients can be produced by circuit breakers with a high current-chopping level and capable of interrupting very high frequency currents (e.g., vacuum circuit breakers), see [58–63]. The characteristics of the transient depend upon the circuit-breaker technology, the operating conditions of the machine, and the characteristics of the system (e.g., the length of the cable connection). Since switching of synchronous generators do not occur very often, and these machines are well protected against lightning, most often these transients are related to switching of large ac motors; in general induction-type machines.

As for transformer windings (see Chapter 4, Section 4.5), the voltage distribution along the machine winding depends greatly on the waveshape of the applied voltage: at low frequencies the distribution is linear along the windings; in case of steeped transients, a larger portion of the voltage applied is distributed on the first few turns of the winding.

Several computational models have been developed to analyze and predict the distribution and magnitude of the stresses originated by fast front or very fast front transients [64–75]. Regardless of the selected model, the representation of inductive, capacitive, and loss components is required.

In the short period of time (i.e., a few microseconds) after the surge arrival to the machine terminals, the flux penetration into the core can be neglected, considering that the core acts as a flux barrier at high or very high frequencies, and the core inductance is represented as a completely linear element. However, this may not be true in some cases where flux penetration dynamics in the core should be taken into account; for instance, when switching a three-phase machine, a suitable representation for the flux penetration into the iron core is needed for longer periods of time in order to predict the effect of the second and the third pole closures [72]. Ref. [71] did show that assuming that slot walls act as flux barriers may be acceptable when the coil structure is embedded in a solid slot core but it is not valid for a laminated slot core, which can allow deeper magnetic flux penetration than a solid slot core. In fact, laminated iron does not act as a flux barrier at frequencies below 20 MHz [76].

Frequency dependence of both series and shunt elements of the windings at high frequencies should be considered: skin and proximity effects produce frequency dependence of winding and core impedances because of the reduced flux penetration; at very high frequencies, the conductance representing the winding capacitive loss also depends on frequency.

A first classification of computational models may be based on the simulation goal: models can be aimed at either determining voltages at machine terminals or predicting surge distribution in machine windings (e.g., interturn voltages, line-end coil voltages). For the first case, a simple lumped-parameter representation of the machine will suffice [77,78]; to analyze surge distribution in machine windings, more sophisticated approaches are required [64–75]. Models developed for surge distribution in windings use either a lumped-parameter equivalent circuit [65,66] or a multiconductor distributed-parameter line model [64,75]. Although there are some differences between multiconductor line models, they have several common features: (1) each coil is divided into two parts (slot and overhang), each part being segmented into sections with uniform geometry (see Figure 5.42); (2) a phase winding may be modeled by cascading coil models.

A short description of some of the proposed models is provided in the following paragraphs.

1. A machine winding is represented in [66] by a lumped-parameter equivalent circuit, as depicted in Figure 5.43. The circuit parameters are grouped to form a

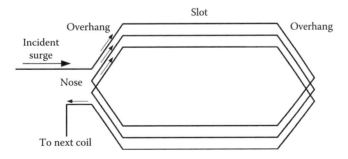

FIGURE 5.42
Rotating machine coil model.

circuit with nodes between adjacent turns at the coil connection end and at the entrance and exit from the coil.

2. Slot and overhang (end) sections are modeled as pi-equivalent circuits with parameters determined by means of an EMTP supporting routine; Ref. [65] uses Cable Constants (see Chapter 3). A coil consists of two slot sections and two end connections. One phase of the armature winding consists of series connection of the phase coils. Finally, the complete machine model consists of the winding models for the three phases.

3. The winding model is divided into two main blocks as shown in Figure 5.44 [75]: (1) the first one is based on a multiconductor frequency-independent distributed-parameter line model and represents capacitances-to-ground, insulation flux and mutual

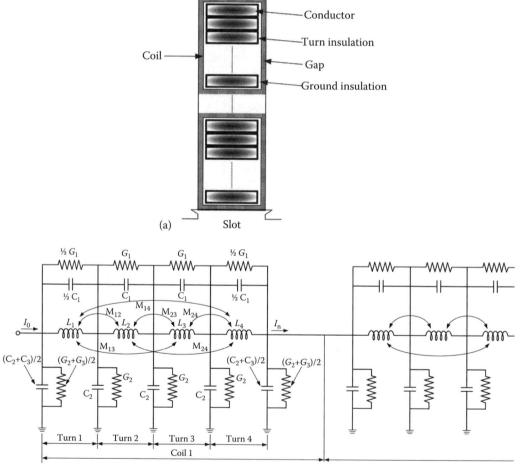

C_1, G_1 = turn-to-turn capacitance and conductance
C_2, G_2 = turn-to-ground capacitance and conductance
C_3, G_3 = additional capacitance and conductance to account for end conditions at the top and the bottom of the coil sides

FIGURE 5.43
Lumped-parameter equivalent circuit of a machine winding. (a) Slot configuration cross coil section; (b) equivalent circuit. (From Rhudy, R.G. et al., *IEEE Trans. Energy Conversion*, 1, 50, 1986. With permission.)

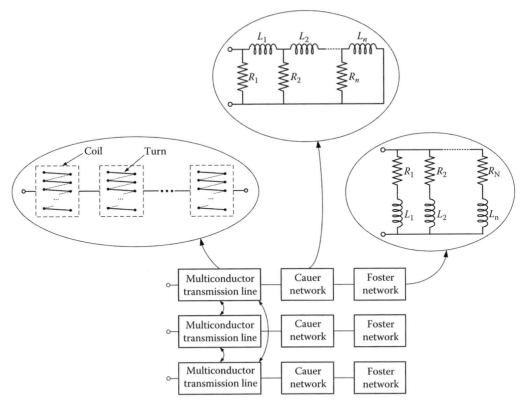

FIGURE 5.44
Multiconductor line model with frequency-dependence of winding parameters. (From Guardado, J.L. et al., *IEEE Trans. Energy Conversion*, 20, 322, 2005. With permission.)

coupling between coils; (2) the second block is based on Cauer and Foster networks that represent respectively the frequency dependence of winding parameters due to flux penetration into the iron core (Cauer network) and the skin effect in the copper conductors (Foster network).

5.10.2 Calculation of Electrical Parameters

Several approaches have been used for the calculation of electrical parameters to be specified in machine winding models adequate for fast front and very fast front transients [68–75]. As an example, the expressions proposed for application in the model depicted in Figure 5.44 are presented in this section, see Ref. [75].

- *Calculation of the parallel admittance*

  The parallel admittance of the slot section is

$$Y = j\omega C \tag{5.160}$$

where the capacitance to ground in the slot section C is calculated by considering the coil walls as parallel plates

$$C = \frac{\varepsilon l p_c}{d} \tag{5.161}$$

ε is the insulation permittivity, l is the total slot length, p_c is the coil perimeter, and d is the insulation thickness.

- *Calculation of the series impedance*

  The expression for the series impedance must account for the flux penetration into the iron core and the skin effect in conductors. The general expression per unit length may be as follows:

$$Z = \left(R_{ic} + R_{sk}\right) + j\omega\left(L_{ic} + L_{sk} + L_{in}\right) \tag{5.162}$$

where the subscripts "*ic*," "*sk*," and "*in*" are used to denote iron core, skin effect, and insulation, respectively.

The self-inductance of the coil due to the flux penetration into the iron core per unit length can be approximated by

$$L_{ic} = \frac{\mu n^2 p}{z} \tag{5.163}$$

where
μ is the iron permeability
n is the number of turns of the coil
z is the slot perimeter
p is the depth penetration, which is given by the following expression:

$$p = \sqrt{\frac{\rho}{j\omega\mu}} \tag{5.164}$$

where ρ is the iron resistivity.
The iron losses are represented by a series resistance whose expression in per unit length is as follows:

$$R_{ic} = \rho\frac{l}{pz} \tag{5.165}$$

where l is again the coil length in the iron.
The contribution due to skin effect in conductors can be calculated by means of the following expression:

$$Z_{sk} = \frac{l}{A\xi\sigma} \tag{5.166}$$

where
A is the conductor cross section
σ is the conductor conductivity
ξ is the complex skin depth given by

$$\xi = \sqrt{\frac{2(1+j)}{\omega\mu\sigma}} \qquad (5.167)$$

Finally, the inductance due to the flux in the insulation (between turns and the slot walls and wedges) is approximated by means of

$$L_{in} = \frac{\mu_o}{2\pi}n^2\ln\left(\frac{2h}{r}\right) \qquad (5.168)$$

where
  μ_o is the permeability in vacuum
  h is half the slot width
  r is the radius of the equivalent circular conductor that substitutes the actual rectangular

This equivalent radius is given by

$$r = 0.2235(a+b) \qquad (5.169)$$

where a and b are the height and the width, respectively.

For more details about the mathematical formulation of the machine winding model, see the original references and also Chapter 4 of this book.

5.11 Synchronous Machine Mechanical System Equations

The single-mass representation of the turbine–generator rotor is adequate when analyzing power system dynamic performance that accounts for the oscillation of the entire turbine–generator rotor with respect to other generators. However, the turbine–generator rotor has a very complex mechanical structure consisting of several rotors with different sizes connected by shafts of finite stiffness. Therefore, when the generator is perturbed, the rotor masses will oscillate relative to one another at one or more of the rotor's mechanical natural frequencies called torsional mode frequencies, depending on the perturbation. Torsional oscillations in subsynchronous range may interact with the electrical system under certain conditions, and detailed representation of turbine–generator rotor is required while studying special problems related to torsional oscillations such as subsynchronous resonance.

Although a continuum model of the rotor is required to account for the complete range of torsional oscillations, a simple lumped multimass model is adequate for studying problems related to torsional oscillations [79]. In this approach, each major rotor element is considered to be a rigid mass connected to adjacent elements by massless shafts with a single equivalent torsional stiffness constant. This lumped multimass model of the rotor has the torsional modes in the frequency range of interest.

5.11.1 Lumped Multimass Model

Figure 5.45 illustrates the structure of a typical lumped multimass model of a generator driven by a tandem compound steam turbine. The six torsional masses represent the rotors

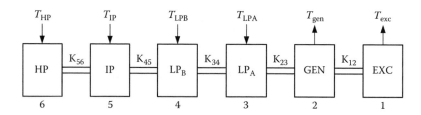

FIGURE 5.45
Structure of a typical lumped multimass shaft system.

of the exciter, the generator, two low-pressure (LP) turbine sections, the intermediate-pressure (IP) turbine section, and the high-pressure (HP) turbine section, respectively.

The shaft system can be considered as a mass–spring–damper system and its characteristics are defined by three set of parameters: the moment of inertia of each individual mass, the torsional stiffness of each shaft section, and the damping coefficients associated with each mass.

There are a number of sources contributing to the mechanical damping of torsional oscillations:

A. Steam forces on turbine blades, which are represented by damping torques proportional to the speed deviations of individual masses from synchronous mechanical speed. The corresponding damping coefficient changes depending on generator loading.

B. Bearing friction and windage on shaft elements, which are represented by damping torques proportional to the angular speed of individual masses.

C. Shaft material hysteresis, which are represented by damping torques proportional to the angular speed difference between the adjacent masses.

From the rotational form of the Newton's second law, the equation for the ith mass connected by elastic shaft sections to mass $(i - 1)$ and $(i + 1)$ is given by

$$J_i \frac{d\omega_i}{dt} + D_{asi}\omega_i + D_{ssi}(\omega_i - \omega_0) + D_{mi,i-1}(\omega_i - \omega_{i-1}) + D_{mi,i+1}(\omega_i - \omega_{i+1})$$

$$+ K_{i,i-1}(\theta_i - \theta_{i-1}) + K_{i,i+1}(\theta_i - \theta_{i+1}) = T_{mi} - T_{ei} \tag{5.170}$$

$$\frac{d\theta_i}{dt} = \omega_i$$

where

J_i is the moment of inertia of the ith mass, in kg m^2

$K_{i,i-1}$ is the torsional stiffness of shaft section between the ith and the $(i - 1)$th mass, in N m

D_{ssi} is the speed deviation self-damping coefficient of the ith mass, in N m s (coefficient corresponding to component (A) of mechanical damping, see above)

D_{asi} is the absolute speed self-damping coefficient of the ith mass, in N m s (coefficient corresponding to component (B) of mechanical damping, see above)

$D_{mi,i-1}$ is the mutual-damping coefficient between the ith and the $(i-1)$th mass, in N m s (coefficient corresponding to component (C) of mechanical damping, see above)

T_{mi} is the mechanical torques developed by the ith turbine sections (T_{HP}, T_{IP}, T_{LPB}, T_{LPA}), in N m

T_{ei} is the electromagnetic torques of either the generator or the exciter (T_{gen}, T_{ex}), in N m

ω_0 is the synchronously rotating reference, in rad/s

ω_i is the speed of ith mass, in rad/s

θ_i is the angular position of ith mass, in radians

The angular position in Equation 5.170 is expressed with respect to a stationary frame. However, in steady state, torsional masses are stationary with respect to synchronously rotating frame. Therefore, it might be more convenient to rewrite the equation of motion in terms of angular position of masses, with respect to a synchronously rotating frame, and of speed deviation of masses. Let $\delta_i = \theta_i - \omega_0 t$, $\Delta\omega_i = \omega_i - \omega_0$; the coefficient D_{si} is the damping coefficient representing the total damping torques due to (A) and (B), which were represented separately with coefficients D_{ssi} and D_{asi} in Equation 5.170, which becomes

$$J_i \frac{d\Delta\omega_i}{dt} + D_{si}\Delta\omega_i + D_{mi,i-1}\left(\Delta\omega_i - \Delta\omega_{i-1}\right) + D_{mi,i+1}\left(\Delta\omega_i - \Delta\omega_{i+1}\right)$$

$$+ K_{i,i-1}\left(\delta_i - \delta_{i-1}\right) + K_{i,i+1}\left(\delta_i - \delta_{i+1}\right) = T_{mi} - T_{ei} \tag{5.171}$$

$$\frac{d\delta_i}{dt} = \Delta\omega_i$$

While analyzing problems related to torsional oscillations, it might be necessary to obtain electrical and mechanical system data on a common base. The shaft system equation in per unit can be written as

$$\frac{1}{\omega_{mB}}\left[\bar{J}\right]\frac{d}{dt}\Delta\bar{\omega} + \left[\bar{D}\right]\Delta\bar{\omega} + \left[\bar{K}\right]\bar{\delta} = \bar{T}_m - \bar{T}_e$$

$$\frac{1}{\omega_{mB}}\frac{d\bar{\delta}_i}{dt} = \Delta\bar{\omega}_i \tag{5.172}$$

or alternatively:

$$\frac{1}{\omega_{mB}^2}\left[\bar{J}\right]\frac{d^2\bar{\delta}}{dt^2} + \frac{1}{\omega_{mB}}\left[\bar{D}\right]\frac{d\bar{\delta}}{dt} + \left[\bar{K}\right]\bar{\delta} = \bar{T}_m - \bar{T}_e \tag{5.173}$$

where the superbar indicates per unit values.

The base quantities of the mechanical system are as follows:

Base mechanical power, $S_{mB} = S_B$, VA

Base mechanical angular frequency, $\omega_{mB} = \omega_B \dfrac{2}{p}$, rad/s

Base mechanical angle, $\theta_{mB} = 1\,\text{rad}$

Base mechanical torque, $T_{mB} = \dfrac{S_{mB}}{\omega_{mB}}$, N·m

Base moment of inertia, $J_{mB} = \dfrac{T_{mB}}{\omega_{mB}^2}$, kg m^2

Base damping coefficient, $D_{mB} = \dfrac{T_{mB}}{\omega_{mB}}$, N·m·s

Base stiffness coefficient, $K_{mB} = \dfrac{T_{mB}}{\theta_{mB}}$, N·m

In Equation 5.173 $[\bar{J}]$ is the diagonal inertia matrix, $[\bar{D}]$ and $[\bar{K}]$ are tridiagonal symmetric matrices of damping and stiffness coefficients, respectively. $\overline{T}_m$ and $\overline{T}_e$ are the vectors of mechanical and electrical torques.

5.11.2 Mechanical System Model in Terms of Modal Parameters

The mechanical damping is determined from decrements tests; therefore, it is available in terms of modal damping. Moreover, the system model in terms of modal parameters is uncoupled, and it might be a more convenient utilization of the modal mass–spring–damper model described in terms of modal inertia, modal stiffness, and modal damping.

The modal parameters can be obtained through transformation of Equation 5.173. By neglecting the damping, that equation reduces to

$$\frac{1}{\omega_{mB}^2}\left[\bar{J}\right]\frac{d^2\bar{\delta}}{dt^2}+\left[\bar{K}\right]\bar{\delta}=\overline{T}_m-\overline{T}_e=\overline{T} \tag{5.174}$$

Let $[Q]$ be a transformation defined by

$$\bar{\delta}=[Q]\bar{\delta}^m \tag{5.175}$$

After substituting Equation 5.175 into Equation 5.174, and multiplying both sides by $[Q]^t$, the following result is derived

$$[Q]^t\left[\bar{J}\right][Q]p^2\bar{\delta}^m+[Q]^t\left[\bar{K}\right][Q]\bar{\delta}^m=[Q]^t\overline{T} \tag{5.176}$$

or

$$\left[\bar{J}^m\right]p^2\bar{\delta}^m+\left[\bar{K}^m\right]\bar{\delta}^m=[Q]^t\overline{T} \tag{5.177}$$

where

$$p=\frac{1}{\omega_{mB}}\frac{d}{dt} \tag{5.178}$$

Both $[\overline{J}^m]$ and $[\overline{K}^m]$ can be written in a diagonal form with proper choice of $[Q]$. The choice of $[Q]$ can be made by considering the properties of eigenvectors of real symmetric matrices. The stiffness matrix $[\overline{K}]$ is a real symmetric matrix having only real eigenvalues, which are mutually orthogonal. Hence, if $[C]$ is a matrix whose columns are eigenvectors of a real symmetric matrix $[\overline{K}']$ defined by

$$\left[\overline{K}'\right] = \left[\overline{J}\right]^{-\frac{1}{2}}\left[\overline{K}\right]\left[\overline{J}\right]^{-\frac{1}{2}} \tag{5.179}$$

then $[C]^t [C] = [D1]$ is a diagonal matrix.

It can be also shown that

$$\left[\overline{K}'\right][C] = [C][\Lambda] \tag{5.180}$$

where $[\Lambda]$ is the diagonal matrix of eigenvalues of $[\overline{K}']$, which are the squares of torsional mode frequencies in per unit. Multiplying both sides of Equation 5.180 by $[C]^t$, the following equation is obtained

$$[C]^t\left[\overline{K}'\right][C] = [C]^t[C][\Lambda] = [D1][\Lambda] = [\Lambda'] \tag{5.181}$$

Hence, if $[Q]$ is chosen as

$$[Q] = \left[\overline{J}\right]^{-\frac{1}{2}}[C] \tag{5.182}$$

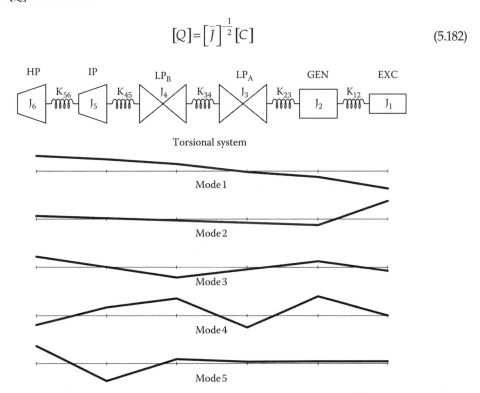

FIGURE 5.46
Typical mode shapes of torsional mechanics.

then

$$[Q]^{\dagger}[\bar{J}][Q]=[C]^{\dagger}[\bar{J}]^{\frac{1}{2}}[\bar{J}][\bar{J}]^{-\frac{1}{2}}[C]=[C]^{\dagger}[C]=[D1]=\left[\bar{J}^{m}\right]$$

(5.183a)

$$[Q]^{\dagger}\left[\bar{K}\right][Q]=[C]^{\dagger}[\bar{J}]^{-\frac{1}{2}}\left[\bar{K}\right][\bar{J}]^{-\frac{1}{2}}[C]=[C]^{\dagger}\left[\bar{K}'\right][C]=[\Lambda']=\left[\bar{K}^{m}\right]$$

(5.183b)

The column vectors of matrix [Q] are also called mode shapes of torsional motion. The elements of each vector indicate the relative rotational displacements of the individual turbine–generator rotor elements when the mechanical system oscillates at the corresponding natural frequency under steady-state conditions. There are several common ways of performing normalization of the mode shapes, and hence modal spring–mass model data is not unique. One method assigns the value unity to the generator mass location. Another assigns values such that one or more values are unity and all others are less than unity. Figure 5.46 illustrates typical mode shapes of a turbine–generator shaft system. Each mode shape is normalized by assigning unity to the largest element.

In general, a turbine–generator rotor with N masses has $N - 1$ torsional modes, and they are numbered sequentially according to mode frequency and number of phase reversals in the mode shape. Mode 1 has the lowest mode frequency and only one polarity reversal in the mode shape. Mode i has the ith lowest torsional frequency and the corresponding mode shape has i polarity reversals. The mode corresponding to the rigid body is Mode 0 and the participation of each mass is equal in this mode.

If each vector of [Q] is normalized by assigning the value unity to the generator mass location, then the linearized mechanical system equation for each mode can be expressed as follows:

$$\bar{J}^{i}p^{2}\Delta\bar{\delta}^{i}+\bar{K}^{i}\Delta\bar{\delta}^{i}=\Delta\bar{T}_{e} \qquad i=0,1,2,\ldots,(N-1)$$

(5.184)

In this equation, the mechanical torque produced by each turbine section is assumed to be constant.

From Equation 5.184, the angular frequency of each torsional mode can be written as

$$\omega^{i}=\sqrt{\frac{\omega_{mB}^{2}\bar{K}^{i}}{\bar{J}^{i}}}$$

(5.185)

It should be noted that, for Mode 0, $\bar{J}^{0}=\sum_{i=1}^{N}\bar{J}_{i}$, $\bar{K}^{0}=0$, and hence $\omega^{0}=0$.

Torsional mode damping quantifies the rate of decay of torsional oscillations at a torsional mode frequency and can be expressed in several ways. The most easily measured quantity is the ratio of successive peaks of oscillation; the natural logarithm of this ratio is known as the logarithmic decrement, or log-dec. For slow decay, the log-dec is approximately equal to the fraction of decay per cycle. A more accurate measure is the time in seconds for the envelope of decay to decrease to the fraction $1/e$ of its value from an earlier point in time. This measure is the time constant of decay. The inverse of the time constant is defined as the decrement factor (σ^i) and is equal to the mode frequency in hertz multiplied by the log-dec [80].

If damping is included, Equation 5.184 can be rewritten as

$$\overline{J}^i p^2 \Delta \overline{\delta}^i + \overline{D}^i p \Delta \overline{\delta}^i + \overline{K}^i \Delta \overline{\delta}^i = \Delta \overline{T}_e \qquad i = 0, 1, 2, ..., (N-1) \tag{5.186}$$

From Equation 5.186, the decrement factor corresponding to Mode i (σ^i), can be written as

$$\sigma^i = \frac{\omega_{mB} \overline{D}^i}{2 \overline{J}^i} \tag{5.187}$$

The mechanical damping is determined from decrements tests; i.e., the decrement factors corresponding to each mode (σ^i), are available. Thus, the modal damping values ($\overline{D}^i$), required for modal mass–spring–damper model, can be found from Equation 5.187.

An alternative approach to the mechanical system model in terms of model parameters has been presented in Ref. [81], in which the decoupled mechanical system model is obtained from an electrical analogy.

Example 5.7

The rotor mass–spring parameters of a 892.4 MVA, 3600 r/min, generation unit taken from Ref. [82] are given in Table 5.11.

Since the rotor is composed of six masses, there are five torsional modes. The mode shapes of torsional motion and torsional frequencies are illustrated in Figure 5.47. As seen from this figure, mode i has the ith lowest torsional frequency and the corresponding mode shape has i polarity reversals. The mode corresponding to rigid body is Mode 0 having 0 Hz torsional frequency, the participation of each mass being equal in this mode.

TABLE 5.11

Example 5.7: Rotor Mass–Spring Parameters of a 892.4 MVA, 3600 r/min Generation Unit

Mass	J (p.u.)	Shaft Section	K (p.u.)
HP	70.0429	HP-IP	19.303
IP	117.312	IP-LPA	34.929
LPA	647.423	LPA-LPB	52.038
LPB	666.684	LPB-GEN	70.858
GEN	654.831	GEN-EX	2.822
EX	25.799		

Source: IEEE Subsynchronous Resonance Task Force of the Dynamic System Performance Working Group, *IEEE Trans. Power Apparatus Syst.*, 96(5), 1565, 1977. With permission.

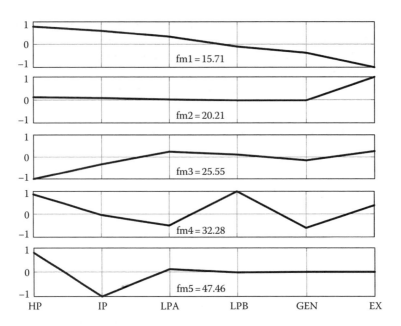

FIGURE 5.47

Example 5.7: Rotor torsional frequencies and mode shapes.

5.11.3 Determination of Mechanical Parameters

As mentioned in the previous section, the single-mass representation is adequate when analyzing power system dynamic performance that accounts for the oscillation of the entire turbine–generator rotor with respect to other generators, and the lumped multimass model is required when studying special problems related to torsional oscillations, such as subsynchronous resonance.

Manufacturers can provide turbine–generator models that are adequate for preliminary studies, but these estimates have shown as much as 10% difference between calculated and measured natural frequencies [83]. Moreover, damping values determine whether a particular unit faces a torsional interaction problem requiring capital expenditures to control. Therefore, identification of torsional parameters of the rotor might be indispensable to provide an accuracy that eliminated uncertainties in the models and study results.

In general, these tests can be divided into two groups [84]:

1. The first group of tests is performed on the turbogenerator at standstill, and includes harmonic torque excitation [85] and engagement of the shaft turning gear. However, supplying an excitation force of sufficient magnitude to a large turbogenerator shaft is not an easy task.

2. The second group of tests uses the turbogenerator's electrical torque to excite the shaft, and is performed with the turbogenerator synchronized to the network. These include manual out-of-phase synchronization, sinusoidal variation of the exciter power, and a system-switching disturbance.

The tests that fall in the first group provide the required parameters of the spring–mass system for the turbine–generator shaft. However, as stated in Section 5.11.1, the mechanical damping varies with the load of the unit and it is necessary to determine the mechanical damping values at different load levels. On the other hand, when one of the tests of the second group is performed, the generator is synchronized and the spring–mass system effectively has an additional spring and damper between the generator and a fixed reference (the infinite bus). This additional spring and damper represent respectively the synchronizing torque coefficient and the electrical damping, and they are functions of load, system impedance, and machine parameters. Therefore, synchronized and unsynchronized generator cases have different sets of mode parameters.

The electrical spring coefficient is much smaller when compared to the mechanical constants of the shaft. Therefore, the torsional frequencies and the mode shapes do not change significantly. The most visible change occurs for Mode 0, which corresponds to the rigid mass. Its value increases to around 1 Hz from 0 Hz when the generator is synchronized. On the other hand, the electrical damping may not be small when compared to the mechanical damping values of the shaft. Therefore, measured damping values, including the combined effects of both the mechanical and the electrical system damping, may be very different when compared to actual mechanical damping values. Many studies related to torsional interactions require mechanical damping values as input data to the study, and hence, some care is required when translating the test data to the studies to prevent representing electrical torsional damping effects twice.

The details regarding these tests, as well as the procedures to be followed, can be found in Refs. [84–87].

5.11.4 Determination of Damping Factors of the Mass–Spring–Damper Model

In general, the mechanical damping is determined from decrements tests, and therefore, it is available in terms of modal damping. However, most EMTP-type programs require the mass–spring–damper model of the shaft system. A trial-and-error method has been proposed in Ref. [88] with a reasonable success. That method takes no-load and full-load decrement factors corresponding to each mode as inputs and outputs of mutual- and self-damping coefficients. As the generator loading increases, the steam forces on turbine blades, and hence the corresponding components of self-damping coefficients increase. There is very little damping present due to steam forces under no-load conditions. Based on this fact, the method assumes that the self-damping coefficients are negligible when compared to the mutual-damping coefficients at no-load conditions. The proposed method may be summarized as follows:

- Step 1: Obtain no-load and full-load test data.
- Step 2: Set self-damping coefficients to zero ($D_{si} = 0$).
- Step 3: Make an initial guess for mutual-damping coefficients (D_{mij}).
- Step 4: Compute eigenvalues of the mechanical system.
- Step 5: Compare the real part of the eigenvalues corresponding to the particular modes with the no-load modal damping test data. If they are sufficiently close, go to Step 6; otherwise, revise mutual-damping coefficients, and go to Step 4.
- Step 6: Use the resulting mutual-damping coefficients and make an initial guess for self-damping coefficients.
- Step 7: Compute eigenvalues of the mechanical system.

- Step 8: Compare the real part of the eigenvalue that corresponds to each particular mode with the full-load modal damping test data. If they are sufficiently close, stop the procedure; otherwise, revise self-damping coefficients, and go to Step 7.

The efficiency of the algorithm can be improved by making the initial guesses for the mutual-damping coefficients proportional to the differences between relative displacements of the masses connected to each end of the particular shaft. The guess for self-damping factors for each turbine can be proportional to the turbine contribution fractions, since the self-damping factor for each turbine is approximately proportional to its MW power divided by the generator MVA output [89].

Appendix A. Third-Order Equivalent Circuits

Data conversion algorithms for parameter determination of third-order *d*- and *q*-axes equivalent circuits are presented in the sections of this appendix. Each section includes the available input information, the scheme of the equivalent circuit that can be determined, and the different steps of the corresponding data conversion algorithm.

5.A.1   Algorithm for Parameter Calculation from $G(s)$ and $L_d(s)$

Data input and equivalent circuit (Figure 5.A.1)

$$G(s) = \frac{L_{ad}}{R_F} \frac{(1+sT_{kd1})(1+sT_{kd2})}{(1+sT'_{d0})(1+sT''_{d0})(1+sT'''_{d0})}$$

$$L_d(s) = L_d \frac{(1+sT'_d)(1+sT''_d)(1+sT'''_d)}{(1+sT'_{d0})(1+sT''_{d0})(1+sT'''_{d0})}$$

Data conversion algorithm
Step 1

$$L_{ad} = L_d - L_{al}$$

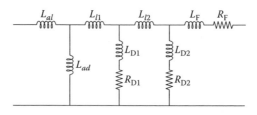

FIGURE 5.A.1
Complete third-order *d*-axis equivalent circuit.

Step 2: Calculate intermediate parameters

$$R_0 = \frac{L_{ad}^2}{L_d \left(T'_{d0} + T''_{d0} + T'''_{d0}\right) - \left(T'_d + T''_d + T'''_d\right)}$$

$$a = \frac{L_d \left(T'_d + T''_d + T'''_d\right) - L_{al} \left(T'_{d0} + T''_{d0} + T'''_{d0}\right)}{L_{ad}}$$

$$b = \frac{L_d \left(T'_d T''_d + T'_d T'''_d + T''_d T'''_d\right) - L_{al} \left(T'_{d0} T''_{d0} + T'_{d0} T'''_{d0} + T''_{d0} T'''_{d0}\right)}{L_{ad}}$$

$$c = \frac{L_d T'_d T''_d T'''_d - L_{al} T'_{d0} T''_{d0} T'''_{d0}}{L_{ad}}$$

$$d = \frac{\left(T'_{d0} T''_{d0} + T'_{d0} T'''_{d0} + T''_{d0} T'''_{d0}\right) - \left(T'_d T''_d + T'_d T'''_d + T''_d T'''_d\right)}{\left(T'_{d0} + T''_{d0} + T'''_{d0}\right) - \left(T'_d + T''_d + T'''_d\right)}$$

$$e = \frac{T'_{d0} T''_{d0} T'''_{d0} - T'_d T''_d T'''_d}{\left(T'_{d0} + T''_{d0} + T'''_{d0}\right) - \left(T'_d + T''_d + T'''_d\right)}$$

Step 3: Calculate parameter L_{l1}

$$L_{l1} = R_0 \frac{c - T_{kd1}\left[b - T_{kd1}\left(a - T_{kd1}\right)\right]}{e + T_{kd1}(T_{kd1} - d)}$$

Step 4: Calculate intermediate parameters

$$A = a - T_{kd1} - \frac{L_{l1}}{R_0}$$

$$B = a - A T_{kd1} - \frac{d L_{l1}}{R_0}$$

Step 5: Calculate parameters R_{D1} and L_{D1}

$$R_{D1} = R_0 \frac{A T_{kd1} - B - T_{kd1}^2}{d T_{kd1} - e - T_{kd1}^2}$$

$$L_{D1} = R_{D1} T_{kd1}$$

Step 6: Calculate intermediate parameters

$$R_p = \frac{R_{D1} R_0}{R_{D1} - R_0}$$

$$T_p = \frac{R_p}{T_{kd1}}\left(\frac{e}{R_0} - \frac{B}{R_{D1}}\right)$$

Step 7: Calculate parameter L_{l2}

$$L_{l2} = R_p \frac{AT_{kd2} - B - T_{kd2}^2}{T_{kd2} - T_p}$$

Step 8: Calculate intermediate parameter

$$T_f = A - T_{kd2} - \frac{L_{l2}}{R_p}$$

Step 9: Calculate the remaining parameters

$$R_{D2} = R_p \frac{T_f - T_{kd2}}{T_p - T_{kd2}}$$

$$L_{D2} = R_{D2}T_{kd2}$$

$$R_F = \frac{R_{D2}R_p}{R_{D2} - R_p}$$

$$L_F = R_F T_f$$

5.A.2 Algorithm for Parameter Calculation from $L_d(s)$ and L_C

Data input and equivalent circuit (Figure 5.A.2)

$$L_d(s) = L_d \frac{(1 + sT_d')(1 + sT_d'')(1 + sT_d''')}{(1 + sT_{d0}')(1 + sT_{d0}'')(1 + sT_{d0}''')}$$

$$L_C$$

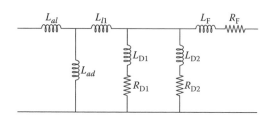

FIGURE 5.A.2
Third-order *d*-axis equivalent circuit.

Data conversion algorithm
Step 1

$$L_{ad} = L_d - L_{al}$$

Step 2

$$L_{l1} = \frac{L_{ad}(L_C - L_{al})}{L_{ad} - L_C + L_{al}}$$

Step 3: Calculate $Z_{eq}(s)$:

$$Z_{eq}(s) = \frac{sL_{ad}[L_d(s) - L_{al}]}{[L_d - L_d(s)]} - sL_{l1}$$

Step 4: Put $Z_{eq}(s)$ in the following form:

$$Z_{eq}(s) = R_p \frac{(1 + sT_{w1})(1 + sT_{w2})(1 + sT_{w3})}{(1 + sT_m)(1 + sT_n)}$$

Step 5: Obtain R_{w1}, R_{w2}, R_{w3}, L_{w1}, L_{w2}, and L_{w3} from the following equations:

$$
\begin{bmatrix}
1 & 1 & 1 \\
T_{w2} + T_{w3} & T_{w1} + T_{w3} & T_{w1} + T_{w2} \\
T_{w2}T_{w3} & T_{w1}T_{w3} & T_{w1}T_{w2}
\end{bmatrix}
\begin{bmatrix}
\dfrac{1}{R_{w1}} \\
\dfrac{1}{R_{w2}} \\
\dfrac{1}{R_{w3}}
\end{bmatrix}
= \frac{1}{R_p}
\begin{bmatrix}
1 \\
T_m + T_n \\
T_m T_n
\end{bmatrix}
$$

$$L_{w1} = R_{w1}T_{w1}$$
$$L_{w2} = R_{w2}T_{w2}$$
$$L_{w3} = R_{w3}T_{w3}$$

Step 6: Identify field and damper winding parameters

$$T_1 = \frac{L_{ad} + L_{l1} + L_{w1}}{R_{w1}}$$

$$T_2 = \frac{L_{ad} + L_{l1} + L_{w2}}{R_{w2}}$$

$$T_3 = \frac{L_{ad} + L_{l1} + L_{w3}}{R_{w3}}$$

The highest time constant (i.e., maximum of T_1, T_2, and T_3) belongs to the excitation winding.

5.A.3   Algorithm for Parameter Calculation from $L_d(s)$

Data input and equivalent circuit (Figure 5.A.3)

$$L_d(s) = L_d \frac{(1+sT'_d)(1+sT''_d)(1+sT'''_d)}{(1+sT'_{do})(1+sT''_{do})(1+sT'''_{do})}$$

Data conversion algorithm
Step 1

$$L_{ad} = L_d - L_{al}$$

Step 2: Calculate $Z_{eq}(s)$

$$Z_{eq}(s) = \frac{sL_{ad}\left[L_d(s)-L_{al}\right]}{\left[L_d - L_d(s)\right]}$$

Step 3: Put $Z_{eq}(s)$ in the following form:

$$Z_{eq}(s) = R_p \frac{(1+sT_{w1})(1+sT_{w2})(1+sT_{w3})}{(1+sT_m)(1+sT_n)}$$

Step 4: Obtain R_{w1}, R_{w2}, R_{w3}, L_{w1}, L_{w2}, and L_{w3} from the following equations:

$$\begin{bmatrix} 1 & 1 & 1 \\ T_{w2}+T_{w3} & T_{w1}+T_{w3} & T_{w1}+T_{w2} \\ T_{w2}T_{w3} & T_{w1}T_{w3} & T_{w1}T_{w2} \end{bmatrix} \begin{bmatrix} \dfrac{1}{R_{w1}} \\ \dfrac{1}{R_{w2}} \\ \dfrac{1}{R_{w3}} \end{bmatrix} = \frac{1}{R_p}\begin{bmatrix} 1 \\ T_m+T_n \\ T_mT_n \end{bmatrix}$$

$$L_{w1} = R_{w1}T_{w1}$$
$$L_{w2} = R_{w2}T_{w2}$$
$$L_{w3} = R_{w3}T_{w3}$$

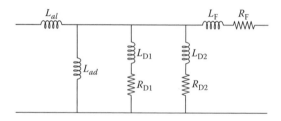

FIGURE 5.A.3
Simplified third-order *d*-axis equivalent circuit.

Step 5: Identify field and damper winding parameters

$$T_1 = \frac{L_{ad} + L_{w1}}{R_{w1}}$$

$$T_2 = \frac{L_{ad} + L_{w2}}{R_{w2}}$$

$$T_3 = \frac{L_{ad} + L_{w3}}{R_{w3}}$$

The highest time constant (i.e., maximum of T_1, T_2, and T_3) belongs to the excitation winding.

5.A.4 Algorithm for Parameter Calculation from $L_q(s)$

Data input and equivalent circuit (Figure 5.A.4)

$$L_q(s) = L_q \frac{\left(1 + sT_q'\right)\left(1 + sT_q''\right)\left(1 + sT_q'''\right)}{\left(1 + sT_{q0}'\right)\left(1 + sT_{q0}''\right)\left(1 + sT_{q0}'''\right)}$$

Data conversion algorithm
Step 1

$$L_{aq} = L_q - L_{al}$$

Step 2: Calculate $Z_{eq}(s)$

$$Z_{eq}(s) = \frac{sL_{aq}\left[L_q(s) - L_{al}\right]}{\left[L_q - L_q(s)\right]}$$

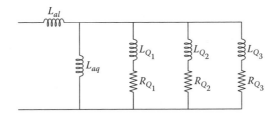

FIGURE 5.A.4
Third-order *q*-axis equivalent circuit.

Step 3: Put $Z_{eq}(s)$ in the following form:

$$Z_{eq}(s) = R_p \frac{\left(1+sT_{Q_1}\right)\left(1+sT_{Q_2}\right)\left(1+sT_{Q_3}\right)}{\left(1+sT_m\right)\left(1+sT_n\right)}$$

Step 4: Obtain R_{Q1}, R_{Q2}, R_{Q3}, L_{Q1}, L_{Q2}, and L_{Q3} from the following equations:

$$\begin{bmatrix} 1 & 1 & 1 \\ T_{Q_2}+T_{Q_3} & T_{Q_1}+T_{Q_3} & T_{Q_1}+T_{Q_2} \\ T_{Q_2}T_{Q_3} & T_{Q_1}T_{Q_3} & T_{Q_1}T_{Q_2} \end{bmatrix} \begin{bmatrix} \dfrac{1}{R_{Q_1}} \\ \dfrac{1}{R_{Q_2}} \\ \dfrac{1}{R_{Q_3}} \end{bmatrix} = \frac{1}{R_p} \begin{bmatrix} 1 \\ T_m+T_n \\ T_m T_n \end{bmatrix}$$

$$L_{Q_1} = R_{Q_1} T_{Q_1}$$
$$L_{Q_2} = R_{Q_2} T_{Q_2}$$
$$L_{Q_3} = R_{Q_3} T_{Q_3}$$

References

1. CIGRE WG 33.02, Guidelines for Representation of Network Elements when Calculating Transients, CIGRE Brochure no. 39, 1990.
2. ANSI/IEEE Std 115-1995, IEEE Guide: Test Procedures for Synchronous Machines.
3. IEC 34-4, Rotating electrical machines—Part 4: Methods for determining synchronous machine quantities from tests, 1985.
4. I. Boldea, *Synchronous Generators*, CRC Press, Boca Raton, FL, 2006.
5. P.C. Krause, O. Wasynczuk, and S.D. Sudhoff, *Analysis of Electric Machinery*, IEEE Press, New York, 1995.
6. P.M. Anderson and A.A. Fouad, *Power System Control and Stability*, IEEE Press, Piscataway, NJ, 1994.
7. P. Kundur, *Power System Stability and Control*, McGraw-Hill, New York, 1994.
8. E.W. Kimbark, *Power System Stability: Synchronous Machine*, John Wiley & Sons, New York, 1956.
9. R.H. Park, Two reaction theory of synchronous machines—Part I, *AIEE Trans.*, 48, 716–730, July 1929.
10. R.H. Park, Two reaction theory of synchronous machines—Part II, *AIEE Trans.*, 52, 352–355, June 1933.
11. M.R. Harris, P.J. Lawrenson, and J.M. Stephenson, *Per-Unit Systems with Special Reference to Electric Machines*, IEE Monograph, Cambridge University Press, London, 1970.
12. ANSI/IEEE Std 1110-1991, IEEE Guide for Synchronous Generator Modeling Practices in Stability Analysis.
13. J. Mahseredjian, *EMTP-RV Documentation*, 2007.
14. I.M. Canay, Extended synchronous machine model for the calculation of transient processes and stability, *Electric Machines and Electromagnetics*, 1, 137–150, 1977.

15. I.M. Canay, Causes of discrepancies on calculation of rotor quantities and exact equivalent diagrams of the synchronous machine, *IEEE Transactions on Power Apparatus and Systems*, 88(7), 1114–1120, July 1969.

16. K. Prubhaskankar and W. Janischewskyi, Digital simulation of multimachine power systems for stability studies, *IEEE Transactions on Power Apparatus and Systems*, 87(1), 73–80, January 1968.

17. A.W. Rankin, Per-unit impedances of synchronous machines, *AIEE Transactions*, 64, Part (i), 569–572; Part (ii), 839–841, 1945.

18. Symposium on Synchronous Machine Modeling for Power Systems Studies, IEEE Power Engineering Society, Publication 83THO101-6-PWR, 1983.

19. S.K. Sen and B. Adkins, The application of the frequency-response method to electrical machines, *Proceedings of the IEEE*, Pt. C, 103, 378–391, 1956.

20. P.M. Anderson, B.L. Agrawal, and J.E. Van Ness, *Subsynchronous Resonance in Power Systems*, IEEE Press, New York, 1990.

21. A. Walton, A systematic method for the determination of the parameters of synchronous machines from results of frequency response tests, *IEEE Transactions on Energy Conversion*, 15(2), 218–223, June 2000.

22. S.D. Umans, J.A. Malick, and G.L. Wilson, Modeling of solid rotor turbogenerators. Part I: Theory and techniques, *IEEE Transactions on Power Apparatus and Systems*, 97(1), 269–277, January/February 1978.

23. S.D. Umans, J.A. Malick, and G.L. Wilson, Modeling of solid rotor turbogenerators. Part II: Example of model derivation and use in digital simulation, *IEEE Transactions on Power Apparatus and Systems*, 97(1), 278–291, January/February 1978.

24. J.D. Hurley and H.R. Schwenk, Standstill frequency response modeling and evaluation by field tests on a 645 MVA turbine generator, *IEEE Transactions on Power Apparatus and Systems*, 100(2), 1637–1645, April 1978.

25. P.L. Dandeno and A.T. Poray, Development of detailed turbogenerator equivalent circuits from standstill frequency response measurements, *IEEE Transactions on Power Apparatus and Systems*, 100(4), 1646–1655, April 1981.

26. P.L. Dandeno, P. Kundur, A.T. Poray, and M.E. Coultes, Validation of turbogenerator stability models by comparison with power system tests, *IEEE Transactions on Power Apparatus and Systems*, 100(4), 1637–1645, April 1981.

27. M.E. Coultes and W. Watson, Synchronous machine models by standstill frequency response tests, *IEEE Transactions on Power Apparatus and Systems*, 100(4), 1480–1489, April 1981.

28. Symposium on Synchronous Machine Modeling for Power System Studies, IEEE Special Publication, 83THO101-6-PWR, 1983.

29. E. Eitelberg and R.G. Harley, Estimating synchronous machine electrical parameters from frequency response tests, *IEEE Transactions on Energy Conversion*, 2(1), 132–138, March 1987.

30. J.C. Balda, M.F. Hadingham, R.E. Fairbairn, R.G. Harley, and E. Eitelberg, Measurement of synchronous machine parameters by a modified frequency response method. Part I: Theory, *IEEE Transactions on Energy Conversion*, 2(4), 646–651, December 1987.

31. J.C. Balda, M.F. Hadingham, R.E. Fairbairn, R.G. Harley, and E. Eitelberg, Measurement of synchronous machine parameters by a modified frequency response method. Part II: Measured results, *IEEE Transactions on Energy Conversion*, 2(4), 652–657, December 1987.

32. F.P. de Mello, L.N. Hannett, and J.R. Willis, Determination of synchronous machine stator and field leakage inductances standstill frequency response tests, *IEEE Transactions on Power Systems.*, 3(4), 1625–1632, November 1988.

33. A. Keyhani, S. Hao, and G. Dayal, Maximum likelihood estimation of solid-rotor synchronous machine parameters from SSFR test data, *IEEE Transactions on Energy Conversion*, 4(3), 551–558, September 1989.

34. A. Keyhani, S. Hao, and G. Dayal, The effects of noise on frequency-domain parameter estimation of synchronous machine models, *IEEE Transactions on Energy Conversion*, 4(4), 600–607, December 1989.

35. Y. Jin and A.M. El-Serafi, A three transfer functions approach for the standstill frequency response test of synchronous machines, *IEEE Transactions on Energy Conversion*, 5(4), 740–749, December 1990.
36. A. Keyhani, S. Hao, and R.P. Schulz, Maximum likelihood estimation of generator stability constants using SSFR test data, *IEEE Transactions on Energy Conversion*, 6(1), 140–154, March 1991.
37. R.W. Merchant and M.J. Gibbard, Identification of synchronous machine parameters from standstill frequency response tests using recursive estimation with the bilinear operator, *IEEE Proceedings.*, 139, 157–165, March 1992.
38. H. Bissig, K. Reichert, and T.S. Kulig, Modelling and identification of synchronous machines. A new approach with an extended frequency range, *IEEE Transactions on Energy Conversion*, 8(2), 263–271, June 1993.
39. I.M. Canay, Determination of the model parameters of machines from the reactance operators $x_d(p)$, $x_q(p)$ (Evaluation of standstill frequency response test), *IEEE Transactions on Energy Conversion*, 8(2), 272–279, June 1993.
40. A. Keyhani and H. Tsai, Identification of high-order synchronous generator models from SSFR test data, *IEEE Transactions on Energy Conversion*, 9(3), 593–603, September 1994.
41. N.J. Bacalao, P. de Arizon, and R.O. Sanchez, A model for the synchronous machine using frequency response measurements, *IEEE Transactions on Power Systems.*, 10(1), 457–464, February 1995.
42. I. Kamwa and P. Viarouge, On equivalent circuit structures for empirical modeling of turbine-generators, *IEEE Transactions on Energy Conversion*, 9(3), 579–592, September 1994.
43. S. Henschel and H.W. Dommel, Noniterative synchronous machine parameter identification from frequency response tests, *IEEE Transactions on Power Systems.*, 14(2), 553–560, May 1999.
44. G. Slemon and M.H. Awad, On equivalent circuit modeling for synchronous machines, *IEEE Transactions on Energy Conversion*, 14(4), 982–988, December 1999.
45. J. Verbeeck, R. Pintelon, and P. Lataire, Influence of saturation on estimated synchronous machine parameters in standstill frequency response tests, *IEEE Transactions on Energy Conversion*, 15(3), 377–283, September 2000.
46. D.Y. Park, H.C. Karmaker, G.E. Dawson, and A.R. Eastham, Standstill frequency-response testing and modeling of salient-pole synchronous machines, *IEEE Transactions on Energy Conversion*, 13(3), 230–236, September 1998.
47. R.M. Sanders, Standstill frequency-response methods and salient pole synchronous machines, *IEEE Transactions on Energy Conversion*, 14(4), 1033–1037, December 1999.
48. IEEE PES WG 12 (Chair: P.L. Dandeno), Experience with standstill frequency response (SSFR) testing and analysis of salient pole synchronous machines, *IEEE Transactions on Energy Conversion*, 14(4), 1209–1217, December 1999.
49. E. da Costa Bortoni and J.A. Jardini, A standstill frequency response method for large salient pole synchronous machines, *IEEE Transactions on Energy Conversion*, 19(4), 687–691, December 2004.
50. J.A. Martinez, B. Johnson, and C. Grande-Moran, Parameter determination for modeling system transients. Part IV: Rotating machines, *IEEE Transactions on Power Delivery*, 20(3), 2063–2072, July 2005.
51. P.L. Dandeno, P. Kundur, A.T. Poray, and H.M. Zein El-din, Adaptation and validation of turbogenerator model parameters through on-line frequency response measurements, *IEEE Transactions on Power Apparatus and Systems*, 100(4), 1656–1664, April 1981.
52. H. Tsai, A. Keyhani, J. Demcko, and R.G. Farmer, On-line synchronous machine parameter estimation from small disturbance operating data, *IEEE Transactions on Energy Conversion*, 10(1), 25–36, March 1995.
53. R. Wamkeue, I. Kamwa, X. Dai-Do, and A. Keyhani, Iteratively reweighted least squares for maximum likelihood identification of synchronous machine parameters from on-line tests, *IEEE Transactions on Energy Conversion*, 14(2), 159–166, June 1999.
54. M. Karrari and O.P. Malik, Identification of physical parameters of a synchronous generator from online measurements, *IEEE Transactions on Energy Conversion*, 19(2), 407–415, June 2004.

55. E. Kyriakides and G.T. Heydt, An observer for the estimation of synchronous generator damper currents for use in parameter identification, *IEEE Transactions on Energy Conversion*, 18(1), 175–177, March 2003.
56. E. Kyriakides, G.T. Heydt, and V. Vittal, On-line estimation of synchronous generator parameters using a damper current observer and a graphic user interface, *IEEE Transactions on Energy Conversion*, 19(3), 499–507, September 2004.
57. E. Kyriakides, G.T. Heydt, and V. Vittal, Online parameter estimation of round rotor synchronous generators including magnetic saturation, *IEEE Transactions on Energy Conversion*, 20(3), 529–537, September 2005.
58. B.K. Gupta, D.K. Sharma, and D.C. Bacvarov, Measured propagation of surges in the winding of a large ac motor, *IEEE Transactions on Energy Conversion*, 1(1), 122–129, March 1986.
59. E. Colombo, G. Costa, and L. Piccarreta, Results of an investigation on the overvoltages due to a vacuum circuit breaker when switching an HV motor, *IEEE Transactions on Power Delivery*, 3(1), 205–213, January 1988.
60. J.G. Reckleff, J.K. Nelson, R.J. Musil, and S. Wenger, Characterization of fast rise-time transients when energizing large 12.3 kV motors, *IEEE Transactions on Power Delivery*, 3(2), 627–636, April 1988.
61. J.G. Reckleff, E.N. Fromholtz, R.J. Musil, and S. Wenger, Measurements of fast rise-time transients switching large 12.3 kV motors, *IEEE Transactions on Power Delivery*, 3(3), 1022–1028, July 1988.
62. H.G. Tempelaar, Determination of transient overvoltages caused by switching of high voltage motors, *IEEE Transactions on Energy Conversion*, 3(4), 806–814, December 1988.
63. E.P. Dick, B.K. Gupta, P. Pillai, A. Narang, T.S. Lauber, and D.K. Sharma, Prestriking voltages associated with motor breaker closing, *IEEE Transactions on Energy Conversion*, 3(4), 855–863, December 1988.
64. P.G. McLaren and H. Oraee, Multiconductor transmission line model for the line end coil of large ac machines, *Proceedings of the IEEE*, Pt. B, 130(3), 149–156, May 1985.
65. D.C. Bacvarov and D.K. Sarma, Risk of winding insulation breakdown in large ac motors caused by steep switching surges. Part I: Computed switching surges, *IEEE Transactions on Energy Conversion*, 1(1), 130–139, March 1986.
66. R.G. Rhudy, E.L. Owen, and D.K. Sharma, Voltage distribution among the coils and turns of a form wound ac rotating machine exposed to impulse voltage, *IEEE Transactions on Energy Conversion*, 1(2), 50–60, June 1986.
67. P.G. McLaren and M.H. Abdel-Rahman, Modeling of large AC motor coils for steep-fronted surge studies, *IEEE Transactions on Industry Applications*, 24(3), 422–426, May/June 1988.
68. J.L. Guardado and K.J. Cornick, A computer model for calculating steep-fronted surge distribution in machine windings, *IEEE Transactions on Energy Conversion*, 4(1), 95–101, March 1989.
69. A. Narang, B.K. Gupta, E.P. Dick, and D.K. Sharma, Measurement and analysis of surge distribution in motor stator windings, *IEEE Transactions on Energy Conversion*, 4(1), 126–134, March 1989.
70. E.P. Dick, R.W. Cheung, and J.W. Porter, Generator models for overvoltages simulations, *IEEE Transactions on Power Delivery*, 6(2), 728–735, April 1991.
71. W.W.L. Keerthipala and P.G. McLaren, Modeling of effects of lamination on steep fronted surge propagation in large ac motor coils, *IEEE Transactions on Industry Applications*, 27(4), 640–644, July/August 1991.
72. J.L. Guardado and K.J. Cornick, Calculation of machine winding electrical parameters at high frequencies for switching transient studies, *IEEE Transactions on Energy Conversion*, 11(1), 33–40, March 1996.
73. J.L. Guardado, K.J. Cornick, V. Venegas, J.L. Naredo, and E. Melgoza, A three-phase model for surge distribution studies in electrical machines, *IEEE Transactions on Energy Conversion*, 12(1), 24–31, March 1997.

74. B.S. Oyegoke, A comparative analysis of methods for calculating the transient voltage distribution within the stator winding of an electric machine subjected to steep-fronted surge, *Electrical Engineering*, 82, 173–182, 2000.

75. J.L. Guardado, J.A. Flores, V. Venegas, J.L. Naredo, and F.A. Uribe, A machine winding model for switching transient studies using network synthesis, *IEEE Transactions on Energy Conversion*, 20(3), 322–328, June 2005.

76. P.J. Tavner and R.J. Jackson, Coupling of discharge currents between conductors of electrical machines owing to laminated steel core, *IEEE Proceedings.*, Pt. B, 135(6), 295–307, November 1988.

77. E.P. Dick, B.K. Gupta, P. Pillai, A. Narang, and D.K. Sharma, Equivalent circuits for simulating switching surges at motor terminals, *IEEE Transactions on Energy Conversion*, 3(3), 696–704, September 1988.

78. E.P. Dick, B.K. Gupta, P. Pillai, A. Narang, and D.K. Sharma, Practical calculation of switching surges at motor terminals, *IEEE Transactions on Energy Conversion*, 3(4), 864–872, December 1988.

79. R.G. Ramey, A.C. Sismour, and G.C. Kung, Important parameters in considering transient torques on turbine-generator shaft systems, *IEEE Transactions on Power Apparatus and Systems*, 99(1), 311–317, January/February, 1980.

80. IEEE Subsynchronous Resonance Working Group of the System Dynamic Performance Subcommittee, Terms, definitions and symbols for subsynchronous oscillations, *IEEE Transactions on Power Apparatus and Systems*, 104(6), 1326–1334, June 1985.

81. K.R. Padiyar, *Analysis of Subsynchronous Resonance in Power Systems*, Kluwer Academic Publishers, Boston, MA, 1999.

82. IEEE Subsynchronous Resonance Task Force of the Dynamic System Performance Working Group, First benchmark model for computer simulation of subsynchronous resonance, *IEEE Transactions on Power Apparatus and Systems*, 96(5), 1565–1572, September/October 1977.

83. IEEE Committee Report, Reader's guide to subsynchronous resonance, *IEEE Transactions on Power Systems.*, 7(1), 150–157, February 1992.

84. R.E. Fairbairn, G.D. Jennings, and R.G. Harley, Turbogenerator torsional mechanical modal parameter identification from on-line measurements, *IEEE Transactions on Power Systems.*, 6(4), 1389–1395, November 1991.

85. R. Bigret, D.J. Coetzee, D.C. Levy, and R.G. Harley, Measuring the torsional frequencies of a 900 MW turbogenerator, *IEEE Transactions on Energy Conversion*, 1(4), 99–107, December 1986.

86. D.N. Walker, C.E.J. Bowler, R.L. Jackson, and D.A. Hodges, Results of subsynchronous resonance test at Mohave, *IEEE Transactions on Power Apparatus and Systems*, 94(5), 1878–1889, September/October 1975.

87. D.E. Walker and A.L. Schwalb, Results of Subsynchronous Resonance Test at Navajo, IEEE PES Special Publication 76CH1066 + PWR.

88. G. Gross and M.C. Hall, Synchronous machine and torsional dynamics simulation in the computation of electromagnetic transients, *IEEE Transactions on Power Apparatus and Systems*, 97(4), 1074–1086, July/August 1978.

89. T.J. Hammons, Electrical damping and its effect on accumulative fatigue life expenditure of turbine-generator shafts following worst-case supply system disturbance, *IEEE Transactions on Power Apparatus and Systems*, 102(6), 1552–1565, June 1983.

6

Surge Arresters

Juan A. Martinez-Velasco and Ferley Castro-Aranda

CONTENTS

6.1 Introduction 352
6.2 Valve-Type Arresters...353
    6.2.1 Gapped Silicon Carbide Arresters...353
    6.2.2 Gapless Metal Oxide Arresters 354
    6.2.3 Gapped Metal Oxide Arresters 355
6.3 Metal Oxide Surge Arresters Requirements..357
    6.3.1 Introduction: Arrester Classification...357
    6.3.2 Durability/Capability Tests 359
    6.3.3 Protective Characteristic Tests 359
    6.3.4 Energy Capabilities..360
6.4 Models for Metal Oxide Surge Arresters 362
    6.4.1 Introduction 362
    6.4.2 Models Mathematics 364
    6.4.3 Models for Low-Frequency and Slow-Front Transients....................365
    6.4.4 Models for Fast Transients..366
        6.4.4.1 Introduction 366
        6.4.4.2 Arrester Models...367
        6.4.4.3 Parameter Determination from Field Measurements.........374
    6.4.5 Models for Very Fast Front Transients 374
6.5 Electrothermal Models 375
    6.5.1 Introduction 375
    6.5.2 Thermal Stability of Metal Oxide Arresters 376
    6.5.3 Analog Models 378
        6.5.3.1 Equivalent Circuits 378
        6.5.3.2 Determination of Thermal Parameters 379
        6.5.3.3 Power Input due to Solar Radiation.....................................383
        6.5.3.4 Electric Power Input 383
    6.5.4 Use of Heat Sinks 385
6.6 Selection of Metal Oxide Surge Arresters 385
    6.6.1 Introduction 385
    6.6.2 Overvoltages..386
        6.6.2.1 Temporary Overvoltages 386
        6.6.2.2 Switching Overvoltages 386
        6.6.2.3 Lightning Overvoltages 386
    6.6.3 Protective Levels 387
    6.6.4 Procedure for Arrester Selection...387

        6.6.5   Phase-to-Phase Transformer Protection..400
        6.6.6   Arrester Selection for Distribution Systems.....................................401
6.7   Arrester Applications...403
        6.7.1   Introduction...403
        6.7.2   Protection of Transmission Systems..404
        6.7.3   Protection of Distribution Systems...422
References...438

6.1   Introduction

Surge arresters are connected across an apparatus to provide a low-resistance path and to limit the various types of transient voltages below the corresponding insulation level of the apparatus. A surge arrester should act like an open circuit during normal operation of the system, limit transient voltages to a safe level, and bring the system back to its normal operational mode as soon as the transient voltages are suppressed [1]. Therefore, a surge arrester must have an extremely high resistance during normal system operation and a relatively low resistance during transient overvoltages; that is, its voltage–current (V–I) characteristic must be nonlinear.

There are two main types of surge arresters: the expulsion type and the valve type. There might be expulsion arresters still in use, but they are no longer manufactured, and they are not reviewed in this chapter.

Valve arresters consist of nonlinear resistors which act like valves when voltages are applied to them. There are two types of valve arresters: silicon carbide (SiC) and metal oxide (MO).

Early protective devices used spark gaps connected in series with disks made with a non-linear SiC material. The spark gaps provide the high impedance during normal conditions while the SiC disks impede the flow of current following sparkover. The V–I characteristic of SiC-type surge arresters are a combination of both the SiC disk and the gap behavior.

The MO varistor material used in modern high voltage surge arresters has a highly nonlinear V–I characteristic. Varistors are made up of zinc oxide (ZnO) powder and traces of oxides of other metals bound in a ceramic mold. Their characteristic avoids the need for series spark gaps. Therefore, the electrical behavior is determined solely by the properties of the MO blocks.

The V–I characteristic of both SiC and ZnO valve elements are shown in Figure 6.1. Note the difference between leakage currents in both valve types for a given voltage. The series gap is essential for a SiC valve arrester to prevent thermal runaway during normal operation, while gapless operation is possible with ZnO arresters because of the low leakage current during normal operation.

Both types of arresters are reviewed below, but the rest of the chapter is dedicated to the MO surge arrester, since it is the most common type that is presently installed.

The goal of this chapter is to provide procedures for development of models and selection of MO surge arresters. The chapter includes a summary of arrester requirements, as established in standards, a description of MO surge arrester models for time-domain simulation, an introduction to models for thermal analysis, detailed procedures to select arresters for transmission and distribution levels (>1 kV), application guidelines, and some examples that will illustrate how to select and develop models for various applications at different voltage levels.

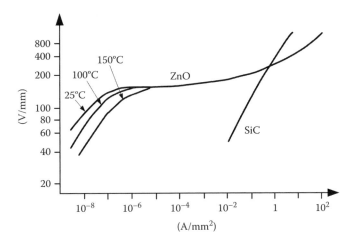

FIGURE 6.1
V–I characteristic of ZnO and SiC elements.

A comparison of the main characteristics and behavior of SiC valve arresters and both gapless and gapped MO valve arresters is presented in the next section.

International Electrotechnical Commission (IEC) and Institute of Electrical and Electronics Engineers (IEEE) standards are a very valuable source of information about the main aspects (ratings, requirements, and test designs, classification, selection, application) of any type of arresters [2–10]; readers are encouraged to consult standards, which are periodically reviewed and updated.

6.2 Valve-Type Arresters

6.2.1 Gapped Silicon Carbide Arresters

SiC gapped arresters consist of series spark gaps with or without series blocks of nonlinear resistors which act as current limiters. The current-limiting block has nonlinear resistance characteristics. The nonlinear resistors are built from powdered SiC mixed with a binding material, molded into a circular disk and baked. The disk diameter depends on its energy rating, and the thickness on its voltage rating. The *V–I* characteristic of a SiC nonlinear valve block has a hysteresis-type loop, being the resistance higher during the rising part of the impulse current than during the tail of the current wave, as shown in Figure 6.2 [1,11,12]. The nonlinear properties are due to the resistance–temperature properties of the junction between SiC crystals.

The function of the air gap is to isolate the current-limiting block from the power-frequency voltage under normal operating conditions. Without a gap, a power-frequency leakage current would constantly flow through the valve block, which would overheat the block, and could eventually lead to thermal runaway. The series gap sparks over if the magnitude of the applied voltage exceeds a preset level. The sparkover voltage depends upon the voltage waveshape: the steeper the voltage rise, the higher the sparkover voltage. Once the energy in the transient voltage is dissipated, power-follow current will flow through the valve elements. A relatively high current will continue to flow through the

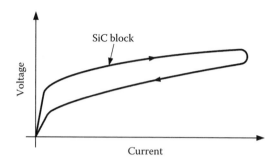

FIGURE 6.2
V–I characteristics of linear resistor and SiC block.

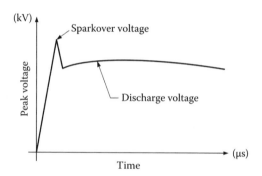

FIGURE 6.3
Sparkover and discharge characteristic of a SiC arrester.

arrester, after sparkover and after the overvoltage has gone down, if the power voltage across the arrester is high enough, since the gap arc has a low resistance and the current is primarily determined by the *V–I* characteristic of the SiC block. The voltage drop across the valves rises with the discharge current that flows through them. However, due to the nonlinearity of the valve blocks, the flow of current will be so low that the arc across the series gap elements will become unstable and be quenched, thus isolating the power-frequency source from the arrester. The combined operation of the gap and the current-limiting block results in the voltage–current characteristic illustrated in Figure 6.3 [1]. Note that the sparkover and the discharge voltage are not equal.

   To ensure arc extinction and resealing, a mechanism is provided for the control of the arc. An important effect of this control is that for fast transients almost the entire voltage appears across the current-limiting block.

6.2.2   Gapless Metal Oxide Arresters

A MO varistor is formed from a variety of materials via a manufacturing process which provides the desired electrical properties to the varistor. The typical structure of an MO varistor consists of highly conductive particles of MO, usually ZnO, suspended in a semi-conducting material. The manufacturing process determines the size of the MO particles as well as the thickness and resistivity of the semiconducting material. In a gapless arrester the MO block is continuously subjected to the power-frequency voltage, but the power-frequency leakage current through an MO arrester is so small that there is no danger of a thermal runaway.

The *V–I* characteristic of an MO surge arrester exhibits a knee for small currents (in the milliampere region). The voltage–current characteristic of an MO block in the protection region (high current flow) is insensitive to the temperature of the block, but the voltage–current characteristic near the knee is temperature-sensitive, see Figure 6.1. For an applied voltage near nominal, the electric current through the arrester is mostly capacitive and the low value on the order of a milliampere. As the voltage increases, the current increases much faster. The increase of the current occurs in the component which is in phase with the voltage while the capacitive component remains constant.

The arrester discharge voltage for a given current magnitude is directly proportional to the height of the valve element stack and is more or less proportional to the arrester rated voltage. The operation of an arrester is sensitive to the rate of rise of the incoming surge current: the higher the rate of rise of the current, the more the arrester limiting voltage rises. It has been suggested that the higher discharge voltage of the MO block for higher rates of current rise is caused by the negative temperature coefficient of the valve resistance, see Figure 6.1. The instantaneous temperature and resistance of a valve block is a function of the energy dissipated in the block up to that instant. However, the energy dissipated in a valve block for any specific current level on the front of the current wave is smaller for a faster rising current than that for a slower rising current at the same current level. Then, the instantaneous resistance of a valve block will be higher for a faster rising current than for a slower rising current. Hence, the discharge voltage for a faster rising current will be higher.

In the earlier construction the MO elements were surrounded by a gaseous medium and the end fittings were generally sealed with rubber rings. With time in service, especially in hostile environments, the seals tended to deteriorate allowing the ingress of moisture. In the 1980s polymeric-housed arresters were developed. In their design the surface of the MO elements column is bonded homogeneously with glass fiber-reinforced resin. This construction is void free, gives the unit a high mechanical strength, and provides a uniform dielectric at the surface of the MO column. The housing material is resistant to tracking and suitable for application in polluted regions. The advantages of the polymeric-housed arresters over their porcelain-housed equivalents are many and include light modular assembly, no risk to personnel or adjacent equipment during fault current operation, or decreased pollution flashover problems.

To reduce and simplify maintenance, arresters built into equipment (e.g., line insulators) for both distribution and transmission levels have been developed [13,14].

6.2.3 Gapped Metal Oxide Arresters

The MO arrester has several advantages in comparison with the SiC arrester (simplicity of design, decreased protective characteristics, increased energy absorption capability) [15,16]; however, the power-frequency voltage is continuously resident across the MO. High currents can result from temporary overvoltages (TOVs), such as faults or ferroresonance, and produce heating; if the TOVs are sufficiently large in magnitude and long in duration, temperatures may increase sufficiently so that thermal runaway and failure occur. In addition, the discharge voltage increases as the arrester discharge current increases.

The performance of MO blocks at higher discharge currents can be improved by equipping them with a shunt gap, which is designed to sparkover whenever the discharge current through the arrester exceeds a certain value; e.g., 10 kA. A typical shunt-gapped arrester *V–I* characteristic is illustrated in Figure 6.4 [5]. Initially, the arrester voltage increases with the surge current following the range A–B on the *V–I* characteristic. When the surge current magnitude reaches 250–500 A (range B to C), sparkover of a gap connected in parallel with a few MO valve elements results in a shunting of the current around these valve elements, and a lower discharge voltage (in the range D to E). With a further current increase, the voltage increases according to the characteristic E–F. As the surge current decreases, the arrester voltage decreases following the characteristic F–G until the shunt gaps extinguish at a low level of current. After the extinction of the arrester leakage current, the arrester operation returns to point A.

Some distribution-class arresters use series gaps that are shunted by a linear component impedance network. A typical *V–I* characteristic is illustrated in Figure 6.5 [5]. The arrester voltage begins to rise with the surge current in the range A–B. At about 1 A (depending on rate of rise in the range B to C), the gaps sparkover and the arrester voltage is reduced to the discharge voltage of the MO valve elements only. For further increase in surge current, the voltage increases according to the characteristic D–E–F. As the surge current decreases, the arrester voltage decreases accordingly, following the characteristic F–G until the series gaps extinguish at a low level of current.

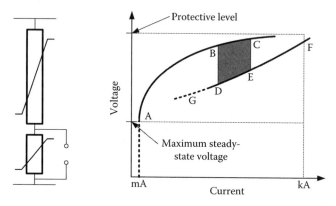

FIGURE 6.4
Characteristic of shunt-gapped MO surge arrester. (From IEEE Std C62.22-1997, IEEE guide for the application of metal-oxide surge arresters for alternating-current systems. With permission.)

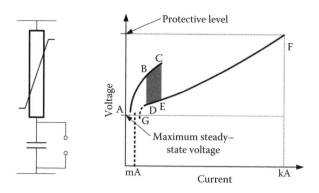

FIGURE 6.5
Characteristic of series-gapped MO arrester. (From IEEE Std C62.22-1997, IEEE guide for the application of metal-oxide surge arresters for alternating-current systems. With permission.)

The advantages of series-gapped MO surge arresters for different types of applications have been discussed in several papers, see [17,18].

6.3 Metal Oxide Surge Arresters Requirements

6.3.1 Introduction: Arrester Classification

Some information (voltage ratings, class or discharge current, frequency) is needed for identification of an MO surge arrester. These values are obtained from tests established and detailed in standards.

IEC standards classify arresters according to the nominal discharge current and line discharge class. Discharge current and discharge class cannot be selected independently of each other [15].

IEEE standards classify arresters into three primary *durability* or *capability* classes [4]: (1) *station*, used primarily in HV and EHV systems; (2) *intermediate*, used between station and distribution; and (3) *distribution*, used in distribution systems, and further divided into *heavy duty*, *normal duty*, and *light duty*. In addition, specific arresters are produced for distribution systems: the *riser pole* arrester for cables, the *dead front* arrester for padmount transformers, and the *liquid immersed* arrester used internally in a transformer. Also arresters are manufactured for use on transmission or distribution lines. They are placed across the line insulation to improve the lightning performance.

IEC standards for high voltage (>1 kV) arresters are denoted IEC 60999; for gapless metal oxide arresters, Part 4 defines the tests and ratings [8], while Part 5 details the application recommendations [9]. Main IEEE standards for MO surge arresters are C62.11, which defines tests and ratings [4], and C62.22, which provides the application guide [5].

While in service an MO surge arrester must withstand the maximum rms value of power-frequency voltage that may appear across its terminals and be capable of operating under the maximum TOV that can occur at its location during the length of time that such overvoltage will exist. In addition, the arrester should have an energy capability greater than the energy associated with the expected switching surges on the system.

The ratings of an MO surge arrester in the IEC Application Guide are the continuous operating voltage (COV), the rated voltage (TOV capability at 10 s), the nominal discharge current, the line discharge class, and the pressure relief class. The nominal discharge current is selected by calculation or estimation of the lightning current discharge by the arrester. The line discharge class is selected by comparison of the arrester energy capability with the energy discharge required. The pressure relief class is selected by comparison to the system fault current.

The selection of an MO arrester according to the IEEE standard is based on similar information. The IEC-rated voltage is similar to the IEEE duty-cycle voltage except that in IEEE this voltage is not defined in terms of the TOV capability. Therefore, to determine arrester ratings, the following rules are to be considered:

1. The steady-state voltage that the arrester can support indefinitely, known as the maximum continuous operating voltage (MCOV) in IEEE and COV in IEC, must be equal to or greater than the maximum line-to-ground system voltage.
2. The TOV across the arrester must be less than the arrester TOV capability.

3. Switching surge energy discharged by the arrester must be less than the energy capability.

4. The pressure relief current must be equal to or greater than the fault current.

Each class and type of arrester is subjected to a series of tests, which may be divided between those that serve to define the arrester ability to protect itself (durability/capability) and those that define the arrester ability to protect the equipment to which it is applied (protective characteristics). In addition, manufacturers provide energy capabilities to define the arrester ability to discharge the energy in a switching surge or a lightning discharge. Tests to establish these energies are not yet included in standards. As an example, test requirements specified in IEEE standard are shown in Table 6.1.

TABLE 6.1

IEEE Arrester Classification and Test Requirements

Design Test	Station and Intermediate Class		Distribution Class			
	Porcelain	Polymer	Porcelain	Polymer	Liquid-Immersed	Dead Front
Arrester insulation withstand test	X	X	X	X	X	X
Discharge–voltage characteristics test	X	X	X	X	X	X
Accelerated aging procedure	X	X	X	X	X	X
Accelerated aging test of polymer-housed arresters	NR	X	NR	X	NR	NR
Contamination test	X	X	X	X	NR	NR
Distribution arrester seal integrity design test	NR	NR	X	X	NR	NR
Radio-influence voltage test	X	X	X	X	NR	NR
Partial discharge test	X	X	X	X	X	X
High-current short-duration withstand test	X	X	X	X	X	X
Low-current long-duration withstand test	X	X	X	X	X	X
Duty-cycle test	X	X	X	X	X	X
TOV test	X	X	X	X	X	X
Pressure relief test for station and intermediate arresters	X	X	NR	NR	NR	NR
Short-circuit test	NR	NR	X	X	NR	NR
Failure mode	NR	NR	NR	NR	X	X
Distribution arrester disconnector test	NR	NR	X	X	NR	NR
Maximum design cantilever load (MDCL) and moisture ingress test for polymer-housed	NR	X	NR	X	NR	NR
Ultimate mechanical strength-static (UMS-static) test for porcelain-housed arresters	X	NR	X	NR	NR	NR

Source: IEEE Std C62.11-2005, IEEE standard for metal-oxide surge arresters for ac power circuits (>1 kV). With permission.

6.3.2 Durability/Capability Tests

IEC and IEEE specify a set of withstand tests for surge arresters. Table 6.2 summarizes the purpose of the major tests as prescribed in the IEEE standard. The main differences with respect to the IEC standard are [15,19]:

- IEC discharge energies are about three times the IEEE values for heavy-duty and two times for normal- and light-duty. That is, the IEC test is more severe.
- The IEC standard has two types of duty-cycle tests, switching impulse and lightning impulse (high current), while only one type of test is listed in IEEE.
- The TOV capability tests in IEC and IEEE are almost identical except for the 10 kA, Classes 2 and 3, and the 20 kA, Classes 4 and 5. The IEC tests may be viewed as with prior energy, whereas the IEEE tests are with and without prior energy. The IEC rated voltage is the TOV capability at 10 s.
- There is no contamination test in IEC.

6.3.3 Protective Characteristic Tests

The purpose of the discharge–voltage characteristic tests is to verify the voltage level appearing across an arrester under specific surge conditions. The protective characteristics

TABLE 6.2

Durability/Capability Design Tests

Design Test	Purpose
Arrester insulation withstand test	Verify that, under usual system conditions, arrester housings will not flashover under defined impulse, switching, and power-frequency conditions.
Contamination test	Demonstrate, through examination of thermal stability and insulation withstand, the ability of the arrester to withstand electrical stresses on the arrester housing caused by contamination.
High-current short-duration withstand test	Demonstrate that under severe high-current impulse conditions, the arrester will remain functional. The high-current short-duration test indirectly evaluates the dielectric strength of the material as well as the thermal withstand capability for impulses that resemble lightning.
Low-current long-duration withstand test	Demonstrate that under severe low-current impulse conditions, the arrester will remain functional. There are two test methods: (a) A transmission-line discharge test for station and intermediate arresters. (b) A rectangular wave test for distribution arresters.
Duty-cycle test	Verify that the arrester can withstand multiple lightning-type impulses without causing thermal instability or dielectric failure.
TOV test	Demonstrate the TOV capability of the arrester. The TOV is strictly a power-frequency overvoltage for time periods from 0.01 to 10,000 s. Manufacturer data shall include curves with abscissa scaled in time and ordinate in per unit of MCOV. In addition, the manufacturer shall publish a table of TOV values listed in per unit of MCOV to three significant digits, for times 0.02, 0.1, 1, 10, 100 s, and 1000 s. The table values shall be taken from the curves and shall include data for "No Prior Duty" and for "Prior Duty".
Pressure relief test for station and intermediate arresters	Demonstrate that arresters remain intact, or fall in a small area around the unit, for given fault currents during an end-of-life event. Arresters shall also be tested to demonstrate pressure relief capability for low levels of fault current.
Short-circuit test	Demonstrate that arresters will withstand fault-current levels claimed by the arrester manufacturer.

are voltages across the arrester for a specified discharge current magnitude and shape. Both IEC and IEEE standards identify three characteristics tests whose objectives are to determine the following voltages [5,8,15,16]:

1. *Steep current impulse discharge voltage*: This voltage is known as *front-of-wave protective level* (FOW) in IEEE. In the IEC standard, it is the discharge voltage obtained by discharging a current having a 1 μs front and a crest current equal to the nominal discharge current. According to the IEEE standard, it is the voltage across the arrester having a time-to-crest of 0.5 μs when discharging the lightning impulse current. This discharge voltage is obtained by using different times to crest (usually 1, 2, and 3 μs) and plotting the voltage as a function of the time-to-crest of the voltage.

2. *Lightning impulse protection level* (LPL): It is the highest of the discharge voltages obtained across the arrester for arrester discharge currents having an 8/20 μs waveshape and crest magnitudes equal to 0.5, 1.0, and 2.0 times the nominal discharge current in IEC, or values of 1.5, 3.0, 5, 10, and 20 kA in IEEE. For arresters applicable to 500 kV systems, IEEE also specifies the 15 kA discharge. Manufacturers may also provide the 40 kA discharge voltage.

3. *Switching impulse protection level*: This is the voltage across the arrester when discharging a current impulse with a front greater than 30 μs and a tail less than 100 μs in IEC standard, or having a 45–60 μs time-to-crest in IEEE standard. The magnitude is determined according to Table 6.3. This test is not required for distribution arresters.

The discharge voltage magnitude and time-to-crest are functions of the time-to-crest and magnitude of the discharge current, see Figure 6.6.

6.3.4 Energy Capabilities

The energy that an arrester can absorb during an overvoltage is known as *energy withstand capability*. This capability is often expressed in terms of kJ/kV of arrester MCOV/COV or per kV of duty-cycle rating. Because it is dependent on the specific form (magnitude, waveshape, and duration) of the overvoltage, the energy-handling capability cannot be

TABLE 6.3

Switching Impulse Currents

Standard	Arrester Class	Crest Current (A)	Waveshape
IEC	20 kA, classes 4 and 5	500 and 2000	
			Time-to-crest > 30 μs
	10 kA, class 3	250 and 1000	
			Tail time < 100 μs
	10 kA, classes 1 and 2	125 and 500	
IEEE	Station 3–150 kV	500	Time-to-crest: 45–60 μs
	151–325 kV	1000	
	326–900 kV	2000	
	Intermediate	500	

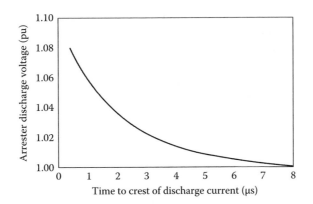

FIGURE 6.6
Effect of time-to-crest of arrester current. (From Hileman A.R., *Insulation Coordination for Power Systems*, Marcel Dekker, New York, 1999. With permission.)

expressed by a single value of kJ/kV. Manufacturers typically publish some information on energy handling capability, but there are no standardized tests to determine the energy handling capability of arresters.

The energy capability for various waveforms and current magnitudes has been the subject of many investigations [15,20–26]. The switching surge energy capability is of importance when selecting the arrester ratings; on the other hand, with the increased use of arresters for protection of transmission and distribution lines, the energy capability in the lightning region is also essential.

The capability of discharging the energy contained in a switching surge is partially determined by the low-current, long-duration test. This energy is the energy from multiple discharges, distributed over 1 min, in which the arrester current is less than a specified magnitude. Thus it becomes evident that the energy capability depends on the rate at which energy is discharged by the arrester.

Some reports have shown that when arresters are tested until failure, the energy capability shows a probabilistic behavior [15,22,24]. Assuming that the Weibull cumulative distribution function can be used to model the energy characteristic and setting the discharge energy for the standard tests at the mean −4 standard deviations, the probability of arrester failure is given by the following equations:

$$P_F = 1 - 0.5^{\left(\frac{Z}{4}+1\right)^5} \tag{6.1}$$

where

$$Z = \frac{W_C - \mu}{\sigma} = \frac{(W_C/W_R) - 2.5}{0.375} \tag{6.2}$$

where
  μ is the average or mean energy capability
  σ is the standard deviation
  W_C is the energy capability
  W_R is the rated energy from the standard tests; i.e., that provided by the manufacturer

The equation for Z assumes a standard deviation of 15% of the mean.

An estimation of the energy discharged by a surge arrester assumes that the entire line is charged to the prospective switching surge voltage at the arrester location and discharged during twice the travel time of the line [5,15].

The current though the arrester is obtained from the expression

$$I_d = \frac{V_S - V_d}{Z} \tag{6.3}$$

where

V_S is the switching surge overvoltage
V_d is the arrester discharge voltage
Z is the single-phase surge impedance of the line

The energy discharged by the arrester is then

$$W_C = 2V_d I_d T_L \tag{6.4}$$

where T_L is the travel time of the line.

A conservative estimate of the switching surge energy discharged by an arrester is [15]:

$$W_C = \frac{1}{2}CV_S^2 \tag{6.5}$$

where C is the total capacitance of the line. The peak voltage will be usually above 2 pu of the peak line-to-neutral voltage, although it may be higher for cases like capacitor bank restrikes and line reclosing. The total capacitance may be that of an overhead line, a cable, or a capacitor bank.

The lightning surge energy discharge may be described as a charge duty [16], being the values of the wave front unimportant for this evaluation. If one arrester discharges all of the lightning stroke current, a conservative estimate of the energy duty is

$$W_C = QV_d \tag{6.6}$$

where

Q is the total stroke charge
V_d is the arrester discharge voltage at the peak lightning stroke current

In practice, the actual energy discharge is reduced by the nonlinear arrester characteristics and by sharing from nearby arresters.

A better quantification of the energy discharge may be obtained through time-domain simulation.

6.4 Models for Metal Oxide Surge Arresters

6.4.1 Introduction

The nonlinear *V–I* characteristic of an arrester valve (SiC or ZnO) is given by

$$I = kV^\alpha \tag{6.7}$$

The parameter k depends upon the dimensions of the valve block, while α, which describes the nonlinear characteristic, depends upon the valve-block material. For a SiC block, α is typically 5, whereas for a ZnO block α is greater than 30.

The *V–I* characteristic of MO arresters can be divided into three regions (see Figure 6.7) [15]:

1. In region 1, I is less than 1 mA and is primarily capacitive.
2. In region 2, I is from 1 mA to about 1000 or 2000 A and is primarily a resistive current.
3. In region 3, I is from 1 to 100 kA. For very large currents, the characteristic approaches a linear relationship with voltage; i.e., the MO varistor becomes a pure resistor.

Coefficient α is variable for MO varistors, reaching a maximum of about 50 in the first region and decreasing to about 7–10 in the third region. Thus, for MO varistors, α is primarily used to indicate the flatness of the characteristic and should not be employed to model the arrester. However, in some cases, it is convenient to use an α value within a limited range to assess the arrester performance; e.g., the TOV capability of the arrester.

As mentioned above, the *V–I* characteristic depends upon the waveshape of the arrester current, with faster rise times resulting in higher peak voltages. Table 6.4 shows modeling guidelines derived from CIGRE WG 33-02 [27]. The commonly used frequency-independent surge arrester model is appropriate for simulations containing low and most switching frequencies (Groups I and II). However, a frequency-dependent model should be used when very high frequencies are simulated (Groups III and IV); such a model has to incorporate the inherent inductance of the surge arrester. A lumped inductance of about 1 μH/m for the ground leads should also be included in models for high frequencies.

MO surge arresters with series or parallel spark gaps can also be represented with the model described below where the device has two curves: (1) prior to sparkover and (2) after sparkover.

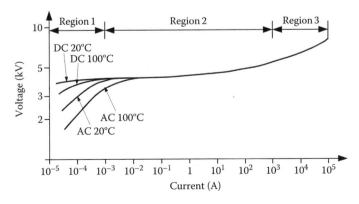

FIGURE 6.7
Typical characteristics of an MO arrester disk.

TABLE 6.4

Guidelines to Represent Metal Oxide Surge Arresters

Model Characteristics	Low Frequency Transients	Slow Front Transients	Fast Front Transients	Very Fast Front Transients
Temperature-dependent V–I characteristic	Important	Negligible	Negligible	Negligible
Frequency-dependent V–I characteristic	Negligible	Negligible	Important	Very important
MOV block inductance	Negligible	Negligible	Important	Very important
Ground lead inductance	Negligible	Negligible	Important	Very important

Sources: CIGRE WG 33.02, Guidelines for Representation of Network Elements when Calculating Transients, CIGRE Brochure 39, 1990. With permission.

6.4.2   Models Mathematics

The V–I characteristic of a surge arrester has several exponential segments [28], and each segment can be approximated by the following formula:

$$i = p \left(\frac{v}{V_{\text{ref}}} \right)^q \tag{6.8}$$

where
  q is the exponent
  p is the multiplier for that segment
  V_{ref} is an arbitrary reference voltage that normalizes the equation

The first segment can be approximated by a linear relationship to avoid numerical underflow and speed up the simulation. The resistance of this first segment should be very high since the surge arrester should have little effect on the steady-state solution of the network. Remember that surge arrester currents during normal steady-state operation are less that 0.1 A. The second segment is defined by the parameters p, q, and V_{min}, the minimum voltage for that segment. Multiple segments are typically used to enhance the accuracy of the model since the exponent decreases as the current level increases. Each segment has its own values for p, q, and V_{min}.

Manufacturers test each disk with a current pulse (typically a current pulse with a 10 kA peak) and record a reference voltage. The resulting peak voltage is the reference voltage V_{10}, the voltage at 10 kA for single-column surge arrester. The V–I curves often use the V_{10} value as the 1.0 per unit value. The V–I curve can be determined by multiplying the per unit arrester voltages by the V_{10} for that rating.

To construct an arrester model, the following information must be selected [29]:

- A reference voltage proportional to the arrester rating V_{10}
- The number of parallel columns of disks
- The V–I characteristic in per unit of the reference voltage

The choice of arrester *V–I* characteristic depends upon the type of transient being simulated since current waveshapes with a faster rise times will result in higher peak voltages. Manufacturers publish several curves:

- The 8/20 µs characteristic, which applies for typical lightning surge simulations
- The front-of-wave (FOW) characteristic, which applies for transients with current rise times less than 1 µs
- The switching discharge voltage for an associated switching surge current, which applies to switching surge simulations
- The 1 ms characteristic, which applies to low frequency phenomena

In addition, manufacturers may supply minimum and maximum curves for each test waveshape. The maximum curve is generally used since it results in the highest overvoltages and the most conservative equipment insulation requirements. The minimum curves are used to determine the highest energy levels absorbed by the arrester.

A supporting routine available in most transients programs allows users to convert the set of manufacturer's *V–I* points to a set of p, q, and V_{min} values. A different curve should be created for each waveshape and manufacturing tolerance (maximum or minimum). The voltages are usually given in a per unit fashion where the reference voltage (1.0 per unit) is either the voltage rating or V_{10}, the peak voltage for a 10 kA, 8/20 µs current wave.

The subsequent sections present the rationale for models representing MO surge arresters in low- (Groups I and II) and high-frequency (Groups III and IV) transients simulations. Main characteristics and parameter determination are discussed for each model.

6.4.3 Models for Low-Frequency and Slow-Front Transients

To construct a surge arrester model for these frequency ranges, data to be obtained from the manufacturer's literature include ratings and characteristics, as well as *V–I* curves. The arrester model is then edited following the guidelines presented in Table 6.4 and the supporting routine mentioned above. The following example is used to demonstrate the role of a surge arrester.

Example 6.1

Figure 6.8a shows the test circuit used in this example. The 350 Ω resistance represents the surge impedance of a transmission overhead line. Figure 6.8b shows the *V–I* characteristic of the surge arrester, which has a rating adequate for 220 kV applications, with a V_{10} of 437 kV. The surge voltage has a triangle waveshape that peaks at 600 kV. Figure 6.8c shows the surge voltage, the arrester voltage, and the arrester current.

The current through the surge arrester is very small until the voltage across its terminal reaches about 300 kV. Until that moment, the surge arrester voltage is approximately the same as the surge voltage because the voltage drop across the surge impedance of the line is nearly zero. As the current increases, the voltage across the surge impedance increases, resulting in a lower voltage at the arrester. At 30 µs, a peak current of 667 A results in a voltage across the resistor of 233.45 kV, being the maximum arrester voltage 366.55 kV. Since the instantaneous voltage for the period between 0 and 30 µs is the same as the period between 30 and 60 µs, these waveshapes have symmetry about the 30 µs point in time.

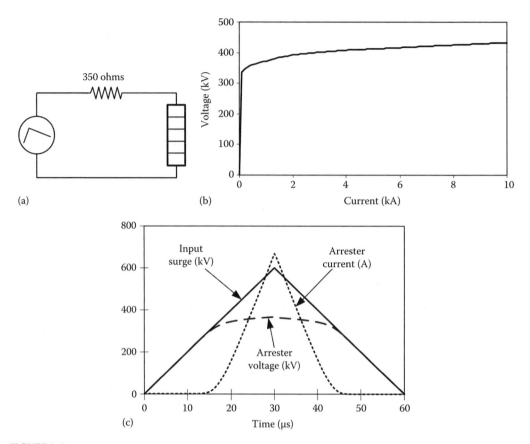

FIGURE 6.8
Example 6.1: Test circuit and simulation results. (a) Test circuit; (b) surge arrester V–I characteristic; (c) voltages and current.

6.4.4 Models for Fast Transients

6.4.4.1 Introduction

The surge arrester model described above does not incorporate time or frequency dependence. Actually, the surge arrester waveshapes would be skewed if they were physically measured in a laboratory, and the peak of the arrester voltage would occur before the peak of the current. For a given peak current, the peak voltage increases as the front time decreases. However, the percentage increase is only slightly proportional to the current magnitude. The fast front phenomenon appears to be an inductive effect, but it is not that of a simple linear inductance.

The dynamic performance of MO surge arresters was described in late 1970s [30]. Since then several models have been developed to account for a frequency-dependent behavior [31–40]. Basically, all these models incorporate a nonlinear resistor to account for the V–I characteristic of MO varistor materials, and an inductor to include frequency-dependent behavior.

6.4.4.2 Arrester Models

A summary of some of the most popular models proposed during the last years to reproduce the performance of an MO surge arrester in high-frequency transients is presented
in the following paragraphs.

1. CIGRE model

The equivalent circuit of a gapless MO surge arrester should include the possible time delays
for the change in the conduction mechanism from thermal to tunnel effects, the capacitance
formed by the parallel/series connection of the granular layers, and the inductance of the
varistor elements, determined by the geometry of the current path in the varistor [35,36].

Real tests show that for fast-front currents (i.e., lightning currents), the discharge voltage
of an MO block reaches the peak prior to the current peak. Furthermore, an increase of this
voltage can be observed with decreasing current. However, for two different discharge
current shapes, the discharge voltages differ in the current front, but they approach in
the current tail. In addition, the difference between voltages remains constant during the
current front. This conduction time delay can be represented by a resistance in series with
the conventional steady-state nonlinear resistance, which reproduces the behavior of MO
blocks for low-frequency discharge currents.

To describe the dynamic performance of an MO surge arrester, the equivalent circuit
shown in Figure 6.9 was proposed [35]. The current-dependent resistance of the granular layer is subdivided into the steady-state current-dependent resistance $R(i)$, the turn-on
resistance $R(dv/dt, V, \tau)$, and the temperature-dependent resistance $R(\theta)$, all of them represent the behavior of the arrester for low frequency. The elements R and L represent the
ZnO grain, whereas the other circuit elements are related to the grain boundaries.

This model was later simplified and adopted by a CIGRE WG [36,41]. The temperature-
dependent resistance $R(\theta)$ and the capacitance C can be neglected, and the current-dependent resistance $R(i)$ and the ZnO resistance can be combined to a single resistance R_l. The
resulting equivalent circuit is then reduced to a series combination of a nonlinear resistance

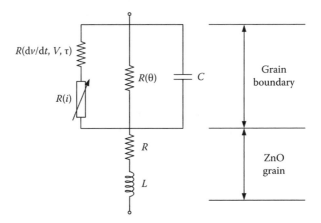

FIGURE 6.9
Equivalent circuit for an MO arrester block. (From Schmidt, W. et al., *IEEE Trans. Power Deliv.*, 4(1), 292, 1989.
With permission.)

R_I, the turn-on linear resistance R_T, and the inductance of the current path L, see Figure 6.10.

The resistance R_I can be determined from the discharge voltage for 8/20 μs currents with various peak values. The turn-on resistance can be described by a set of curves or obtained from the following differential equation [36]:

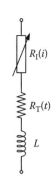

$$\frac{dG}{dt} = \frac{G_{ref}}{T} \cdot \left(1 + \frac{G}{G_{ref}}\right)\left(1 + \frac{G}{G_{ref}}\left(\frac{i}{I_{ref}}\right)^2\right) \cdot e^{\frac{v}{V_{ref}}} \qquad (6.9)$$

where

$$G_{ref} = \frac{34}{V_{10}} \qquad (6.10)$$

$$V_{ref} = kV_{10} \qquad (6.11)$$

FIGURE 6.10
MO surge arrester model for fast front surges. (From Hileman, A.R. et al., *Electra*, no. 133, 132, 1990. With permission.)

where
G is the turn-on conductance, in mho
i is the current through the element, in kA
v is the voltage across the turn-on element, in kV
V_{10} is the discharge voltage for a 10 kA, 8/20 μs current, in kV
G_{ref} is the reference conductance, in mho
V_{ref} is the reference voltage, in kV
I_{ref} is the reference current, which is equal to 5.4 kA
k is a constant ranging between 0.3 and 0.5
T is the reference time, which is equal to 80 μs

Ref. [36] proposes other expression for V_{ref} in addition to Equation 6.11.

The element L can be represented either as an inductance or as an ideal line with a surge impedance and a travel time estimated as follows [36]:

- For outdoor arresters   Inductance: $L = 1\,\mu H/m$ of arrester length
  Ideal line: $Z = 300\ \Omega$; travel time = 3.33 ns/m of arrester length.
- For GIS   Inductance: $L = 0.33\,\mu H/m$ of arrester length
  Ideal line: $Z = 100\ \Omega$; travel time = 3.33 ns/m of arrester length.

2. IEEE model

It is shown in Figure 6.11; it incorporates two time-independent nonlinear resistors (A_0 and A_1), a pair of linear inductors (L_0 and L_1) paralleled by a pair of linear resistors (R_0 and R_1) and a capacitor C. The V–I characteristic of A_1 is slightly less than the 8/20 μs curve while A_0 is 20% to 30% higher. L_1 and R_1 form a low pass filter that sees a decaying voltage across it. A lumped inductance of about $1\,\mu H/m$ for the ground leads should also be included in series with the model. In transients simulations the nonlinear resistors should be modeled as exponential segments as described above. This model was proposed by D.W. Durbak [32] and adopted by the IEEE, including committee papers [37] and standards [5].

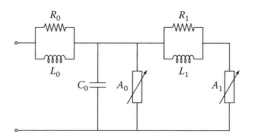

FIGURE 6.11
IEEE MO surge arrester model for fast front surges. (From IEEE Working Group on Surge Arrester Modeling, *IEEE Trans. Power Deliv.*, 7, 302, 1992. With permission.)

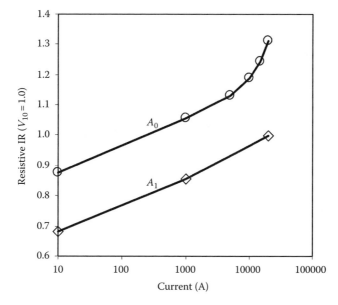

FIGURE 6.12
V–I characteristics for nonlinear resistors. (From Durbak, D.W., *EMTP Newslett.*, 5, 1985.)

For low-frequency surges, the impedance of the filter R_1 and L_1 is very low and A_0 and A_1 are practically in parallel. During high-frequency transients, the impedance of the filter becomes very high and the discharge current is distributed between the two nonlinear branches.

Figure 6.12 shows V–I characteristics of A_0 and A_1, see also Table 6.5, where voltage values are in per unit of V_{10}. A_0 is presented as 5 segment and A_1 as 2 segments.

Formulas to calculate parameters of the circuit shown in Figure 6.11 were initially suggested in [32]. They are based on the estimated height of the arrester, the number of columns of MO disks, and the curves shown in Figure 6.12.

The information required to determine the parameters of the fast front model is as follows:

- d is the height of the arrester, in m
- n is the number of parallel columns of MO disks

TABLE 6.5

Values for A_0 and A_1 in Figure 6.12

Current (kA)	Voltage (per unit of V10)	
	A_0	A_1
0.01	0.875	0.681
1	1.056	0.856
5	1.131	
10	1.188	
15	1.244	
20	1.313	1.000

Source: Durbak, D.W., *EMTP Newslett.*, 5, 1985.

- V_{10} is the discharge voltage for a 10 kA, 8/20 μs current, in kV
- V_{ss} is the switching surge discharge voltage for an associated switching surge current, in kV

Linear parameters are derived from the following equations [37]:

$$L_0 = 0.2 \frac{d}{n} \ (\mu H) \quad R_0 = 100 \frac{d}{n} \ (\Omega) \tag{6.12}$$

$$L_1 = 15 \frac{d}{n} \ (\mu H) \quad R_1 = 65 \frac{d}{n} \ (\Omega) \tag{6.13}$$

$$C = 100 \frac{n}{d} \ (pF) \tag{6.14}$$

These formulas do not always give the best parameters, but provide a good starting point. The procedure proposed by the IEEE WG to determine all parameters can be summarized as follows [37]:

1. Determine linear parameters (L_0, R_0, L_1, R_1, C) from the previously given formulas, and derive the nonlinear characteristics of A_0 and A_1.
2. Adjust A_0 and A_1 to match the switching surge discharge voltage.
3. Adjust the value of L_1 to match the V_{10} voltages.

3. Hysteretic model

The discharge voltage of an MO surge arrester reaches the peak before the discharge current reaches its own peak. This dynamic frequency-dependent behavior can be included by adding an inductance in series with the conventional nonlinear resistance, which reproduces the behavior of MO blocks for low-frequency discharge currents. The *V–I* characteristic of an MO surge arrester with a stepped discharge current has a looping hysteretic tendency. Therefore, the series inductance must be also nonlinear. The equivalent circuit of an MO surge arrester can consist of a series combination of a nonlinear resistor and a nonlinear inductor, as shown in Figure 6.13.

FIGURE 6.13
MO surge arrester model for fast front surges. (From Kim, I. et al., *IEEE Trans. Power Deliv.*, 11, 834, 1996. With permission.)

The principles of the procedure for calculating the nonlinear inductance from the original hysteresis loop were presented in [38]. A previous version of this model was presented in [31].

4. *Simplified model*

A simplified version of the IEEE model was proposed in [39]. According to the authors of this model, the capacitance C in the model shown in Figure 6.11 can be eliminated, since its effect is negligible, and the two resistances in parallel with the inductances can be replaced by a single resistance R, of about $1\,M\Omega$, placed between model terminals to avoid numerical problems, see Figure 6.14.

This model does not take into consideration any physical characteristic of the arrester, and its operating principle is similar to that of the IEEE model:

- The definition of nonlinear resistors characteristics A_0 and A_1 is the same as that for the IEEE model.

- The two inductances are calculated by means of the following equations:

$$L_0 = \frac{1}{12} \cdot (K-1) \cdot V_n \ (\mu H)$$

(6.15)

$$L_1 = \frac{1}{4} \cdot (K-1) \cdot V_n \ (\mu H)$$

(6.16)

where

$$K = \frac{V_{1/T2}}{V_{10}}$$

(6.17)

where
   V_n the arrester rated voltage (in kV)
   V_{10} the discharge voltage for a $10\,kA$, $8/20\,\mu s$ current (in kV)
   $V_{1/T2}$ the discharge voltage for a $10\,kA$ steep current pulse (in kV)

The decrease time T_2 in $V_{1/T2}$ is not specified since it can vary between 2 and $20\,s$ and each manufacturer can choose the preferred value.

The value $V_{1/T2}$ is similar to the front-of-wave (*FOW*) discharge voltage defined in the IEEE standard [5].

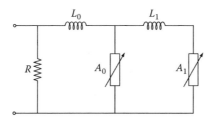

FIGURE 6.14
Simplified MO surge arrester model for fast front surges. (From Pinceti, P. and Giannettoni, M., *IEEE Trans. Power Deliv.*, 14, 393, 1999. With permission.)

The model was refined in a later work [40]. It was proved that whenever $V_{1/T2}$ is not available and the factor K is more than 1.18, the inductance parameters can be calculated as follows:

$$L_0 = 0.01 \cdot V_n (\mu H) \tag{6.18}$$

$$L_1 = 0.03 \cdot V_n (\mu H) \tag{6.19}$$

The following example uses the model adopted by the IEEE [37].

Example 6.2

The procedure to calculate parameters of the fast front model will be applied to a one-column arrester, with an overall height of 1.45 m, being $V_{10} = 248$ kV, and $V_{ss} = 225$ kV for a 3 kA, 300/1000 μs current waveshape. This example was presented in Ref. [37], a summary of the procedure taken from this reference is presented below; readers are referred to the original paper for more details.

- Initial values: the initial parameters that result from using Equations 6.12 to 6.14 are

$$R_0 = (100 \cdot 1.45) / 1 = 145 \ (\Omega)$$

$$L_0 = (0.2 \cdot 1.45) / 1 = 0.29 \ (\mu H)$$

$$R_1 = (65 \cdot 1.45) / 1 = 94.25 \ (\Omega)$$

$$L_1 = (15 \cdot 1.45) / 1 = 21.75 \ (\mu H)$$

$$C = 100 \cdot \frac{1}{1.45} = 68.97 \ (pF)$$

- Adjustment of A_0 and A_1 to match switching surge discharge voltage: the arrester model was tested to adjust the nonlinear resistances A_0 and A_1. A 3 kA, 300/1000 μs double-ramp current was injected into the initial model. The result was a 224.87 kV voltage peak that matches the manufacturer's value within an error of less than 1%.
- Adjustment of L_1 to match lightning surge discharge voltage: the model was tested to match the discharge voltage for a 10 kA, 8/20 μs current. The resulting procedure, similar to that applied in [37], is presented in Table 6.6.

The final model is that shown in Figure 6.15.
Figures 6.16 and 6.17 present the performance of the model. Simulated results depict the discharge voltage developed across the MO surge arrester for different discharge current waveshapes.

TABLE 6.6

Example 6.2: Adjustment of Surge Arrester Model Parameters

Run	L_1 (μH)	Simulated V_{10} (kV)	Difference (%)
1	21.750	262.5	3.67
2	10.875	252.1	0.52
3	5.4375	246.0	0.15
4	7.2500	248.1	

Simulations with a concave current source were performed by using the Heidler model [42]. If the definition of the lightning surge parameters is according to Ref. [43], an adjustment of values is previously needed to derive the parameters of the concave waveshape. Therefore, some care is advisable when using this waveshape. Voltage waveshapes shown in Figure 6.17 were deduced by assuming that the time-to-crest of the current surges were 8 and 2 μs, respectively. However, this time should not be used as a measurement of the rise time [29]. Figure 6.18 shows the results obtained respectively with a time-to-crest of 2 μs and a rise time $(=1.67(t_{90} - t_{30}))$ of 5 kA/μs; one can observe that the difference between voltage peaks is about 2%.

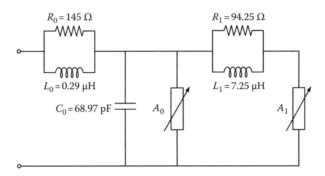

FIGURE 6.15
Example 6.2: Fast front arrester model with IEEE example parameters.

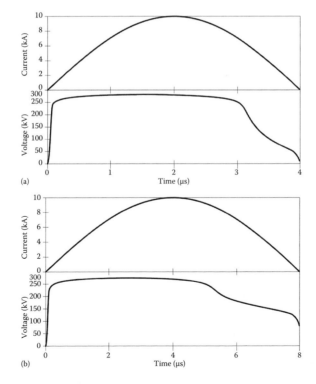

FIGURE 6.16
Example 6.2: Test results with a sinusoidal surge current waveshape. (a) 10 kA, 2 μs time-to-crest surge; (b) 10 kA, 4 μs time-to-crest surge. (From Martinez, J.A. and Durbak, D., *IEEE Trans. Power Deliv.*, 20, 2073, 2005. With permission.)

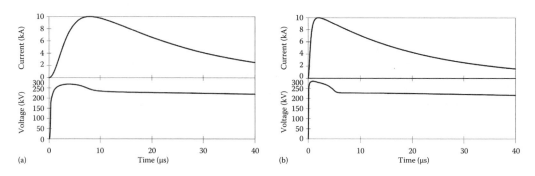

FIGURE 6.17

Example 6.2: Test results with a concave lightning surge current waveshape. (a) 10 kA, 8/20 μs time-to-crest current surge; (b) 10 kA, 2/20 μs time-to-crest current surge. (From Martinez, J.A. and Durbak, D., *IEEE Trans. Power Deliv.*, 20, 2073, 2005. With permission.)

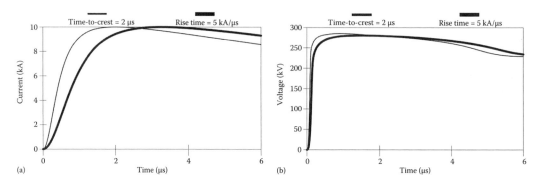

FIGURE 6.18

Example 6.2: Influence of the current slope. (a) Current surges; (b) arrester voltages. (From Martinez, J.A. and Durbak, D., *IEEE Trans. Power Deliv.*, 20, 2073, 2005. With permission.)

6.4.4.3 Parameter Determination from Field Measurements

The estimation of parameters to be specified in the above arrester models is usually based on manufacturer's data and arrester geometry. A technique based on the measured residual voltage derived from 8/20 μs input current test was presented in [44]. The proposed optimization procedure was applied to models presented in [37–39], see Figures 6.11, 6.13, and 6.14.

6.4.5 Models for Very Fast Front Transients

Limited experience is presently available on modeling and validation of MO surge arrester for very fast front transients simulation. An interesting discussion on this subject was presented in Ref. [35]. Basically, the recommended models are similar to those proposed for modeling surge arresters in fast front surge simulations, but representing frequency-dependent behavior by means of a distributed parameter lossless line, see Section 6.7.

6.5 Electrothermal Models

6.5.1 Introduction

Models presented in the previous section were developed to determine residual voltages between arrester terminals. The energy absorption capability is another important issue when selecting an arrester. The thermal behavior can be therefore of concern. Several models based on an analog electrical equivalent circuit have been proposed to include the thermal part of the arrester, see for instance [45–48].

MO varistor materials have a temperature dependence that is significant only at low current densities. Temperature dependence does not need to be represented in simulations for typical overvoltage studies where the arrester currents exceed 1 A; it is of concern when selecting arrester ratings for steady state and TOVs. The temperature-dependent *V–I* characteristic is important only for the evaluation of energy absorbed by the surge arrester, and should not influence the insulation protective margins.

When an MO surge arrester is connected to an electric power system, a very small resistive leakage current will flow through the arrester. When a large surge energy is absorbed, the temperature of ZnO elements increases, causing the resistance value to decrease. As a result, the leakage current becomes greater and the heat generation increases. But, even if the heat generation increases, the temperature of ZnO elements may be below a limit intrinsic to the surge arrester. This occurs when the amount of heat generation is smaller than that of heat dissipation; in such case the temperature of the ZnO elements gradually drops and returns to that prior to the absorption of surge energy. However, when the temperature rise exceeds the temperature limit, the amount of heat generation becomes greater than the heat dissipation, and a thermal runaway occurs. Therefore, this temperature limit is a very important factor.

Failure of MO surge arresters may occur due to breakdown of elements and degradation of elements [26,47]:

1. Arresters may absorb an excessive amount of energy adiabatically (i.e., in a duration shorter than 1 ms). The energy absorption limit is determined by one of two factors: *adiabatic heating*, which means that the energy is nonuniformly absorbed by local areas within the disk before temperature gradient causes excessive stresses leading to cracking; or *puncture* by melting of a region where current is concentrated. A few seconds are required to achieve a uniform temperature distribution. Manufacturers specify a delay between energy discharges to avoid these types of failures. Each rated energy discharge will translate into a sudden (adiabatic) temperature rise of the whole disk. If the rated energy is exceeded, disk failure may result.

2. Arresters may reach excessive temperatures (above 100°C) and fail by thermal runaway (in min or h) at voltages as low as the normal COV. If MO disks are brought up to excessive temperatures and submitted to their MCOV, their leakage current and temperature will increase boundlessly up to breakdown. This is a longer duration phenomenon and results from an unbalance between the heat generated inside the surge arrester and the heat dissipated through their housing. It is a function of MO disk characteristics, arrester design, and ambient temperature.

6.5.2 Thermal Stability of Metal Oxide Arresters

The thermal behavior of MO arresters is influenced by the performance of ZnO elements at higher temperatures and the features of arrester assembly. Field arresters are exposed to stresses, including environmental conditions, such as ambient temperature, solar radiation, and pollution [45], which essentially result in heating up of the ZnO elements.

The thermal stability of an MO arrester is defined by its heat loss-input balance diagram, as shown in Figure 6.19. The electrical power dissipation of a valve element at a constant power-frequency voltage is temperature-dependent (solid line in Figure 6.19), and can be expressed as [49]:

$$P = Ae^{-(W/kT)} \tag{6.20}$$

where
A is a factor that depends on the applied voltage and the valve element material
W is the activation energy
k is the Boltzmann constant
T is the temperature of the valve elements

The heat output to ambient of the valve element is nearly proportional to its temperature rise above the ambient (broken line in Figure 6.19) [49]:

$$Q = B(T - T_a) \tag{6.21}$$

where
T is the temperature of the valve elements
T_a is the ambient temperature
B is the thermal dissipation factor

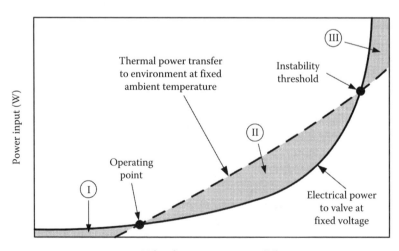

FIGURE 6.19
Heat loss-input diagram for steady-state operation of a ZnO surge arrester. (From Lat, M.V., *IEEE Trans. Power Apparatus Syst.*, 102, 2194, 1983. With permission.)

Thermal stability is achieved when the electrical power dissipation is balanced against heat output to the environment. As shown in Figure 6.19, there are two intersection points: one at low temperature, known as *operating point*, and other at high temperature, known as *instability threshold*. If the power input exceeds heat transfer to environment from the element, (regions I and III in Figure 6.19) then excess energy is stored in the element and its temperature increases. Inversely, if the heat transfer exceeds power input, temperature of the element decreases, (region II in Figure 6.19). Consequently, the valve element temperature will settle at the stable operating point between regions I and II, if the initial valve temperature does not exceed the instability threshold. As the two characteristics diverge beyond the instability threshold point, a thermal runaway will result from temperature excursions above this point.

The heat loss-input diagram can only be used to evaluate thermal instability arising from steady-state conditions; that is, when there has been a sufficient amount of time allowed to stabilize both valve element and arrester housing temperatures.

The valve element operating voltage may be expressed in terms of the applied voltage ratio, which is defined as the ratio of the peak power-frequency voltage to the DC voltage that produces a DC current of 1 mA through the valve element at a temperature of 20°C–25°C. Figure 6.20 shows that a change in the ambient temperature or in operating voltage will shift both the operating and the stability limit temperatures. In a limiting situation, the heat loss curve may become tangential to the power dissipation curve, thus making operating and stability limit temperatures equal. At voltages or ambient temperatures above this point the arrester becomes inherently unstable. Generally, voltages in this range (i.e., TOVs) can be permitted only for limited time duration.

The absorption of relatively low surge energy on a repetitive basis is a nonadiabatic process. A significant fraction of surge energy is dissipated into the environment during intervals between consecutive surges. Cooling between discharges can significantly reduce the overall energy stress. Since most MO arrester applications involve nonadiabatic energy absorption processes, it is evident that a more complex model is required to represent the electrothermal behavior.

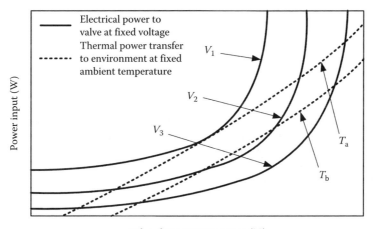

FIGURE 6.20
Influence of the operating voltage ($V_1 > V_2 > V_3$) and the ambient temperature ($T_b > T_a$) on the operating point of a ZnO surge arrester.

The thermal stability limit temperature determined by this method is only valid for conditions where the valve element and the housing have reached thermal equilibrium (steady state). Stability limits for events that occur in short time spans may be up to 40°C higher than the steady-state limits. However, the steady-state limits are used for most applications, because they give conservative design margins and they can be readily derived from the heat loss–gain diagram.

Real world conditions that could cause thermal runaway include long-term overvoltages and deterioration of valve elements. Therefore, this diagram is of limited use, mainly when thermal instability due to sudden energy input, switching, or lightning surges, or even TOV must be considered. A thermal model, capable of representing the steady-state as well as transient operation of an MO surge arrester, is needed to analyze such conditions.

6.5.3 Analog Models

6.5.3.1 Equivalent Circuits

Several approaches have been developed and used to analyze the electrothermal behavior of MO surge arresters [45–56]. They use an analog equivalent or solve the heat transfer equations of an arrester by using either the method of finite differences [52,53], or the finite element method [55]. Only analog models are described in this chapter. For more details about the other approaches, readers are referred to the above references.

Analog models use an equivalent electric circuit to represent the thermal behavior of the surge arrester. In these models, electric currents, voltages, resistances, and capacitances represent heat flows, temperature, thermal resistances, and thermal capacitances, respectively. One of the first approaches for analyzing the thermal behavior of surge arresters was proposed by Lat [45]; it included only a thermal analog whose electrical parameters were derived from experimental measurements of the temperatures after de-energization; the model was validated for distribution surge arresters. This model was later refined [46], and a new model to analyze the electrical behavior of an MO surge arrester was proposed and assembled to the previous thermal analog.

Another approach to analyze the electrothermal behavior of an MO surge arrester was presented in [47]; it used the thermal model proposed by Lat, which was assembled to a new electrical model that provided power input in the form of Watt–loss curve derived from experimentally derived *V–I* curves at different temperatures. The model could be used to predict two values: the instantaneous temperature rise of the MO disks during adiabatic (short duration) heating and the steady-state temperature that the disks would reach to dissipate the power loss. The first value is quite uniform for all disks, while the second one is a function of the arrester design and can vary significantly.

The electrical analog model presented in this section is basically the model presented and validated in Refs. [45] and [46]. The electrical equivalency is based on representing power flow (W) as current and temperature (°C) as voltage. The equivalent units of resistance and capacitance are then °C/W and W·s/°C, respectively. The electrical analog circuit is shown in Figure 6.21, where

C_E, C_H are the thermal capacities of the valve element and adjacent housing, respectively (W·s/°C)

R_{EH} is the thermal resistance from element to external surface of the housing (°C/W)

R_{HAO} is the thermal resistance from housing to ambient, radiation, and natural convection components (°C/W)

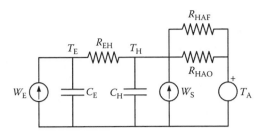

FIGURE 6.21
Electrical analog circuit for the thermal analysis of an MO surge arrester. (From Lat, M.V., *IEEE Trans. Power Apparatus Syst.*, 102, 2194, 1983. With permission.)

R_{HAF} is the thermal resistance from housing to ambient, forced convection component (°C/W)
T_E is the valve element temperature (°C)
T_H is the housing temperature (°C)
T_A is the ambient temperature (°C)
W_E is the power input (W)
W_S is the heat input due to solar radiation (W)

The model can be used to analyze transient or steady-state conditions. In steady state, C_E and C_H can be omitted. The analog circuit uses only lumped components of thermal resistance and heat capacity, although these properties are in reality distributed.

The thermal energy, released within the MO element assembly is transferred to the environment by means of radiation, conduction, and convection. Although heat loss occurs in radial and axial directions, the model assumes that in the middle section of the surge arrester the axial heat flow is minimal and most of the energy transfer takes place radially. Laboratory experiments show that this assumption is valid about three valve element heights from the end of the assembly [45]. Consequently, model parameters derived on the basis of radial heat loss only can accurately represent valve element assemblies consisting of more than seven elements.

The complete electrothermal model is the combination of two parts: (1) an electrical model, to obtain the power dissipated into the MO surge arrester; (2) a thermal model, to obtain the temperature rise of the MO disks. The power input due to solar radiation must be also calculated. The determination of parameters of the thermal model and the calculation of power input due to both solar radiation and to any type of voltage stress are presented in the following subsections.

6.5.3.2 Determination of Thermal Parameters

Thermal resistances: If it is accepted that the heat flow from a valve element is mainly directed radially outward, thermal resistances will be a function of the radial cross-section design layout and of materials involved in the heat transfer path. In air gaps between the valve element and the housing, heat is transferred by radiation, conduction, and convection (although convection may be negligible for very small air gaps). In solid materials, such as porcelain housings or heat transfer pads, the heat is transferred by conduction. On the surface of the arrester housing, heat is lost by radiation and convection. Additional losses may be provided by wind-induced forced convection.

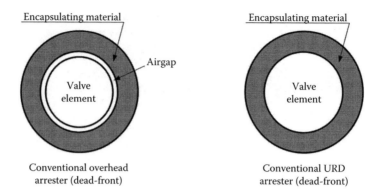

FIGURE 6.22
Some cross-sectional geometries of MO surge arresters.

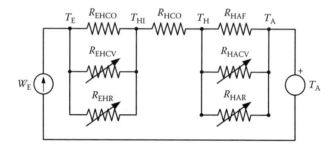

FIGURE 6.23
Electrical analog circuit for thermal steady-state analysis. (From Lat, M.V., *IEEE Trans. Power Apparatus Syst.*, 102, 2194, 1983. With permission.)

Assume that the arrester geometry is one of those shown in Figure 6.22. The components necessary to determine steady-state thermal operation are shown in Figure 6.23. The procedures for deriving the values of all of the components are detailed below.
Heat transfer by radiation: The expressions needed to obtain thermal resistances involved in heat transfer by radiation are presented below [45].

 a. R_{EHR}—thermal resistance between element and housing.

$$R_{\text{EHR}} = \frac{T_{\text{E}} - T_{\text{HI}}}{\varepsilon_{\text{EH}} \cdot K \cdot \left(T_{\text{E}}^4 - T_{\text{HI}}^4\right) \cdot \pi \cdot D_{\text{E}} \cdot h} \tag{6.22}$$

 b. R_{HAR}—thermal resistance between housing and ambient.

$$R_{\text{HAR}} = \frac{T_{\text{H}} - T_{\text{A}}}{\varepsilon_{\text{H}} \cdot k \cdot \left(T_{\text{H}}^4 - T_{\text{A}}^4\right) \cdot \pi \cdot D_{\text{max}} \cdot h} \tag{6.23}$$

where

T_{HI} is the temperature of the internal surface of the housing
T_E is the temperature of the internal surface of the housing
T_H is the temperature of the external surface of the housing
T_A is the ambient temperature
k is the Stefan–Boltzmann constant ($=5.72 \times 10^{-12}$ W/cm^{2}K^4)
ε_{EH} is an equivalent emissivity between the element and the housing

$$\varepsilon_{EH} = \frac{1}{\dfrac{1}{\varepsilon_H} + \dfrac{1}{\varepsilon_E} - 1} \tag{6.24}$$

ε_E is the emissivity of the valve element
ε_H is the emissivity of the housing
h is the arrester height
D_E is the arrester diameter
D_{max} is the maximum housing diameter

Heat transfer by conduction: Heat transfer by conduction can be treated according to expressions for either uniform cross sections (thin heat transfer pads and air gaps) or nonuniform cross sections (arrester housings).

a. R_{EHCO}—thermal resistance of air gap between elements and housing.

$$R_{EHCO} = \frac{D_{HIN} - D_E}{\lambda_A \cdot \pi \cdot (D_{HIN} + D_E) \cdot h} \tag{6.25}$$

b. R_{HCO}—thermal resistance of arrester housing.

$$R_{HCO} = \frac{\ln\left(\dfrac{D_{HOA}}{D_{HIN}}\right)}{2 \cdot \lambda_H \cdot \pi \cdot h} \tag{6.26}$$

where

D_{HIN} is the average internal diameter of the housing
D_{HOA} is the average external diameter of the housing
λ_A is air conductivity at 20°C
λ_H is housing conductivity at 20°C

Heat transfer by natural convection: It is a complex phenomenon which depends on the physical parameters of the transfer medium, in this case air. The analysis can be simplified with certain assumptions.

a. R_{EHCV}—thermal resistance between element and housing due to natural convection.

$$R_{EHCV} = \frac{1}{2.3 \times 10^{-4} \cdot (T_E - T_{HI})^{0.25} \cdot \pi \cdot D_E \cdot h} \tag{6.27}$$

b. R_{HACV}—thermal resistance between housing and ambient due to natural convection

$$R_{HACV} = \frac{1}{2.3\times10^{-4} \cdot (T_H - T_A)^{0.25} \cdot \pi \cdot D_{HOA} \cdot h \cdot \varphi} \tag{6.28}$$

where

$$\varphi = \frac{\text{Shed contour length}}{\text{Distance between sheds}} \tag{6.29}$$

Heat transfer by forced convection: Heat loss by wind-induced forced convection from the arrester surface may be a major factor in determining its performance under field conditions. The thermal resistance due to forced convection can be determined by means of the following expression:

$$R_{HAF} = \frac{1}{\lambda_A \cdot C \cdot (Re)^P \cdot \pi \cdot D_{HOA} \cdot h \cdot \varphi} \tag{6.30}$$

where C and P are empirical constants developed for smooth horizontal cylinders

- For $100 < Re < 5000$ $P = 0.485$ and $C = 0.55$
- For $5000 < Re < 50000$ $P = 0.650$ and $C = 0.13$

and Re is the Reynolds number, which is a function of wind speed (v_A, in km/h)

$$Re = v_A \cdot D_{HOA} \cdot 14.6 \tag{6.31}$$

Some thermal resistances may be combined to yield overall resistances, see Figures 6.21 and 6.23

$$R_{EH} = R_{EHR} \mathbin{/\mkern-6mu/} R_{EHCV} \mathbin{/\mkern-6mu/} R_{EHCO} + R_{HCO}$$

$$R_{HAO} = R_{HAR} \mathbin{/\mkern-6mu/} R_{HACV} \tag{6.32}$$

A more complex equivalent analog circuit could be needed for other arrester housing configurations, see [45].

The solution of this circuit is complicated because of temperature-dependent (nonlinear) behavior of the radiation and convection components. An iterative solution of the different expressions for a fixed power input is usually required, since the values of R_{EH} and R_{HAO} are different at different temperatures.

Thermal capacitances: A capacitance represents the amount of energy that can be stored in a particular component of the arrester. It is a physical property of a particular material, and depends on the total mass and the temperature. The heat capacity of any particular component of an arrester can be determined as follows:

$$C = V \cdot (c + \alpha T) \tag{6.33}$$

where

V is the volume of arrester component, in cm^3
T is the temperature, in °C
c is the specific heat constant, in J/(cm^3 °C)
α is the temperature coefficient, in J/(cm^3 °C^2)

6.5.3.3 Power Input due to Solar Radiation

Solar radiation can account for a significant power input to the arrester. The power (W_S) is injected at the housing surface. The power input to a vertically positioned arrester is relatively constant over a period of 8 h in the day [45]. The power input density to the arrester surface is then approximately 60%–75% of maximum solar radiation intensity (measured at noon) so that the power input to the arrester model can be expressed as follows:

$$W_S = 0.068 \cdot D_{max} \cdot h \tag{6.34}$$

Assuming that power density at the housing surface is uniform, temperature rise of the housing can be obtained from the following expression:

$$\Delta T_{HS} = R_{HAO} \cdot W_S \tag{6.35}$$

This equation gives maximum temperature rises of 13°C–18°C. In reality only the surface zone exposed to the sun receives solar radiation, which results in a nonuniform temperature distribution along the housing circumference.

6.5.3.4 Electric Power Input

An electrothermal model of a surge arrester should be capable of simulating the following situations [46]:

1. Arrester operation under continuous power-frequency voltage with diverse variations of applied voltage, ambient temperature, and valve element degradation.
2. Low magnitude TOV performance prediction. Long-duration, low-magnitude TOV is characterized by power-frequency peak voltages below the arrester voltage that corresponds to 1 A current. Generally, the arrester valve element is the current limiting impedance in the circuit.
3. Adiabatic energy absorption under duty cycle conditions, involving various voltage sources such as lightning surges, capacitor switching surges, line switching surges, and short-duration, high-magnitude TOV (characterized by arrester currents above 1 A).

A typical *V–I* characteristic of a valve block was shown in Figure 6.7. The high-current characteristic is associated with adiabatic energy absorption processes, such as surge energy absorption or large TOV conditions. The low-current characteristics show very significant variations with respect to temperature and waveshape. This behavior determines the response of the arrester to nonadiabatic processes such as power dissipation due to COV and due to long-term, low-magnitude TOV.

- In the high-current region the element voltage drop is mainly governed by current magnitude, and consequently a relatively simple mathematical expression may be used to represent the *V–I* characteristic. The calculation of energy absorption is based on the solution of a model such as that presented above.

- Deriving an explicit mathematical model for the low-current region is a difficult task. (Nonlinear dependence of current on voltage and temperature, capacitive currents, hysteresis effect in AC resistive current waveform, and inconsistency between AC resistive and DC current characteristics.) An empirical approach can be more appropriate; that is, this region can be represented by a look-up table of empirical values of power dissipation at different voltages and temperatures [45,50].

Figure 6.24 shows the power dissipation characteristics of a typical MO valve element. The simulation of the arrester thermal performance requires power dissipation input at rated frequency as a continuous function of temperature at a constant voltage level. The empirical data may be in the form of discrete points at randomly distributed voltage levels, a format which is not directly compatible with the input requirements of the thermal model [46].

A two-step interpolation may be used to convert the test data into a suitable format [46]:

1. Data is interpolated between two adjacent points on each isothermal line shown in Figure 6.24 to obtain power dissipation at the exact operating voltage level. Extrapolation may be used to obtain operating points at power dissipation levels outside of the measured range. The interpolation is based on the following equation:

$$P(V) = \beta e^{\gamma V} \tag{6.36}$$

where
$P(V)$ is the power dissipation
V is the peak rated-frequency voltage
β and γ are interpolation constants

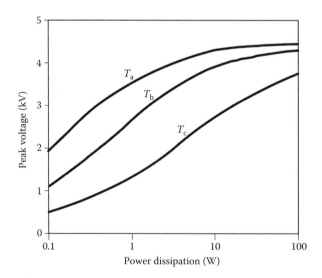

FIGURE 6.24
Power dissipation characteristics of a typical MO valve element ($T_a < T_b < T_c$).

A single value of power dissipation is obtained for each discrete temperature level, representing the constant voltage operation of the valve element.

2. The power dissipation in between temperature levels is also calculated by interpolation, using the following expression:

$$P(T) = \phi e^{\lambda T} \tag{6.37}$$

where
   $P(T)$ is the power dissipation
   T is the temperature of valve element
   ϕ and λ are interpolation constants

Equation 6.37 is then used in a piecewise method to determine power dissipation in the valve element over the full range of temperatures.

6.5.4 Use of Heat Sinks

One method for improving thermal stability is to place heat sinks between the MO elements [56,57]. This has three beneficial effects: (1) in the short term, heat sinks absorb heat from the MO elements to lower their temperature and reduce power dissipation; (2) in the longer term, the elements conduct heat from the large heat sink interfaces to the heat sink–gas interface, thereby increasing heat transfer from the elements to the surrounding gas; and (3) since the greatest reduction in MO temperature occurs near the heat sink surface, this decreases the MO conductivity in that region compared to the middle of the element, which results in the maximum power dissipation occurring near the heat sink interface where heat can be removed from the MO element most efficiently.

6.6 Selection of Metal Oxide Surge Arresters

6.6.1 Introduction

Arresters must be selected to withstand the voltages and the resulting currents through them with a sufficiently high reliability taking into account pollution and other environmental stresses. The selection of an adequate arrester constitutes an optimization process, which has to consider system and equipment parameters.

For a given application, the selection of an appropriate arrester involves considerations of MCOV, protective characteristics, durability, service conditions, and pressure relief requirements [5,9]. Durability and protective level considerations will primarily determine the class of arrester selected. The above information plus arrester geometry will suffice for developing an accurate surge arrester model and validating its applicability. For real applications, other information could appear on the arrester nameplate; e.g., contamination withstand level of the enclosure, type, and identification of the arrester, the year of manufacture. For some studies additional information, such as energy withstand capability, may be required.

This section is aimed at presenting procedures proposed in standards for arrester selection. A summary of the overvoltages that can stress arrester operation and the protective levels are presented prior to the selection procedure.

Although the basic principles of arrester selection and application are similar for transmission and distribution system arresters, there are specific differences that require special consideration for distribution arresters: (1) distribution lines are generally not shielded and therefore are particularly susceptible to direct lightning strokes; (2) systems to be protected by distribution arresters are either three-wire (wye or delta, high- or low-impedance grounded at the source), or four-wire multigrounded wye; (3) distribution equipment includes underground cable systems. On the other hand, arresters installed phase-to-ground may not provide adequate phase-to-phase protection for delta-connected transformer windings. A solution is to apply phase-to-phase arresters. The last subsections discuss the selection of MO surge arresters to protect delta-connected transformers and distribution level equipment.

6.6.2 Overvoltages

The causes of overvoltages in power systems may be external (e.g., lightning) or internal (e.g., switching maneuvers, faults, ferroresonance, and load rejection), and may also occur as a combination of several events (e.g., fault and load rejection). Since their magnitude can exceed the maximum permissible levels of some equipment insulation, it is fundamental to either prevent or to reduce them [5,9,15,58–60]. A short summary of the most important overvoltages is presented in the following paragraphs.

6.6.2.1 Temporary Overvoltages

TOVs are lightly damped power-frequency voltage oscillations of relatively long duration. Proper application of MO surge arresters requires knowledge of the most dangerous and frequent forms of TOV, as well as their duration. When detailed system studies are not available, the overvoltages due to line-to-ground faults are generally used for arrester selection. Arresters on a grounded system are exposed to lower magnitude TOV during single-line-to-ground faults than on a system which is either ungrounded or grounded through an impedance.

6.6.2.2 Switching Overvoltages

Switching overvoltages are of concern in systems above 220 kV and in all systems where the effective surge impedance as seen from the arrester location is low (e.g., cable and capacitor bank circuits). The switching surge duty of MO surge arresters applied on overhead transmission lines increases as the voltage and the length of the line to be protected increase, see Section 6.3.4. In general, transients caused by high speed reclosing impose greater duty than energizing. The application of arresters to protect shunt capacitor banks or cables may require a detailed system study because of the likelihood of high-discharge currents, which may force to select arresters of higher energy capability or to install parallel arrester setups [5].

6.6.2.3 Lightning Overvoltages

Lightning surge voltages that arrive at a station, traveling along a transmission line, are caused by a lightning stroke terminating either on a shield wire, a tower, or a phase conductor. The voltage magnitude and shape of these voltages are functions of the magnitude, polarity, and shape of the lightning stroke current, the line surge impedance, the tower surge impedance and footing impedance, and the lightning impulse critical flashover voltage (CFO, U_{50} in IEC) of the line insulation. For lines that are effectively shielded, the surge voltages caused by a backflash are usually more severe than those caused by a shielding failure; that is, they have greater steepness and greater crest voltage. They are, therefore, the only ones generally considered for analysis of station protection [15,58]. Although simplified calculations can be performed, a detailed transient analysis may be needed to accurately asses the lightning performance of the transmission lines and to obtain the characteristics of the incoming surge.

6.6.3 Protective Levels

As defined in Section 6.3.3, the protective level of an arrester is the maximum crest voltage that appears across the arrester terminals under specified conditions of operation. For gapless MO arresters, the protective level is the arrester discharge voltage for a specified discharge current. Using the IEEE terminology, the protective levels specified in standards are:

1. *Front-of-wave* (FOW) *protective level*: It is the crest discharge voltage resulting from a current wave through the arrester of lightning impulse classifying current magnitude with a rate of rise high enough to produce arrester crest voltage in 0.5 μs. IEC recommend 1 μs.
2. *Lightning impulse protective level* (LPL): It is the highest discharge voltage obtained by test using 8/20 μs discharge current impulses for specified surge voltage waves.
3. *Switching impulse protective level* (SPL): It is the highest discharge voltage measured with a switching impulse current through the arrester according to Table 6.3.

6.6.4 Procedure for Arrester Selection

IEC and IEEE standards propose procedures for selection of MO arresters [5,9]. As an example, Figure 6.25 shows a summary of the steps required to select arresters as suggested in IEEE Std C62.22-1997. Although the sequence proposed in IEC 60099-5 is not the same, both procedures are based on the same calculations, and may be reduced to three main steps:

a. Select an arrester and determine its protective characteristics.
b. Select (or determine) the insulation withstand.
c. Evaluate the insulation coordination.

The whole procedure may be iterative. Since this chapter is intended for application of MO arrester models suitable for time-domain simulations, only the first step is analyzed. The determination of the insulation withstand voltages of protected equipment is out of the scope of this chapter, while insulation coordination evaluation will be analyzed in some illustrative test cases included at the end of this chapter. Readers are referred to the specialized literature [15,58,59].

Remember that using the MO surge arrester model supported by IEEE, the required information is the height of the arrester, the number of parallel columns of MO disks, the discharge voltage for a 10 kA, 8/20 μs current, and the switching surge discharge voltage for an associated switching surge current. This information can be obtained from manufacturer's catalog once the arrester selection has been made using the information discussed in the following paragraphs.

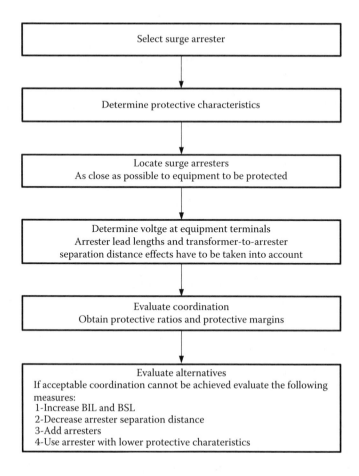

FIGURE 6.25
Procedure for arrester selection. (Reproduced from IEEE Std C62.22-1997, IEEE guide for the application of metal-oxide surge arresters for alternating-current systems.)

1. *Maximum continuous operating voltage* (MCOV–COV): This rating is the maximum designated rms value of power-frequency voltage that may be applied continuously between the terminals of the arrester. Consequently, the arrester rating must equal or exceed the maximum expected steady-state voltage of the system. Proper application requires that the system configuration (single-phase, delta, or wye) and the arrester connection (phase-to-ground, phase-to-phase, or phase-to-neutral) be evaluated. Remember that some insulation (e.g., delta-connected windings) may be exposed to phase-to-phase voltages, see Section 6.6.5.

2. *Temporary overvoltage capability* (TOV): MO surge arresters are capable of operating for limited periods of time at power-frequency voltages above their MCOV rating. The overvoltage that an arrester can successfully tolerate depends on its duration. Manufacturers provide the arrester overvoltage capability in the form of a curve that shows temporary power-frequency overvoltage versus allowable time. Figure 6.26 is a typical TOV curve for station and intermediate class arresters. These curves are sensitive to ambient temperature and prior energy input. The upper curve shows the time the arrester withstands overvoltages; the lower curve is similar to the upper one, but it must be used when the arrester has absorbed

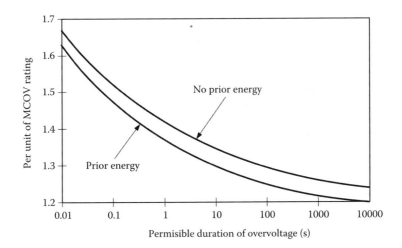

FIGURE 6.26
Typical arrester TOV data. (Reproduced from IEEE Std C62.22-1997, IEEE guide for the application of metal-oxide surge arresters for alternating-current systems.)

prior energy. As already mentioned, the maximum TOV can result from several situations (e.g., overvoltage on an unfaulted phase during a phase-to-ground fault, switching transients, and ferroresonance). During phase-to-ground fault, the surge arrester is subjected to a TOV whose duration is a function of the protective relay settings and fault interrupting devices; therefore, the MCOV of the selected arrester should be high enough so neither the magnitude nor the duration of the TOV exceeds the capability of the arrester.

TOV on unfaulted phases caused by faults to ground can be estimated by using either the *coefficient of grounding* or the *earth-fault factor* (EFF), which are defined as follows [5,9]:

- The coefficient of grounding (COG) is the ratio of the line-to-ground rms voltage on the unfaulted phase to the nominal system line-to-line rms voltage (i.e., the prefault voltage).
- The EFF is the ratio of the line-to-ground rms voltage on the unfaulted phase to the normal system line-to-ground rms voltage.

Obviously

$$EFF = \sqrt{3} \cdot COG \tag{6.38}$$

Given the zero- and positive-sequence impedances at the fault location, the COG can be estimated from the following expressions:

- Single-line-to-ground fault

$$COG = -\frac{\sqrt{3}}{2}\left[\frac{\sqrt{3}K}{2+K} \pm j1\right] \tag{6.39}$$

$$K = \frac{Z_0 + R_f}{Z_1 + R_f} \quad (6.40)$$

- Double-line-to-ground fault

$$COG = \frac{3K}{2K+1} \quad (6.41)$$

$$K = \frac{Z_0 + 2R_f}{Z_1 + 2R_f} \quad (6.42)$$

where
$Z_0 = R_0 + jX_0$
$Z_1 = R_1 + jX_1$

IEEE standards provide COG curves for several relationships of R_1/X_1 [5], while IEC standards provide EFF curves [9,59].

Other causes of TOVs that need consideration are load rejection, resonance effects (e.g., when charging long unloaded lines or when resonances exist between systems), voltage rise along long lines (Ferranti effect), harmonic overvoltages (e.g., when switching transformers), accidental contact with conductors of higher system voltage, backfeed through interconnected transformer windings (e.g., dual transformer station with common secondary bus during fault clearing or single-phase switched three-phase transformer with an unbalanced secondary load), or loss of system grounding.

Table 6.7 shows typical TOV values and the estimated duration due to some of the causes mentioned above. The overvoltage duration depends for many situations on relaying practices. TOVs due to ferroresonance should not be the basis for surge arrester selection, and any possibility of ferroresonant overvoltages should be prevented or eliminated (see Section 6.6.6).

If a combination of causes (e.g., ground faults caused by load rejection) are considered sufficiently probable and they may result in higher TOV values than the single events, the overvoltage factors for each cause have to be multiplied, taking into account the actual system configuration.

3. *Switching surge durability*: The amount of energy absorbed by a surge arrester depends upon the switching surge magnitude and its waveshape, the system impedance and topology, the arrester voltage–current characteristics, and the number of operations (single/multiple events), see Section 6.3.4. The selected arrester should have an energy capability greater than the energy associated with the expected switching surges at the arrester location. The actual amount of energy discharged by an MO surge arrester during a switching surge can be determined through detailed system studies, or estimated by following the procedures presented in Section 6.3.4. Since arresters are constructed with series repeated sections, the energy can be presented in per-unit of MCOV/COV. The number of discharges allowed in a short period of time (1 min or less) is the arrester energy capability divided by the energy per discharge.

TABLE 6.7

Magnitude and Duration of TOVs

Overvoltage Cause	COG (pu)	Duration
Faults to ground		
Grounded systems		
• High short-circuit capacity	0.69–0.80	1 s
• Low short-circuit capacity	0.69–0.87	1 s
• Low impedance	0.80–1.0	1 s
Resonant grounded systems		
• Meshed network	1.0	Several hours
• Radial lines	1.0–1.15	Several hours
Isolated systems		
• Distribution	1.0–1.04	With fault clearing: several seconds Without fault clearing: several hours
Load rejection		
• Extended systems	0.69–0.87	>10 s
• Generator—step-up transformer	0.64–0.87	<3 s
Resonance		
• Saturation phenomena	1.15–1.70	0.5–10 s
• Coupled circuits	1.70–2.90	0.5–10 s
Unloaded long lines		
• High short-circuit capacity	0.58–0.64	>10 s
• Low short-circuit capacity	0.58–0.69	>10 s

Sources: IEEE Std C62.22-1997, IEEE guide for the application of metal-oxide surge arresters for alternating-current systems; Hileman, A.R., *Insulation Coordination for Power Systems*, Marcel Dekker, Newyork, 1999. With permission.

4. *Selection of arrester class*: It should be selected taking into account the importance of the equipment to be protected, the required level of protection, and the following information: available voltage ratings; pressure relief current limits, which should not be exceeded by the available short-circuit current and duration at the arrester location; durability characteristics that are adequate for systems requirements.

Other aspects to be accounted for are discussed in the following paragraphs [5].

A. *Coordinating currents*: They are needed to determine the protective levels of the arrester. A distinction between coordinating currents for lightning surges and for switching surges is to be made.

1. *Lightning surges*: Factors that affect this selection include the importance and the degree of protection desired, line insulation, ground flash density, and the lightning performance of the line. The appropriate coordinating current for lightning surges depends strongly on the line shielding:

• *Completely shielded lines*: The lightning performance of shielded lines is based on their shielding failure and backflashover rates. A line is considered effectively shielded when the number of line insulation flashovers due to shielding failures is negligible, and backflashover is the predominant mechanism of line insulation flashover.

- *Partially shielded lines*: They are generally shielded for a short distance adjacent to the station. In these cases, higher arrester discharge currents may be required. In assessing the arrester discharge current, it is necessary to consider the ground flash density, the probability of strokes to the line exceeding a selected value, and the percentage of total stroke current that discharges through the arrester.

- *Unshielded lines:* This situation is usually limited to either lower voltage lines (i.e., 34.5 kV and below) and/or to lines located in areas of low lightning ground flash density. When located in areas of high lightning ground flash density, the coordinating current should not be less than 20 kA. In severe thunderstorm areas, higher levels should be considered. For lines located in areas of low lightning ground flash density, coordinating currents may be similar to those for effectively shielded lines in areas of high lightning ground flash density.

2. *Switching surges*: The current an arrester conducts during a switching surge is a function of the arrester ratings and its location. The effective impedance seen by the arrester during a switching surge can vary from several hundred ohms for an overhead transmission line to tens of ohms for arresters connected near cables and large capacitor banks. In these two cases, the arrester current and the resulting arrester energy vary significantly for a switching surge of a given magnitude and duration.

    Guidelines to select the coordinating current are provided in standards [5,9]. As for other aspects, a better quantification may be obtained through accurate time-domain simulation.

B. *Location*: As a general rule, the voltage at the protected equipment in a station is higher than the arrester discharge voltage due to the separation distance between arrester and equipment. It is always good practice to reduce this separation to a minimum. A major factor in locating arresters within a station is the shielding of lines and the station, since the probability of high voltages and steep fronts within the station is reduced by the station shielding. However, the majority of strokes are to the incoming lines.

1. *Unshielded lines*: They are subjected to the highest lightning currents and voltage rates-of-rise. The minimum possible separation is recommended for installations where complete shielding is not used. With a single unshielded incoming overhead line, the arrester should be located as near as possible to the terminals of the equipment (usually a transformer) to be protected. When several unshielded incoming overhead lines meet in the station, the incoming overvoltage waves are reduced by refraction. However, it is important to consider the case when one or more of the lines may be out of service. In these cases, the reduction by refraction is lower, and insulation in equipment such as circuit breakers, potential transformers, or current transformers connected on the line side might be damaged; so arresters may be required at the respective line entrances.

2. *Shielded lines*: Incoming voltages from shielded lines are lower in amplitude and steepness than voltages from unshielded lines. In many cases, this allows

some separation between the arresters and the insulation to be protected. In a single shielded overhead line scenario, one set of arresters may suffice to provide protection to all equipment. At stations with multiple shielded incoming overhead lines, arresters are not always required at the terminals of every transformer. Important installations may justify a detailed transient study.

3. *Cable-connected equipment*: The apparatus within the station are connected to the cable, which in turn is connected to an overhead line. The overhead line may or may not be shielded at the line–cable junction. It is always preferable to install arresters at the overhead line–cable junction. In the case of unshielded overhead lines, it may be recommendable to install additional protective devices a few spans before the junction. If it is impossible or undesirable to install arresters at the protected equipment terminals, it is then necessary to evaluate the protection that can be obtained with an arrester at the junction. Arresters installed at the line–cable junction should be grounded to the station ground through a low-impedance path.

The following example presents how to select MO surge arresters for protection of an air insulated substation.

Example 6.3

Consider the diagram shown in Figure 6.27. It corresponds to a 50 Hz, 220 kV single-line substation. The objective is to select the MO surge arresters in order to obtain a specified MTBF. The information required for arrester selection is as follows:

- Power system: Frequency = 50 Hz
                  Rated voltage = 220 kV
                  Grounding = Low-impedance system, EFF = 1.4
                  Duration of temporary overvoltage = 1 s

- Line: Length = 80 km
                  Span length = 250 m
                  Insulator string strike distance = 2.0 m
                  Positive polarity CFO (U50 in IEC) = 1400 kV
                  Conductor configuration = 1 conductor per phase
                  Capacitance per unit length = 10.5 nF/km

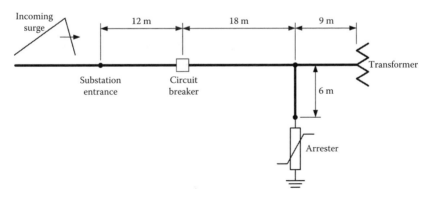

FIGURE 6.27
Example 6.3: Diagram of a 220 kV substation.

Backflashover rate (BFR) = 2.0 flashovers/100 km year
Ground wire surge impedance = 438 Ω
High-current grounding resistance = 25 Ω
Maximum switching surge at the remote end = 520 kV (peak value)
- Substation: Single-line substation in an area of high lightning activity
MTBF = 100/200 years
Transformer capacitance = Between 2 and 4 nF.

Assume that the surge impedance of each substation section involved in calculations is the same as that of the surge impedance of the line, 400 Ω.

The substation is located at sea level, and calculations are performed assuming standard atmospheric conditions (see Chapter 2).

A. *Arrester selection*

In IEC the rated voltage is the TOV capability at 10 s with prior energy, TOV_{10}, whose value can be obtained by means of the following expression [9]:

$$TOV_{10} = TOV_C \cdot \left(\frac{t}{10}\right)^m \qquad (6.43)$$

where
TOV_C is the representative temporary overvoltage
t is the duration of this overvoltage
m is a factor, for which the recommended value is 0.02

The standard maximum system voltage for 220 kV is U_m = 245 kV in IEC, and 242 kV in IEEE. Taking into account that there is a low-impedance grounding system with an EFF of 1.4 and the duration of the TOV of 1 s, using the IEC standard maximum voltage, the following values are obtained:

- MCOV–COV

$$COV = \frac{245}{\sqrt{3}} = 141.45 \, kV$$

- TOVC

$$TOV_C = 1.4 \cdot \frac{245}{\sqrt{3}} = 198.0 \, kV$$

- TOV_{10}

$$TOV_{10} = 1.4 \cdot \frac{245}{\sqrt{3}} \cdot \left(\frac{1}{10}\right)^{0.02} = 189.1 \, kV$$

Table 6.8 presents ratings and protective data of surge arresters for this voltage level guaranteed by one manufacturer. The nominal discharge current is 20 kA according to IEC and 15 kA

TABLE 6.8

Example 6.3: Surge Arrester Protective Data

Max. System Voltage U_m (kV$_{rms}$)	Rated Voltage U_r (kV$_{rms}$)	MCOV		TOV Capability 10 s (kV$_{rms}$)	Maximum Residual Voltage with Current Wave	
		IEC U_c (kV$_{rms}$)	IEEE MCOV (kV$_{rms}$)		30/60 μs 3 kA (kV$_{peak}$)	8/20 μs 10 kA (kV$_{peak}$)
245	192	154	154	211	398	437
	198	156	160	217	410	451
	210	156	170	231	435	478
	214	156	173	235	445	488
	216	156	175	237	448	492

according to IEEE. Note that a 20 kA rated current should be used for a higher rated voltage according to IEC. The arrester housing is made of silicone.

The switching surge energy discharged by the arrester is estimated using Equation 6.5. The result is 113.6 kJ, or less than 1 kJ/kV for any arrester of Table 6.8, which is lower than the usual switching impulse energy ratings of station-class arresters.

Since the first arresters have values of MCOV/COV and TOV too close to the expected values in the system, the selected arrester in this example is the third one.

Therefore, the ratings of the arrester will be as follows:

- Rated voltage (rms): $U_r = 210$ kV
- MCOV (rms): $U_c = 156$ kV (170 kV according to IEEE)
- TOV capability at 10 s: $TOV_{10} = 231$ kV
- Nominal discharge current (peak): $I = 20$ kA (15 kA according to IEEE)
- Line discharge class: Class 4.

From the manufacturer data sheets, the height and the creepage distance of the selected arrester are respectively 2.105 and 7.250 m.

B. *Arrester model*

The procedure to calculate parameters of the fast front model will be applied to a one-column arrester, with an overall height of 2.105 m, being $V_{10} = 478$ kV, and $V_{ss} = 435$ kV for a 3 kA, 30/60 μs current waveshape.

- Initial values: The initial parameters that result from using Equations 6.12 to 6.14 are

$$R_0 = 100 \cdot 2.105 = 210.5 \ (\Omega)$$

$$L_0 = 0.2 \cdot 2.105 = 0.4210 \ (\mu H)$$

$$R_1 = 65 \cdot 2.105 = 136.825 \ (\Omega)$$

$$L_1 = 15 \cdot 2.105 = 31.575 \ (\mu H)$$

$$C = 100 \cdot \frac{1}{2.105} = 47.51 \ (pF)$$

TABLE 6.9

Example 6.3: Adjustment of Surge Arrester Model Parameters

Run	L_1 (mH)	Simulated V_{10} (kV)	Difference (%,	Next Value of L_1
1	31.57	496.7	3.76	15.78
2	15.78	480.0	0.42	14.20
3	14.20	478.2	0.04	–

- Adjustment of A_0 and A_1 to match switching surge discharge voltage: the arrester model was tested to adjust the nonlinear resistances A_0 and A_1. A 3 kA, 30/60 μs double-ramp current was injected into the initial model. The result was a 436.1 kV voltage peak that matches the manufacturer's value within an error of less than 1%.

- Adjustment of L_1 to match lightning surge discharge voltage: next, the model was tested to match the discharge voltages for an 8/20 μs current. This is now made by modifying the value of L_1 until a good agreement between the simulation result and the manufacturer's value is achieved. The resulting procedure is presented in Table 6.9.

C. Incoming surge

A conservative estimate of the crest voltage for the incoming surge is to assume it is 20% above the CFO [58]. For the line under study, the crest voltage is 1.2*1400 = 1680 kV.

The distance to flashover is obtained by using the following expression:

$$d_m = \frac{1}{n \cdot (\text{MTBF}) \cdot \text{BFR}} \tag{6.44}$$

where

  n is the number of lines arriving to the substation
  MTBF is the mean time between failures of the substation
  BFR is the backflashover rate of the line

The substation is designed for a given MTBF and two values are considered: 100 and 200. Since the BFR of the line is 2.0 flashovers/100 km years, the span length is 250 m, and it is a single-line substation (n = 1), the distance to flashover for each case is

- MTBF = 100 years $d_m = \dfrac{1}{100 \cdot (2/100)} = 0.50$ (km)

- MTBF = 200 years $d_m = \dfrac{1}{200 \cdot (2/100)} = 0.25$ (km)

These distances coincide with a tower location, so they do not have to be modified. That is, $d = d_m$.

The steepness, S, and tail time, t_h, of the incoming surge are calculated by using the following expressions [58]:

$$S = \frac{K_S}{d} \tag{6.45}$$

$$t_h = \frac{Z_g}{R_i} t_s \tag{6.46}$$

TABLE 6.10

Corona Constant

Conductor	K_S (kV·km)/µs
Single conductor	700
Two conductor bundle	1000
Three or four conductor bundle	1700
Six or eight conductor bundle	2500

where
  K_S is the corona constant, obtained from Table 6.10
  d is the backflashover location
  Z_g is the shield (ground) wire surge impedance
  R_i is the high-current resistance
  t_s is the travel time of one line span

For a single conductor line $K_S = 700$ (kV km)/µs. The other values to be used in the above expressions are $Z_g = 438\,\Omega$, $R_i = 25\,\Omega$, and $t_s = 0.833\,µs$. The incoming surge corresponding to each MTBF will have the following characteristic values:

- MTBF = 100 years $S = 1400$ (kV/µs) $t_f = 1.20$ (µs)
- MTBF = 200 years $S = 2800$ (kV/µs) $t_f = 0.60$ (µs)

where t_f is the front time. The tail time is in both cases the same, $t_h = 14.6\,µs$.

D. Simulation results
The voltage at the different equipment locations (station entrance, circuit breaker, arrester-bus junction, transformer) must be calculated taking into account the power-frequency voltage at the time the incoming surge arrives to the substation. IEEE Std 1313.2 recommends a voltage of opposite polarity to the surge, equal to 83% of the crest line-to-neutral power-frequency voltage [58]. In the case under study, the value of this voltage is 149 kV.

Figure 6.28 shows the results obtained with the incoming surge deduced for a MTBF = 200 years. Table 6.11 shows the simulation results obtained with the two values of the MTBF and the two transformer capacitances.

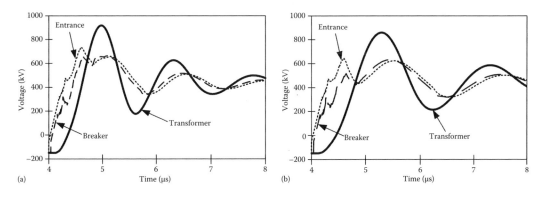

FIGURE 6.28
Example 6.3: Simulation results—MTBF = 200 years. (a) Transformer capacitance = 2 nF; (b) transformer capacitance = 4 nF.

TABLE 6.11

Example 6.3: Voltage at Substation Equipment (kV)

MTBF	Location	Transformer Capacitance	
		2 nF	4 nF
100 years	Station entrance	819	663
	Circuit breaker	764	635
	Arrester–bus junction	643	630
	Transformer	752	785
200 years	Station entrance	737	645
	Circuit breaker	669	632
	Arrester–bus junction	684	657
	Transformer	917	858

E. Estimating substation equipment BIL

The BIL of non-self-restoring equipment (e.g., transformer, breaker) is estimated applying a safety margin. Although there are some discrepancies between IEEE standards [5,58], this safety margin should be in all cases equal or greater than 15%. IEEE standards take into account the time-to-crest of the overvoltage, although the border value is different in each standard. IEEE Std 1313.2 also recommends to account for the tail time constant. Another aspect to be considered is that IEC standards do not consider the front-of-wave (FOW) protective level.

IEEE Std 1313.2 recommends to select the BIL for transformers and other equipment as follows:

- *Transformers*: Apply a safety margin of 15% taking into account voltage and time-to-crest values

$$\text{BIL} = 1.15 \cdot V_t \qquad \text{if the time-to-crest value is more than } 3\,\mu s \qquad (6.47a)$$

$$\text{BIL} = 1.15 \cdot \frac{V_t}{1.10} \qquad \text{if the time-to-crest value is less than } 3\,\mu s \qquad (6.47b)$$

where V_t is the peak voltage-to-ground at the transformer.

- *Other equipment*: Apply the following expressions:

$$\text{BIL} = \frac{V_e}{\delta} \qquad \text{if } \frac{V_e}{V_r} \leq 1.15 \qquad (6.48a)$$

$$\text{BIL} = \frac{V_e}{1.15\delta} \qquad \text{if } \frac{V_e}{V_r} \geq 1.15 \qquad (6.48b)$$

where

V_e is the peak voltage-to-ground at the equipment
V_r is the arrester discharge voltage
δ is the relative air density

For this example, $\delta = 1$, and only the second expression must be applied.

Table 6.12 shows the standardized values of withstand voltages according to IEEE and IEC for the rated voltage. Considering the recommendations presented above and using the largest voltage

TABLE 6.12

Standard Insulation Levels

Standard	Maximum System Voltage (kV$_{rms}$)	Standard Rated Short-Duration Power-Frequency Withstand Voltage (kV$_{rms}$)	Standard Rated Lightning Impulse Withstand Voltage (kV$_{peak}$)
IEEE	242	275	650
		325	750
		360	825
		395	900
		480	975
			1050
IEC	245	(275)	(650)
		(325)	(750)
		360	850
		395	950
		460	1050

Sources: IEEE Std 1313-1996, IEEE Standard for Insulation Coordination—Definitions, Principles and Rules; IEC 60071-1, Insulation Co-ordination, Part 2: Definitions, principles and rules, 2006. With permission.

Note: Values in brackets are considered insufficient by IEC to prove that the required phase-to-phase withstand voltages are met.

TABLE 6.13

Example 6.3: BIL Selection

MTBF	Equipment	Voltage (kV)	BIL Required (kV)	BIL Selected (kV)
100 years	Station entrance	819	712	850–825
	Circuit breaker	764	664	850–825
	Arrester–bus junction	643	559	850–825
	Transformer	785	903	1050
200 years	Station entrance	737	641	850–825
	Circuit breaker	669	582	850–825
	Arrester–bus junction	684	595	850–825
	Transformer	917	1055	–

obtained by simulations with both capacitance values, the BIL required and the BIL selected for each equipment/location are shown in Table 6.13 [61, 62].

The selected arrester is adequate for an MTBF of 100 years, but it cannot guarantee an adequate protection of the transformer if a 200 MTBF is desired. In this second case some of the measures indicated in Figure 6.25 (e.g., an arrester with lower protective characteristics) should be evaluated. For instance, there are other arresters that could be considered for selection from Table 6.8 with a lower residual voltage.

Another aspect to be mentioned is that the positive polarity CFO is unusually high for a 220 kV line. Thus, this case illustrates a problem that can occur in insulation coordination studies: increasing the withstand voltage of one equipment can translate the insulation problem to other part of the system. In this case the increase beyond normal of the insulator string lengths could avoid achieving a 200 MTBF at the substation because the incoming surge will arrive with a too high voltage. Of course, a more detailed simulation of the study zone would be advisable to support this conclusion.

The example is aimed at illustrating how to select arrester ratings and evaluate insulation coordination. More aspects must be considered for a real substation (e.g., substation air clearances) [5]. In addition, a detailed transient study of the transmission line lightning performance might be required if the line is not effectively shielded. In such case, the polarity of the lightning return strokes should be considered as well as the different behavior of the line insulation in front of the impulse polarity, see Chapter 2.

6.6.5 Phase-to-Phase Transformer Protection

Phase-to-phase overvoltages exceeding transformer insulation withstand can result from the following causes [5]:

a. *Switching surges:* High phase-to-phase switching overvoltages may occur due to capacitor bank switching or misoperation of capacitor bank switching devices [63,64].

b. *Lightning surges:* Lightning initiates current and voltage waves which propagate along the struck phase conductor and also induce voltage on the other phase conductors. At the struck location, the induced voltages have the same polarity as the struck phase voltage, and the phase-to-phase voltage is just the difference of the struck and the induced phase voltages. However, due to the propagation, it is possible for the voltage waveforms to become of opposite polarity, and the maximum phase-to-phase overvoltage could be as high as the sum of the absolute values of the peak voltages on the struck and the induced phases. In such case, the resulting phase-to-phase overvoltage can exceed a delta-connected transformer insulation withstand level [65].

c. *Surge transfer through transformer windings:* Lightning surges entering a transformer terminal can excite the natural frequencies of delta-connected windings resulting in phase-to-phase overvoltages in excess of the transformer insulation withstand level [65].

When surge arresters are installed phase-to-ground at each terminal of the delta-connected transformer windings, each winding is protected by two arresters connected in series through their ground connection. The protective level of these two arresters may not provide the recommended protective ratios for the transformer insulation. The protection of delta-connected transformer windings can be accomplished by either of the arrangements shown in Figure 6.29: Figure 6.29a represents a six-surge arrester arrangement,

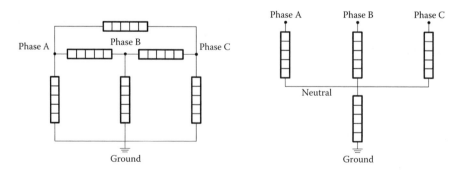

FIGURE 6.29
Phase-to-phase protection of delta winding. (From Keri, A.J.F. et al., *IEEE Trans. Power Deliv.*, 9, 772, 1994. With permission.)

consisting of three phase-to-phase and three phase-to-ground arresters; Figure 6.29b represents a four-legged surge arrester arrangement, consisting of three surge arresters connected from three phases to common neutral, and one arrester connected from the common neutral to ground.

The following procedures are recommended for selection of surge arrester ratings and capabilities [5,65]:

  a. Six-surge arrester arrangement: (1) the traditional selection process should be used to select the phase-to-ground surge arresters; (2) the phase-to-phase surge arrester MCOV rating should be equal to or slightly greater than the maximum phase-to-phase system voltage.

  b. Four-legged surge arrester arrangement:

   • Phase-to-neutral arrester MCOV should be equal to or slightly greater than the maximum phase-to-phase system voltage divided by square root of three. These phase-to-neutral arresters must be also matched to avoid overstressing the neutral-to-ground surge arrester.

   • Neutral-to-ground arrester MCOV can be selected by using the following procedure: (1) determine the minimum required phase-to-ground MCOV based on the traditional phase-to-ground requirements. (2) Subtract the phase-to-neutral arrester MCOV from the minimum required phase-to-ground MCOV obtained just before. (3) Select a surge arrester MCOV rating equal to or slightly greater than this value. If the phase-to-neutral arrester MCOV rating had to be increased to match available MCOV ratings, then the neutral-to-ground arrester MCOV may be reduced, provided the conditions discussed above are met. This iteration will permit the lowest ratings to be used. Proper insulation coordination should be established between the series combination of the phase-to-neutral and neutral-to-ground arrester, and the transformer phase-to-ground insulation.

6.6.6 Arrester Selection for Distribution Systems

Distribution arresters are usually selected so that they can be used anywhere on a system rather than for a particular location. Insulation coordination for distribution systems is in general based on lightning surge voltages, although severe arrester stresses can occur when arresters are installed to protect switching capacitor banks, or when arresters are subjected to ferroresonant or backfeed overvoltages. A summary of the procedure recommended in IEEE standard is presented below [5].

General procedure: As for transmission systems, proper application of MO surge arresters on distribution systems requires knowledge of the maximum normal operating voltage and the magnitude and duration of TOVs. For arrester application on distribution systems, the TOV is usually based on the maximum phase-to-ground voltage that can occur on unfaulted phases during single-line-to-ground faults. This calculation will depend on the distribution system configuration:

  1. *Three-wire, low-impedance grounded systems* (*grounded at source only*): A general practice is to assume the TOV on the unfaulted phases rises to 1.4 pu, although the voltage rise on the unfaulted phases can exceed this value if the system is grounded through an impedance. The voltage on the unfaulted phases can rise

to an order of 80% and be unequal due to the ground resistance at the point of fault. Where it is possible to backfeed a portion of the circuit which has been disconnected from the source through transformers or capacitors that are connected to that part of the circuit, the TOV should be assumed to be equal to the maximum phase-to-phase voltage. If TOV duration cannot be determined, the arrester should be selected so that its MCOV rating equals or exceeds the maximum system phase-to-phase voltage.

2. *Three-wire, high-impedance grounded systems, or delta-connected systems*: During a single line-to-ground fault, the line-to-ground voltage on the unfaulted phases rises to line-to-line values. Because fault current values are very low, relaying schemes could allow this type of fault to exist for a very long time. A general practice is to choose an arrester with an MCOV rating greater than the maximum system phase-to-phase voltage, although a lower MCOV rating might be used when relaying schemes limit the duration.

3. *Four-wire, multigrounded-wye systems*: It can be assumed that the TOV on unfaulted phases exceeds the nominal line-to-ground voltage by a factor of 1.25 when the line-to-ground resistance is low (i.e., less than 25 Ω) and the neutral conductor size is at least 50% of the phase conductor. The factor can exceed 1.25 when smaller size neutral conductors are used.

There may also be seasonal effects; that is, systems which may be effectively grounded when there is a high soil water content may change to a noneffectively grounded condition as the water content is reduced. A system is considered to be noneffectively grounded when the coefficient of grounding exceeds 80%. High-impedance arcing faults on an ungrounded system can result in high overvoltages, greater than phase-to-phase voltage, and cause arrester failure.

Large voltage variations are allowed within the distribution system as long as the voltage at the customer point is within limits. Special care is required to make sure that the MCOV rating of MO surge arresters is not exceeded where regulators are used to control system voltage, since voltage swings may result when three single-phase voltage regulators are installed at a location where the neutral is permitted to float.

Normal-duty versus heavy-duty surge arresters: The selection of surge arrester duty is more a choice of the user than a decision based on performance data. Heavy-duty arresters should be used when a greater than normal withstand capability is desired or required; for instance, in areas with a high keraunic level, or to discharge surges generated by switching large capacitor banks. It is important to keep in mind that: (1) heavy-duty arresters generally have a lower discharge voltage characteristic than normal-duty arresters; (2) the total lightning current is unlikely to be discharged by a single arrester, and the amount of current discharged by an arrester depends on the distance between the point of impact and the arrester, the presence of other arresters, and the level of line insulation.

Overvoltages caused by ferroresonance: Ferroresonant overvoltages result when a saturable inductance is placed in series with a capacitance in a lightly damped circuit. This circuit topology can result when one or two transformer phases are disconnected from the source. The capacitance may be provided by overhead lines, underground cables, the internal capacitance of the transformer windings, or by shunt capacitor banks [60,66]. Overvoltage magnitudes depend on the transformer primary winding connection and the amount of capacitance present compared to the transformer characteristics. For transformers with ungrounded primary connections, ferroresonant overvoltages can easily exceed 3–4 pu [67,68]. Short underground cable lengths may be sufficient to cause crest voltages of this severity [69]. Internal transformer capacitance may also suffice to cause ferroresonance in ungrounded primary transformers and banks. The ferroresonant overvoltage can persist for as long as the open-phase condition, which may be intentional or due to the operation of a protective device such as a fuse. Single-phase protective devices should be avoided if their operations can result in an open-phase condition present for a long period of time.

Although ferroresonant overvoltages can result in arrester failure, the ferroresonant circuit is a high-impedance source and gapless MO surge arresters limit the voltage while discharging relatively small currents; consequently, accumulation of energy is usually relatively slow and an arrester can often withstand exposure to ferroresonant overvoltages for a period of minutes or longer. Arrester TOV curves do not accurately reflect the ability of MO surge arresters to withstand ferroresonant overvoltage duty. If the arrester valve elements are raised to an excessive temperature due to a ferroresonant overvoltage, arrester failure may not occur until the open phase, to which the arrester is connected, is reclosed into the low-impedance system source. In many cases, the arrester is not overheated by ferroresonance during the brief time required to complete a switching operation. In fact, MO surge arresters can be used to limit ferroresonant overvoltages in situations where the ferroresonance cannot be easily avoided [69].

Overvoltages caused by backfeed: They can occur with configurations such as ungrounded wye-delta banks or in dual-transformer stations [5]. Voltages exceeding 2.5 pu can appear on the open primary of an ungrounded wye-delta bank by feedback from the secondary when a three-phase secondary load is removed, leaving a single-phase load connected. Several alternative practices may be more recommended than installing higher rated MO surge arresters on these wye-delta banks [5]: (1) balancing the load so that the load on each phase of the delta is no more than four times that on each of the other two phases; (2) grounding the wye; (3) closing the disconnect last on the phase that has the largest single-phase load; (4) applying a grounding resistor or reactor in the neutral of the ungrounded-wye windings; (5) closing a neutral grounding switch during the energization of the phases and opening it after all three phases have been closed; (6) placing arresters on the source side instead of the load side.

6.7 Arrester Applications

6.7.1 Introduction

Previous sections have shown that MO surge arresters are installed to protect power equipment against switching and lightning overvoltages, but they are selected from the estimated TOV that can be caused at arrester locations. Insulation withstand voltages may be selected to match the characteristics of certain arresters, or arresters may be matched to available insulation. In general: (a) surge arrester ground terminals are connected to the grounded parts of the protected equipment; (b) both line and ground surge arrester connections are as short as practical; (c) stations and transmission lines are shielded against direct strokes, while distribution lines are not.

The rest of this section is dedicated to discussing the different arrester protection scenarios in both transmission and distribution levels, and how to solve the main problems that can arise when protecting power equipment at both transmission and distribution voltage levels.

6.7.2 Protection of Transmission Systems

Protection of transmission lines. Transmission line insulators are usually protected from lightning flashover by overhead shield wires. The application of surge arresters may significantly reduce the flashover rate of a transmission line. Line surge arresters may be installed phase-to-ground, either in parallel with the line insulators, or built into the insulators. Their protective level should be greater than the protective levels of the adjacent substation arresters. This will reduce the energy absorbed by the line arresters due to switching surges and therefore reduce the possibility of a line arrester failure. The appropriate location of surge arresters depends on many factors including lightning ground stroke density, exposure, span length, conductor geometry, footing resistance, insulation level, and desired line performance. In general, the more arresters are installed, the better the line performance.

The first line surge arresters were installed in the 1980s [70,71]. Since then, MO surge arrester design and reliability have been improved [72–75]. In the beginning arresters were seen as a solution for improving the performance of shielded transmission lines with a poor lightning performance, mostly due to the impossibility of achieving a low enough grounding impedance. Presently, arresters are also seen as a solution to obtain a good lightning performance of unshielded lines.

The energy to be discharged by line surge arresters includes the corresponding to the first and to the subsequent strokes [76–78]. For shielded lines, the required energy withstand capability is in general within arrester capability since strokes with the highest peak current magnitudes terminate on a tower or a shield wire and most of the stroke energy is discharged through the tower grounding impedance. And only strokes with low peak current magnitudes will impact phase conductors. For unshielded lines, the energy capability is critical since strokes with any peak current magnitudes may reach a phase conductor [15]. Studies related to the application of line surge arresters were detailed in Refs. [79–93].

The following example presents an illustrative case of line arrester application.

Example 6.4

Figure 6.30 shows the tower design of the test transmission line with two conductors per phase and one shield wire, whose characteristics are presented in Table 6.14.

The goal of this example is to select the ratings of the arresters and prove that they will significantly improve the performance of the line. Data relevant for this example follows:

- Power system: Frequency = 50 Hz
  Rated voltage = 400 kV
  Grounding = Low-impedance system, EFF = 1.4
  Duration of temporary overvoltage = 1 s
- Line: Length = 45 km
  Span length = 390 m
  Strike distance of insulator strings = 3.066 m
  Conductor configuration = 2 conductor per phase
  Low-current low-frequency grounding resistance = 50 Ω
  Soil resistivity = 200 Ω m.

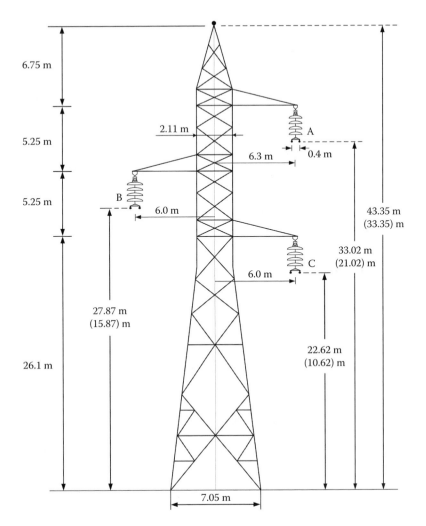

FIGURE 6.30
Example 6.4: 400 kV line configuration (values in brackets are midspan heights).

TABLE 6.14

Example 6.4: Conductors Characteristics

	Type	Diameter (mm)	Resistance (Ω/km)
Phase conductors	Cardinal	30.35	0.0586
Shield wire	7N8	9.78	1.4625

The first step is to check the lightning performance of the test line. A poor lightning performance can be due to an inadequate design of the line (e.g., poor grounding, inadequate shielding) and to a very high atmospheric activity.

An unacceptable lightning flashover rate can be due either to a very high BFR, or to a very high shielding failure flashover rate (SFFR), or to both. A reason for a very high BFR is a poor grounding, while the main reason for a very high SSFR is a poor shielding. An additional reason that can affect both rates is a too short strike distance of insulator strings.

The grounding resistance specified above is not small, but it is not as high as to significantly increase the lightning flashover rate of the line. Soil ionization may significantly decrease the effective grounding resistance value, so if the line model includes a high-current grounding model (see Chapter 2), the lightning performance of the test line is not so affected by the average grounding resistance value assumed for this line. However, the strike distance of insulator strings is rather short for the rated voltage of this line, so it certainly affects the line behavior. Finally, it is advisable to analyze the shielding provided to this line by the single sky wire.

The study has been organized as follows. The initial step assesses the performance of the shielding design, the goal is to apply an incidence model and obtain the maximum peak current magnitude of lightning strokes that can reach a phase conductor. An analysis of the lightning performance including the calculation of both BFR and SFFR will follow. The rest of the study is dedicated to select the MO surge arresters, to analyze the energy withstand capability required for a correct protection, and to obtain the lightning flashover rate that may be achieved after arrester installation considering different options.

A. Line shielding

The incidence model used in this example is the electrogeometric model proposed by Brown and Whitehead [94]. According to this model, the striking distances are calculated as follows:

$$r_c = 7.1 I^{0.75} \tag{6.49a}$$

$$r_g = 6.4 I^{0.75} \tag{6.49b}$$

where
   r_c is the striking distance to both phase conductors and shield wires
   r_g is the striking distance to earth
   I is the peak current magnitude of the lighting return stroke current

Figure 6.31 shows the situation used to define the maximum shielding failure current; that is, the situation where all striking distances (from ground, from the shield wire, from the phase conductor for phase A) coincide at a single point Q.

Using r_{cm} and r_{gm} to denote the striking distances from the conductor and the ground that correspond to the maximum current I_m, then

$$r_{gm} = h_s + r_{cm} \sin(\alpha - \beta) \tag{6.50a}$$

$$\alpha = \tan^{-1} \frac{a}{h_s - h_c} \tag{6.50b}$$

$$\beta = \sin^{-1} \frac{h_s - h_c}{2 r_{cm} \cos\alpha} \tag{6.50c}$$

where
   a is the horizontal separation between the shield wire and the phase conductor
   h_s is the average height of the shield wire
   h_c is the average height of the phase conductor

The average height of a wire or conductor is the height at the tower minus two-thirds of the mid-span sag.

The value of the maximum shielding failure current I_m is that for which Equation 6.50a holds.

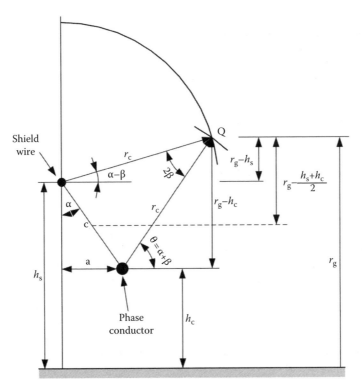

FIGURE 6.31
Finding the maximum shielding failure current.

The most exposed phase conductor in the test line is the upper one (phase A) in Figure 6.30. Therefore, the values from the geometry of the test line to be used in these equations are $a = 6.3\,m$, $h_s = 36.68\,m$, and $h_c = 26.35\,m$. The resulting value for I_m is 26.3 kA. Although it is not a low value, this peak current is not so far from the values usually encountered for transmission lines whose rated voltage is similar to that of the test line. However, it will be important to consider this maximum value at the time of selecting the arrester energy withstand capability.

B. Lightning performance of the test line without arresters
Modeling guidelines: Models used to represent the different parts of a transmission line are detailed in the following paragraphs [27,95–98]. See Chapter 2 for more details on these models.

1. The line (shield wires and phase conductors) is modeled by means of four spans at each side of the point of impact. Each span is represented as a multiphase untransposed constant- and distributed-parameter line section, whose parameters are calculated at 500 kHz.
2. The line termination at each side of the above model, needed to avoid reflections that could affect the simulated overvoltages around the point of impact, is represented by means of a long-enough section, whose parameters are also calculated at 500 kHz.
3. A tower is represented as an ideal single-conductor distributed-parameter line. The surge impedance of the tower is calculated according to CIGRE recommendations [99].
4. The representation of insulator strings relies on the application of the leader progression model [99–101]. The leader propagation is deduced from the following equation:

$$\frac{dl}{dt} = k_l V(t) \left[\frac{V(t)}{g - l} - E_{l0} \right] \qquad (6.51)$$

where
  $V(t)$ is the voltage across the gap
  g is the gap length
  l is the leader length
  E_{l0} is the critical leader inception gradient
  k_l is a leader coefficient

5. The footing impedance is represented as a nonlinear resistance whose value is approximated by the following expression [58,59]:

$$R_T = \frac{R_0}{\sqrt{1 + I/I_g}} \qquad \left(I_g = \frac{E_0 \rho}{2\pi R_0^2} \right) \tag{6.52}$$

where
  R_0 is the footing resistance at low current and low frequency
  I is the stroke current through the resistance
  I_g is the limiting current to initiate sufficient soil ionization, being ρ the soil resistivity
    (Ω m) and E_0 the soil ionization gradient (about 400 kV/m [102])

6. A lightning stroke is represented as an ideal current source with a concave waveform. In this work return stroke currents are represented by means of the Heidler model [42]. A return stroke waveform is defined by the peak current magnitude, I_{100}, the rise time, $t_f (= 1.67 (t_{90} - t_{30}))$, and the tail time, t_h, that is the time interval between the start of the wave and the 50% of peak current on tail.

Monte Carlo procedure: The main aspects of the Monte Carlo method used in this study are summarized below [103]:

a. The calculation of random values includes the parameters of the lightning stroke (peak current, rise time, tail time, and location of the vertical leader channel), phase conductor voltages, the footing resistance, and the insulator strength.
b. The last step of a return stroke is determined by means of the electrogeometric model. As mentioned in the previous section, the model used in this study is that proposed by Brown and Whitehead [94].
c. Overvoltage calculations are performed once the point of impact has been determined.
d. If a flashover occurs in an insulator string, the run is stopped and the flashover rate updated.
e. The convergence of the Monte Carlo method is checked by comparing the probability density function of all random variables to their theoretical functions; the procedure is stopped when they match within the specified error.

The following probability distributions are used:
• Lightning flashes are assumed to be of negative polarity, with a single stroke and with parameters independently distributed, being their statistical behavior approximated by a lognormal distribution, whose probability density function is [43]:

$$p(x) = \frac{1}{\sqrt{2\pi} x \sigma_{\ln x}} \exp\left[-0.5 \left(\frac{\ln x - \ln x_m}{\sigma_{\ln x}} \right)^2 \right] \tag{6.53}$$

where
  $\sigma_{\ln x}$ is the standard deviation of $\ln x$
  x_m is the median value of x

TABLE 6.15

Example 6.4: Statistical Parameters of Negative Polarity Return Strokes

Parameter	x	$\sigma_{\ln x}$
I_{100}, kA	34.0	0.740
t_f, μs	2.0	0.494
t_h, μs	77.5	0.577

Source: IEEE TF on Parameters of Lightning Strokes, *IEEE Transactions on Power Delivery,* 20(1), 346, 2005. With permission.

Table 6.15 shows the values selected for each parameter.

- The power-frequency reference angle of phase conductors is uniformly distributed between 0° and 360°.
- Insulator string parameters are determined according to a Weibull distribution. The mean value and the standard deviation of E_{l0} are 570 kV/m and 5%, respectively. The value of the leader coefficient is $k_l = 1.3E{-}6\,m^2/(V^2{\cdot}s)$ [99].
- The footing resistance has a normal distribution with a mean value of 50 Ω and a standard deviation of 5 Ω. The value of the soil resistivity was 200 Ω·m.
- The stroke location, before the application of the electrogeometric model, was generated by assuming a vertical path and a uniform ground distribution of the leader. Only flashovers across insulator strings are assumed.

Simulation results: After 20,000 runs, the flashover rates due to backflashovers and to shielding failures are respectively 1.665 and 0.075 per 100 km year. Therefore, the total flashover rate is 1.74 per 100 km year. These values were obtained with a ground flash density of $N_g = 1$ fl/km² year. That is, if $N_g = 3$ fl/km² year, the total rate will be about 5.2 flashovers per 100 km year. These flashover rate values are in general too high for a transmission line.

Figure 6.32 shows some simulation results that will help to understand the lightning performance of this line. A backflashover with a peak current magnitude of 80 kA or below is not unusual, but it can justify the backflashover rate of 1.665, which is a rather high value for a transmission line. This is mainly but not only due to the short strike distance of the insulator strings. As deduced from Figure 6.32a, a not-so-small percentage of backflashovers were caused by strokes with a peak current magnitude above 300 kA. These high values result from the theoretical statistical distribution of stroke parameters, but they will not frequently occur in reality, although peak current magnitudes higher than 800 kA have been measured [104]. One can observe that the highest peak current magnitude that causes shielding failure flashover is below the value deduced above (i.e., 26.3 kA). This is an indication that more runs should be performed to obtain a better accuracy with the Monte Carlo method. Take into account that only 19 strokes out of 20,000 impacted a phase conductors, and 15 of them caused flashover.

C. Arrester selection

The procedure followed for selecting the line arresters is initially similar to that used in Example 6.3. There is, however, an important aspect that has to be carefully analyzed when arrester are

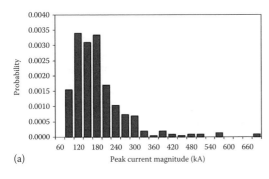

(a)

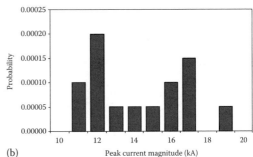

(b)

FIGURE 6.32

Example 6.4: Distribution of stroke currents. (a) Strokes to shield wires that caused flashover; (b) strokes to phase conductors that caused flashover.

installed to protect overhead lines: their energy withstand capability must be selected taking into account the actual energy stresses.

The corresponding standard maximum system voltage specified for 400 kV in IEC is $U_m = 420$ kV. Since the EFF is 1.4, the following values are obtained, assuming that the duration of the TOV is 1 s:

MCOV–COV

$$COV = \frac{420}{\sqrt{3}} = 242.49 \text{ kV}$$

TOVC

$$TOV_C = 1.4 \cdot \frac{245}{\sqrt{3}} = 339.48 \text{ kV}$$

TOV$_{10}$

$$TOV_{10} = 1.4 \cdot \frac{245}{\sqrt{3}} \cdot \left(\frac{1}{10}\right)^{0.02} = 324.20 \text{ kV}$$

Given the length of the test line, the switching surge energy discharged by arrester is much lower than the usual switching impulse energy ratings of station-class arresters.

All the above values are rms values. Table 6.16 presents ratings and protective data of surge arresters for this voltage level guaranteed by one manufacturer. The nominal discharge current is 20 kA according to IEC and 15 kA according to ANSI/IEEE. The arrester housing is made of silicone.

The only arrester that provides a safety margin of 20% in both MCOV and TOV capability is the last one, which is the selected arrester in this example. Therefore, the ratings of the arrester will be as follows

- Rated voltage (rms): $U_r = 360$ kV
- Maximum continuous operating voltage (rms): $U_c = 267$ kV (291 kV according to IEEE)
- TOV capability at 10 s: TOV$_{10} = 396$ kV
- Nominal discharge current (peak): $I = 20$ kA (15 kA according to IEEE)
- Line discharge class: Class 4.

From the manufacturer data sheets, the height and the creepage distance of the selected arrester are respectively 3.216 and 10.875 m.

TABLE 6.16

Example 6.4: Surge Arrester Protective Data

Max. System Voltage U_m (kV$_{rms}$)	Rated Voltage U_r (kV$_{rms}$)	MCOV		TOV Capability 10 s (kV$_{rms}$)	Maximum Residual Voltage with Current Wave	
		IEC U_c (kV$_{rms}$)	IEEE MCOV (kV$_{rms}$)		30/60 µs 3 kA (kV$_{peak}$)	8/20 µs 10 kA (kV$_{peak}$)
420	330	264	267	363	684	751
	336	267	272	369	696	765
	342	267	277	376	709	779
	360	267	291	396	746	819

D. Arrester model

The procedure to calculate parameters of the fast front transient model will be applied to a one-column arrester, with an overall height of 3.216 m, being $V_{10} = 819$ kV, and $V_{ss} = 746$ kV for a 3 kA, 30/60 μs current waveshape.

- Initial values: The initial parameters that result from using Equations 6.12 to 6.14 are

$$R_0 = 100 \cdot 3.216 = 321.6 \ (\Omega)$$

$$L_0 = 0.2 \cdot 3.216 = 0.6432 \ (\mu H)$$

$$R_1 = 65 \cdot 3.216 = 209.04 \ (\Omega)$$

$$L_1 = 15 \cdot 3.216 = 48.24 \ (\mu H)$$

$$C = 100 \cdot \frac{1}{3.216} = 31.095 \ (pF)$$

- Adjustment of A_0 and A_1 to match switching surge discharge voltage: the arrester model was tested to adjust the nonlinear resistances A_0 and A_1. A 3 kA, 30/60 μs double-ramp current was injected into the initial model. The result was a 751.1 kV voltage peak that matches the manufacturer's value within an error of less than 1%.
- Adjustment of L_1 to match lightning surge discharge voltage: To match the discharge voltages for an 8/20 μs current L_1 is modified until a good agreement between the simulation result and the manufacturer's value is achieved. The resulting procedure is presented in Table 6.17.

E. Arrester energy analysis

Some modeling guidelines used in the calculation of the flashover rate are no longer valid when the main goal is to estimate the energy discharged by arresters. The most important differences can be summarized as follows [98,105]:

- Spans must be represented as multiphase untransposed frequency-dependent distributed-parameter line sections, since the calculations with a constant parameter model can produce wrong results during the stroke tail, when the steepness of the current is variable and lower than during the front of the wave.
- No less than seven spans at both sides of the point of impact have to be included in the model for arrester energy evaluation. This is mainly due to the fact that the discharged energy is shared by arresters located at adjacent towers. However, it is worth mentioning that such number of spans is needed because corona effect is not included in the calculations, so the damping and energy losses caused during propagation are not as high as when corona is included, see Chapter 2.
- The tail time of the return stroke current has a strong influence, being the effect of the rise time very small, or even negligible for low peak current values. Therefore, the simulation time must be always longer than the tail time of the lightning stroke current.

TABLE 6.17

Example 6.4: Adjustment of Surge Arrester Model Parameters

Run	L_1 (mH)	Simulated V_{10} (kV)	Difference (%)	Next Value of L_1
1	48.24	844.7	3.04	24.12
2	24.12	819.9	0.11	–

The effect of phase conductor voltages can be more important for arrester energy evaluation than for flashover rate calculations [105]. However, the effect of the arrester leads will be neglected since it is assumed that arresters are in parallel to insulator strings. Given the difference of lengths, no effect on either the voltage or the energy should be expected.

A line model with the guidelines discussed above was created to estimate the configuration (i.e., line phases at which arresters are installed) with which the energy discharged by arresters could reach the maximum value when the stroke hits either a tower or a phase conductor. Table 6.18 shows some results obtained with a grounding resistance model for which $R_0 = 50\,\Omega$ and $\rho = 200\,\Omega\cdot m$. The calculations were made with arresters installed in all towers and assuming that the reference phase angle (phase A angle) was $0°$.

As shown in the table, the peak current magnitude used to obtain the maximum discharged energy when the stroke hits a tower is different from that assumed when the stroke hits a phase conductor. Note, on the other hand, that the placement of a single arrester at phase C, the bottom phase in Figure 6.30, was not studied since, according to the electrogeometric model, a vertical-path stroke will never impact that phase.

From the results shown in the table, one can conclude that differences between the maximum energy stresses deduced with each configuration are rather small. That is, when the stroke hits a phase conductor, the maximum energy discharged by the arrester installed at the struck phase is very similar in all configurations. The maximum energy stress in arresters corresponds to a return stroke that impacts a phase conductor, although the peak current magnitude is much lower.

A conclusion from the results shown in the table is that the estimation of the maximum energy stress can be performed by assuming that arresters are installed at all the phases, although it is not the most onerous case.

Plots of Figure 6.33 show the results from a sensitivity study aimed at estimating the maximum energy discharged by arresters considering a different range of peak current values for strokes to a tower or to a phase conductor. Although the peak current values of return strokes that can hit a tower can reach values larger than 300 kA, one can conclude from the first plot that the impact of a stroke to a tower will not cause arrester failure. The results presented in the second plot, when the stroke hits a phase conductor, are different, but the maximum energy discharged by arresters for a peak current magnitude of 50 kA is below the maximum energy withstand capability of the selected arresters.

When a lightning stroke with the same parameters used in Table 6.18 (30 kA, 2/50 μs) impacts a phase conductor at the midspan, the resulting energy discharged by arresters installed at the first tower is, from Equation 6.6, about 600 kJ. The value shown in Table 6.18 is less than 50%, which can be seen as a proof that the energy discharge is shared by arresters at the adjacent towers.

It is important to keep in mind that the main stroke parameters when calculating the energy discharged by arresters are the peak current magnitude and the tail time. The influence of the rise time on the energy discharged by the arrester is negligible. Since the total charge of a double-ramp

TABLE 6.18

Example 6.4: Energy Discharged by Surge Arresters (in kJ)

Arresters per Tower	Stroke to a Tower[a]	Stroke to a Phase Conductor[b]
A–B–C	141.9	264.3
A–B	140.6	264.3
B–C	159.2	259.1
C–A	174.0	264.3
A	140.8	264.3
B	178.4	259.1

[a] Waveform of the stroke to a tower = 150 kA, 2/50 μs.
[b] Waveform of the stroke to a conductor = 30 kA, 2/50 μs.

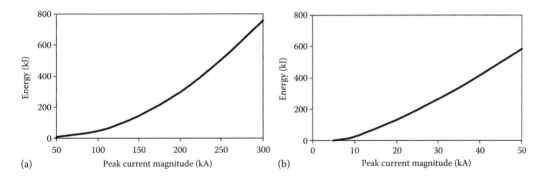

FIGURE 6.33
Example 6.4: Energy discharged by surge arresters. Stroke waveform = 2/50 μs; footing resistance: $R_0 = 50 \Omega$, $\rho = 200 \Omega \cdot m$. (a) Strokes to a tower; (b) strokes to a phase conductor.

waveshape lightning stroke is approximately linear with respect to the tail time, using Equation 6.6 one can deduce that the energy discharged by arresters is linearly dependent on the peak current magnitude and on the tail time. Therefore, the maximum energy discharged by the selected arresters when a lightning stroke with a peak current magnitude of 200 kA and a tail time of 100 μs impacts a tower would be about 600 kJ, which is less than 2 kJ/kV(U_r); that is, well below the usual energy withstand capability of the selected arrester. Since the probability of a stroke with these parameters is very low, the conclusion is that the probability of arrester failure caused by a high energy stress is virtually negligible.

The footing resistance parameters could be different from those used to obtain Figure 6.33b, but their influence, when the stroke hits a phase conductor, is unimportant. That is, values with a different footing resistance will be very similar to those depicted in Figure 6.33b. However, a more detailed study would be justified with higher grounding resistance values because then arrester energy stress might significantly increase.

F. Lightning flashover rate with arresters

From the results presented and discussed in the previous sections one can deduce that the line has a poor lightning performance and the probability of arrester failure caused by a stroke to a tower or a phase conductor will be very low. Only lightning strokes with a too high peak current magnitude and a too long tail time could case arrester failure.

The goal of the new study is to estimate the improvement of the flashover rate that can be achieved by installing surge arresters at all towers of the test line, but not necessarily at all phases.

The flashover rate of the test line with the different combinations of arresters previously analyzed (see Table 6.18) was estimated. The new simulations were performed without measuring the energy discharged by arresters, so the line model, as well as return stroke parameters, were those detailed before. Table 6.19 shows a summary of the new results.

The following conclusions are derived from these results:

- When arresters are installed at all phases, the total flashover rate is reduced to zero, as expected.
- To obtain a good performance, when arresters are installed at two phases, one arrester should be installed at the upper phase. Note that when two arresters are installed, the best results in both BFOR and SFFOR are achieved when the upper phase (phase A) is protected.
- Shielding failure flashovers can be avoided by installing only two arresters, at phases A and B, per tower. This is due to the protection provided by the phase A, which will avoid that a return stroke reaches phase C. In fact, SFFOR is also negligible if phase B is not protected; although this does not mean that lightning strokes do not reach phase B, but the maximum peak current magnitude is not high enough to cause flashover.

TABLE 6.19

Example 6.4: Flashover Rate with Arresters (per 100 km Year)

Arrester Protection	BFOR	SFFOR	Total Flashover Rate
A–B–C	0.000	0.000	0.000
A–B	0.200	0.000	0.200
B–C	0.575	0.075	0.650
C–A	0.495	0.000	0.495
A	0.615	0.000	0.615
B	0.925	0.075	1.000
C	1.065	0.075	1.140

- There is a significant difference between the rates obtained with only one arrester. The best location is the upper phase, and the improvement is even more significant for the SFFOR. This behavior can be justified by analyzing the electrogeometric model as applied in this work.

G. Remarks

1. Modeling guidelines different from those used in this example have been proposed. The representation of the footing impedances is a critical aspect. A constant resistance model is frequently used to represent this part of the line [106]. The differences between the results derived with a constant resistance model and those with a nonlinear resistance would be very important for both flashover rates and arrester energies, mainly when the parameter R_0 is very high. A constant resistance will always provide more conservative values; i.e., higher BFR values.

2. All the simulations were performed without including corona effect in the line model. As shown in Chapter 2, surge propagation with corona is highly damped; therefore, results presented in this example, mainly those related to arrester energy assessment, are conservative. In fact, a lower number of line spans is required for a proper arrester energy assessment if corona effect is included in the model.

3. Several incidence models have been proposed and developed to date [15]. Arrester selection for the test line has been performed from results derived from the application of the model proposed by Brown and Whitehead [94]. The application of other electrogeometric model might provide different results and even suggest the selection of a different arrester (e.g., higher duty class).

4. Calculations were made by assuming that return strokes had a vertical path. Flashovers caused by strokes with a nonvertical or even a horizontal path have been reported [107]. The highest flashover rate is usually obtained by assuming vertical paths; however, the assumption of nonvertical paths will have a nonnegligible effect in some studies [108]. In addition, they could also impact the bottom phase of the test line.

5. The lognormal distribution used to characterize the statistical variation of the peak current magnitude of a return stroke can provide unrealistic values of this parameter; e.g., $I_{100} > 1000$ kA. Although strokes with a peak current magnitude above 800 kA have been recorded [104], they are very rare. Therefore, the calculation of the flashover rate could have been performed by truncating the peak current magnitude; e.g., at 400 kA. However, one can observe that this would have hardly affected the flashover rate, since the percentage of return strokes that caused backflashover with a peak current magnitude above 400 kA was less than 0.05%, see Figure 6.32a.

Protection of transformers. In addition to the practices discussed in the previous sections, there are some additional situations that must be considered [5]: (1) surge protection across series windings could be required; however, when arresters are connected in parallel with the series winding, it is necessary to insulate both arrester terminals from ground, in such case the arrester must be installed at or close to the terminals of the equipment. (2) In some cases, multiwinding transformers have connections brought out to external bushings that do not have lines connected. Arresters should always be connected at or close to the terminals of such bushings. (3) Isolated neutral terminals are subjected to TOVs caused by line-to-ground faults and to surge voltages as a result of overvoltages at the line terminals propagating through the windings, so they may require arrester protection.

Protection of shunt capacitor banks. Shunt capacitor banks are usually installed at stations, wye-connected, with or without grounded neutrals. The increase in capacitor bank voltage due to an incoming lightning surge on any connected transmission line does not depend on the rate of rise, but on how much charge is absorbed. If the charge results in excessive overvoltages, surge arresters should be installed to discharge energy and limit the overvoltage level. Due to the low surge impedance of shunt capacitor banks, adding additional surge arresters beyond those that already exist at a station may not be necessary.

Certain capacitor switching operations can produce potentially dangerous overvoltages, not only to the capacitor bank but to other station equipment [60]. Arresters installed to protect transformers and other station equipment can be subjected to severe energy absorption duty during capacitor switching because of the large energy stored in the capacitor bank. The capability of all nearby surge arresters to withstand the energies dissipated during capacitor switching is an important aspect.

Overvoltage protection should be considered at some locations [5]: (1) on the capacitor primary and backup switchgear to limit transient recovery voltages (TRVs) when shunt capacitors are being switched out; (2) at the end of transformer-terminated lines to limit phase-to-phase overvoltages resulting from capacitor switching or line switching in the presence of shunt capacitor banks; (3) on transformers when energized in the presence of shunt capacitor banks; (4) on lower voltage systems that are inductively coupled through transformers to higher voltage systems with shunt capacitor banks; (5) on the neutrals of ungrounded shunt capacitor banks.

Protection of series capacitor banks. Under fault conditions a varistor protects capacitor units by commutating a portion of the capacitor current. Thus, the varistor must be capable of withstanding the currents and energies present until the fault is removed or a bypass takes place. To control varistor duty, the protection system monitors varistor and, in some cases, capacitor currents. Protective functions usually provided are fast bypass, thermal bypass, and imbalance. Depending on the series capacitor location and the available fault current, the varistor may be protected by a bypass device such as a breaker, a gap, or a thyristor. Protective level, energy handling capability, current sharing, or pressure relief are performance characteristics that should be considered for a proper varistor application [5].

Protection of underground cables. Many of the concerns identified in protecting shunt capacitor banks should be considered also for high-voltage cable installations. In addition, overvoltage protection of the junction between overhead lines and cables should be evaluated [111,112]. Lightning may also be an important consideration at cable terminals. Cables may require further consideration because of traveling-wave phenomena.

- *Cable insulation.* Cables have low surge impedance, which means that surges incoming from overhead lines will be reduced significantly at the line–cable junction. In turn, surges originated at a station will increase in voltage at the cable–line connection due to the much higher surge impedance of the line. Since there is little surge attenuation in cables and the ratio of surge impedances is so large, it is common for the reflected wave plus the oncoming wave to cause a voltage doubling at the cable–line connection. MO surge arresters should be capable of absorbing the high energy that can be stored in a cable when subjected to an overvoltage that causes the arrester to discharge. For complex cable and overhead line systems, optimum protection can be determined by carrying out a detailed study [113–115].

- *Sheath and joint insulation.* High-voltage power cables are provided with metallic sheaths to give a uniform field distribution to the solid dielectric, to protect it from external damage and to provide a return path for fault current, see Chapter 3. To ensure safety and to avoid the losses associated with circulating currents requires special bonding and grounding of the metallic sheath circuits. Common sheath bonding systems are single-point bonding and cross bonding. A disadvantage of both methods, however, is that a change in surge impedance occurs at the ungrounded terminals of the cable sheath and at the sheath sectionalizing insulators; as a result, all traveling-wave surges entering the cable system due to lightning, switching operations, or faults will be subjected to partial reflection and refraction at these locations, and dangerous overvoltages can be developed across the sheath joint insulators and sheath jacket insulation [5]. MO surge arresters installed to protect cable sheath and joint bonding should be suitable for continuous operation during normal and emergency conditions, withstand power-frequency overvoltages resulting from faults, limit surge voltages below the withstand strength of the jacket and sheath joint insulators, and absorb currents and energies associated to lightning, switching, or fault initiation [116].

Protection of gas-insulated substations (GIS). A GIS is more sensitive to overvoltages than an air-insulated station (AIS). This is a result of the high electrical stress placed on relatively small geometries. The volt–time characteristics of GIS equipment are independent of the atmospheric conditions and much flatter than for atmospheric pressure air or for solid dielectrics, especially for fast front transients. This means that any incoming surge having sufficiently high peak and rate-of-rise values is likely to cause breakdown before flashing over any coordinating air gaps. In general, GIS with connections to overhead lines will need arresters on each line entrance. One of the most common questions is related to the location and type of the surge arrester within the GIS system. A detailed study is usually recommended to determine whether additional surge arresters are needed within the station or not [15,109].

The effect of very fast front transients on equipment, such as transformers connected to the high voltage side, should be taken into account. However, arrester protection within the GIS against very fast front overvoltages is difficult to achieve due to the steep front of wave and to the delay of the conduction mechanism in MO varistor blocks.

Protection of cable-connected stations. A station is in general well protected by arresters installed at the line–cable transition, although it is not possible to establish a general recommendation [5,110]. Those arresters may suffice when more than one line is connected to the station through underground insulated cables. If cables can function with a terminal open, then arresters could be also required at that terminal.

The insulation coordination of a GIS connected to an overhead transmission line through an underground cable is a particular problem for which the selection of the highest standard rated lightning impulse voltage is recommended [110]. In addition, surge arresters should be installed at the line–cable junction. The following example analyzes the effectiveness of this protection scheme and checks whether additional protection may be needed.

Example 6.5

Figure 6.34 shows a schematic diagram of a cable-connected GIS. The objective is to decide where to install MO surge arresters and what ratings should be selected for them to obtain a given MTBF. The information required for arrester selection is as follows:

- Power system: Frequency = 50 Hz
  Rated voltage = 220 kV
  Grounding = Low-impedance system, EFF = 1.4
  Duration of temporary overvoltage = 1 s
- Line: Length = 60 km
  Span length = 250 m
  Insulator string strike distance = 1.78 m
  Positive polarity CFO (U_{50} in IEC) = 1246 kV
  Conductor configuration = 2 conductor per phase
  Capacitance per unit length = 11.0 nF/km
  BFR = 2.2 flashovers/100 km year
  Ground wire surge impedance = 480 Ω
  Phase conductor surge impedance = 400 Ω
  High-current grounding tower resistance = 20 Ω
  Maximum switching surge at the remote end = 520 kV (peak value)
- Cable: Length = Variable
  Surge impedance = Between 25 and 50 Ω
  Propagation velocity = 150 m/μs

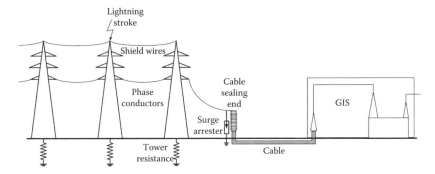

FIGURE 6.34
Example 6.5: Diagram of a 220 kV cable-connected GIS.

Rated short-duration power-frequency withstand voltage = 460 kV
Rated lightning impulse withstand voltage = 1050 kV
- GIS: Single-line substation in an area of high lightning activity
                        MTBF = 1000 years
                        Bus duct length = 30 m
                        Bus duct surge impedance = 105 Ω
                        Propagation velocity = 285 m/μs
                        Rated short-duration power-frequency withstand voltage = 460 kV
                        Rated lightning impulse withstand voltage = 1050 kV
                        Transformer capacitance = Neglected.

The substation is located at sea level, and calculations are performed assuming standard atmospheric conditions. Rated withstand voltages are measured phase-to-ground.

A. Modeling guidelines

The overvoltage protection provided by MO surge arresters is studied considering an idealized representation for components: the line, the cable, and the GIS are represented by means of nondissipative single-phase traveling-wave models; an ideal current source with a double-ramp waveshape is used to represent the lightning stroke.

The factors that should be considered for analysis are many (e.g., tower grounding resistance, cable length and surge impedance, cable bonding practice, surge arrester location, GIS topology, and surge impedance of GIS bus ducts). This example analyzes how the length and the surge impedance of the cable can affect the protection scheme to be implemented. The idealized model of the cable implies that its sheath is bonded to the substation grounding grid and to the tower grounding at the line end. The system model assumes that there is a common grounding for shield wires, surge arresters, and cable screens, and it is represented as a constant resistance of 5 Ω. The effect of the substation grounding impedance at the cable–GIS junction is neglected. GIS topology may be rather complex; in this example it is represented by a single bus duct with an open terminal.

The study has been divided into two parts. First, the simplified representation of power components, including the MO surge arrester, is used to gain insight into the physical phenomena and into the performance of different protection schemes. The more accurate model for MO surge arresters is used in the second part.

Figure 6.35 shows the diagram of the test system with the initial configuration, in which both the cable and the GIS are protected by the arresters installed at the line–cable junction only.

B. Arrester selection and model

Ratings of the power system for this example are the same that were used in Example 6.3; therefore, the procedure to select surge arrester ratings and to develop the arrester model will be also the same.

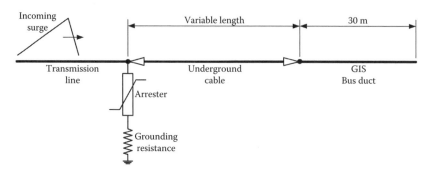

FIGURE 6.35
Example 6.5: Simplified diagram of test system.

As in Example 6.3, the maximum system voltage is 245 kV, the EFF is 1.4, and the duration of the TOV of 1 s. The resulting values are then the same:

- MCOV–COV = 141.45 kV
- TOV_C = 198.0 kV
- TOV_{10} = 189.1 kV.

Arrester ratings are selected again from Table 6.8. The ratings of the arrester selected for this example are as follows:

- Rated voltage (rms): U_r = 198 kV
- MCOV (rms): U_c = 156 kV (160 kV according to IEEE)
- TOV capability at 10 s: TOV_{10} = 217 kV
- Nominal discharge current (peak): I = 20 kA (15 kA according to IEEE)
- Line discharge class: Class 4.

The height and the creepage distance of the selected arrester are respectively 2.105 and 7.250 m.

The procedure to calculate parameters of the fast front model is again that used with the previous examples. The procedure will be applied to a one-column arrester, with an overall height of 2.105 m, being V_{10} = 451 kV, and V_{ss} = 410 kV for a 3 kA, 30/60 μs current waveshape. After adjustment of L_1, the parameters to be specified in the model depicted in Figure 6.11 are R_0 = 210.5 Ω, L_0 = 0.4210 μH, R_1 = 136.825 Ω, L_1 = 15.788 μH, and C = 47.51 pF.

C. Incoming surge

The crest voltage of the incoming surge is, using the approach recommended in IEEE Std 1313.2 [58], 1.2*1246 = 1495.2 kV.

Since only one line arrives to the substation, and it is designed for a MTBF of 1000 years, the distance to flashover, taking into account the lightning performance of the transmission line, is

$$d_m = \frac{1}{n \cdot (MTBF) \cdot BFR} = \frac{1}{1 \cdot 1000 \cdot (2.2/100)} = 0.045 \text{ (km)}$$

This distance is midway between the first and the second tower, so it is increased to the second tower location, and it is assumed that the backflashover occurs one span from the line–cable transition; that is 250 m away from the cable entrance.

Equations 6.45 and 6.46 are used to obtain respectively the steepness, S, and tail time, t_h, of the incoming surge. Taking into account that for a two-conductor bundle, the corona constant K_s is 1400 (kV·km)/μs, see Table 6.10, the resulting characteristics are S = 4000 kV/μs and t_h = 20 μs.

D. Simulation results

The first simulations are aimed at justifying the need for surge arrester protection at the line–cable junction. Initially, an ideal behavior is assumed for MO surge arresters; that is, an arrester behaves as an infinite impedance for surge voltages lower than its discharge voltage, and keeps constant its terminal voltage once the discharge voltage has been reached. Arrester leads are assumed very short and their effect is neglected.

As in Example 6.3, the voltage at the different system locations is calculated taking into account the power-frequency voltage at the time the incoming surge arrives to the substation. Using again IEEE Std 1313.2 recommendation, a voltage of opposite polarity to the surge, equal to 83% of the crest line-to-neutral power-frequency voltage is used [58]. In the case under study, the value of this voltage is 149 kV.

The arrester discharge voltage used in simulations is the value of V_{10} of the selected arrester; that is, 451 kV.

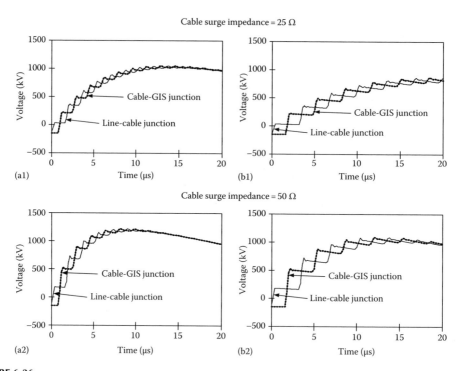

FIGURE 6.36
Example 6.5: Simulation results with arrester protection at the line–cable junction. (a) Cable length = 120 m; (b) cable length = 240 m.

Figure 6.36 shows some results obtained with two different values of both the length and the surge impedance of the cable. It is obvious from these results that some protection is required. Although the first wave that propagates along the cable is significantly reduced at the line–cable junction due to the negative reflection, the subsequent reflections at both cable ends are all positive and there is voltage escalation at both junctions that can cause dangerous stresses for the cable and the GIS. One can observe that the shorter the cable the faster the voltage escalation phenomenon. On the other hand, it is also evident that the surge impedance of the cable is another important factor: the highest peak voltage value is reached with the highest value of cable surge impedance.

Figure 6.37 shows some results that correspond to the same values used in the previous study when surge arresters are installed at the line–cable junction. This time, irrespectively of the length and the surge impedance of the cable, the overvoltages are limited within an acceptable range of values for the insulation of both the cable and the GIS bus duct because no voltage does exceed 800 kV. Since the surge impedance values used in the simulations are within realistic ranges for lines, cables, and GIS ducts, a provisional conclusion of this study is that both the cable and the GIS can be protected by installing arresters at the line–cable junction only.

A further study is made by using an advanced MO surge arrester model and considering only the parameters for which the most onerous overvoltages result. The study includes also the analysis of the energy discharged by surge arresters. From the simulation results shown in Figure 6.38 one can deduce that the differences with respect to the ideal arrester model are unimportant and the arrester energy stress for this example is not of concern.

An important aspect to be analyzed is the limitation of the component models used in simulations. The model used to represent a single-phase distributed-parameter cable or GIS bus duct does not allow users to measure voltages between conductors and screens or enclosures.

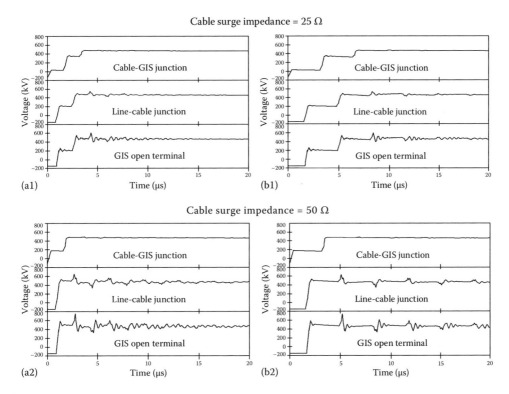

FIGURE 6.37
Example 6.5: Simulation results with arrester protection at the line–cable junction. (a) Cable length = 120 m; (b) cable length = 240 m.

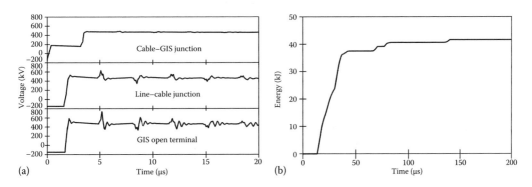

FIGURE 6.38
Example 6.5: Simulation results with advanced arrester model. Cable parameters: length = 240 m, surge imped-ance = 50 Ω. (a) Overvoltages; (b) energy discharged by arresters.

This is why the grounding resistance model at the cable–GIS junction has not been included. Since the terminals of the arrester at the line–cable junction must be connected between the conductor and the screen of the cable, the wave propagation along the cable and the GIS duct obtained from the simulation of this example is not strictly correct. However, the influence of the grounding resistance at the line–cable junction is negligible, so it would be the influ-ence of the grounding resistance at the cable–GIS junction. There are several reasons for this conclusion. First, the current through the arrester at the line–cable junction is very low before the arrester discharge voltage is reached, and once this voltage has been reached the voltage drop across the grounding resistance is rather low and its effect negligible. Second, the arrester current is always below 10 kA; as a consequence the arrester energy discharge is also small, as shown in Figure 6.38b.

In fact, this result is a consequence of a basic assumption applied in this example: the incoming surge is caused by a transmission line backflashover, which means that most of the lightning stroke energy has been bypassed to ground before the incoming surge reaches the line–cable transition. This might not be always true, and a more detailed study could be needed if the incoming surge could be caused by a shielding failure due to a poor transmission line shielding or when the transmission line is unshielded (see Section 6.6.4), although this later case is very rare.

Protection of circuit breakers. The use of MO surge arresters can be a more economical solution than increasing the number of interrupting chambers to withstand circuit breaker TRVs. If the surge arresters are mounted across the interrupters of multichambered circuit breakers, some attention must be paid to the interruption of fault currents. A reignition of a single chamber can result in fault currents flowing through the surge arrester across the nonreignited chambers [5]. When reclosing, arresters should be able to withstand the difference of the power-frequency out-of-phase voltages on each side of the open circuit breakers during the time required for reclosing. Surge arresters can also be installed phase-to-ground at either or both sides of the circuit breakers. However, this may require detailed computer studies, which should include coordination with any other surge arresters installed in the station or on shunt reactors.

6.7.3 Protection of Distribution Systems

Introduction: A usual problem for distribution systems is the large number of locations, overvoltages, and durations to be considered in their selection, mainly if the goal is to select a single arrester rating to be used on an entire feeder. Some practices to be considered when installing MO surge arresters for distribution equipment protection are discussed in the following paragraphs [5]. For selection and application of MO surge arresters in distribution environments see [21,117–122].

- Distribution arresters should be placed as close as possible to protected equipment. This is particularly important where only one arrester is used to protect equipment that is connected to a line that runs in two directions from the tap point, since surges approaching from the unprotected side can exceed the protective level of the arrester. Surges approaching from the arrester side are limited by arrester action, but the separation distance effect can be significant.

- The discharge of lightning currents through arrester lead wires produces a voltage that adds to the arrester discharge voltage only during the rise of the discharge current. The lead can be represented as an inductance whose vale is basically a function of its length, being the effect of the conductor diameter relatively minor. A value of 1–1.3 μH/m is representative of typical applications. The discharge voltage, without lead length effects, should be coordinated with the full-wave withstand of the protected equipment, although it may be recommendable to coordinate the sum of the lead voltage and the arrester discharge voltage with the chopped-wave withstand of the protected equipment [5].

- Placing an arrester on the source side of a fused cutout often results in very long arrester lead lengths. Since it requires that the fuse carries arrester discharge current, nuisance fuse blowing or fuse damage can result. Alternatives that allow both overload protection of small transformers and avoid excess arrester lead lengths are dual-element fuses installed in cutouts ahead of tank-mounted or internal arresters, or internal fuses located between the transformer winding and a tank-mounted or internal arrester. The coordination of MO surge arresters and fuses to achieve a proper protection of distribution transformers and avoid nuisance fuse operation have been analyzed in several works, see for instance [123–125].

- Primary and secondary grounds of distribution transformers must be interconnected with the arrester ground terminal. Ground connections should be made to the tanks of transformers, reclosers, capacitor support frames, and all hardware associated with the protected equipment. If regulation does not allow grounding of equipment support structures, protective gaps should be connected between the arrester ground terminal and the structure. Transformer-mounted arresters are grounded to the transformer tank, and the tank can be isolated from ground by inserting the protective gap between the transformer tank and ground.

Protection of distribution lines. Distribution lines are generally not shielded and particularly susceptible to direct lightning strokes. The effectiveness of MO surge arrester protection has been the subject of works with different objectives [126–128], and analyzed in IEEE Std 1410 [129]. Arresters can be used, instead of overhead shield wires, to protect the distribution lines from lightning flashovers. The level and frequency of occurrence of discharge currents depends to a great extent on the exposure of the distribution system and the ground flash density. Arresters applied on exposed lines (few trees and buildings) and located in areas of high ground flash density will see large magnitude currents more often than arresters in shielded locations. Although the primary cause of flashover is in many distribution lines a surge induced by a nearby stroke to ground, when the line is protected by arresters, only direct strokes to unshielded lines are of concern, since only in such cases the energy discharged by arresters may be larger than energy withstand capability. Multiple-stroke flashes are another factor to be considered. Since strokes with any peak current magnitudes may reach a phase conductor of a distribution line, heavy-duty class arresters are in general used to protect these lines [15]. The energy discharged by arresters installed on distribution lines has been analyzed in a number of works [129–132]. Arrester damage caused by direct lightning strokes can be effectively prevented by installing overhead ground wires or by increasing arrester withstand capability. Lowering the grounding resistance may also help, but it is not a decisive factor [131].

Protection of transformers. The maximum protection is provided by shortening to minimum arrester lead lengths and by installing secondary side arresters [133,134]. Secondary surges can be produced by lightning strokes to either the primary or the secondary side. Strokes to the primary system elevate the transformer secondary neutral potential above the neutral potential at the customer service, forcing surge current into the transformer. Surges impressed on the primary winding of the transformer also are transferred to the secondary by inductive and capacitive coupling, although this is not usually of concern to the integrity of the secondary winding insulation. Secondary surges may be caused by direct strokes to the secondary service conductors or due to induced voltages caused by lightning strokes to nearby objects, such as trees and structures, or to ground. Where possible, the secondary neutral terminal should be bonded to the primary neutral which also should be bonded to the tank.

The use of secondary arresters at the distribution transformer, may increase the surge voltage that reaches the customer service connection; consequently, additional secondary protection at the customer service entrance may be required.

Protection of capacitor banks. Pole-mounted shunt capacitor banks may be protected by line-to-ground connection of arresters mounted on the same pole. Grounded-wye capacitor banks, when protected by MO surge arresters, can only be charged to the protective level of the arrester, and the stroke current is shared by the arrester and the bank. Arrester operation on ungrounded banks is usually caused by a high transient voltage transmitted from the line to the bank; however, little of the transient energy is added to the stored energy in the capacitors and no special high-energy capability is required for arresters protecting ungrounded capacitor banks against lightning surges. If a capacitor bank is switched, arresters having high energy absorption capability may be required regardless of the circuit configuration, since arresters can be exposed to high-energy surges if restriking of the switching device occurs when the bank is being de-energized. In the case of an ungrounded capacitor bank, a two-phase restrike can cause excessive current to flow in the arresters associated with the restruck phases. Arresters on either side of the switching device can experience high-energy switching transients.

Protection of switches, reclosers, and sectionalizers. If a device is in the closed position, a good protection may be obtained by applying one arrester from line to ground on the source side; however, there is some risk of lightning damage when the device is open. Switches, reclosers, or sectionalizers are best protected by installing arresters on both the source and load side. Some reclosers are designed with a built-in bypass protector across the series coils.

Protection of regulators. Line voltage regulators should be protected on both line and load sides. For the most effective protection, the arrester should be mounted on the tank with the arrester ground connected to the tank. Bus voltage regulators are often protected by station or intermediate class arresters on the station bus or on the station transformer low-voltage bushings, and by distribution arresters adjacent to the substation on the outgoing feeders.

Protection of equipment on underground systems. An underground system can basically be seen as a cable terminated by an open point: the magnitude of surge voltages that enter the underground system from the overhead feeder at the riser pole is limited by the arrester on the riser pole; however, surge voltage above the protective level of the riser pole arresters can occur on the cable and at equipment locations far from the riser pole because of reflection from the open point. Different alternatives for protection of underground systems are summarized below.

- When arrester protection is provided at the riser pole only, the voltage at this location is the sum of the arrester discharge voltage and the voltage drop in the arrester connecting leads. This voltage propagates into the cable and can approximately double its value because of the reflections at points such as open switches and terminating transformers.

- When a second arrester is placed at the remote end of the cable, the voltage at this end is limited to the discharge voltage of the remote arrester at a current usually smaller than the current through the riser pole arrester. Because the remote arrester appears as an open circuit until the arrester becomes conductive, a portion of the incoming wave front is reflected and superimposed on the approaching surge voltage wave. Therefore, the voltage at intermediate points in the cable system will usually be higher than at either end, see Example 6.6. The maximum voltage at intermediate points will be the protective level of the riser pole arrester (discharge voltage plus lead voltage drop) plus some fraction of the discharge voltage of the remote arrester.

- Equipment protection can be improved by installing an additional arrester at a location upstream from the open-point termination. This midcircuit arrester will suppress the reflected surge as it is being superimposed on the incoming surge voltage wave; however, equipment connected between the remote arrester and the midcircuit arrester may need individual arrester protection, see Example 6.6.

- The most effective protection method is to install arresters at the riser pole, at the open point, and at each underground equipment location. The voltage on each piece of equipment will be held to the low-current discharge voltage of its arrester, and only the section of cable between the open point and the first upstream arrester will see a higher surge voltage.

Arresters installed on underground equipment may be either elbow arresters (for dead front equipment) or base/bracket-mounted arresters. Liquid-immersed arresters are also available mounted inside the transformers [5].

The following example analyzes the protection of an underground system and justifies some of the above statements.

Example 6.6

Figure 6.39 shows the scheme of an overhead-underground transition, located in an area of low lightning activity. The surge voltage caused by a lightning stroke that hits the unshielded distribution line may damage the non-self-restoring cable insulation. The objective of this example is to justify a protection scheme that will prevent any disruptive discharge in the cable and select the characteristics of the MO surge arresters. Known data for this example are as follows:

- Distribution system: Frequency = 50 Hz
  Rated voltage = 25 kV
  Maximum system voltage = 27.5 kV
  Grounding = High-impedance system, EFF = 1.7
  Duration of temporary overvoltage = 60 s

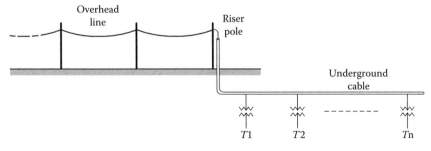

FIGURE 6.39
Example 6.6: Overhead-underground transition.

- Line: Three-wires
                                      Length = 10 km
                                      Average span length = 30 m
                                      Surge impedance = 360 Ω
                                      Propagation velocity = 300 m/μs
- Cable: Length = 600 m
                                      Surge impedance = 30 Ω
                                      Propagation velocity = 150 m/μs
                                      Rated short-duration power-frequency withstand voltage = 70 kV
                                      Rated lightning impulse withstand voltage = 170 kV

The corresponding standard maximum phase-to-phase voltage specified for 25 kV in IEC is 36 kV, while in IEEE is 26.2 kV. Since the voltage regulation in the distribution line may increase the voltage above 26.2 kV, the next value (i.e., 36.2 kV) should be considered. In that case the IEEE standard rated voltages would be: (a) rated short-duration power-frequency withstand voltage = 70 kV; (b) rated lightning impulse withstand voltage = 200 kV. Note that the rated short-duration power-frequency withstand voltage is the same in both standards, but the rated lightning impulse withstand voltage is higher in IEEE than in IEC.

Lightning overvoltages on the cable can be produced under several circumstances:

a. The return stroke hits a line phase conductor several spans from the transition point. The surge voltage that reaches the cable is below the lightning CFO of the line due to damping and corona effect during propagation. The surge voltage that will pass to the cable after reflection will be about 10% of the incoming surge. Since the CFO of the distribution line may be between 200 and 300 kV, the event should not cause any damage to the cable regardless of the scheme selected to protect the cable, or even if no protection was installed.

b. The return stroke impacts ground in the neighborhood of the line. As in the previous case, the maximum incoming surge voltage that reaches the cable is limited by the CFO of the line. Even if the point of impact was close to the transition point, the surge voltage that would arrive to that point would not exceed 300 kV [134], and the surge voltage transmitted to the cable would be reduced to a nondangerous level for the cable.

c. The return stroke hits the line span close to the transition point. In most cases, the surge voltage that will propagate along the cable, provided that no flashover is produced across air between line conductors, will be so high that, even after being reduced a 90%, it could damage the insulation of the cable if this is not protected.

The study that follows is based on the assumption that the lightning strokes hits the line span close to the transition overhead-underground and no flashover across air (i.e., between conductors) is produced. To simplify the analysis it is also assumed that line insulators do not flashover.

The actual phenomena are much more complicated than those derived from these models and several additional aspects should be considered for a very accurate simulation. When the lightning stroke impacts a conductor there will be a voltage raise not only at the struck conductor but also at the other two phases, consequently the lowest withstand voltage may be at a cross arm [135]. In general, damage occurs on the conductor, not far from insulators, or on the insulators, not at the point of impact. Surge propagation along bare conductors is, on the other hand, affected by corona, although this effect will not be very significant if the case under study assumes that the impact is on the span close to the line–cable junction. In addition, there will be a flow of predischarge currents between the struck phase and conductors of the other phases, which can reduce the steepness of the arrester current [136]; however, this effect can be generally ignored [137].

The effectiveness of the different protection schemes is studied considering an idealized representation for components. That is, lines and cables are represented by means of nondissipative single-phase traveling-wave models. Transformers can be represented as a capacitance-to-ground (see Example 6.3), although in this study they are represented as open circuits. As for Example

6.5, an ideal gapless MO surge arrester behaves as an infinite impedance for surge voltages lower than its discharge voltage, and keeps constant its terminal voltage once the discharge voltage is exceeded. Arrester leads are modeled as lumped-parameter reactances.

The study has been divided into two parts. A simplified representation of power components is used to gain insight into the physical phenomena and the performance of the different protection schemes. More accurate model for surge arresters and lightning stroke current are used in the last part. First of all, arrester selection and model development, following procedures similar to those used in the previous examples, are presented.

A. Arrester selection and model
The following values are obtained from the data specified for this example:
- MCOV–COV

$$COV = \frac{27.5}{\sqrt{3}} \approx 15.9 \, kV$$

- TOV_C

$$TOV_C = 1.7 \cdot \frac{27.5}{\sqrt{3}} \approx 27.0 \, kV$$

- TOV_{10}

$$TOV_{10} = 1.7 \cdot \frac{27.5}{\sqrt{3}} \cdot \left(\frac{60}{10}\right)^{0.02} \approx 28.0 \, kV$$

Table 6.20 presents ratings and protective data of surge arresters for this voltage level guaranteed by one manufacturer. The nominal discharge current is 10 kA and the arrester housing is made of silicone.

The ratings of the arrester initially selected are as follows:

- Rated voltage (rms): $U_r = 30 \, kV$
- MCOV (rms): $U_c = 24.4 \, kV$

TABLE 6.20

Example 6.6: Surge Arrester Protective Data

			Maximum Residual Voltage with Current Wave	
			30/60 μs	8/20 μs
	MCOV	**TOV Capability**		
Rated Voltage (kV$_{rms}$)	**MCOV (kV$_{rms}$)**	**10 s (kV$_{rms}$)**	**2 kA (kV$_{peak}$)**	**10 kA (kV$_{peak}$)**
18	15.3	21.9	46.7	52.9
21	17.0	24.3	49.6	56.2
24	19.5	27.9	57.7	65.5
27	22.0	31.5	65.0	73.8
30	24.4	34.9	72.6	82.5
33	27.0	38.6	80.1	90.9
36	29.0	41.5	85.0	97.6

- TOV capability at 10 s: $TOV_{10} = 34.9\,kV$
- Nominal discharge current (peak): $I = 10\,kA$
- Line discharge class: Class 2—Heavy-duty riser pole

From the manufacturer data sheets, the height and the creepage distance of the selected arrester are respectively 0.320 and 0.846 m.

The procedure to calculate parameters of the fast front model is again that used with the previous examples. The procedure is applied to a one-column arrester, with an overall height of 0.320 m, being $V_{10} = 82.5\,kV$, and $V_{ss} = 72.6\,kV$ for a 2 kA, 30/60 μs current waveshape. After adjustment of L_1, the parameters to be specified in the model depicted in Figure 6.11 are $R_0 = 32.0$ Ω, $L_0 = 0.064\,\mu H$, $R_1 = 20.8\,\Omega$, $L_1 = 2.4\,\mu H$, and $C = 312.5\,pF$.

B. Simplified analysis

The way in which an ideal gapless MO surge arrester behaves can be summarized as follows. If the maximum value of the wave that arrives to an arrester is larger than the discharge voltage, the maximum value of the wave that propagates further will be limited to the discharge voltage, see Figure 6.40a. This effect can be analyzed by assuming that the wave is not distorted, but an additional wave, hereafter known as the relief wave, is generated and the effect of both waves matches the effect of the actual wave, see Figure 6.40b [138].

Once the arrester voltage has equaled its discharge value, this voltage remains unchanged unless its value decreases below the discharge voltage, and the arrester node behaves as a short-circuit for reflected waves that arrive to this location.

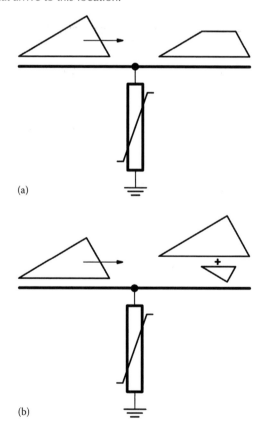

(a)

(b)

FIGURE 6.40
Ideal gapless arrester behavior. (a) Limitation of lightning overvoltages; (b) generation of the relief wave. (From Martinez, J.A. and Gonzalez-Molina, F., *IEEE Trans. Power Deliv.*, 15, 756, 2000. With permission.)

Several factors—front wave steepness of the lightning current, arrester lead length, and system voltage—can affect the maximum overvoltage in the cable. In order to distinguish which is the effect that each of these factors can have, the study is gradually made:

- Initially, the length of lead wires and the system voltage are ignored;
- In a second step, the arrester leads at the riser poles are taken into account;
- Finally, the effect of the system voltage is also included.

1. *The arrester lead and the system voltage are ignored*

Several schemes have been proposed to protect underground cables [139–141]. In this first study, it is assumed that there are arresters at the riser pole only.

When a lightning stroke hits a line conductor, two traveling waves are set up. Once the wave that travels to the cable meets the arrester, reflected and refracted waves are initiated. The refracted wave, which travels to the open terminal, is reflected with sign unchanged and returns to the transition point, where new reflected and refracted waves will be initiated.

Assume that the traveling wave that arrives to the transition point has a rate of rise S_v. The reflection coefficient at this point is given by the following expression

$$\gamma = \frac{Z_c - Z_l}{Z_c + Z_l} \tag{6.54}$$

where Z_l and Z_c are the surge impedances of the overhead line and the underground cable, respectively. Since $Z_c < Z_l$, this coefficient will be always negative. The analysis is performed using the opposite value $\Gamma\ (= -\gamma)$.

The lattice diagram of this case, while the voltage across the arrester is lower than its discharge voltage, is shown in Figure 6.41 [138]. The initial slope of the waves shown in the diagram is S_v times that depicted in the figure.

A detailed analysis of this case could provide information about the overvoltages to be produced at certain point along the cable. However, for most actual cases it can be simplified. Assume that the cable length is such that the voltage at the transition point equals the arrester discharge voltage before the first reflection from the open terminal reaches this transition point.

The time needed to equal the discharge voltage at the transition point can be calculated using the following expression:

$$V_{dis} = S_v \left(1 - \Gamma\right) t_r \tag{6.55}$$

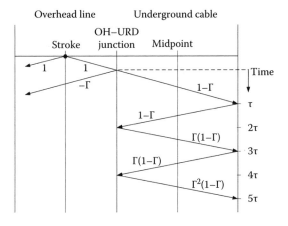

FIGURE 6.41
Example 6.6: Lattice diagram (arrester leads are ignored).

from where

$$t_r = \frac{V_{dis}}{S_v\left(1-\Gamma\right)} \tag{6.56}$$

If the transition time of the cable is τ, the critical length, l_{cr} is defined as the cable length for which $2\tau = t_r$; that is the length for which the first reflected wave from the open end reaches the transition point just when the voltage across the arrester equals the discharge voltage. Therefore:

$$l_{cr} = \tau v = \frac{V_{dis}}{2S_v\left(1-\Gamma\right)}v \tag{6.57}$$

where v is the propagation velocity in the cable.

For a cable length equal or greater than l_{cr} the maximum voltage at the open end will be twice the arrester discharge voltage.

Assume that the discharge voltage of the arrester placed at the riser pole is 82.5 kV, and the peak current magnitude of the lightning stroke is 10 kA, with a front time of 2 μs; that is $S_i = 5$ kA/μs. The rate of the voltage rise S_v depends on the rate of the current rise S_i, and can be approached by:

$$S_v = \frac{S_i}{2}\,Z_l \tag{6.58}$$

In this case, $S_v = 1800$ kV/μs and, according to Equation 6.57, the critical length of this cable would be 22.3 m.

The concept of critical length will be also very useful to analyze a cable protected by arresters at the riser pole and the open terminal. The lattice diagram for this new case before any arrester equals its discharge voltage is still that shown in Figure 6.41. Assume that the discharge voltage is the same for both arresters and that the cable length is the critical one. The voltage at the transition point equals the discharge voltage when the first reflected wave from the open end reaches the transition point

$$V_{rp} = S_v\left(1-\Gamma\right)2\tau \tag{6.59}$$

As the reflection coefficient at the open terminal is unity, the voltage at this point will rise at twice the voltage rise of the wave coming from the transition point, therefore the open-point arrester will equal the discharge voltage in half the time that the arrester at the transition point

$$V_{op} = 2S_v\left(1-\Gamma\right)\tau \tag{6.60}$$

Since this wave is originated with a τ delay, both arresters equal their discharge voltage simultaneously.

The maximum voltage at the point equidistant from both cable terminals can be estimated. At this point there will be the initial wave $S_v(1 - \Gamma)$ during a time τ; then the reflected wave from the open end will travel back and superimpose on the initial wave. After an additional τ delay, relief waves from both ends will reach this point. The maximum voltage stress at this point will be then

$$V_{max} = S_v\left(1-\Gamma\right)\tau + 2S_v\left(1-\Gamma\right)\tau = 3S_v\left(1-\Gamma\right)\tau \tag{6.61}$$

Taking into account Equations 6.59 and 6.60 one can deduce that the maximum voltage stress at this position will be 1.5 times the arrester discharge voltage [139,141]. It is important to notice that this maximum value will be the same for any cable length longer than the critical length.

The lattice diagram, before any arrester equals its discharge voltage, when a third arrester is placed at the cable midpoint is again that shown in Figure 6.41. If all arresters have the same discharge voltage and the length of both cable sections is given by Equation 6.57, one can conclude that:

- The arrester voltage at the transition point equals its discharge value before the first backward reflected wave reaches the transition;
- The arrester voltages at the midpoint and at the open end equal the discharge value at the same time, just when the first reflected wave from the open end reaches the midpoint;
- The maximum voltage stress in the first section will not exceed the discharge voltage of arresters, while the second section will be exposed to 1.5 times the discharge voltage.

From these results it is obvious that the second section should be shorter than the critical length to achieve an acceptable protective margin. However, shortening the second section implies a higher voltage stress in the first section, because then there will be an interval while the reflected wave from the open end will reach the first section before the relief wave from the midpoint arrester is generated. The goal now is to determine the maximum voltage stress produced in the first section as a function of the second section length.

Due to the fact that the second section is shorter than the critical length, the arrester at the midpoint will need some time in excess of the transit time of this second section to equal its discharge voltage

$$V_{dis} = S_v \left(1 - \Gamma\right) 2\tau_2 + 2S_v \left(1 - \Gamma\right) t_r \tag{6.62}$$

being τ_2 the transit time of the second section. Besides, the maximum voltage stress in the first section will be given by

$$V_{max1} = V_{dis} + S_v \left(1 - \Gamma\right) t_r \tag{6.63}$$

From Equation 6.62 one can deduce the following forms:

$$t_r = \frac{V_{dis}}{2S_v \left(1 - \Gamma\right)} - \tau_2 \tag{6.64}$$

$$V_{max1} = 1.5 V_{dis} - S_v \left(1 - \Gamma\right) \tau_2 \tag{6.65}$$

This result shows that the maximum voltage stresses will never be higher than 1.5 times the arrester discharge voltage and that with a length of the second section shorter than the critical length ($\tau_2 > 0$ and $\tau_2 \, v < l_{cr}$) the voltage stress at both sections will be lower than $1.5 V_{dis}$ [138].

The analysis of the three protection schemes has shown how to obtain the maximum overvoltages and the effect that every new arrester has on the maximum value. Roughly speaking, one can assume that the maximum overvoltage when arresters are placed at the riser poles only will be twice the arrester discharge voltage. This value will be limited to 1.5 times the discharge voltage when arresters are placed at the open point too. However, with arresters at a cable midpoint the maximum value could also be 1.5 times the discharge voltage if the second section is longer than the critical length.

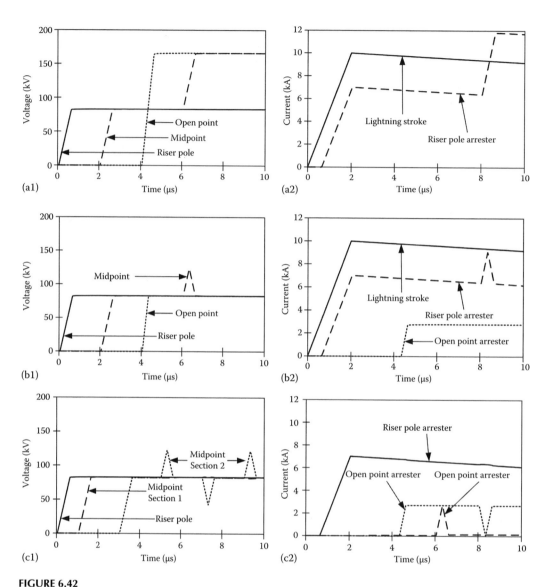

FIGURE 6.42

Example 6.6: Performance of different protection schemes. Lightning stroke: 10 kA crest, front time = 2 μs. Voltages (left side) and currents (right side). (a) Arresters at riser pole only; (b) arresters at riser pole and open point; (c) arresters at riser pole, mid-point and open point.

Figure 6.42 shows some simulation results with the three protection schemes, in which the discharge voltage of the ideal MO surge arresters was in all cases the V_{10} value of the selected arrester; that is, 82.5 kV. These results are according to the previous analysis. With arresters at the riser pole only, the maximum value at any point of the cable would be 165 kV; that is below the rated lightning impulse withstand voltage, 170 kV. However, the protective margin would be very small, and would not cover any deviation from the rated lightning withstand voltage, something that which could be due to ageing. Therefore, a more convenient protection scheme would consider arresters at the open terminal, which would ideally limit the maximum overvoltage to 50% above the discharge voltage of the surge arresters.

Several studies dedicated to determine and measure the current magnitude through distribution arresters have been presented over the years [21,142–146]. According to Ref. [143], from a total of 357 lightning-caused arrester discharge currents (many of which were due to multistroke flashes), about 95% were less than 2 kA, only three were larger than 10 kA, and one had a crest magnitude of 28 kA. Results, presented in Ref. [145] show that the mean of the peak current magnitudes of strokes measured at an experimental line were always below 20 kA. Therefore, the choice of a 10 kA peak current magnitude for representing the lightning stroke that impacts the test line is reasonable, although higher values could be justified.

2. *The effect of the arrester leads is included*
Lightning currents discharging through arrester lead wires produce a voltage that must be added to the arrester voltage. Lead wires may be represented as lumped inductors or distributed-parameters lines (see Example 6.3). In this example they are represented as lumped inductors, whose total length must include distances from arrester terminals to line and to protected equipment ground. When the effect of the arrester lead length is included, the voltage at the transition point should be increased. Assume now that the cable is much longer than the critical length defined above; the maximum voltage at the transition point will be then

$$V_{dis} + L\frac{di}{dt} \tag{6.66}$$

This value will remain constant until the discharge current reaches its peak value. After this, the voltage will decrease, reaching a final value equal to the arrester discharge voltage.

Depending on the protection scheme, the maximum overvoltages produced at a cable will be different

- With an arrester at the riser pole only, the maximum overvoltage is originated at the open point, and its value will be twice the quantity calculated from Equation 6.66.
- If a second arrester is placed at the open point, the analysis is similar to that presented above, and the maximum overvoltage will be 1.5 times the quantity calculated from Equations 6.66; however, this maximum could be originated at a point different from the midpoint.
- When an additional arrester is placed at the midpoint, the wave propagated to the second section is limited to a maximum of V_{dis}, thus the maximum overvoltage at each section will be different:

> $V_{dis} + L$ di/dt between the riser pole and the midpoint   (6.67a)
> $1.5V_{dis}$ between the midpoint and the open point   (6.67b)

For all cases the wave that propagates from the transition point has a peak value given by Equation 6.66.

The analysis is not so simple for the most general case. Several aspects should be considered:

- If the cable length is shorter than the critical length, a reflected wave will reach the transition point when the voltage across the arrester is lower than its discharge voltage;
- If the cable is a little bit longer than the critical length, the reflected wave might reach the transition point before the voltage at this point has decreased to the discharge voltage, then Equation 6.66 will not be correct;
- Even if the cable length is much longer than the critical length, the peak voltage at the transition point could be higher than that calculated from Equation 6.66, although this peak value will never be the highest in the cable;
- If the cable is protected by three arresters and the second section is shorter than the critical length, the maximum voltage at the first section will have a peak value higher than that calculated from Equation 6.66.

Although the analysis is very difficult if every case has to be covered, from the definition of the critical length and the values of this length which are calculated from Equation 6.57, one can conclude that the performance of most actual distribution cables can be analyzed with the hypotheses above assumed.

3. The system voltage is taken into account
Consider that a lightning stroke hits the overhead line at a time that the system voltage has an opposite polarity. The arrester will not go into conduction before the lightning voltage has compensated for the system voltage. Therefore, the wave which propagates from the transition point will be higher than the peak voltage at this point. Obviously the worst case is produced when the stroke hits at the time that the system voltage is at a peak.

The analysis taking into account this effect is rather simple if the cable length is longer than the critical length. The peak value at the transition point is always ($V_{dis} + Ldi/dt$), while the peak value of the positive wave which propagates from the transition point is given by ($V_{sys} + V_{dis} + Ldi/dt$), where V_{sys} is the system voltage at the time the stroke hits the line. The main conclusions for each protection scheme are as follows:

- If no arrester is placed at the open point, the peak voltage will be doubled; however, the peak voltage at the open point will be now

$$2\left(V_{sys} + V_{dis} + L\frac{di}{dt}\right) - V_{sys} \tag{6.68}$$

- If an arrester is placed at the open point, the peak voltage at this point will be the discharge voltage, and the maximum peak will be originated at a certain point, not necessarily equidistant from both terminals, reaching a value given by

$$1.5\left(V_{sys} + V_{dis}\right) + L\frac{di}{dt} - V_{sys} \tag{6.69}$$

- If an additional arrester is placed at the midpoint, the voltage at both the midpoint and the open-point arresters will be the discharge voltage; however, the peak voltages at each section will be different

➢ ($V_{sys} + V_{dis} + Ldi/dt$) − V_{sys} on the first section $\qquad$ (6.70a)
➢ $1.5V_{dis}$ on the second one $\qquad$ (6.70b)

Figure 6.43 shows the simulation results when both the arrester lead ($L = 1\,\mu H$) and the system voltage effects are included. The simulations were performed assuming that the lightning stroke hits the line when the system voltage is at the negative peak value.

The differences with respect to the previous results are evident: the arrester leads add 5 kV at the riser pole and the system voltage increases even more the voltage at the open terminal, when arresters are installed at the riser pole only. However, when more arresters are used (open terminal, midpoint), the voltages at other locations of the cable are basically the same as that were obtained before. As for the discharged energies, the conclusion is the expected one: the lightning stroke current is shared by arresters and the discharged energy by one arrester decreases as more arresters are installed. It is interesting to note that with the values used in this example, the maximum energy is not discharged by the arrester installed at the riser pole when the cable is protected by more than one arrester.

Table 6.21 shows a summary of the results derived from this study for each protection scheme.

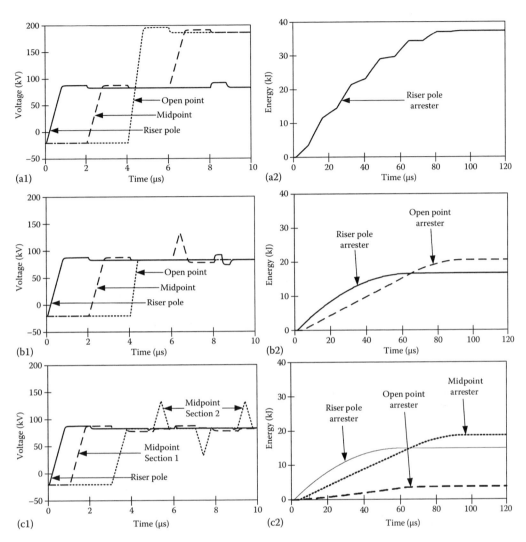

FIGURE 6.43

Example 6.6: Arrester lead and system voltage effects are included. Arrester lead: $L = 1\,\mu H$—lightning stroke: 10 kA crest, front time = 2 μs. Voltages (left side) and energies (right side). (a) Arresters at riser pole only; (b) arresters at riser pole and open point; (c) arresters at riser pole, mid-point and open point.

C. Advanced arrester model

Two important features are changed for the new analysis: arresters are now represented according to the IEEE model and the lightning stroke current is represented by means of a concave waveform, using the Heidler model [42].

The critical length of the cable will be again a useful concept; however, this value must be now calculated from the value of the maximum slope of the current waveshape along the front time. The conclusions will be the same that were derived from a double-ramp representation, provided that the cable length is longer that the critical length.

The new study has been performed considering the three protection schemes and including the effect of the lead length of the arrester placed at the riser pole only.

TABLE 6.21

Example 6.6: Overvoltages Originated with Each Protection Scheme

Protection Scheme	Overvoltages
	Riser pole: $V_{dis} + L\,di/dt$ Open point: $2\,(V_{dis} + L\,di/dt) + V_{sys}$
	Riser pole: $V_{dis} + L\,di/dt$ Midpoint: $1.5V_{dis} + L\,di/dt + 0.5V_{sys}$ Open point: V_{dis}
	Riser pole: $V_{dis} + L\,di/dt$ Midpoint—Section 1: $V_{dis} + L\,di/dt$ Midpoint—cable: V_{dis} Midpoint—Section 2: $1.5V_{dis} + 0.5V_{sys}$ Open point: V_{dis}

Source: Martinez, J.A. and Gonzalez-Molina, F., *IEEE Trans. Power Deliv.*, 15, 756, 2000. With permission.

The maximum voltage at the transition point, if the cable length is longer than the critical length, can be approximated by $(V_{dis} + Ldi/dt)$. However, the value of V_{dis} is not now constant and depends on the discharge current waveshape. Besides, the di/dt value to be used is that corresponding to the maximum slope of the current along its front.

The conclusions derived from digital simulations are summarized as follows:

1. The discharge currents across arresters placed at a midpoint and the open point do not develop a discharge voltage greater than that at the riser pole.
2. The system voltage when the stroke hits the line has no effect at those points where the arresters are placed.
3. If the cable is protected by an arrester at the riser pole only, the maximum overvoltage will be again twice that originated at the riser pole plus the system voltage, see Table 6.21.
4. When an additional arrester is placed at the open terminal, the maximum overvoltage is originated again at a midpoint of the cable. However, the maximum value should be calculated taking into account that the voltage across the arrester placed at the open point is generally lower than that at the riser pole. An acceptable approximation would be:

$$V_{max} < 1.5V_{dis} + L\frac{di}{dt} + 0.5V_{sys} \tag{6.71}$$

where V_{dis} is the discharge voltage at the riser pole arrester.

5. The point where the maximum overvoltage is produced, when a third arrester is placed at a midpoint of the cable, depends on several factors (arrester discharge voltage, discharge

current waveshape across the riser pole arrester, length of each cable section). If both cable sections are much longer than the critical length which corresponds to the lightning current, the maximum overvoltage at each section can be approximated as follows:

$$V_{max1} = V_{dis} + L\frac{di}{dt} \tag{6.72a}$$

$$V_{max2} < 1.5V_{dis} + 0.5V_{sys} \tag{6.72b}$$

The wave originated at the transition point has a maximum value given by $(V_{dis} + V_{sys} + Ldi/dt)$. While this wave propagates along the first section, the maximum overvoltage at each point of this section is that given by Equation 6.72a, which results after subtracting V_{sys} from the expression given above, and equals the voltage at the transition point. The maximum overvoltage at the midpoint arrester is limited to a value close to V_{dis}, so the wave which propagates along the second section has a maximum given by $(V_{dis} + V_{sys})$. As the maximum overvoltage at the open point is also limited by V_{dis}, due to the small discharge current across the arrester placed at this terminal, the analysis for the second section could be made assuming an ideal performance of the arresters. Therefore, the maximum overvoltage in this section can be approached by Equation 6.72b, see also Table 6.21. As the length of this second section is longer than the critical one, when the wave reflected in the open point reaches the midpoint, the arrester at this place has reached the discharge voltage and prevents the wave could pass to the first section, whose maximum voltage is no longer increased.

Figure 6.44 shows some simulation results with the three protection schemes. In all cases the cable has been protected using the selected arrester. The derivative of the discharge current across the arrester placed at the riser pole reached a maximum lower than $10\,\text{kA}/\mu\text{s}$. The simulations were performed again when the system voltage is at a negative peak value.

Theoretical and simulation results have been derived assuming a very simple representation for lines and cables. This is generally acceptable because single-core cables are practically decoupled at lightning frequencies and the surge attenuation is not significant for lengths shorter than 1 km [147,148]. However, the representation of transformers is still very simple.

An obvious conclusion from the results presented above is that the more arresters are placed the lower the overvoltages generated along the cable. However, unless the arresters were placed very close to each other, one can also conclude that very few advantages are derived from installing more than three arresters. When a cable is protected by means of arresters at both cable terminals and at a midpoint, an approximated calculation of the protective margin and the probability of failure are not very difficult tasks. Indeed, the maximum overvoltage generated along the cable for a given lightning stroke waveshape is deduced from Equations 6.72a and 6.72b, while the probability of failure can also be obtained if reliable information about ground flash density and peak current distribution are available.

The calculations have been made assuming that lightning strokes are monopolar, either positive or negative. Although the majority of lightning strokes are monopolar, bipolar strokes can also be encountered [149].

Protection of distributed generators. Overvoltages can be originated when a generator and part of the distribution network are separated from the utility, although protection schemes used at the generation site would be expected to sense the islanding condition and disconnect the generator from the system in a few seconds. These overvoltages are mainly due to ungrounded transformer connections, self-excitation, or ferroresonance problems [150]. MO surge arresters may not be able to survive the sustained overvoltages caused by the presence of a generator. As the number and size of distributed generators increase, these overvoltages could become a serious problem. On systems with a high penetration of distributed generation, utilities may consider using higher rated surge arresters.

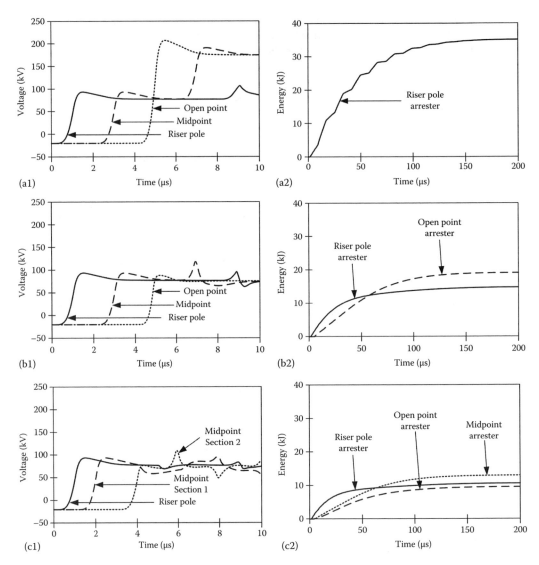

FIGURE 6.44
Example 6.6: Simulation results—advanced arrester model. Arrester lead: $L = 1\,\mu H$; lightning stroke: concave waveform, 10 kA crest, front time = 2 μs. Voltages (left side) and energies (right side). (a) Arrester at riser pole only; (b) arrester at riser pole and open point; (c) arresters at riser pole, mid-point and open point.

References

1. P. Chowdhuri, *Electromagnetic Transients in Power Systems*, RSP-Wiley, Taunton, U.K., 1996.
2. IEEE Std C62.1-1989, IEEE standard for gapped silicon-carbide surge arresters for AC power circuits.
3. IEEE Std C62.2-1987, IEEE guide for the application of gapped silicon-carbide surge arresters for alternating current systems.

4. IEEE Std C62.11-2005, IEEE standard for metal-oxide surge arresters for ac power circuits (>1 kV).

5. IEEE Std C62.22-1997, IEEE guide for the application of metal-oxide surge arresters for alternating-current systems.

6. IEC Std 60099-1, Surge arresters—Part 1: Non-linear resistor type gapped surge arresters for a.c. systems, edition 3.1, 1999.

7. IEC Std 60099-3, Surge arresters—Part 3: Artificial pollution testing of surge arresters, 1st edn., 1990.

8. IEC Std 60099-4, Surge arresters—Part 4: Metal-oxide surge arresters without gaps for a.c. systems, edition 2.1, 2006.

9. IEC Std 60099-5, Surge arresters—Part 5: Selection and application recommendations, edition 1.1, 2000.

10. IEC Std 60099-6, Surge arresters—Part 6: Surge arresters containing both series and parallel gapped structures—Rated 52 kV and less, 1st edn., 2002.

11. D.P. Carroll, R.W. Flugum, J.W. Kalb, and H.A. Peterson, A dynamic surge arrester model for use in power system transient studies, *IEEE Transactions on Power Apparatus and Systems*, 92(3), 1057–1066, May/June 1972.

12. A.P. Sakis Meliopoulos, Lightning and overvoltage protection, Section 27 of *Standard Handbook for Electrical Engineers*, D.G. Fink and H. Wayne Beaty (Eds.), 14th edn., McGraw-Hill, 2000.

13. N. Fujiwara et al., Development of a pin-post insulator with built-in metal oxide varistors for distribution lines, *IEEE Transactions on Power Delivery*, 11(2), 824–833, April 1996.

14. Y. Harumoto et al., Evaluation for application of built-in type zinc oxide gapless surge arresters for power system equipments, *IEEE Transactions on Power Delivery*, 2(3), 750–757, July 1987.

15. A.R. Hileman, *Insulation Coordination for Power Systems*, Marcel Dekker, New York, 1999.

16. T.E. McDermott, Surge Arresters, Chapter 13 of *Power Systems*, L.L. Grigsby (Ed.), CRC Press, New York, 2007.

17. S. Gu et al., Development of surge arresters with series gap against lightning breakage of covered conductors on distribution lines, *IEEE Transactions on Power Delivery*, 22(4), 2191–2198, October 2007.

18. G.L. Goedde, Lj.A. Kojovic, M.B. Marz, and J.J. Woodworth, Series graded gapped arrester provides reliable overvoltage protection in distribution networks, *IEEE PES Winter Meeting*, 28 January–1 February, 2001, Columbus, OH.

19. J. Osterhout, Comparison of IEC and IEEE standards for metal oxide surge arresters, *IEEE Transactions on Power Delivery*, 7(4), 2002–2006, October 1992.

20. S. Kojima, S. Nishiwaki, and T. Sato, Switching surge duty of metal oxide surge arresters in multiphases system and its equivalent conditions in a test, *IEEE Transactions on Power Delivery*, 2(4), 117–1123, October 1987.

21. E.C. Sakshaug, J.J. Burke, and J.S. Kresge, Metal oxide arresters on distribution systems. Fundamental considerations, *IEEE Transactions on Power Delivery*, 4(4), 2076–2089, October 1989.

22. K.G. Ringler, P. Kirkby, C.C. Erven, M.V. Lat, and T.A. Malkiewicz, The energy absorption capability and time-to-failure of varistors used in station-class metal-oxide surge arresters, *IEEE Transactions on Power Delivery*, 12(1), 203–212, January 1997.

23. P. Kirkby, C.C. Erven, and O. Nigol, Long-term stability and energy discharge capacity of metal oxide valve elements, *IEEE Transactions on Power Delivery*, 3(4), 1656–1665, October 1988.

24. M.L.B. Martinez and L.C. Zanetta, On modeling and testing metal oxide resistors to evaluate the arrester withstand energy, *Dielectric Materials, Measurements and Applications Conference*, Edinburg, U.K., Paper 473, pp. 247–252, IEE 2000.

25. M.L.B. Martinez and L.C. Zanetta, A proposal to evaluate the energy withstand capacity of metal oxide resistors, *High Voltage Engineering Symposium*, London, U.K., Paper 467, August 1999.

26. M. Bartkowiak, M.G. Comber, and G.D. Mahan, Failure modes and energy absorption capability of ZnO varistors, *IEEE Transactions on Power Delivery*, 14(1), 152–162, January 1999.

27. CIGRE WG 33.02, Guidelines for Representation of Network Elements when Calculating Transients, CIGRE Brochure 39, 1990.
28. H.W. Dommel, *Electromagnetic Transients Program Reference Manual* (*EMTP Theory Book*), Bonneville Power Administration, Portland, August 1986.
29. J.A. Martinez and D. Durbak, Parameter determination for modeling systems transients. Part V: Surge arresters, *IEEE Transactions on Power Delivery*, 20(3), 2073–2078, July 2005.
30. E.C. Sakshaug, Influence of rate-of-rise on distribution arrester protective characteristics, *IEEE Transactions on Power Apparatus and Systems*, 98(2), 519–526, March/April 1979.
31. S. Tominaga et al., Protective performance of metal oxide surge arrester based on the dynamic V–I characteristics, *IEEE Transactions on Power Apparatus and Systems*, 98(6), 1860–1871, November/December 1979.
32. D.W. Durbak, Zinc-oxide arrester model for fast surges, *EMTP Newsletter*, 5(1), 1–9, January 1985.
33. C. Dang, T.M. Parnell, and P.J. Price, The response of metal oxide surge arresters to steep fronted current impulses, *IEEE Transactions on Power Delivery*, 1(1), 157–163, January 1986.
34. D.W. Durbak, The choice of EMTP surge arrester models, *EMTP Newsletter*, 7(3), 14–18, September 1987.
35. W. Schmidt et al., Behavior of MO-surge-arrester blocks to fast transients, *IEEE Transactions on Power Delivery*, 4(1), 292–300, January 1989.
36. A.R. Hileman, J. Roguin, and K.H. Weck, Metal oxide surge arresters in AC systems. Part V: Protection performance of metal oxide surge arresters, *Electra*, 133, 132–144, December 1990.
37. IEEE Working Group on Surge Arrester Modeling, Modeling of metal oxide surge arresters, *IEEE Transactions on Power Delivery*, 7(1), 302–309, January 1992.
38. I. Kim et al., Study of ZnO arrester model for steep front wave, *IEEE Transactions on Power Delivery*, 11(2), 834–841, April 1996.
39. P. Pinceti and M. Giannettoni, A simplified model for zinc oxide surge arresters, *IEEE Transactions on Power Delivery*, 14(2), 393–398, April 1999.
40. M. Caserza Magro, M. Giannettoni, and P. Pinceti, Validation of ZnO surge arresters model for overvoltage studies, *IEEE Transactions on Power Delivery*, 19(4), 1692–1695, October 2004.
41. CIGRE WG 33.06, Metal Oxide Arresters in AC Systems, Technical Brochure 60, 1991.
42. F. Heidler, J.M. Cvetic, and B.V. Stanic, Calculation of lightning current parameters, *IEEE Transactions on Power Delivery*, 14(2), 399–404, April 1999.
43. IEEE TF on Parameters of Lightning Strokes, Parameters of lightning strokes: A review, *IEEE Transactions on Power Delivery*, 20(1), 346–358, January 2005.
44. H.J. Li, S. Birlasekaran, and S.S. Choi, A parameter identification technique for metal-oxide surge arrester models, *IEEE Transactions on Power Delivery*, 17(3), 736–741, July 2002.
45. M.V. Lat, Thermal properties of metal oxide surge arresters, *IEEE Transactions on Power Apparatus and Systems*, 102(7), 2194–2202, July 1983.
46. M.V. Lat, Analytical method for performance prediction of metal oxide surge arresters, *IEEE Transactions on Power Apparatus and Systems*, 104(10), 2665–2674, October 1985.
47. G. St-Jean and A. Petit, Metal-oxide surge arrester operating limits defined by a temperature-margin concept, *IEEE Transactions on Power Delivery*, 5(2), 627–633, April 1990.
48. A. Petit, X.D. Do, and G. St-Jean, An experimental method to determine the electro-thermalt model parameters of metal oxide arresters, *IEEE Transactions on Power Delivery*, 6(2), 715–721, April 1991.
49. A. Haddad, ZnO Surge Arresters, Chapter 5 of *Advances in High Voltage Engineering*, A. Haddad and D. Warne (Eds.), The Institution of Electrical Engineers, 2004.
50. S. Nishiwaki, H. Kimura, T. Satoh, H. Mizoguchi, and S. Yanabu, Study of thermal runaway/equivalent prorated model of a ZnO surge arrester, *IEEE Transactions on Power Apparatus and Systems*, 103, 413–421, February 1984.
51. M. Kan, S. Nishiwaki, T. Sato, S. Kojima, and S. Yanabu, Surge discharge capability and thermal stability of a metal oxide surge arrester, *IEEE Transactions on Power Apparatus and Systems*, 102(2), 282–289, February 1983.

52. F.R. Stockum, Simulation of the nonlinear thermal behavior of metal oxide surge arresters using a hybrid finite difference and empirical model, *IEEE Transactions on Power Delivery*, 9(1), 306–313, January 1994.

53. E. Guedes da Costa, S.R. Naidu, and A. Guedes da Lima, Electrothermal model for complete metal-oxide surge arresters, *IEE Proceedings—Generation, Transmission and Distribution*, 148(1), 29–33, January 2001.

54. O. Nigol, Methods for analyzing the performance of gapless metal oxide surge arresters, *IEEE Transactions on Power Delivery*, 7(3), 1256–1264, July 1992.

55. S. Boggs, J. Kuang, H. Andoh, and S. Nishiwaki, Electro-thermal-mechanical computations in ZnO arrester elements, *IEEE Transactions on Power Delivery*, 15(1), 128–134, July 2000.

56. Z. Zheng, S. Nishiwaki, S. Boggs, and T. Fukano, Effects of heat sinks in metal-oxide surge arresters on ZnO element power dissipation and temperature distribution, *IEEE Transactions on Power Delivery*, 20(2), 1402–1408, April 2005.

57. Z. Zheng et al. Investigation of heat-of-fusion heat sinks for polymer-type metal-oxide surge arresters, 2006 *IEEE International Symposium on Electrical Insulation*, Tuscaloosa, AL, 2005.

58. IEEE Std 1313.2-1999, IEEE Guide for the Application of Insulation Coordination.

59. IEC 60071-2, Insulation Co-ordination, Part 2: Application Guide, 1996.

60. A. Greenwood, *Electrical Transients in Power Systems*, John Wiley, New York, 1971.

61. IEEE Std 1313-1996, IEEE Standard for Insulation Coordination—Definitions, Principles and Rules.

62. IEC 60071-1, Insulation Co-ordination, Part 2: Definitions, principles and rules, 2006.

63. R.A. Jones and H.S. Fortson Jr., Consideration of phase-to-phase surges in application of capacitor banks, *IEEE Transactions on Power Delivery*, 1(3), 240–244, July 1986.

64. R.P. O'Leary and R.H. Harner, Evaluation of methods for controlling the overvoltages produced by the energization of a shunt capacitor bank, *CIGRE Session*, Paper No. 13–05, 1988, Paris.

65. A.J.F. Keri, Y.I. Musa, and J.A. Halladay, Insulation coordination for delta connected transformers, *IEEE Transactions on Power Delivery*, 9(2), 772–780, April 1994.

66. IEEE Task Force on Slow Transients, Modeling and analysis guidelines for slow transients—Part III: The study of ferroresonance, *IEEE Transactions on Power Delivery*, 15(1), 255–265, January 2000.

67. R.H. Hopkinson, Ferroresonance during single-phase switching of 3-phase distribution transformer banks, *IEEE Transactions on Power Apparatus and Systems*, 84(4), 289–293, April 1965.

68. F.S. Young, R.L. Schmid, and P.I. Fergestad, A laboratory investigation of ferroresonance in cable-connected transformers, *IEEE Transactions on Power Apparatus and Systems*, 87(5), 1240–1249, May 1968.

69. R.A. Walling et al., Performance of metal-oxide arresters exposed to ferroresonance in padmount transformers, *IEEE Transactions on Power Delivery*, 9(2), 788–795, April 1994.

70. R.E. Koch et al., Design of zinc oxide transmission line arresters for application on 138 kV towers, *IEEE Transactions on Power Apparatus and Systems*, 104(10), 2675–2680, October 1985.

71. C.H. Shih et al., Application of special arresters on 138 kV lines of Appalachian Power Company, *IEEE Transactions on Power Apparatus and Systems*, 104(10), 2857–2863, October 1985.

72. S. Furukawa, O. Usuda, T. Isozaki, and T. Irie, Development and application of lightning arresters for transmission lines, *IEEE Transactions on Power Delivery*, 4(4), 2121–2129, October 1989.

73. K. Ishida et al., Development of a 500 kV transmission line arrester and its characteristics, *IEEE Transactions on Power Delivery*, 7(3), 1265–1274, July 1992.

74. T. Yamada et al., Development of suspension-type arresters for transmission lines, *IEEE Transactions on Power Delivery*, 8(3), 1052–1060, July 1993.

75. J.L. He, S.M. Chen, R. Zeng, J. Hu, and C.G. Deng, Development of polymeric surge ZnO arresters for 500-kV compact transmission line, *IEEE Transactions on Power Delivery*, 21(1), 113–120, January 2006.

76. R.A. Sargent, G.L. Dunlop, and M. Darveniza, Effects of multiple impulse currents on the microstructure and electrical. Properties of metal-oxide varistors, *IEEE Transactions on Electrical Insulation*, 27(3), 586–592, June 1992.

77. M. Darveniza, D. Roby, and L.R. Tumma, Laboratory and analytical studies of the effects of multipulse lightning current on metal oxide arresters, *IEEE Transactions on Power Delivery*, 9(2), 764–771, April 1994.
78. M. Darveniza, L.R. Tumma, B. Richter, and D. Roby, Multipulse lightning currents and metal oxide arresters, *IEEE Transactions on Power Delivery*, 12(3), 1168–1175, July 1997.
79. Y. Matsumoto et al., Measurement of lightning surges on test transmission line equipped with arresters struck by natural and triggered lightning, *IEEE Transactions on Power Delivery*, 11(2), 996–1002, April 1996.
80. T.A. Short et al., Application of surge arresters to a 115 kV circuit, *IEEE PES Transmission and Distribution Conference and Exposition*, Los Angeles, September 1996.
81. S. Sadovic, R. Joulie, S. Tartier, and E. Brocard, Use of line surge arresters considering the total lightning performance of 63 kV and 90 kV shielded and unshielded transmission lines, *IEEE Transactions on Power Delivery*, 12(3), 1232–1240, July 1997.
82. T. Kawamura et al., Experience and effectiveness of application of arresters to overhead transmission lines, *CIGRE Session*, Paper 33–301, 1998, Paris.
83. E.J. Tarasiewicz, F. Rimmer, and A.S. Morched, Transmission line arresters energy, cost, and risk of failure analysis for partially shielded transmission lines, *IEEE Transactions on Power Delivery*, 15(3), 919–924, July 2000.
84. T. Wakai, N. Itamoto, T. Sakai, and M. Ishii, Evaluation of transmission line arresters against winter lightning, *IEEE Transactions on Power Delivery*, 15(2), 684–690, April 2000.
85. L.C. Zanetta and C.E. de Morais Pereira, Application studies of line arresters in partially shielded 138-kV transmission lines, *IEEE Transactions on Power Delivery*, 18(1), 95–100, January 2003.
86. L.C. Zanetta, Evaluation of line surge arrester failure rate for multipulse lightning stresses, *IEEE Transactions on Power Delivery*, 18(3), 706–801, July 2003.
87. L. Stenström, Required energy capability based on total charge for a transmission line arrester for protection of a compact 420 kV line for Swedish conditions, *CIGRE International Colloquium on Insulation Coordination*, September 2–3, 1997, Toronto.
88. L. Stenström and J. Lundquist, Energy stress on transmission line arresters considering the total lightning charge distribution, *IEEE Transactions on Power Delivery*, 14(1), 148–151, January 1999.
89. CIGRE WG 33.11, Application of metal oxide surge arresters to overhead lines, *Electra*, no. 186, 82–112, October 1999.
90. M.A. Ismaili, P. Bernard, R. Lambert, and A. Xémard, Estimating the probability of failure of equipment as a result of direct lightning strikes on transmission lines, *IEEE Transactions on Power Delivery*, 14(4), 1394–1400, October 1999.
91. M.S. Savic, Estimation of the surge arrester outage rate caused by lightning overvoltages, *IEEE Transactions on Power Delivery*, 20(1), 116–122, January 2005.
92. T. Hayashi, Y. Mzuno, and K. Naito, Study on transmission-line arresters for tower with high footing resistance, *IEEE Transactions on Power Delivery*, 23(4), 2456–2460, October 2008.
93. A. Xemard, S. Sadovic, I. Uglesic, L. Prikler, and J.A. Martinez, Developments on the Line Surge Arresters Application Guide prepared by the CIGRE WG C4 301, *CIGRE Symposium on Transients in Large Electric Power Systems*, April 18–21, 2007, Zagreb.
94. G.W. Brown and E.R. Whitehead, Field and analytical studies of transmission line shielding: Part II, *IEEE Transactions on Power Apparatus and Systems*, 88(3), 617–626, May 1969.
95. IEEE TF on Fast Front Transients, Modeling guidelines for fast transients, *IEEE Transactions on Power Delivery*, 11(1), 493–506, January 1996.
96. *Modeling and Analysis of System Transients Using Digital Programs*, A.M. Gole, J.A. Martinez-Velasco and A.J.F. Keri (Eds.), IEEE PES Special Publication, TP-133-0, 1999.
97. IEC TR 60071-4, Insulation Co-ordination Part 4: Computational guide to insulation co-ordination and modelling of electrical networks, 2004.
98. J.A. Martinez-Velasco and F. Castro-Aranda, Modelling guidelines for overhead transmission lines in lightning studies, *CIGRE Symposium on Transients in Large Electric Power Systems*, April 18–21, 2007, Zagreb.

99. CIGRE WG 33.01, Guide to Procedures for Estimating the Lightning Performance of Transmission Lines, Technical Brochure 63, 1991.

100. CIGRE WG 33.01, Guide for the Evaluation of the Dielectric Strength of External Insulation, Technical Brochure 72, 1992.

101. A. Pigini et al., Performance of large air gaps under lightning over-voltages: Experimental study and analysis of accuracy of pre-determination methods, *IEEE Transactions on Power Delivery*, 4(2), 1379–1392, April 1989.

102. A.M. Mousa, The soil ionization gradient associated with discharge of high currents into concentrated electrodes, *IEEE Transactions on Power Delivery*, 9(3), 1669–1677, July 1994.

103. J.A. Martinez and F. Castro-Aranda, Lightning performance analysis of overhead transmission lines using the EMTP, *IEEE Transactions on Power Delivery*, 20(3), 2200–2210, July 2005.

104. W.A. Lyons, M. Uliasz, and T.E. Nelson, Large peak current cloud-to-ground lightning flashes during the summer months in the contiguous United States, *Monthly Weather Review*, 126, 2217–2233, 1998.

105. J.A. Martinez and F. Castro-Aranda, Modeling of overhead transmission lines for line arrester studies, *IEEE PES General Meeting*, June 2004, Denver.

106. A. Ametani and T. Kawamura, A method of a lightning surge analysis recommended in Japan using EMTP, *IEEE Transactions on Power Delivery*, 20(2), 867–875, April 2005.

107. V.A. Rakov and M.A. Uman, *Lightning. Physics and Effects*, Cambridge University Press, Cambridge, U.K., 2003.

108. J.A. Martinez and F. Castro-Aranda, Influence of the stroke angle on the flashover rate of an overhead transmission line, *IEEE PES General Meeting 2006*, June 2006, Montreal.

109. CIGRE 33/13.09, *Monograph on GIS Very Fast Transients*, July 1989.

110. M.M. Osborne, A. Xemard, L. Prikler, and J.A. Martinez, Points to consider regarding the insulation coordination of GIS substations with cable connections to overhead lines, *Proceedings of the 7th International Conference on Power Systems Transients*, 4–7, June 2007, Lyon.

111. CIGRE WG B1.05, *Transient Voltages Affecting Long Cables*, Technical Brochure 268, 2005.

112. T. Henriksen, B. Gustavsen, G. Balog, and U. Baur, Maximum lightning overvoltage along a cable protected by surge arresters, *IEEE Transactions on Power Delivery*, 20(2), 859–866, April 2005.

113. E.W. Greenfield, Transient behavior of short and long cables, *IEEE Transactions on Power Apparatus and Systems*, 103(11), 3193–3203, November 1984.

114. L. Marti, Simulation of transients in underground cables with frequency-dependent modal transformation matrices, *IEEE Transactions on Power Delivery*, 3(3), 1099–1110, July 1988.

115. H. van der Merwe and F.S. van der Merwe, Some features of traveling waves on cables, *IEEE Transactions on Power Delivery*, 8(3), 789–797, July 1993.

116. W.E. Reid et al., MOV arrester protection of shield interrupts on 138 kV extruded dielectric cables, *IEEE Transactions on Power Apparatus and Systems*, 103, 3334–3341, November 1984.

117. M.V. Lat and J. Kortschinski, Distribution arrester research, *IEEE Transactions on Power Apparatus and Systems*, 100(7), 3496–3505, July 1981.

118. S.S. Kershaw, G.L. Gaibrois, and K.B. Stump, Applying metal-oxide surge arresters on distribution systems, *IEEE Transactions on Power Delivery*, 4(1), 301–307, January 1989.

119. M.V. Lat, Determining temporary overvoltage levels for application of metal oxide surge arresters on multigrounded distribution systems, *IEEE Transactions on Power Delivery*, 5(2), 936–946, April 1990.

120. J.J. Burke, V. Varneckas, E. Chebli, and G Hoskey, Application of MOV and gapped arresters on non-effectively grounded distribution systems, *IEEE Transactions on Power Delivery*, 6(2), 794–800, April 1991.

121. R.T. Mancao, J.J. Burke, and A. Myers, The effect of distribution system grounding on MOV selection, *IEEE Transactions on Power Delivery*, 8(1), 139–145, January 1993.

122. T.A. Short, J.J. Burke, and R.T. Mancao, Application of MOVs in the distribution environment, *IEEE Transactions on Power Delivery*, 9(1), 293–305, January 1994.

123. F.J. Muench and J.P. Dupont, Coordination of MOV type lightning arresters and current limit-ing fuses, *IEEE Transactions on Power Delivery*, 5(2), 966–971, April 1990.
124. A. Hamel, G. St-Jean, and M. Paquette, Nuisance fuse operation on MV transformers during storms, *IEEE Transactions on Power Delivery*, 5(4), 1866–1874, October 1990.
125. W. Plummer et al., Reduction in distribution transformer failure rates and nuisance outages using improved lightning protection concepts, *IEEE Transactions on Power Delivery*, 10(2), 768–777, April 1995.
126. S. Yokoyama, Distribution surge arrester behavior due to lightning induced voltages, *IEEE Transactions on Power Delivery*, 1(1), 171–178, January 1986.
127. T.E. McDermott, T.A. Short, and J.G. Anderson, Lightning protection of distribution lines, *IEEE Transactions on Power Delivery*, 9(1), 138–152, January 1994.
128. T. Short and R.H. Ammon, Monitoring results of the effectiveness of surge arrester spacings on distribution line protection, *IEEE Transactions on Power Delivery*, 14(3), 1142–1150, July 1999.
129. IEEE Std 1410-1997, IEEE Guide for Improving the Lightning Performance of Electric Power Overhead Distribution Lines.
130. K. Nakada, T. Yokota, S. Yokoyama, A. Asakawa, M. Nakamura, H. Taniguchi, and A. Hashimoto, Energy absorption of surge arresters on power distribution lines due to direct lightning strokes-effects of an overhead ground wire and installation position of surge arresters, *IEEE Transactions on Power Delivery*, 12(4), 1779–1785, October 1997.
131. K. Nakada, S. Yokoyama, T. Yokota, A. Asakawa, and T. Kawabata, Analytical study on preven-tion methods for distribution arrester outages caused by winter lightning, *IEEE Transactions on Power Delivery*, 13(4), 1399–1404, October 1998.
132. K. Nakada et al., Distribution arrester failures caused by lightning current flowing from cus-tomer's structure into distribution lines, *IEEE Transactions on Power Delivery*, 14(4), 1527–1532, October 1999.
133. M. Darveniza and D.R. Mercer, Lightning protection of pole mounted transformers, *IEEE Transactions on Power Delivery*, 4(2), 1087–1095, April 1989.
134. Lightning Protection Manual for Rural Electric Systems, NRECA, RER Project 92–12, 1993.
135. B. Wareing, *Wood Pole Overhead Lines*, The IEE, 2005.
136. C.F. Wagner and A.R. Hileman, Effect of predischarge current upon line performance, *IEEE Transactions on Power Apparatus and Systems*, 82(65), 117–131, April 1963.
137. G.W. Brown and S. Thunander, Frequency of distribution arrester discharge currents due to direct strokes, *IEEE Transactions on Power Apparatus and Systems*, 95(5), 1571–1578, September/October 1976.
138. J.A. Martinez and F. Gonzalez-Molina, Surge protection of underground distribution cables, *IEEE Transactions on Power Delivery*, 15(2), 756–763, April 2000.
139. J.J. Burke, E.C. Sakshaug, and S.L. Smith, The application of gapless arresters on distribution systems, *IEEE Transactions on Power Apparatus and Systems*, 100(3), 1234–1243, March 1981.
140. W.D. Niebuhr, Application of metal-oxide-varistor surge arresters on distribution systems, *IEEE Transactions on Power Apparatus and Systems*, 101(6), 1711–1715, June 1982.
141. M.V. Lat, A simplified method for surge protection of underground distribution systems with metal oxide arresters, *IEEE Transactions on Power Delivery*, 2(4), 1110–1116, October 1987.
142. G.L. Gaibrois, Lightning current surge magnitude through distribution arresters, *IEEE Transactions on Power Apparatus and Systems*, 100(3), 964–970, March 1981.
143. P.P. Baker, R.T. Mancao, D.J. Kvaltine, and D.E. Parrish, Characteristics of lightning surges mea-sured at metal oxide distribution arresters, *IEEE Transactions on Power Delivery*, 8(1), 301–310, January 1993.
144. M.I. Fernandez, K.J. Rambo, V.A. Rakov, and M.A. Uman, Performance of MOV arresters dur-ing very close, direct lightning strikes to a power distribution system, *IEEE Transactions on Power Delivery*, 14(2), 411–418, April 1999.
145. J. Schoene et al., Direct lightning strikes to test power distribution lines—Part I: Experimental and overall results, *IEEE Transactions on Power Delivery*, 22(4), 2236–2244, October 2007.

146. J. Schoene et al., Direct lightning strikes to test power distribution lines—Part II: Measured and modeled current division among multiple arresters and ground, *IEEE Transactions on Power Delivery*, 22(4), 2245–2253, October 2007.
147. R.E. Owen and C.R. Clinkenbeard, Surge protection of UD cable systems. Part I: Cable attenuation and protection constraints, *IEEE Transactions on Power Apparatus and Systems*, 97(4), 1319–1327, July/August 1978.
148. R.C. Dugan and W.L. Sponsler, Surge protection of UD cable systems. Part II: Analytical models and simulations, *IEEE Transactions on Power Apparatus and Systems*, 97(5), 1901–1909, September/October 1978.
149. P.P. Baker, Voltage quadrupling on a UD cable, *IEEE Transactions on Power Delivery*, 5(1), 498–501, January 1990.
150. W.E. Feero and W.B. Gish, Overvoltages caused by DSG operation: Synchronous and induction generators, *IEEE Transactions on Power Delivery*, 1(1), 258–264, January 1986.

7

Circuit Breakers

Juan A. Martinez-Velasco and Marjan Popov

CONTENTS

7.1 Introduction 448
7.2 Principles of Current Interruption...449
    7.2.1 Introduction..449
    7.2.2 The Electric Arc 450
    7.2.3 Thermal and Dielectric Characteristics.....................................452
    7.2.4 Arc–Power System Interaction: Transient Recovery Voltage 453
        7.2.4.1 Current Interruption under Normal Operating Conditions 454
        7.2.4.2 Current Interruption under Fault Conditions 463
7.3 Breaking Technologies...469
    7.3.1 Introduction..469
    7.3.2 Oil Circuit Breakers 472
    7.3.3 Air-Blast Circuit Breakers 473
    7.3.4 SF_6 Circuit Breakers 473
    7.3.5 Vacuum Circuit Breakers 476
7.4 Standards and Ratings...477
    7.4.1 Introduction..477
    7.4.2 Voltage-Related Ratings..477
        7.4.2.1 Maximum Operating Voltage...................................477
        7.4.2.2 Rated Dielectric Strength 477
        7.4.2.3 Rated Transient Recovery Voltage...........................479
    7.4.3 Current-Related Ratings 479
        7.4.3.1 Rated Continuous Current 479
        7.4.3.2 Rated Short-Circuit Current 480
    7.4.4 Additional Requirements..481
7.5 Testing of Circuit Breakers...482
    7.5.1 Introduction..482
    7.5.2 Direct Tests...483
    7.5.3 Synthetic Tests 483
7.6 Modeling of Circuit Breakers 486
    7.6.1 Introduction..486
    7.6.2 Circuit Breaker Modeling during Opening Operations...........487
        7.6.2.1 Current Interruption..487
        7.6.2.2 Circuit Breaker Models 488
        7.6.2.3 Gas-Filled Circuit Breaker Models 489
        7.6.2.4 Vacuum Circuit Breaker Models..............................498
        7.6.2.5 Examples 510

        7.6.3   Circuit Breaker Modeling during Closing Operations 518
                7.6.3.1   Statistical Switch Models 520
                7.6.3.2   Prestrike Models 526
        7.6.4   Discussion 529
7.7   Parameter Determination 531
        7.7.1   Introduction 531
        7.7.2   Gas-Filled Circuit Breakers.. 531
                7.7.2.1   Introduction 531
                7.7.2.2   Parameter Determination Methods.. 532
                7.7.2.3   Validation 536
                7.7.2.4   Conclusions.. 540
        7.7.3   Vacuum Circuit Breakers 541
References..545

7.1   Introduction

A circuit breaker is a mechanical switching device, capable of making, carrying, and breaking currents under normal circuit conditions and also making, carrying for a specified time, and breaking currents under specified abnormal circuit conditions such as those of short circuit. In normal operating conditions, a circuit breaker is in the closed position and some current flows through the closed contacts. The circuit breaker opens its contacts when a tripping signal is received. The performance of an ideal circuit breaker may be summarized as follows [1]:

1. When closed, it is a good conductor and withstands thermally and mechanically any current below or equal to the rated short-circuit current.
2. When opened, it is a good insulator and withstands the voltage between contacts, the voltage to ground or to the other phases.
3. When closed, it can quickly and safely interrupt any current below or equal to the rated short-circuit current.
4. When opened, it can quickly and safely close a shorted circuit.

In a real circuit breaker, an electric arc is formed after contacts start separating; it changes from a conducting to a nonconducting state in a very short period of time. The arc is normally extinguished as the current reaches a natural zero in the alternating current cycle; this mechanism is assisted by drawing the arc out to maximum length, increasing its resistance and limiting its current. Various techniques are adopted to extend the arc; they differ according to size, rating, and application. In high-voltage circuit breakers (>1 kV), the current interruption is performed by cooling the arc. Power circuit breakers are categorized according to the extinguishing medium in which the arc is formed.

The main consideration in the selection of a circuit breaker for a particular system voltage is the current it is capable of carrying continuously without overheating (the rated normal current) and the maximum current it can withstand, interrupt, and make onto under fault conditions (the rated short-circuit current). Power system studies will quantify these values. Ratings can be chosen from tables of preferred ratings in the appropriate standard.

The interruption of current has to be considered in close relationship to the operating voltage of the system, to its structure and to the nature of the system components. The operating voltage itself affects the type of circuit breaker to be selected, while the power system components and structure will influence the design of the circuit breaker unit because of the voltage transients produced during the current interruption process. Standards also specify overvoltages that circuit breakers must withstand. In addition, the circuit breaker has to provide sufficient isolating distance to allow personnel to work safely on the part of the system that has been disconnected.

The preferred ratings, the performance parameters, and the testing requirements of circuit breakers are covered by standards [2–6]. The most recognized and influential circuit breaker standards are supported by IEEE/ANSI (American National Standards Institute) and the IEC. The most significant differences between both standards will be discussed in the appropriate section where the generic circuit breaker requirements are reviewed.

Standards require that circuit breakers must be designed and manufactured to satisfy test specifications with respect to its insulating capacity, switching performance, protection against contact, current-carrying capacity, and mechanical function. Evidence of this is obtained by testing a circuit breaker prototype or sample. Type or design tests are performed on a single unit to prove that the device is capable of performing the rated switching and withstand duties without damage, and that it will provide a satisfactory service life within the limits of specified maintenance. The main type tests are dielectric, temperature rise, mechanical endurance, short-circuit, and switching tests. Routine or production tests are performed on each individual manufactured item to ensure that the construction is satisfactory and that the operating parameters are similar to the unit that was subjected to the type test. These tests include power-frequency voltage-withstand tests on the main circuit, voltage tests on the control and auxiliary circuits, measurement of the resistance of the main circuit, and mechanical operation tests [4,6].

The development of circuit breakers was initially based on practical experience. Although some previous work on circuit breaker modeling had been performed, it was in 1939 and 1943, when the models proposed by A. M. Cassie [7] and O. Mayr [8] were presented, that a significant progress in understanding arc–circuit interaction was made. Much effort has been done afterwards to refine those works and confirm their physical validity through practical measurements. See Chapter 1 of [9] for an historical analysis of circuit breaker models.

The main goal of this chapter is to present the different circuit breaker models proposed and applied to date in opening and closing operations, as well as the procedures that can be used for estimating the parameters to be specified in these models. To better understand the principles of these models, the chapter includes some sections that summarize the physical phenomena involved in circuit interruption and the main breaking technologies used in modern circuit breakers. For more details on each subject, readers are referred to the specialized literature [1,9–31]. Sections 7.4 and 7.5 are dedicated respectively to review circuit breaker standards and to describe testing techniques.

7.2 Principles of Current Interruption

7.2.1 Introduction

All methods of interrupting current in high-voltage systems introduce a nonconducting gap into a conducting medium. This can be achieved by mechanically separating the metallic contacts so that the gap formed is either automatically filled by a liquid, a gas, or even vacuum. However, insulating media may sustain electrical discharges which can prevent electrical isolation from being achieved.

During the current interruption process there is a strong interaction between these physical processes and the power system. The current is determined by the driving voltage and the series impedance. The interruption of a resistive load current is not usually a problem; when the circuit breaker interrupts the current, the voltage rises slowly from its zero to its peak following its natural power-frequency shape, the build up of voltage across the opening contacts is relatively moderate, and it can be sustained as the contact gap increases to its fully open position. However, in many circuits the inductive component of current is much higher than the resistive component; as the breaker contacts open and the current is extinguished, the voltage tends to rise to its peak value. This can result in a high rate of rise of voltage across the contacts, aiming for a peak recovery voltage which may be much higher than the normal peak system voltage; under these circumstances the arc may restrike even though the contacts are separating.

Circuit interruption technology is concerned, on the one hand, with the control and extinction of the discharges that may occur, while on the other hand, it relates to the connected system and the manner in which it produces postcurrent interruption voltage waveforms and magnitudes.

The following Sections 7.2.2–7.2.4 detail the post-arc current phenomenon, the basic characteristics of a circuit breaker during opening, and the recovery voltage caused across circuit breaker terminals by the interaction with the power system.

7.2.2 The Electric Arc

The arc is a plasma channel formed after a gas discharge in the extinguishing medium. The plasma state is reached when the increase in temperature gives the molecules so much energy that they dissociate into atoms; if the energy level is increased even more, electrons of the atoms dissociate into free moving electrons, leaving positive ions.

Since the plasma channel is highly conducting, the current continues to flow after contact separation. For higher temperatures, the conductivity increases rapidly. The thermal ionization, as a result of the high temperatures in the electric arc, is caused by collisions between the fast-moving electrons and photons, the slower positively charged ions and the neutral atoms. At the same time, there is also a recombination process when electrons and positively charged ions recombine to form a neutral atom. When there is a thermal equilibrium, the rate of ionization is in balance with the rate of recombination [1].

The physical mechanisms produced during the electron–ion recombination process have time constants in the order of 10–100 ns, which are much shorter than the rate of change of the electrical phenomena caused in the power system during the current interruption. For this reason, the circuit breaker arc can be assumed to be in thermal-ionization equilibrium for all transient phenomena in the power system [1].

The electric arc can be generally divided into three regions—the middle column, the cathode region, and the anode region, as shown Figure 7.1. The figure also shows a typical potential distribution along the arc channel between the breaker contacts. The potential gradient is a function of the arc current, the physical properties of the surrounding medium, the pressure, the flow velocity, the energy exchange between the plasma channel, and the surrounding medium. There are no space charges in the arc column; the current flow is maintained by electrons, and there is a balance between the electron charges and the positive ion charges [11].

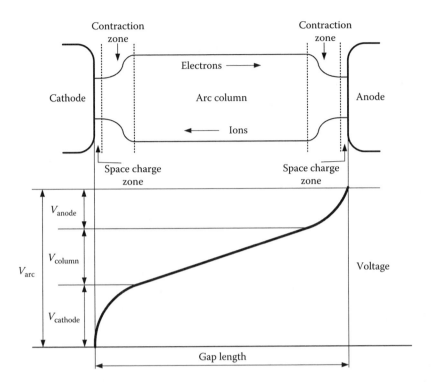

FIGURE 7.1
Potential distribution along the regions of an arc channel.

The cathode emits electrons. A cathode made from refractory material with a high boiling point (e.g., carbon, tungsten, or molybdenum) starts with the emission of electrons when heated to a temperature below the evaporation temperature, the so-called thermionic emission [1]. The cooling of the heated cathode spot is relatively slow compared with the rate of change of the transient recovery voltage (TRV), which appears across the breaker contacts after the arc has extinguished and the current has been interrupted. A cathode made from nonrefractory material with a low boiling point (e.g., copper or mercury) experiences significant material evaporation. These materials emit electrons at temperatures too low for thermionic emission and the emission of electrons is due to field emission. This type of cathode is used in vacuum breakers, for which the current density in the cathode region is much higher than the current density in the arc column itself; this results in a magnetic field gradient that accelerates the gas flow away from the cathode.

The anode can function in either passive or active mode [1]. In passive mode, the anode collects the electrons that leave the arc column; in active mode the anode evaporates, and may supply positive ions to the arc column. At the current-zero crossing, active anode spots in vacuum arcs do not stop emitting ions, whose heat capacity enables the spots to evaporate anode material even when the power input is zero preventing the arc from extinction.

When the arc ignites just after contact separation, evaporation of contact material is the main source of charged particles. When the contact distance increases, the evaporation of contact material becomes the main source of charged particles for the vacuum arcs. For high-pressure arcs burning in air, oil, or sulfur hexafluoride (SF_6), the effect of contact material evaporation is minimal, and the plasma depends mainly on the surrounding medium.

7.2.3 Thermal and Dielectric Characteristics

The arc voltage after maintaining a constant value during the high current period increases to a peak value and then drops to zero with a very steep dv/dt. The current approaches zero crossing with a more or less constant di/dt, but it can be slightly distorted under the influence of the arc voltage. The arc is resistive, so arc voltage and current reach the zero crossing simultaneously. After current interruption, the gas between the breaker contacts is stressed by the rate of rise of the recovery voltage (RRRV), so the present charged particles start to drift and cause a post-arc current, which results in energy input in the gas channel. The problem of current interruption then transforms into one of quenching the discharge against the capability of the system voltage of sustaining a current flow through the discharge. This phase of the process is governed by a competition between the electric power input due to the recovery voltage and the thermal losses from the electric arc; it is known as the thermal recovery phase and has duration of a few microseconds. If the energy input is such that the gas molecules dissociate into free electrons and positive ions, the plasma state is created again and current interruption fails. This is called a thermal breakdown. If the current interruption is successful, the gas channel cools down and the post-arc current disappears. However, a later *dielectric failure* can still occur when the dielectric strength of the gap between the breaker contacts is not sufficient to withstand the TRV.

This performance can be summarized as follows [10,16]:

1. The thermal recovery phase is characterized by rate of rise of recovery voltage (dV/dt) versus rate of current decay (di/dt) diagram (Figure 7.2a). The point of intersection gives the limit of the thermal current. This performance may be improved by increasing the pressure of the gas, the nature of the gas, or the geometry of the interrupter head.

2. For the dielectric recovery phase the characteristic is represented by the boundary that separates fail and clear conditions on a maximum restrike voltage (V_{max}) versus rate of current decay (di/dt) diagram (Figure 7.2b). The point of intersection gives the limit of the dielectric current. This recovery performance may be improved by increasing the number of contact gaps (interrupter units) connected in series.

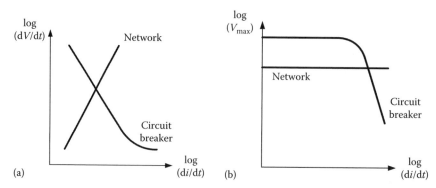

FIGURE 7.2
Circuit breaker performance. (a) Thermal recovery characteristic. (b) Dielectric recovery characteristic.

7.2.4 Arc–Power System Interaction: Transient Recovery Voltage

Transients caused by switching operations in linear systems can be analyzed by using the superposition principle, see Figure 7.3. The switching process caused by an opening operation is obtained by adding the steady-state solution, which exists prior to the opening operation, and the transient response of the system that results from short-circuiting voltage sources and open-circuiting current sources to a current injected through the switch contacts. Since the current through the switch terminals after the operation will be zero, the injected current must equal to the current that was flowing between switch terminals prior to the opening operation. When the contacts of a switch start to open a transient voltage is developed across them. This voltage, known as TRV, is present immediately after the current zero and in actual systems its duration is in the order of milliseconds.

To obtain the TRV waveform, the analysis of the transient response may suffice, since this voltage is zero during steady state; i.e., prior to the opening operation, see Figure 7.3. However, it is important to keep in mind that the recovery voltage will consist of two components: a transient component, which occurs immediately after a current zero, and a steady-state component, which is the voltage that remains after the transient dies out. The actual waveform of the voltage oscillation is determined by the parameters of the power system. Its rate of rise and amplitude are of vital importance for a successful operation of the interrupting device. If the rate of recovery of the contact gap at the instant of current zero is faster than the RRRV, the interruption is successful in the thermal region (i.e., first 4–8 μs of the recovery phase). It may be followed by a successful recovery voltage withstand in the dielectric region (above 50 μs) and then by a full dielectric withstand of the AC recovery voltage. If, however, the RRRV is faster than the recovery of the gap, then failure will occur either in the thermal region or in the dielectric region.

A good understanding of the transient phenomena associated with circuit breaker operations in power systems has led to improved testing practice and resulted in more reliable switchgear. Recommended characteristic values for simulation of the TRV are fixed in standards [4,6,32]. To understand the different requirements applicable to circuit breakers, the most frequent and important cases of current interruption will be analyzed. Switching conditions for TRV analysis have been divided into two groups corresponding to the interruption of current under fault and normal operating conditions, respectively.

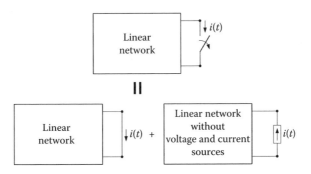

FIGURE 7.3
Application of the principle of superposition to an opening operation.

7.2.4.1 Current Interruption under Normal Operating Conditions

Figure 7.4 shows the recovery voltage across the circuit breaker terminals when interrupting the current in very simple circuits. Observe the different waveshape that appear in each case. The representation of each circuit is that depicted in the figure, except for the interruption of the inductive current since in this case the current zero occurs when the voltage across the inductor terminals is maximum, and a capacitive element is needed to account for the trapped charge. The latter oscillation is caused by the energy transfer between the inductor and the capacitor.

Although real systems are much more complex than the circuits analyzed above, these cases show that switching under normal operating conditions can be categorized as resistive, inductive, and capacitive. Resistive switching is the easiest to clear, while inductive and capacitive are more difficult since the recovery voltage is on the order of twice that for a resistive case, and even higher in three-phase systems (see below). In addition, inductive switching may produce a high-frequency TRV, which is associated to a premature current zero (known as current chopping), and may cause a severe TRV. A short description of the processes originated with the interruption of inductive and capacitive currents, as well as some of the problems that they can cause, is presented in the following paragraphs.

Interruption of small inductive currents: The interruption of small currents can lead to situations that are known as current chopping and virtual chopping [1]. If the current is interrupted at current zero, the interruption is normal and the TRVs are usually within the specified values. However, if premature interruption occurs, due to current chopping, the interruption will be abnormal and it can cause high-frequency reignitions and overvoltages. When the breaker chops the peak current, the voltage increases almost instantaneously, if this overvoltage exceeds the specified dielectric strength of the circuit breaker, reignition takes place. When this process is repeated several times, due to high-frequency reignitions, the voltage increase continues with rapid escalation of voltages. The high-frequency oscillations are governed by the electrical parameters of the concerned circuit, circuit configuration and interrupter design, and result in a zero crossing before the actual power-frequency current zero.

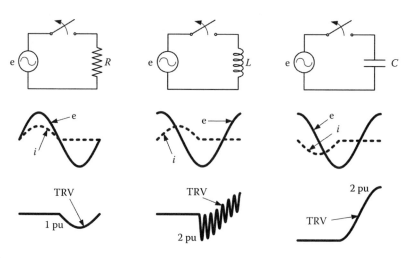

FIGURE 7.4
Transient recovery voltage of simple circuits under normal operating conditions.

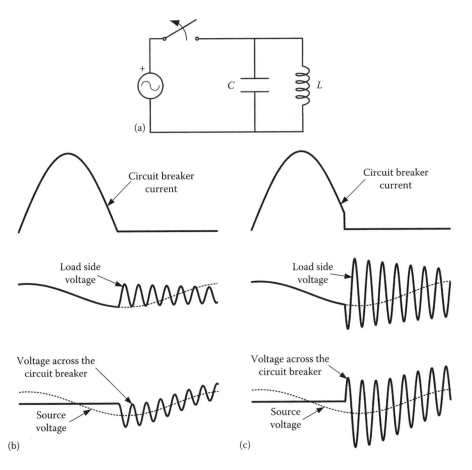

FIGURE 7.5
Transient recovery voltage during interruption of a small inductive current. (a) Equivalent circuit. (b) Arc interruption at current zero. (c) Arc interruption before current zero.

Figure 7.5 compares the load side voltage and the TRVs that are generated when arc interruption takes place at current zero and before current zero (current chopping), respectively. It is obvious from this example that the second case is more severe.

The importance of current chopping can be easily understood by neglecting the influence of losses at the load side. After current interruption at current zero, the energy stored in the load side is the energy stored at the capacitance, whose voltage is at the maximum. This energy is transferred to the inductance and vice versa, and the following expression holds:

$$\frac{1}{2}CV_{max}^2 = \frac{1}{2}LI_{max}^2 \tag{7.1}$$

where V_{max} and I_{max} are the maximum values of the voltage across the capacitance and the current through the inductance, respectively. At the time the current is interrupted, V_{max} is equal to the voltage at the source side.

The frequency of oscillation for both voltage and current is

$$f = \frac{1}{2\pi\sqrt{LC}} \tag{7.2}$$

When premature current interruption occurs, the energy stored at the load side is stored in both the inductance and the capacitance, and it is equal to

$$\frac{1}{2}LI_0^2 + \frac{1}{2}CV_0^2 \tag{7.3}$$

where V_0 and I_0 are the values of the voltage across the capacitance and the current through the inductance at the moment the current is interrupted.

In this case, the frequency of the oscillations will be the same as that for the previous case, but the maximum voltage at the capacitance is obtained from the following equation:

$$\frac{1}{2}CV_{\text{max}}^2 = \frac{1}{2}LI_0^2 + \frac{1}{2}CV_0^2 \tag{7.4}$$

where

$$V_{\text{max}} = \sqrt{\frac{L}{C}I_0^2 + V_0^2} \tag{7.5}$$

This equation can be normalized with respect to the voltage at the source side. Since the current chop occurs with a voltage across the capacitance very close to that of the source side, the following expression results:

$$V_{\text{max(pu)}} \approx \sqrt{1 + \frac{L}{C}I_0^2} \tag{7.6}$$

The interruption at current zero is just a particular case in which $I_0 = 0$ and $V_{\text{max}} = V_0$ (or $V_{\text{max(pu)}} = 1$). In many real cases, the value of L is large and the value of C is small; therefore, regardless of the value of I_0, the maximum voltage that can be caused at the load side, and consequently the TRV across the circuit breaker when current chops, can be much larger than when the current is interrupted at current zero, as illustrated in Figure 7.5.

In the case of current chopping, the instability of the arc around current zero causes a high-frequency transient current to flow in the neighboring network elements. This high-frequency current superimposes on the power-frequency current whose amplitude is small and which is actually chopped to zero. In the case of virtual chopping, the arc is made unstable through a superimposed high-frequency current caused by oscillations with the neighboring phases in which current chopping took place. Virtual chopping has been observed for gaseous arcs in air, SF_6, and oil. Vacuum arcs are also very sensitive to current chopping.

The phenomena of chopping and reignition, with associated high-frequency oscillatory overvoltages, are attributed to the design of the circuit breaker. Circuit breakers are designed to cope with high fault currents. If a design is concentrated only on an efficient performance for high currents, it will be also efficient for small current and will try to interrupt before the natural current zero. This may produce current chopping and reignitions with adverse consequences. The breaker design should incorporate features to cope equally well with small and high currents.

The following practical case analyzes a realistic interruption of inductive current with a premature current interruption [33].

Example 7.1

Figure 7.6 shows the simplified circuit of an induction motor fed through a cable. The motor is a highly inductive load, and it might cause overvoltages when disconnected during startup. Parameters for this case are

Source	Peak voltage = 5144 V
Cable	Series parameters: $R_s = 8.26\ \Omega$, $L_s = 50\ \mu H$
	Parallel parameters: $C_p = 15\ nF$, $R_p = 0.33\ \Omega$
Motor	Stator parameters: $R_1 = 0.204\ \Omega$, $L_1 = 10\ mH$
	Magnetizing inductance: $L_m = 302\ mH$
	Rotor parameters: $R_2 = 2.588\ \Omega$, $L_2 = 8.2\ mH$

The resistance R in parallel with $R_s - L_s$ is needed to stabilize simulation results. The selected value for this resistance is 100 Ω.

During the opening process an arc is drawn between the contacts that help the current to flow until the first current zero. It is assumed that the arc voltage drop does not vary in a wide range of the current change, and its value is a few tens of volts, which is a realistic assumption for a vacuum breaker.

When the contacts open much before current zero the gap is large enough to withstand the TRV and the arc can be cleared even with current chopping and a few reignitions or without any reignition.

In this example, a constant current value of 4 A is assumed for the chopping current level. Figures 7.7 and 7.8 show two different cases of current chopping and arc reignition during an interval of 1 ms [33]:

- In the first case (Figure 7.7), the circuit breaker contacts open approximately 150 μs before current zero and the arc is unsuccessfully cleared at first current zero. Due to the short arcing time, the gap is short when the TRV begins to rise, so the dielectric characteristic rises slower than the TRV. Reignition does not occur at the first local maximum of the TRV, but at the second one. The voltage escalates with the same frequency as the reignited current. The high-frequency components of both current and voltage depend on the circuit parameters (motor, cable).
- In the second case (Figure 7.8), the contacts begin to open approximately 250 μs before the first current zero and the arc is successfully cleared after three reignitions. The frequency of the voltage at the motor side after current interruption depends on the motor parameters.

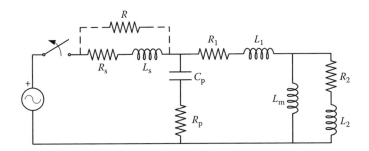

FIGURE 7.6
Example 7.1: Disconnection of an induction motor during startup.

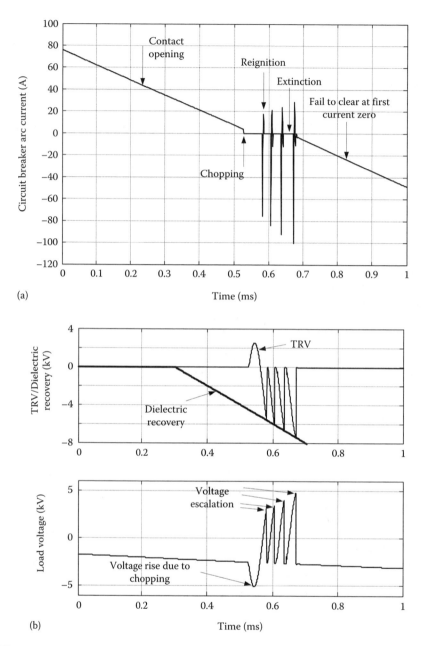

FIGURE 7.7
Example 7.1: Chopping current and reignition—Contacts open at 150 µs before current zero. (a) Circuit breaker current. (b) TRV and load side voltage.

A successful clearing can be achieved by decreasing the inductance between the circuit breaker and the motor. In practice, this inductance is mainly from the cable and the bus bar.

By applying a simple analysis, amplitudes and frequencies of voltages and currents can be easily estimated. According to the previous analysis, the maximum voltage that occurs at motor terminals can be approximated by the following expression:

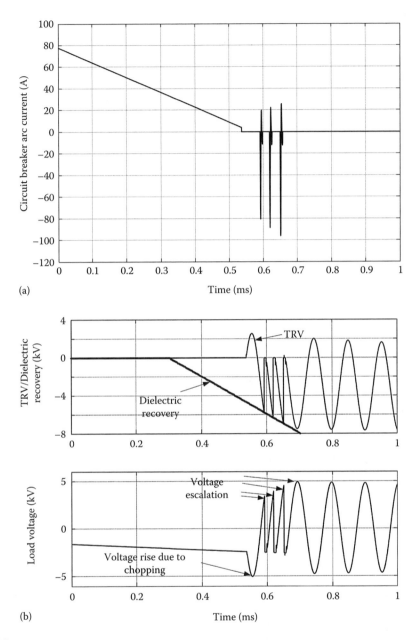

FIGURE 7.8
Example 7.1: Chopping current and reignition—Contacts open at 250 μs before current zero. (a) Circuit breaker current. (b) TRV and load side voltage.

$$V_{\mathrm{m}} = \sqrt{\frac{L_{\mathrm{eq}}}{C_{\mathrm{p}}} i_{\mathrm{ch}}^2 + V_0^2}$$ (7.7)

where
   V_0 is the voltage at the instant of current chopping
   i_{ch} is the current chopping value

L_{eq} is the equivalent inductance, which in this case is

$$L_{eq} = L_1 + \frac{L_2 L_m}{L_2 L_m} \tag{7.8}$$

From the parameters of this example and taking into account that $i_{ch} = 4$ A and $L_{eq} = 18$ mH, one obtains $V_m = 5044$ kV, which is very close to the value shown in Figure 7.7b.

When a restrike occurs, the current contains a high-frequency component which is restrained in the loop source–R_s–L_s–R_p–C_p–source. This frequency can be estimated as

$$f = \frac{1}{2\pi\sqrt{L_s C_p}} \tag{7.9}$$

The resulting value from this expression is 184 kHz, which is close to the measured value 196 kHz.

The amplitude of the first reignition can be estimated as

$$\hat{I} = \left| \frac{V_{first}}{Z_{loop}} \right| \tag{7.10}$$

where
Z_{loop} is

$$Z_{loop} = \frac{R(R_s + j\omega L_s)}{R + R_s + j\omega L_s} + R_p + \frac{1}{j\omega C_p} \tag{7.11}$$

V_{first} is the loop voltage, which is the source voltage at the instant the first restrike takes place

In this example $Z_{loop} = 31.23 - j12.95$ Ω where ω is roughly calculated at 200 kHz, and $V_{first} = 2558$ V. Applying Equation 7.10, the peak restrike current of the first restrike is 75.64 A.

The second restrike occurs at a higher TRV and its amplitude is approximately

$$\hat{I} = \left| \frac{V_{second}}{Z_{loop}} \right| \tag{7.12}$$

where the loop voltage V_{second} is the value of the source voltage increased by the initial capacitor voltage because of the trapped charge at the instant of restrike. V_{second} is equal to the difference between the value of the withstand voltage capability and the capacitor voltage at that instant.

Table 7.1 shows a summary of the successive restrike current amplitudes and the associated loop voltages.

In case of successful interruption and after the restrikes are eliminated, the voltage starts oscillating at a frequency which can be estimated as

TABLE 7.1

Example 7.1: Currents and Loop Voltage during Restrikes

	First Restrike	Second Restrike	Third Restrike	Fourth Restrike
Loop voltage (V)	2558	2756	3051	3429
Restrike current—calculated (A)	75.64	81.53	90.26	101.00
Restrike current—simulated (A)	77.34	84.10	91.91	100.76

$$f = \frac{1}{2\pi\sqrt{L_{eq}C_p}} \tag{7.13}$$

from which a value of 9.69 kHz is obtained.

Interruption of capacitive currents: This scenario usually corresponds to the disconnection of capacitor banks and unloaded lines. In a solid-grounded circuit, the line-to-ground voltage stored in the capacitor is equal to 1.0 per unit (pu) at the time the current is zero. Since half cycle later the source voltage reaches its peak with opposite polarity, the total voltage across the contacts reaches a value of 2.0 pu. If the circuit has an isolated neutral connection, then the voltage trapped in the capacitor for the first phase to clear has a line-to-ground value of 1.5 pu and the total voltage across the contacts half cycle later will be equal to 2.5 pu.

If the circuit breaker restrikes in any of these cases, there will be an inrush current flow which will force the voltage in the capacitor to oscillate with respect to the instantaneous system voltage to a peak value that is approximately equal to the initial value at which it started but with a reversed polarity. If the restrike happens at the peak of the system voltage, then the capacitor voltage will reach a value of 3.0 pu. Under these conditions, if the high-frequency inrush current is interrupted at the zero crossing, then the capacitor will be left with a charge corresponding to a voltage of 3.0 pu and half cycle later there will be a voltage of 4.0 pu applied across the circuit breaker contacts, as shown in Figure 7.9. If the sequence is repeated, the capacitor voltage will reach a 5.0 pu value [9]. If damping is ignored, there could be a theoretical unlimited voltage escalation across the capacitor.

The characteristics of the transient processes originated when interrupting a circuit under normal operating conditions can be summarized as follows [34]:

- Under fault conditions, the interrupted currents are much lower than the currents that exist. In some cases, the current can be very small.
- The interruption of a normal load is made with a high and inductive power factor, and causes a low to moderate TRV, since it is not strongly driven during the transient period.
- The interruption of a shunt reactor is made with an almost zero power factor and it may cause high transient overvoltages, since the following transient occur when recovery voltage conditions are at a maximum.

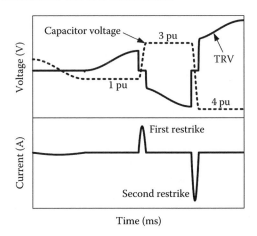

FIGURE 7.9
Voltage escalation during the interruption of a capacitive current.

- The interruption of an unloaded transformer will be likely premature, because of low magnitude current; that is, the current may chop. But the energy available is mostly retained within the core due to residual magnetism and will not contribute to driving the TRV.

- The interruption of capacitor banks, unloaded lines, and cables is made with an almost zero power factor. As with the interruption of shunt reactors, the recovery voltage conditions are at a maximum, but in this case postinterruption transient is moderate due to the voltage retention of the capacitive circuit. The magnitude of the recovery voltage will depend on the capacitive bank connection and the system grounding.

Table 7.2 provides a picture of the main cases that can be considered when the current interruption is performed under normal operating conditions. The plots presented in the

TABLE 7.2

Transient Recovery Voltage under Normal Operating Conditions

Case	Diagram	Typical TRV Waveshape
Normal loads		<1 pu
Shunt reactors		<2 pu
Unloaded transformers		>1 pu
Capacitor banks, unloaded lines, and cables	Grounded capacitances	>2 pu
	Ungrounded capacitances	>2.5 pu

table correspond only to natural TRVs, without reignitions and restrikes. For more details and a deeper analysis of these cases see Ref. [34].

7.2.4.2   Current Interruption under Fault Conditions

Although the cases to be analyzed are many, they present some common characteristics: the current is high, much above normal load currents; the power factor of the circuit is low and inductive; after the current becomes zero, the voltage across the circuit breaker is near the maximum value, so the TRV is strongly driven. TRV waveshapes caused by fault current interruption are usually classified into three types: exponential (also known as ex-cos), oscillatory (also known as 1-cos), and sawtoothed. Figure 7.10 shows the three waveshapes and the typical fault conditions that can cause each of them [35,36].

A short description of some processes originated with the interruption under fault conditions is presented in the following paragraphs.

Short-circuit current interruption: The simplest case is the interruption of a symmetrical rated frequency short-circuit current since the current reduces naturally to zero once every half cycle (Figure 7.11). This represents the minimum natural rate of current decay (di/dt) so that for conventional power systems, which are inherently inductive, the induced voltage following current interruption is minimized.

However, other current interruption situations may exist in which the sinusoidal waveform is superimposed on a damped unidirectional current to form an asymmetric wave

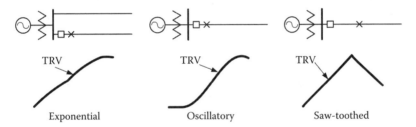

Exponential   Oscillatory   Saw-toothed

FIGURE 7.10
Characteristic transient recovery voltages under fault conditions.

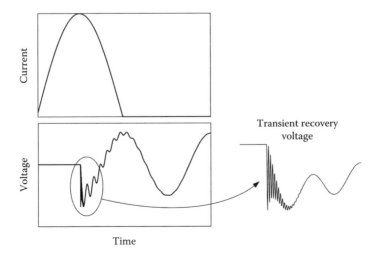

FIGURE 7.11
Voltage and current waveforms during current interruption.

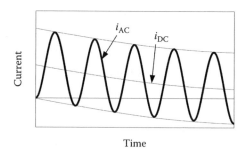

FIGURE 7.12
Short-circuit current waveform with a DC component.

(see Figure 7.12), which will cause different circuit breaker stress. A related condition, which occurs in generator faults, corresponds to the power-frequency wave superimposed on an exponentially decaying component so that zero current crossing may be delayed for several half cycles. Standards distinguish between fault current close to a generator and fault current far from a generator [37].

The following example presents a case in which the TRV across the circuit breaker terminals is caused by a three-phase fault on the line side of the breaker (i.e., a bus fault). The TRV will be analyzed by using the current-injection method, whose principle has been schematized in Figure 7.3. After arc interruption at current zero, the recovery voltage is proportional to the fault current (as a function of time) and the impedance is seen from the current breaker location.

Example 7.2

Figure 7.13 shows a substation layout with a transformer and some transmission lines connected to the lower voltage bus. One line is open at the remote end. Relevant system data follows:

Rated frequency	50 Hz
Short-circuit capacity at node 1	(400 kV) 12,000 MVA
Substation transformer substation	400/220 kVA, 300 MVA, 8%
Substation capacitance (measured at 220 kV)	4 nF
Short-circuit capacity at nodes 4 and 5 (220 kV)	6000 MVA
Lines	Surge impedance: 400 Ω
	Propagation velocity: 3.0E + 5 km/s

All specified voltage values are rms phase-to-phase values.

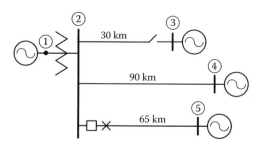

FIGURE 7.13
Example 7.2: Scheme of the test system.

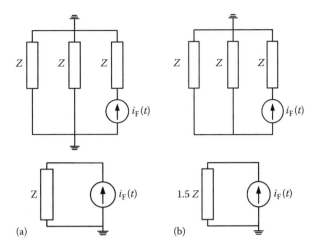

FIGURE 7.14
Equivalent circuits to calculate the TRV during three-phase faults. (a) Grounded fault. (b) Ungrounded fault.

The faulted transmission line is disconnected, and the objective of the study is to estimate the TRV across the circuit breaker terminals by the most onerous three-phase fault.

To estimate the TRV that can be caused by a three-phase fault the circuits shown in Figure 7.14 are used. They show how superposition can be applied to obtain the TRV caused by both grounded and ungrounded three-phase faults. One can observe that the ungrounded fault is more severe and increases the TRV by 1.5 times with respect to what would be caused by a grounded fault. Therefore, only the ungrounded case is used in calculations.

A simplified lossless single-phase equivalent circuit will be used to obtain the TRV across the circuit breaker caused by an ungrounded three-phase fault using superposition. The one-line equivalent circuit is shown in Figure 7.15.

- The current injected at the substation node is the fault current, which has been derived by using a transients program. The value shown in the figure accounts for the increase to be considered in ungrounded faults.
- Transmission lines are represented as single-phase ideal lines with constant and distributed parameters, and characterized by means of a surge impedance and a travel time.

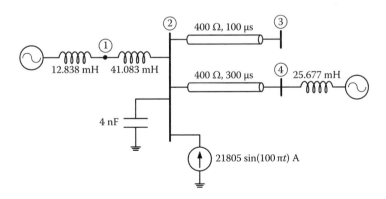

FIGURE 7.15
Example 7.2: One-line equivalent circuit for TRV calculation.

- System equivalents at nodes 1 and 4, as well as the substation transformer, are replaced by the inductances whose values are obtained as follows:

$$\text{Node 1} \quad L = \frac{1}{2\pi 50} \cdot \frac{220^2}{12,000} \equiv 12.838\,\text{mH}$$

$$\text{Transformer} \quad L = \frac{1}{2\pi 50} \cdot 0.08 \cdot \frac{220^2}{300} \equiv 41.083\,\text{mH}$$

$$\text{Node 4} \quad L = \frac{1}{2\pi 50} \cdot \frac{220^2}{6000} \equiv 25.677\,\text{mH}$$

To obtain the initial recovery voltage, traveling-wave models for lines are replaced by resistances whose values are equal to their surge impedance. A simplified circuit, shown in Figure 7.16, can be applied before the first reflection from any remote line end arrives to the substation. In this circuit $R = Z/n$, where Z is the surge impedance of all lines and n is the number of parallel lines at the substation node.

The analysis of this circuit, after neglecting the effect of the capacitance, provides the following approximated expression for the voltage across the circuit breaker [38]:

$$v(t) \approx 1.5 \cdot (\sqrt{2}I) \cdot \omega L \cdot (1 - e^{-(Z/nL)t}) \tag{7.14}$$

where I is the rms fault current measured at the fault location. In this case $I \approx 10.278$ (kA).

Figure 7.17 shows the simulation results. This transient voltage across the circuit breaker was obtained by simulating the opening of the switch (not using the current-injection method). One can observe that the initial rise follows an exponential function, which ends at the time of the first reflection from the open end, node 3, arrives at the substation; this occurs after a round-trip travel (i.e., 200 μs after arc interruption). At 200 μs, the value of the maximum recovery voltage obtained from expression (Equation 7.7) is 193.5 kV, which agrees with the value shown in Figure 7.17.

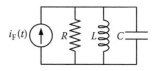

FIGURE 7.16
Example 7.2: Simplified circuit for calculation of the initial TRV ($R = 200\,\Omega$, $L = 53.921\,\text{mH}$, $C = 4\,\text{nF}$).

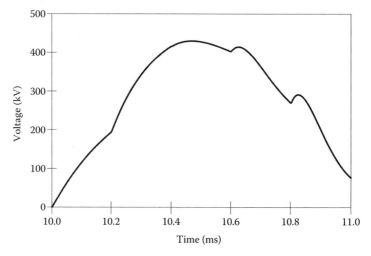

FIGURE 7.17
Example 7.2: Transient recovery voltage.

Since the new wave is reflected with a positive coefficient, it increases the TRV. This trend remains until a new reflected wave from the same open end arrives at the substation 200 μs later, after which the TRV still continues increasing until about 470 μs. The first reflection from node 4 arrives at 600 μs and originates a small increase of the TRV, although it is below the maximum, which was reached at about 470 μs.

The initial rate of rise is an important parameter for the selection of the circuit breaker. It can be estimated from Equation 7.7 by taking the derivative and setting $t = 0$. The result is

$$RRRV \approx 1.5 \cdot (\sqrt{2}I) \cdot \omega \frac{Z}{n} \tag{7.15}$$

The resulting value for this example is 1.37 kV/μs, which is below the value fixed in standards for this class of circuit breaker.

Short-line fault current interruption: A fault on a transmission line close to the terminals of a high-voltage circuit breaker is known as a short-line fault. The clearing of a short-line fault puts a high thermal stress on the arc channel in the first few microseconds after current interruption due to the electromagnetic waves' reflection from the short circuit back to the terminals of the circuit breaker which can result in a TRV with a rate of rise from 5 to 10 kV/μs. Figure 7.18 shows the typical sawtooth shape of the recovery voltage during a short-line fault clearing.

For some kinds of circuit breakers, the initial TRV is the most critical period, and the stress caused by a short-line fault may be the most severe one. The value of the rate of rise at the line side depends on the interrupted short-circuit current and the characteristic impedance of the overhead transmission line, as analyzed in the following example.

Example 7.3

Assume that a grounded three-phase short circuit occurs at a 400 kV transmission line close to the circuit breaker. The single-phase equivalent circuit of this case may be represented as in Figure 7.18, where the network equivalent at the breaker location has been replaced by a series inductance and a shunt capacitance. Consider the following parameters:

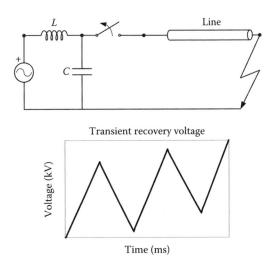

FIGURE 7.18
TRV during a short-line fault.

High-voltage network:	Rated rms phase-to-phase voltage	400 kV
	Frequency	50 Hz
	Short-circuit capacity at breaker location	10,000 MVA
	Equivalent capacitance to ground	3 μF
Line:	Surge impedance	400 Ω
	Propagation velocity	300 m/μs

The analysis of a short-line fault is usually concentrated in the first few microseconds which correspond to a round-trip travel time on the faulted line section, after which the TRV will start decreasing. For that reason the equivalent circuit can be that shown in Figure 7.19.

As for the previous example, this case can be also resolved by applying the principle of superposition. Assuming that the phase-to-ground voltage is $\sqrt{2}V \sin\omega t$, with $\omega = 2\pi 50$, the following approximated expression for the recovery voltage across the circuit breaker is obtained:

$$v(t) \approx \sqrt{2}V \cdot \left[\frac{Z}{L_{eq}} \cdot t + (1 - \cos\omega t) \right] \tag{7.16}$$

The first term of this expression is the voltage that develops at the line side, while the second is the contribution of the source side.

Taking into account that

$$Z = \frac{\left(\sqrt{3}V\right)^2}{SIL} \qquad L_{eq} = \frac{\left(\sqrt{3}V\right)^2}{\omega S_{sh}} \tag{7.17}$$

where
  SIL is the surge impedance load of the transmission line
  S_{sh} is the short-circuit capacity at breaker location, it results

$$v(t) \approx \sqrt{2}V \cdot \left[\omega \frac{S_{sh}}{SIL} \cdot t + (1 - \cos\omega t) \right] \tag{7.18}$$

The maximum value of this expression is reached after two trips of the traveling waves. During such short interval, the voltage at the source side can be assumed constant, so the term $(1 - \cos)$ can be neglected, and the following expression yields for the voltage across the circuit breaker:

$$v(t) \approx \sqrt{2}V\omega \frac{S_{sh}}{SIL} t \tag{7.19}$$

whose maximum is reached after $t = 2\tau$, where τ is the travel time between the breaker and the fault location. Remember that the equivalent circuit used for this analysis is only valid for an interval of time equal to 2τ.

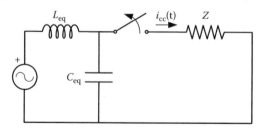

FIGURE 7.19
Example 7.3: Single-phase equivalent circuit for ITRV calculation.

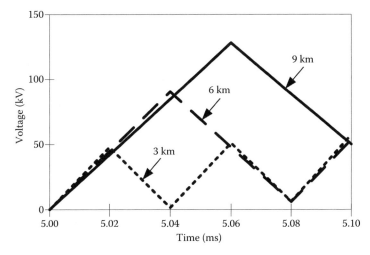

FIGURE 7.20
Example 7.3: TRV during a short-line fault.

Since the travel time is a very small value, the maximum value that results from the above expression is not usually too high. However, the parameter of concern is not the maximum TRV but its initial rate of rise. After deriving Equation 7.19, it results:

$$\text{RRRV} = \left.\frac{dv(t)}{dt}\right|_{t=0} \approx \sqrt{2}V\omega\frac{S_{sh}}{\text{SIL}} \tag{7.20}$$

This approximate value does not depend on the distance between the fault location and the circuit breaker, provided it is short enough, and it linearly increases with the rated voltage and the rated frequency. For the case under study, SIL = 400 MW, and

$$\text{RRRV} \approx \sqrt{2}\frac{400}{\sqrt{3}}\cdot 2\pi 50\cdot\frac{10,000}{400} \equiv 2.57\,\text{kV/}\mu\text{s}$$

Figure 7.20 shows the typical sawtooth waveshape of the TRV that results in different distances between the fault location and the circuit breaker in this test case. One can observe that the initial recovery is not the same but very similar for the three cases and the initial slope agrees with that obtained above.

Table 7.3 summarizes the main cases when the current interruption is performed under fault conditions. As for the cases under normal load conditions, the plots presented in the table correspond only to natural TRVs, without reignitions and restrikes. More details on the TRV caused during fault clearing can be found in Ref. [29].

7.3 Breaking Technologies

7.3.1 Introduction

To achieve a successful current interruption, a circuit breaker must fulfill the following conditions: the power dissipated during arcing through the Joule effect must remain less than the cooling capacity of the device, the deionization rate of the medium must be high, and the

TABLE 7.3

Transient Recovery Voltage under Fault Conditions

Case	Diagram	Typical TRV Waveshape
Single-line fed bus fault		
Multiple-line fed bus fault		
Transformer-through fault		
Multiple-line and transformer-fed bus fault		
Short-line fault		

intercontact space must have sufficient dielectric strength. The breaking medium must therefore [24] (1) have high thermal conductivity in the extinction phase to remove the arc thermal energy; (2) recover its dielectric properties as soon as possible in order to avoid restriking; (3) be a good electrical conductor at high temperatures to reduce arc resistance; and (4) be a good electrical insulator at low temperatures to make it easier to restore the voltage.

This insulating quality of the breaking medium is measured by the dielectric strength between the contacts, which depends on the gas pressure and the distance between the electrodes. The breakdown voltage is a function of the interelectrode distance and the pressure. Figure 7.21 shows the voltage ranges in which each breaking technique is currently used [24].

Switchgear for distribution and transmission voltage ranges show similarities, but many functions and practices are different. A particular requirement for transmission is the ability to deal with TRVs generated during interruption of faults on overhead lines when the position of the fault is close to the circuit breaker, which is known as short-line fault condition. A further requirement is the energization of long overhead lines, which can generate large transient overvoltages on the system unless countermeasures are taken.

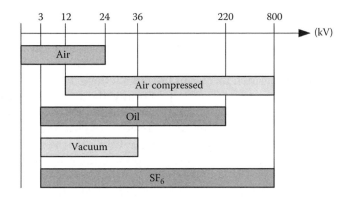

FIGURE 7.21
Voltage range of application of breaking technologies. (Adapted from Theoleyre, S., MV breaking techniques, Schneider Cahier Technique, no. 193, June 1999.)

Early oil designs used plain break contacts in an oil tank capable of withstanding the pressure built up from large quantities of gas generated by long arcs. During the 1920s and 1930s various designs of arc control device were introduced to improve performance. These were designed such that the arc created between the contacts produces enough energy to breakdown the oil molecules, generating gases and vapors which resulted in successful clearance at current zero. The use of oil switchgear has been significantly reduced because of the need for regular maintenance and the risk of fire in the event of failure.

The 1940s saw the development and introduction of air-blast circuit breakers, which used compressed air, not only to separate the contacts but also to cool and deionize the arc. Early interrupter designs were not permanently pressurized and reclosed when the blast was shut off. Isolation was achieved separately by a series-connected switch which interrupted any residual resistive and/or capacitive grading current, and had the full rated fault-making capacity. The increase of system voltages to 400 kV in the late 1950s coincided with the discovery, investigation, and specification of the short-line fault condition. Since air-blast interrupters are particularly susceptible to a high RRRV, parallel-connected resistors are needed to assist interruption. As short-circuit current levels increased, parallel resistors of up to a few hundred ohms per phase were required. To deal with this, permanently pressurized breakers incorporating parallel resistor switching contacts were used, with multiple interrupters per phase in order to achieve the required fault-switching capacity at high transmission voltages [1]. Air-blast circuit breakers with only four or six interrupters per phase have been installed up to the late 1970s.

The next change occurred with the introduction of SF_6 circuit breakers. SF_6 is a very effective gas for arc interruption. The merits of electronegative gases such as SF_6 had long been recognized, but the reliability of seals was a problem. It was during the nuclear reactor development when sufficiently reliable gas seals were developed to enable gas-insulated equipment to be reconsidered. SF_6 circuit breakers were initially introduced and used predominantly in transmission switchgear, but they became popular at distribution voltages from the late 1970s. The first generation of SF_6 circuit breakers were double-pressure devices that were based on air-blast designs, but using the advantages of SF_6. Such design was followed by puffer circuit breakers which use a piston and cylinder arrangement, driven by an operating mechanism. Puffer circuit breakers form the basis of most current transmission circuit breaker designs. Other designs of SF_6 circuit breaker, introduced in the 1980s, made more use of the arc energy to aid interruption, allowing a lighter and cheaper operating mechanism.

The main alternative to the SF_6 circuit breaker in the medium-voltage range is the *vacuum circuit breaker*. Although it is a simple device, comprising only a fixed and a moving contact located in a vacuum vessel, it is also the most difficult to develop. Early work started in the 1920s, but it was not until the 1960s that the first vacuum breakers capable of breaking large currents were developed; commercial circuit breakers followed about a decade later.

Presently, the SF_6 breaking technique is practically the only one used in EHV. All the technologies are still used in MV applications, but SF_6 and vacuum breaking techniques have in fact replaced breaking in air for reasons of cost and space requirements, and breaking in oil for reasons of reliability, safety, and reduced maintenance. Finally, magnetic breaking in air is, with a few rare exceptions, the only technique used in LV applications.

Sections 7.3.2 through 7.3.5 summarize the main characteristics of the breaking technologies presently used for transmission and distribution voltage levels.

7.3.2 Oil Circuit Breakers

Circuit breakers built in the beginning of the twentieth century were mainly oil circuit breakers. Their breaking capacity was sufficient to meet the required short-circuit level in the substations. The first design of an oil circuit breaker was based on an air switch whose all three phases were enclosed within a tank of oil. This circuit breaker was not equipped with any arc quenching device, so arc extinction was determined solely by the oil characteristics and the pressure rise within the circuit breaker tank. For short-circuit currents below 12 kA such circuit breakers operated satisfactorily since high pressures were built up within the circuit breaker tank to assist in the arc extinction process. Under single-phase or low-current fault conditions, however, insufficient pressure may be built up to extinguish the arc and long arcs can be produced which may continue until the circuit breaker contacts reach the end of their travel, after which failure may result.

To overcome these shortcomings, solid-insulated assemblies, known as arc control devices, were developed to enclose the arc produced at each set of interrupter contacts. For a high short-circuit current very high pressure is built up in the top section of the arc control device as a consequence of vaporization of the oil, resulting in rapid arc extinction. For lower fault currents insufficient pressure is produced in the upper section of the device to extinguish the arc, and much longer arcs are drawn out into the lower region; this allows a higher pressure to result in this lower region. There are usually interconnecting chambers within the device which allow a flow of oil upwards and across the upper region of the arc to assist in cooling and arc extinction. This oil flow lends its name to the arc control device, which allows satisfactory high short-circuit current ratings to be achieved and safe interruption of both single-phase and low-current fault conditions (see Figure 7.22).

A step further was the development of circuit breakers where each phase has its own independent oil chamber, which is of relatively small volume. As the chamber is at line potential, it must be supported from earthed metalwork by some form of insulator, while its outgoing HV terminal must be insulated. Such devices are referred to as *minimum oil* circuit breakers. Due to the small oil volume, oil carbonization rapidly occurs on fault interruption and more frequent maintenance is required than for a bulk oil-type circuit breaker. If maintenance is not undertaken at the appropriate time, internal electrical tracking and deterioration can occur on insulating surfaces within the oil chamber which may lead to failure.

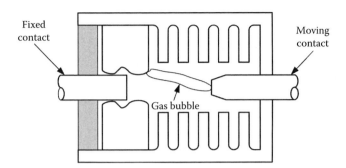

FIGURE 7.22
Arc control device of an oil circuit breaker.

7.3.3 Air-Blast Circuit Breakers

Although air can be used as an extinguishing medium for current interruption, at atmospheric pressure its interrupting capability is limited to low voltage and medium voltage only. For medium-voltage applications up to 50 kV, the breakers are mainly of the *magnetic air-blast type* in which the arc is blown into a segmented compartment by the magnetic field generated by the fault current. In this way, the arc length, the arc voltage, and the surface of the arc column are increased. The arc voltage decreases the fault current, and the larger arc column surface improves the cooling of the arc channel. These circuit breakers provide fault clearance under all fault conditions within one cycle of arcing, and are ideal for providing a means of clearing both high fault current levels and low inductive fault currents. They are rarely found on distribution networks but have been occasionally used for switching MV shunt reactors.

Air-blast breakers operating with compressed air can interrupt higher currents at considerable higher voltage levels. However, they cannot provide permanently open contacts, and during arc extinction process, a mechanically driven sequential disconnector is required to provide an open gap before the contacts reclose. Closure of the circuit breaker is achieved by closing the sequential disconnector. Air-blast breakers using compressed air can be of the *cross-blast* or the *axial-blast* type. The cross-blast type operates similar to the magnetic-type breaker; i.e., compressed air blows the arc into a segmented arc-chute compartment. Because the arc voltage increases with the arc length, this is also called high-resistance interruption; it has the disadvantage that the energy dissipated during the interruption process is rather high. In the axial-blast design, the arc is cooled in axial direction by the airflow and the current is interrupted when the ionization level is brought down around current zero. Because the arc voltage hardly increases, this is called low-resistance interruption. These circuit breakers do not require a series sequential disconnector, since their contacts are held open by pressurized air. They are used only at transmission voltages where very fast fault-clearing times are required.

7.3.4 SF₆ Circuit Breakers

The SF_6 is a gas with features particularly suited to switchgear applications. Its high dielectric withstand characteristic is due to its high electron attachment coefficient. The AC voltage withstand performance of SF_6 gas at 0.9 bar(g) is comparable with that of insulating oil; however, SF_6 arc voltage characteristic is low, so the arc-energy removal requirements are low. At temperatures above 1,000°C, SF_6 gas starts to fragment and at an arc-core temperature of about 20,000°C, the process of dissociation accelerates producing a number of constituent gases, some of which are highly toxic. However, they recombine very quickly as the temperature starts to fall and the dielectric strength of the gap recovers to its original level in microseconds, which allows several interruptions in quick succession.

Early EHV SF$_6$ circuit breakers were based on the two-pressure air-blast technology which was modified to give a closed loop for the exhaust gases. SF$_6$ at high pressure was released by the blast valve through a nozzle to a low-pressure reservoir instead of being exhausted to atmosphere. The gas was recycled through filters, then compressed and stored in the high-pressure reservoir for subsequent operations. The liquefying temperature of SF$_6$ gas depends on the pressure but lies in the range of the ambient temperature of the breaker. This means that the SF$_6$ reservoir should be equipped with a heating element which introduces an extra failure possibility for the circuit breaker: When the heating element does not work, the breaker cannot operate. The cost and complexity of this design led to the development of simpler and more reliable single-pressure puffer-type interrupters.

The principle of a single-pressure puffer-type interrupter is explained by the operation of a cycle pump, where air is compressed by the relative movement of a piston against a cylinder. The opening stroke made by the moving contacts moves a piston, compressing the gas in the puffer chamber and thus causing an axial gas flow along the arc channel. The nozzle must be able to withstand the high temperatures without deterioration, see Figure 7.23.

The puffer interrupter principle has been widely applied in both distribution and transmission circuit breaker applications. However, during an opening operation, the pressure rise generated by compression and heating of SF$_6$ produces retarding forces, acting on the piston surface. High-energy mechanisms are required to overcome these forces and to provide consistent opening characteristics for all short-circuit duties.

Some SF$_6$ circuit breaker designs use rotating arc techniques, one of which utilizes an electromagnetic coil. The arc is rotated at a very high speed by an electromagnetic field, see Figure 7.24. The field is produced by the coil through which the current to be interrupted flows during the period between contact separation and arc extinction. Rotating arc circuit breakers have relatively long arc durations at low levels of fault current, because a lower magnetic field generated by a lower fault current will not cause the arc to rotate very quickly. At lower currents arc rotation may cease and arc interruption may be fortuitous.

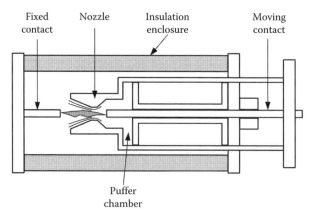

FIGURE 7.23
Operating principle of an SF$_6$ puffer circuit breaker.

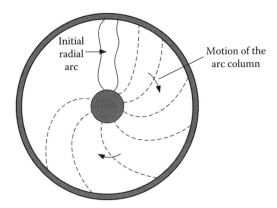

FIGURE 7.24
Operating principle of rotating-arc SF$_6$ circuit breaker.

This may be particularly critical if the circuit breaker is required to switch low energy reactive currents, where very high values of TRV may occur.

Auto-expansion is another interruption technique that uses the power dissipated by the arc to increase pressure within the expansion chamber. The overpressure generated has the double effect of causing strong turbulence in the gas and forcing the gas out of the chamber across the arc as soon as the nozzle starts to open, thus extinguishing the arc. This circuit breaker uses a low energy operating mechanism and produces switching overvoltages up to 2 pu.

The majority of on-site failures are mechanical; therefore, manufacturers have concentrated their efforts on producing simple interrupting devices, and reliable and economical mechanisms. They have been able to develop new interrupters with some improved design features: reduction in energy has been achieved by optimizing the puffer-type interrupter designs, while reduction in drive energy has been achieved by utilizing the arc energy to heat the SF$_6$ gas, thus generating sufficient high pressure to quench the arc and assist the mechanism during the opening stroke.

This principle can be used to produce low energy circuit breakers whose design depends on optimizing the volumes of the two chambers of the interrupter: (1) the expansion chamber, to provide the necessary quenching pressure by heating the gas with arc energy; and (2) the puffer chamber, to provide sufficient gas pressure for clearing the small inductive, capacitive, and normal load currents. This design provides softer interruption, producing low overvoltages for switching of small inductive and capacitive currents, low-energy mechanisms, lighter moving parts, simpler damping devices, long service life, increased reliability, and low-cost devices.

To date no single gas superior to pure SF$_6$ gas in dielectric withstand capability and arc interruption has been found. Several gas mixtures have been studied with the objective of exploiting the properties of the component gases so that they could be used effectively in switchgear. A survey based on published works on gas mixtures has shown that the performance of some gas mixtures is very promising [16].

New developments are concentrated on improving the short-circuit rating, better understanding of the interruption techniques, improving the life of arcing contacts, or reducing the ablation rate of nozzles. Most of the present SF$_6$ circuit breaker designs are virtually maintenance-free.

7.3.5 Vacuum Circuit Breakers

The arc in a vacuum circuit breaker differs from the high-pressure arc because it burns in vacuum (i.e., in the absence of an extinguishing medium). The physical process in the vacuum arc column is a metal surface phenomenon rather than a phenomenon in an insulating medium.

It was already known in the 1920s that contacts opening in a vacuum would not sustain an arc. However, this feature was not employed due to the difficulties of providing a sealed chamber that could hold a vacuum without leakage. It was not until the 1960s when technology made it possible to manufacture gas-free electrodes and the first practical interrupters were built. Post-World War II saw very rapid developments in television technology, which required the tubes to be operated with a sustained vacuum. Television tube technology was thus applied to produce a sealed insulation enclosure for opening the contacts of a vacuum circuit breaker [1].

There is no mechanical way to cool a vacuum arc, and the only possibility to influence the arc channel is by means of a magnetic field. The vacuum arc is the result of a metal-vapor/ion/electron emission phenomenon. To avoid uneven erosion of the surface of the arcing contacts (especially the surface of the cathode), the arc should be kept diffused or in a spiral motion; this can be achieved by making slits in the arcing contacts or by applying horseshoe magnets. The movement of the circuit breaker moving contacts is achieved by means of steel bellows (see Figure 7.25).

Vacuum circuit breakers are prone to current chopping and reignitions because of their excellent interrupting characteristics. When the arc current goes to zero it does so in discrete steps of a few amperes. For the last current step to zero, this can cause a noticeable chopping of the current. Therefore, care should be taken when selecting a vacuum circuit breaker for small inductive and reactor switching duties.

In general, vacuum circuit breakers are compact, safe and reliable, have low contact wear, and require low maintenance. The dielectric withstand voltage of a 1 cm gap in a vacuum of 10^{-6} mm of mercury is about 200 kV, but increases only very slightly with increase in contact gap. This limits the withstand voltage. Although some manufacturers produce vacuum bottles up to 145 kV, they are very expensive and only used on circuit breakers for special duties. For reasons of cost, almost all manufacturers limit the use of vacuum circuit breakers up to a voltage of 36 kV.

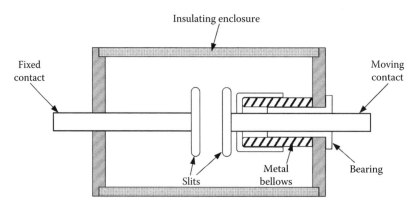

FIGURE 7.25
Cross section of a vacuum circuit breaker.

7.4 Standards and Ratings

7.4.1 Introduction

Standards define and establish the general requirements that are applicable to all types of high-voltage switching equipment [2–6]. The standards assigned are considered to be the minimum designated limits of performance that are expected to be met by the device when operated within specified operating conditions. The power-frequency rating is a significant factor during current interruption, because the rate of change of current at the zero crossing is a meaningful parameter. It should be kept in mind that a 60 Hz current is generally more difficult to interrupt than a 50 Hz current of the same magnitude.

Circuit breaker ratings include, in addition to the fundamental voltage and current parameters, other requirements, which are reviewed together with the most important voltage- and current-related ratings in Sections 7.4.2 and 7.4.3. For more details, see the related standards and Ref. [9].

7.4.2 Voltage-Related Ratings

7.4.2.1 Maximum Operating Voltage

This rating (called rated maximum operating voltage in IEEE/ANSI, and rated voltage in IEC) sets the upper limit of the system voltage for which the circuit breaker application is intended. The practice of using nominal operating voltages is no longer used because it is common practice in other standards for apparatus that are used in conjunction with circuit breakers to refer only to the maximum rated voltage.

7.4.2.2 Rated Dielectric Strength

The rated dielectric strength capability of a circuit breaker is specified in the form of a series of prescribed voltage values, which represent stresses that are produced by power-frequency overvoltages, or caused by switching operations and lightning strokes.

Power-frequency dielectric withstand: IEEE and IEC standards specify voltage and duration values to be considered in dry and wet power-frequency dielectric-withstand tests. In general, power-frequency overvoltages that occur in an electric system are much lower than the power-frequency withstand values that are required by the standards. In general, the basic impulse level (BIL) withstand capability defines the worst-case condition, and when the design of a circuit breaker meets the BIL then it is by default overdesigned for the corresponding standard power-frequency withstand requirement.

Lightning impulse withstand: Overvoltages produced by lightning strokes are one of the primary causes of system outages and dielectric failures of power equipment. Since it would be impractical to design circuit breakers to withstand the highest lightning overvoltage, the specified impulse levels are lower than the expected lightning overvoltage levels. The objective of specifying an impulse withstand level is to define the upper capability limit for a circuit breaker and to define the level of voltage coordination that must be provided. The specified BIL reflects the insulation coordination practices used in the design of power systems, which are influenced by the insulation limits and the protection requirements of power transformers and other apparatuses [9]. For both indoor and outdoor circuit breakers IEEE specifies only one BIL value for each circuit breaker rating with the exception of breakers rated 25.8 and 38 kV where two BIL values are given. For circuit breakers rated 72.5 kV and above IEEE specifies two BIL levels. IEC specifies two BIL ratings for all voltage classes except for 48.3 and 72.5 kV, and for all circuit breakers rated above 100 kV, except for the 245 kV rating where three values are specified [6].

Basic lightning impulse tests are made under dry conditions using both positive and negative impulse waves. The standard lightning impulse wave is defined as a 1.2/50 μs wave. The waveform and the points used for defining the wave are shown in Figure 7.26a. The 1.2 μs value is the front time t_f and is defined as 1.67 times the time interval that encompasses the 30% and 90% points of the voltage magnitude when these two points are joined by a straight line. The 50 μs represents the tail time and is defined as the point where the voltage has decreased to half of its peak value. These values are defined from the virtual origin; i.e., the intercept of the straight line between the 30% and 90% values and the horizontal axis that represents time.

The *chopped wave withstand* is a dielectric requirement specified only in IEEE. It was added to account for the effect of the separation distance from the arrester and to eliminate the need for installation of surge arresters at the line side of the breaker [9]. The 2 μs peak is 1.29 times the corresponding BIL, being this higher voltage intended to account for the additional separation that may exist from the circuit breaker terminals to the arresters in comparison to the transformer–arrester separation. The chopped wave is shown in Figure 7.26b; the front of the wave is defined in the same manner as the lightning impulse wave, but the time t_c represents the chopping time and is defined as the time from the virtual origin to the point of the chopping initiation.

The impulse bias test is an IEC requirement for all circuit breakers rated 300 kV and above to account for the effects produced by an impulse wave upon the power-frequency wave, due to the fact that a lightning stroke may reach the circuit breaker at any time. This requirement has not been established by IEEE.

Switching impulse withstand: This requirement is applicable to circuit breakers rated by IEEE at 362 kV or above, and by IEC at 300 kV and above. There are two sets of voltage levels listed in IEEE and IEC. With the exception of switching surge required for a 362 kV circuit breaker all the higher voltage levels are the same in both standards. The lower values required by IEEE are less than those required by IEC. The required switching surge withstand capability across the isolating gap is higher for IEC except at the 550 kV level were the requirements are the same in both standards. Phase-to-phase requirement is specified by IEC but not by IEEE.

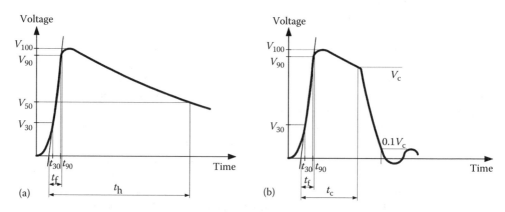

FIGURE 7.26
Standard waveforms for (a) lightning-withstand tests and (b) chopped wave tests.

7.4.2.3 Rated Transient Recovery Voltage

TRVs depend upon the type of fault being interrupted, the configuration of the system, and the characteristics of its components. In addition, a distinction has to be made for terminal faults, short-line faults, and for the initial TRV condition. See Section 7.2.4 for more details. The standards have specific requirements for each one of these conditions.

Both IEEE and IEC have adopted the basic waveforms of TRVs originally used in IEC [6,39].

IEC defines two TRV envelopes, see Figure 7.27. The first one is applicable to circuit breakers rated below 100 kV and is known as the two-parameter method. The second one is applicable to circuit breakers rated 100 kV and above, and is known as the four-parameter method. The two-parameter envelope is defined by (1) U_c, the reference voltage (TRV peak), in kilovolts and (2) t_3, the time to reach U_c, in microseconds. The four-parameter envelope is defined by (1) U_1, the intersection point that corresponds to the first reference voltage in kilovolts; (2) t_1, the time to reach U_1, in microseconds; (3) U_c, the intersection point that corresponds to the second reference voltage (TRV peak), in kilovolts; and (4) t_2, the time to reach U_c, in microseconds.

A capacitance on the source side of the breaker produces a slower rate of rise of the TRV. Consequently, standards specify a time delay that ranges from a maximum of 16 μs down to 8 μs. For circuit breakers rated above 121 kV the delay time specified is a constant with a value of 2 μs.

When the inherent TRV of the system exceeds standard envelopes, there are some alternatives, outside of reconfiguring the system [39]: (1) using a breaker with a higher voltage rating, or a modified circuit breaker; (2) adding capacitance to the circuit breaker terminal(s) to reduce the rate of rise of the TRV; and (3) consulting the manufacturer.

The only important difference between IEEE and IEC for short-line faults is that IEC requires this capability only on circuit breakers rated 52 kV and above, and which are designed for direct connection to overhead lines. IEEE requires the same capability for all outdoor circuit breakers.

7.4.3 Current-Related Ratings

7.4.3.1 Rated Continuous Current

The continuous current rating serves to set the limits for the circuit breaker temperature rise. These limits are chosen to prevent a temperature runaway condition, when the type

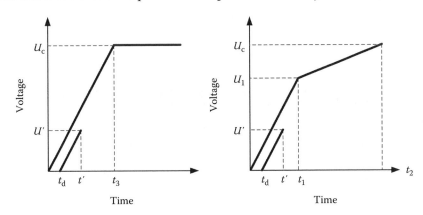

FIGURE 7.27
IEC two- and four-parameter limiting curves.

of material used in the contacts or conducting joints is taken into consideration, and to avoid that the temperature of the conducting parts, which are in contact with insulating materials, exceeds the softening temperature of such material. The temperature limits are given in terms of both the total temperature and the temperature rise over the maximum allowable rated ambient operating temperature. The maximum temperature limits for contacts and conducting joints are based on the knowledge that exists about the change in resistance due to the formation of oxide films, and the prevailing temperature at the point of contact. An additional requirement is the maximum allowable temperature of circuit breaker parts that may be handled by an operator [9].

The preferred continuous current ratings specified by IEEE/ANSI are 600, 1200, 1600, 2000, or 3000 A. The corresponding IEC ratings are 630, 800, 1250, 1600, 2000, 3150, or 4000 A.

7.4.3.2 Rated Short-Circuit Current

This rating corresponds to the maximum value of the symmetrical current that can be interrupted by a circuit breaker. In both standards the preferred values for outdoor circuit breakers are based on the so-called R10 series. Associated with the symmetrical current value, which is the basis of the rating, there are a number of related capabilities, as they are referred by IEEE, or as definite ratings as specified by IEC. The terminology is different, but the meaning of the parameters is the same.

Asymmetrical currents: In most cases, the AC short-circuit current has an additional DC component, whose magnitude is a function of the time constant of the circuit, and of the elapsed time between the initiation of the fault and the separation of the circuit breaker contacts. Standards establish the magnitude of the DC component based on a time interval consisting of the sum of the actual contact opening time plus one-half cycle of rated frequency. The requirements of the two standards are basically the same, except that in IEEE the factors are not exact numbers but an approximation for certain ranges of asymmetry [9].

The interrupting time has a common definition in IEEE and IEC: It consists of the time period from the instant the trip coil is energized until the time the arc extinguishes, see Figure 7.28 [9,40]. Note that the contact parting time is the summation of the relay time plus the opening time.

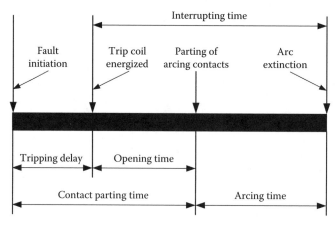

FIGURE 7.28
Interrupting time relationships.

Peak closing current: It is established to define the mechanical capability of the circuit breaker, since its contacts and its mechanism must withstand the maximum electromagnetic forces that can be generated when the circuit breaker closes into a fault. The peak magnitude of the possible fault current against which the circuit breaker is closed is expressed in terms of multiples of the rated symmetrical short-circuit current. IEEE specifies a 2.6 factor, while IEC specifies a 2.5 multiplier. The difference between the two values is due to the difference in the rated power frequencies.

Short time current: It is by definition the rms value of the current that the circuit breaker must carry for a prescribed length of time. The purpose of this requirement is to assure that the short-time heating capability of the conducting parts of the circuit breaker is not exceeded. The magnitude of the current is equal to the rated symmetrical short-circuit current that is assigned. For the required length of time that it must be carried, IEEE specifies 2 s for circuit breakers rated below 100 kV and 1 s for those rated above 100 kV. The IEC requirement is 1 s for all circuit breakers. Nevertheless, IEC recommends a value of 3 s if longer than 1 s periods are required. The 1 s specification of IEC corresponds to the allowable tripping delay, which is referred to as the rated duration of short circuit.

Rated operating duty cycle: It is known as the rated operating sequence by IEC. IEEE specifies the standard operating duty as a sequence consisting of the following operations: O–15 s–CO–3 min–CO; that is, an opening operation followed, after a 15 s delay, by a close–open operation and finally after a 3 min delay by another close–open operation. For circuit breakers intended for rapid reclosing duty the sequence is O–0.3 s–CO–3 min–CO. IEC offers two alternatives: the first one is O–3 min–CO–3 min–CO, while the second alternative is the same duty cycle prescribed by IEEE. However, for circuit breakers that are rated for rapid reclosing duties the time between the opening and the close operation is reduced to 0.3 s. IEEE refers to the reclosing duty as being an O–0 s–CO cycle, which implies that there is no time delay between the opening and the closing operation. However in C37.06 the rated reclosing time, which corresponds to the mechanical resetting time of the mechanism, is given as 20 or 30 cycles, which corresponds respectively to 0.33 or 0.5 s, depending upon the rating of the circuit breaker. When referring to reclosing duties IEEE allows a current derating factor; IEC does not use any derating factor.

Service capability: It defines a minimum acceptable number of times a circuit breaker must interrupt its rated short-circuit current without having to replace its contacts. This capability is expressed in terms of the accumulated interrupted current (a value of 400% is specified for older technologies, being 80% for modern SF_6 and vacuum circuit breakers). Except for the new circuit breaker-type designation created by IEC, see below, this type of requirement is not addressed by IEC. The definition of an IEC Type E_2 circuit breaker does not establish a specific numerical value as in IEEE but when the test duties required are added the accumulated value would be greater than 800%.

Electrical endurance: Up to rated voltages of 52 kV IEC has established two types of HV circuit breakers. Type E_2 is designed so as not to require maintenance of the interrupting parts of the main circuit during its expected operating life, and only minimal maintenance of its other parts. Type E_1 is defined as a circuit breaker that does not fall into the E_2 category.

7.4.4 Additional Requirements

Aside from switching short-circuit currents, circuit breakers must also execute other switching operations and fulfill other requirements. Standards recognize additional needs and define requirements for these operations [41,42]. As with most of all other requirements, there are some differences which are discussed in the following paragraphs.

Capacitance switching: There were some significant differences between IEEE and IEC on this subject, but after harmonization a single standard has been developed. The new standards establish a new classification (C_2) for a circuit breaker that has a *very low probability* of restrike. The old IEEE Class C_1 is now defined as a low probability of restrike. To demonstrate the probabilities a significant amount of data are needed, consequently, the test procedure has also been revised and specific tests are mandated to demonstrate the class assignments. In the new standards single capacitor bank switching and back-to-back capacitor bank switching capabilities are optional ratings, not required for all circuit breakers; line charging requirements are mandatory for all circuit breakers rated 72.5 kV and higher; cable charging is a special case of capacitance switching requirement, which is mandatory for all circuit breakers rated equal or less than 52 kV. Reignitions, restrike, and overvoltages are requirements that serve to define an acceptable performance of a circuit breaker during capacitive current switching. The limitations that have been specified in the standards are aimed at assuring that the effects of restrikes and the potential for voltage escalation are maintained within safe limits. The maximum overvoltage factor is specified because it is recognized that certain types of circuit breakers are prone to have restrikes. This rating allows to restrike only when appropriate means have been implemented within the circuit breaker to limit the overvoltages to the maximum value given in the respective standard. A common method used for voltage control is the insertion of shunt resistors to control the magnitude of the overvoltage.

Mechanical requirements: They include operating life, endurance design tests, and operating mechanism functional characteristics. Mechanical life refers to the number of mechanical operations that can be expected from a circuit breaker. One mechanical operation consists of one opening and one closing operation. For indoor applications, IEEE fixes the number of operations as a function of the current and voltage ratings of the circuit breaker; all outdoor circuit breakers have a common requirement of 2000 operations [3]. In IEC, circuit breakers of the E_1 type have a 2,000 operations requirement; those belonging to the E_2 type require 10,000 operations [43].

7.5 Testing of Circuit Breakers

7.5.1 Introduction

The interrupting process in a circuit breaker is a very complex physical phenomenon that involves thermodynamics, static and dynamic mechanics, fluid dynamics, chemical processes, and electromagnetic fields. Such complexity makes difficult the design and simulation of circuit breakers; therefore, testing becomes an indispensable approach in their development and verification. As summarized in Section 7.4, standard requirements cover short-circuit current making and breaking tests (including terminal fault tests, short-line fault test, and out-of-phase test), inductive and capacitive current switching tests (including line charging, cable charging, single capacitor bank, and back-to-back capacitor bank tests), and mechanical and environmental tests.

Except for the mechanical and environmental tests, type test requirements are designed to prove the interrupting ability of circuit breakers. The testing of circuit breakers can be carried out either in the actual system or in a laboratory. Real system testing offers the advantage that no special investments are necessary for the testing equipment, and the breakers face the fault conditions as they would in service; however, it interferes with normal system operation and security and it cannot easily reproduce the various system conditions that are prescribed by the standards.

In a direct test circuit, both the interrupting current and the recovery voltage are supplied by the same power source and the tests can be simultaneously carried out for the three phases. When the interrupting capabilities of circuit breakers exceeded the power available from direct tests, the limitation could be initially overcome by using the unit testing method, in which units consisting of one or more interrupting chambers were tested separately at a fraction of the rated voltage of a complete breaker pole. Modern circuit breakers have a high interrupting rating per break which surpasses the short-circuit power of the largest high-power laboratories. The synthetic testing method, developed to overcome the problem of insufficient direct test power, applies two source circuits to supply the high current and the recovery voltage separately.

The main characteristics and limitations of direct test and synthetic test circuits are detailed in Sections 7.5.2 and 7.5.3. See Refs. [1,9,44–46] for more details. Other works on circuit breaker testing were presented in Ref. [47–65].

7.5.2 Direct Tests

Direct test circuits can be either fed by specially designed generators or supplied from the network. Figure 7.29 depicts the single-phase equivalent circuit of a direct test, in which TB is the breaker under test. This breaker is in a closed position in a breaking test. After a successful interruption of the short-circuit current, the breaker is stressed by the TRV, whose waveform depends on the adjusting shunt elements together with the inductance L_s, which is formed by the current-limiting reactor, as well as by the generator and the transformer reactances. In making tests, the TB is in open position and must close in on a short circuit. The current trace in a making test will show a DC component, which is determined by the instant of closing. Since the supply circuit is mainly inductive, the DC component is zero and the current is called *symmetrical* when TB closes at the maximum of the supply voltage. However, if TB closes at voltage zero, the current starts with maximum offset and is called *asymmetrical*; the DC component damps out with a time constant that can be adjusted by varying the circuit resistance.

7.5.3 Synthetic Tests

Synthetic testing methods are based on the fact that during the interrupting process, the circuit breaker is stressed by a high current and by a high voltage at different time periods.

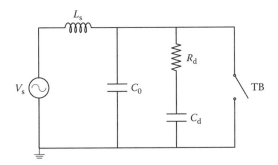

FIGURE 7.29
Single-phase test circuit for a short-circuit test on a HV circuit breaker.

Two separate sources of energy may be used: one source supplies the short-circuit current during the arcing period, while the other source supplies the recovery voltage. The overlap of the current and voltage source takes place during the *interaction interval* around the current zero.

Two different methods, current injection and voltage injection, are used in synthetic testing techniques. With the current-injection method, a high-frequency current is injected into the arc of the test breaker (TB) before the short-circuit current has its zero crossing. There are two possibilities: parallel and series current injection. With the voltage-injection method, a high-frequency voltage is injected across the contacts of the TB after the short-circuit current has been interrupted. There are also two possibilities: a parallel and a series scheme.

Figure 7.30 shows the synthetic test circuit used for the parallel current-injection method. Figure 7.31 depicts the interrupting current of the TB. Before the test, both the auxiliary breaker (AB) and the TB are in a closed position. At t_0, AB and TB open their contacts, and the short-circuit current flows through AB and TB. At t_1, the spark gap is fired and the capacitor bank C_0, which is initially charged, discharges through the inductance L_1 and injects a high-frequency current i_v in the arc channel of TB. During the interval from t_1 to t_2, the current source circuit and the injection circuit are connected in parallel to TB, the current through TB being the sum of the short-circuit current supplied by the generator and the injected current. At t_2, the power-frequency short-circuit current supplied by the generator, i_g, reaches current zero. When TB interrupts the injected current at t_3, it is stressed by the TRV resulting from the voltage that oscillates across the TB.

The shape of the current waveform in the interaction interval around current zero is important, and it is necessary to keep the rate of change of the interrupted current just

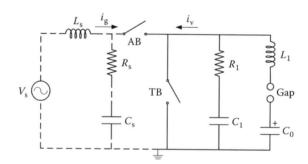

FIGURE 7.30
Parallel current-injection test circuit.

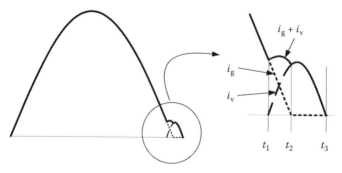

FIGURE 7.31
Current in the parallel current-injection test circuit.

before current zero the same as in the direct test circuit. During the interval around current zero, there is a very strong arc–circuit interaction this time period being very critical for the circuit breaker; the current is approaching zero and at the same time the arc voltage is changing. During the high current interval, when the circuit breaker is in its arcing phase, it is necessary that the energy released in the interrupting chamber is the same as the energy that would be released in a direct test, so that at the beginning of the interaction interval, the conditions in the breaker are equal to those in a direct test. The magnitude of the current is controlled by the reactance of the current circuit and the driving voltage of the generator, and is influenced by the arc voltages of both TB and AB.

Figure 7.32 shows the series voltage-injection synthetic test circuit. The initial part of the TRV comes from the current source circuit; after current zero, the voltage source circuit is connected to supply the rest of the TRV. Before the test, both AB and TB are in a closed position, and the generator supplies the short-circuit current and the post-arc current for the TB. During the whole interruption interval, the arcs in TB and AB are in series. The AB and TB clear almost simultaneously, and the delay for TB to clear depends on the value of the parallel capacitor across AB. Once TB and AB have interrupted the short-circuit current, the recovery voltage of the current source is brought to TB via the parallel capacitor. Just before this recovery voltage reaches its crest value the spark gap is fired, and the voltage oscillation of the injection circuit is added to the recovery voltage of the current source. The resulting TRV is supplied by the voltage circuit only. Because the current injected by the voltage circuit is small, the amount of capacitive energy in the voltage circuit is relatively low. In this test circuits, AB and TB interrupt the short-circuit current simultaneously, and both the current source circuit and the voltage source circuit contribute to it.

The voltage-injection method can use smaller capacitor values than the current-injection method, and is therefore more economical. However, during the interaction interval the circuit parameters of the voltage-injection circuit differ from the circuit parameters of the direct test circuit, which is not the case for the current-injection circuit. In addition, the voltage-injection test circuit requires an accurate timing for the injection of voltage, which is rather difficult to control. For these reasons the current-injection method is more frequently used when the direct test circuit reaches its testing limits.

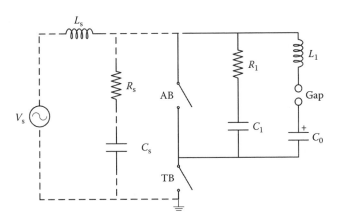

FIGURE 7.32
Series voltage-injection test circuit.

7.6 Modeling of Circuit Breakers

7.6.1 Introduction

Circuit breaker models are needed to analyze both closing and opening operations. Unless an ideal representation is assumed, no single model can be used in either type of operation. In addition, several approaches have been developed to represent circuit breakers in making and breaking currents.

The separation of the contacts of a circuit breaker causes the generation of an electric arc. Although a large number of arc models have been proposed, and some of them have been successfully applied, there is no model that takes into account all physical phenomena in a circuit breaker. Several approaches can be used to reproduce the arc interruption phenomenon in the development, testing, and operation of circuit breakers [66]:

1. Physical arc models, which include the physical process in detail; that is, the overall arc behavior is calculated from conservation laws, gas and plasma properties, and more or less detailed models of exchange mechanisms (radiation, heat conduction, turbulence).
2. Black box models, which consider the arc as a two-pole and determine the transfer function using a chosen mathematical form and fitting the remaining free parameters to measured voltage and current traces.
3. Formulae and diagrams, which give parameter dependencies for special cases and scaling laws. They can be derived from tests or from calculations with the two previous models.

A summary of different applications indicating the approaches that can be used in development, testing, and operation of circuit breakers, is given in Table 7.4 [66].

The most suitable representations of a circuit breaker in a transients program are the so-called black box models [66,67]. The aim of a black box model is to describe the interaction of an arc and an electrical circuit during an interruption process. Rather than internal processes, it is the electrical behavior of the arc which is of importance.

Several models can also be used to represent a circuit breaker in closing operations. As for models used to reproduce opening operations, different approaches and different

TABLE 7.4

Applications of Arc Models

Type of Problem	Development	Testing	Operation
Design optimization	PM		–
Dimensioning of flow and mechanical system, pressure buildup	PM, CF	PM, CF	–
Description of dielectric recovery	PM, CF	CF	CF
Influence of arc asymmetry and delayed current zeros	PM, BB, CF	BB, CF	BB
Small inductive currents	PM, BB, CF	BB, CF	BB, CF
Short-line fault, including TRV	PM, BB, CF	BB, CF	BB, CF
Design and verification of test circuits	CF	BB	–

Source: CIGRE Working Group 13.01, Practical application of arc physics in circuit breakers. Survey of calculation methods and application guide, *Electra*, no. 118, pp. 64–79, 1988. With permission.

Note: BB, Black box arc model; CF, Characteristic curve/formulae; PM, Physical model.

levels of complexity may be considered. When using a black box-type model, the simplest approach assumes that the breaker behaves as an ideal switch whose impedance passes instantaneously from an infinite value to a zero value at the closing time. A more sophisticated approach assumes that there is a closing time from the moment at which the contacts start to close to the moment that they finally make and its withstand voltage decreases as the separation distance between contacts decreases. Finally, a dynamic arc representation can be included in case of breaker prestrike.

Table 7.5 shows modeling guidelines proposed by the CIGRE WG 33-02 for representing circuit breakers in both closing and opening operations [68].

The rest of this section deals with different approaches proposed for representing circuit breakers during both opening and closing operations using black box models. Several test examples are presented to illustrate the scope and limitations of each approach.

7.6.2 Circuit Breaker Modeling during Opening Operations

7.6.2.1 Current Interruption

Circuit breakers accomplish the task of interrupting an electrical current by using some interrupting medium for dissipating the energy input. For most breaking technologies the interrupting process is well understood and can be described with some accuracy.

TABLE 7.5

Modeling Guidelines for Circuit Breakers

Operation		Low-Frequency Transients	Slow Front Transients	Fast Front Transients	Very Fast Front Transients
Closing	Mechanical pole spread	Important	Very important	Negligible	Negligible
	Prestrikes (decrease of sparkover voltage versus time)	Negligible	Important	Important	Very important
Opening	High current interruption (arc equations)	Important only for interruption capability studies	Important only for interruption capability studies	Negligible	Negligible
	Current chopping (arc instability)	Negligible	Important only for interruption of small inductive currents	Important only for interruption of small inductive currents	Negligible
	Restrike characteristic (increase of sparkover voltage versus time)	Negligible	Important only for interruption of small inductive currents	Very important	Very important
	High-frequency current interruption	Negligible	Important only for interruption of small inductive currents	Very important	Very important

Source: CIGRE Working Group 02 (SC 33), Guidelines for representation of network elements when calculating transients, CIGRE Technical Brochure no. 39, 1990. With permission.

The electric arc is a self-sustained discharge capable of supporting large currents with a relatively low voltage drop. The arc acts like a nonlinear resistor: after current interruption, the resistance value does not change rapidly from a low to an infinite value. For several microseconds after interruption, a post-arc current flows. The interruption process can be separated into three periods: the arcing period, the current-zero period, and the dielectric recovery period.

During arcing period, the time constant of the DC component in a short-circuit current becomes smaller because of the arc resistance. As the arc current approaches a current-zero point, the ratio of arc heat loss to electrical input energy increases and the arc voltage rises abruptly. After the arc current is extinguished, the recovery voltage appears. Since the space between contacts does not change to a completely insulating state, a small current, the post-arc current, flows through the breaker as the recovery voltage builds up and soon disappears. If the extinguishing ability of the circuit breaker is small, the post-arc current does not decrease, and an interruption failure can occur; it is known as thermal failure or reignition.

The recovery voltage has two components: a TRV and a power-frequency recovery voltage (PFRV). The TRV has a direct effect on the interrupting ability of a circuit breaker, but the PFRV is also important because it determines the center of the TRV oscillation. The waveform and magnitude of the TRV vary according to many factors, such as system voltages, equipment parameters, and fault types. The waveform determined by the system parameters alone is called the inherent TRV. All regions of the TRV have an effect on the breaking interrupting ability. For some kinds of circuit breakers, the initial TRV is the most critical period.

The circuit interruption is complete only when the circuit breaker contacts have recovered sufficient dielectric strength after arc extinction near current zero. Here, the dielectric strength recovery characteristics play an important role. Depending on the type of circuit breaker, this period is also dangerous. An interruption failure during this period is known as dielectric failure or restrike.

7.6.2.2 Circuit Breaker Models

The main objectives of a circuit breaker model are [69]

- From the system viewpoint, to determine all voltages and currents that are produced within the system as a result of the breaker action
- From the breaker viewpoint, to determine whether the breaker will be successful when operating within a given system under a given set of conditions

The aim of a black box model is to describe the interaction of the switching device and the corresponding electrical circuit during an interruption process. Rather than internal processes, it is the electrical behavior of the circuit breaker which is of importance. Black box models are aimed at obtaining a quantitatively correct performance of the circuit breaker. Several levels of model complexity are possible [69]:

1. The simplest model considers an ideal breaking action that is completely independent of the arc. The breaker is represented as an ideal switch that opens at first current-zero crossing after the tripping signal is given. This model can be used to obtain the voltage across the breaker, this voltage is to be compared with a pre-specified TRV withstand capability for the breaker. This model cannot reproduce any interaction between the arc and the system.

2. A more elaborated model considers the arc as a time-varying resistance or conductance. The time variation is determined ahead of time based on the breaker characteristic and perhaps upon the knowledge of the initial interrupting current. This model can represent the effect of the arc on the system, but requires advanced knowledge of the effect of the system on the arc. Arc parameters are not always easy to obtain and the model still requires the use of precomputed TRV curves to determine the adequacy of the breaker.

3. The most advanced model represents the breaker as a dynamically varying resistance or conductance, whose value depends on the past history of arc voltage and current. This model can represent both, the effect of the arc on the system and the effect of the system on the arc. No precomputed TRV curves are required. These models are generally developed to determine initial arc quenching; that is, to study the thermal period only, although some can be used to determine arc reignition due to insufficient voltage withstand capability of the dielectric between breaker contacts. Their most important application cases are short-line fault interruption and switching of small inductive currents.

Many models for circuit breakers, represented as a dynamic resistance/conductance, have been proposed. A survey on black box models of gas (air, SF_6) circuit breakers was presented in Ref. [67]. Electromagnetic transients program (EMTP) implementation of dynamic arc models, adequate for gas and oil circuit breakers, was presented in Ref. [70]. All those models are useful to represent a circuit breaker during the thermal period; models for representation of SF_6 breakers during thermal and dielectric periods were discussed and used in Refs. [71,72]. For some of the models proposed during the last decades see also Refs. [73–122]. These models are almost exclusively applied to gas-filled circuit breaker. They are less attractive for other types of switching devices (e.g., oil circuit breakers) because the thermal process is less significant in the behavior of these types of circuit breakers.

Vacuum circuit breakers have a different performance, they lack an extinguishing medium and their representation has to consider statistical properties. Vacuum circuit breaker models may include arc voltage characteristic, calculation of mean chopping current value with known di/dt, dielectric breakdown voltage characteristic, contacts separation dynamics, probability of high-frequency arc quenching capability, probability of high-frequency zero current passing. Several models and test studies for this type of breaker have been presented during the last years [123–157].

Sections 7.6.2.3 and 7.6.2.4 present different approaches developed to date for representing both gas-filled and vacuum circuit breakers.

7.6.2.3 Gas-Filled Circuit Breaker Models

7.6.2.3.1 Ideal Switch Models

The simplest approach assumes that the breaker has zero impedance when closed and opens at first current-zero crossing after the tripping signal is given. The breaking action is therefore completely independent of the arc. Two options are analyzed here:

1. The model is used to obtain the voltage across the breaker; this voltage may be later compared with the TRV withstand capability for the breaker. With this approach, the model does not reproduce any interaction between the breaker and the system.

2. The model incorporates the specified TRV curve, allowing in this manner the possibility of reproducing a breakdown and the interaction of the breaker and the system.

The following example shows the simulation of a short-line fault by means of a very simple model, which illustrates the application of both approaches and the influence of some parameters in the recovery voltage across the breaker terminals.

Example 7.4

Figure 7.33 shows the scheme of the test system. It is an overhead line, with a fault at the receiving end, and fed from a HV network represented by its equivalent. The line model chosen for this example is the simplest one, an ideal traveling-wave single-phase model, specified by the surge impedance and the propagation velocity. Consider the following parameters:

Network equivalent:	Voltage	Peak value = 300 kV
	Series inductance	$L = 16.9\,\text{mH}$
	Shunt capacitance	$C = 1.8\,\mu\text{F}$
Line:	Length	$\ell = 4.5\,\text{km}$
	Surge impedance	$Z_c = 380\,\Omega$
	Propagation velocity	$\upsilon = 300\,\text{m/}\mu\text{s}$

Simulation results obtained with the simplest model are shown in Figure 7.34. The results present the typical sawtooth waveform of the TRV caused when clearing a short-line fault. The simulation was performed by simply sending a trip signal to the ideal switch which opened after the first current-zero crossing.

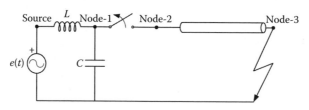

FIGURE 7.33
Example 7.4: Diagram of the test case.

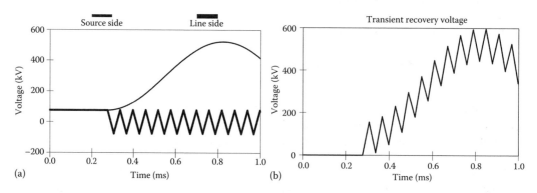

FIGURE 7.34
Example 7.4: TRV across the breaker. (a) Source and line side voltages. (b) Transient recovery voltage.

The same test system is simulated again to illustrate the characteristics of a more elaborated circuit breaker model, in which the specified TRV is included. This allows to compare the resulting recovery voltage to the withstand capability of the breaker and decide whether the operation will be successful or not during the simulation. With the new model, the operating point of the breaker is updated every time step, depending on its past state as well as on the voltage and current applied from the electrical circuit. Breaker states and transitions are calculated using the following rules [158,159]:

1. The breaker can first open from the "normal closed" state after instant "to open" as soon as the measured current becomes lower than the user-defined power-frequency chopping level.
2. From this "tentatively open" state, the arc may reignite whenever the TRV measured across the circuit breaker exceeds the voltage withstand characteristic of the breaker, defined as a function of time by the point-list function characteristic.
3. The arc will be reextinguished as soon as the following conditions are simultaneously true:
   - The breaker has been in the "reignited" state for a minimum amount of time in order to prevent unrealistic opening in the megahertz range.
   - The current is lower than the high-frequency chopping level.
   - The slope of the current is lower than the critical current slope of the breaker.
4. The breaker will move from a "tentatively open" state to a "normal open" state when the time elapsed since the instant of contact separation becomes larger than the defined reignition time-window.

Figure 7.35 shows the transitions between possible states of this model. Figure 7.36 depicts some results of the TRV and the current of the breaker. Two different values of the equivalent inductance were used to compare the influence of the parameter. One can observe that the breaking

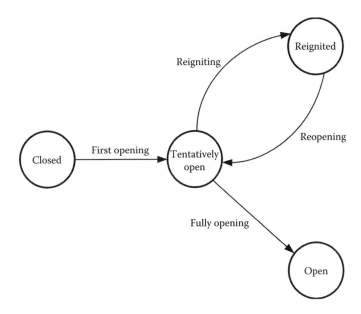

FIGURE 7.35
State transition model of a circuit breaker.

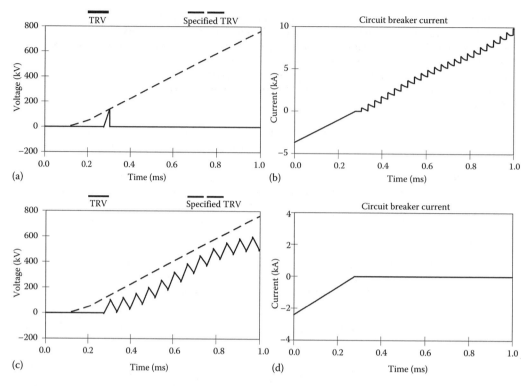

FIGURE 7.36
Example 7.4: Performance of a circuit breaker represented by a precomputed voltage withstand characteristic. (a) TRV across the circuit breaker ($L = 16.9\,\text{mH}$). (b) Circuit breaker current ($L = 16.9\,\text{mH}$). (c) TRV across the circuit breaker ($L = 28.9\,\text{mH}$). (d) Circuit breaker current ($L = 28.9\,\text{mH}$).

operation fails with the lower value of the equivalent inductance, which is in accordance with the analysis presented in Section 7.2.4. The lower the inductance, the higher the short-circuit capacity at the breaker position, which according to Equation 7.20 increases the initial TRV as well as the probability of failure.

7.6.2.3.2 Dynamic Arc Representation

Arc models are mathematically expressed as formulae for the time-varying resistance or conductance as a function of arc current, arc voltage and several time-varying parameters representing arc properties. Several arc models are available in the literature; some are derived from Cassie and Mayr equations, some are a combination of both models [160]:

- The Cassie model is given by the following equation:

$$\frac{1}{g}\frac{dg_c}{dt} = \frac{1}{\tau_c}\left(\left(\frac{v}{v_0}\right)^2 - 1\right) = \frac{1}{\tau_c}\left(\left(\frac{i}{v_0 g_c}\right)^2 - 1\right) \tag{7.21}$$

It assumes an arc channel with constant temperature, current density, and electric field strength. Changes of the arc conductance result from changes of arc cross section; energy removal is obtained by convection.

- The Mayr model is given by the following equation:

$$\frac{1}{g_m}\frac{dg_m}{dt} = \frac{1}{\tau_m}\left(\frac{vi}{P_0}-1\right) = \frac{1}{\tau_m}\left(\frac{i^2}{P_0 g_m}-1\right) \tag{7.22}$$

This model assumes that changes of arc temperature are dominant, and size and profile of the arc column are constant. Thermal conduction is the main mechanism of energy removal.

In these equations, g is the arc conductance, v is the arc voltage, i is the arc current, τ is the arc time constant, P_0 is the steady-state power loss, and v_0 is the constant part of the arc voltage. g_c is in the region of $1\,\mu s$ (SF$_6$) and g_m is between 0.1 and $0.5\,\mu s$ (SF$_6$). These parameters are not strictly constant for an actual arc, but observations indicate that during the brief time around current zero these parameters vary sufficiently slowly to assume them to be constant.

- A combination of both models gives the Cassie–Mayr model:

$$\frac{1}{g} = \frac{1}{g_c} + \frac{1}{g_m} \tag{7.23}$$

This is justified by the fact that at high currents the entire voltage drop takes place in the Cassie equation, but before current zero the contribution from the Mayr equation increases, while the Cassie part goes to zero.

Table 7.6 shows a comparison of the main features of both models. One can observe that both models are defined by only two parameters:

- τ and v_0 for Cassie model
- τ and P_0 for Mayr model

Both models give a qualitative description of an arc behavior; they should be carefully used for quantitative representations. A great number of modifications of these equations

TABLE 7.6

Comparison of Arc Models

	Cassie Model	Mayr Model
Arc conductance	By change of arc diameter $G \propto q$	By change of ionization degree $G \propto \exp\left(\frac{q}{q_0}\right)$
Heat loss	By thermal convection $P \propto q$	By thermal conduction $P = $ constant
Dynamic expression	$\frac{1}{g}\frac{dg}{dt} = \frac{1}{\tau}\left(\left(\frac{v}{v_0}\right)^2-1\right)$	$\frac{1}{g}\frac{dg}{dt} = \frac{1}{\tau}\left(\frac{vi}{P_0}-1\right)$
Suitability	Large current region	Small current region

Source: Adapted from Nakanishi, K. (Ed.), *Switching Phenomena in High-Voltage Circuit Breakers*, Marcel Dekker, New York, 1991. With permission.

have been formulated. These modifications introduce more parameters into the equation or define the equation in a more general form, making models more adaptive.

Other widely used models are presented below [69,160].

1. The model proposed by Avdonin is for air-blast and SF_6 breakers [85]. The arc resistance of this model is expressed by

$$\frac{dr_a}{Et} = \frac{r_a^{1-\alpha}}{A} - v_a r_a \frac{r_a^{1-\alpha-\beta}}{AB} \qquad (7.24)$$

which is derived from the modified Mayr equation

$$\frac{dr_a}{dt} = \frac{r_a}{\tau}\left[1 - \frac{v_a r_a}{P_0}\right] \qquad (7.25)$$

with

$$\tau = A r_a^\alpha \quad P_0 = B r_a^\beta \qquad (7.26)$$

where
   r_a, v_a, and i_a are the arc resistance, voltage, and current, respectively
   τ is the arc time constant
   P_0 is the breaker cooling power

An extra parameter is needed to specify the number of breaks per pole.

This model can be used to represent thermal failure near current interruption and conductivity in the post-arc region. Resistance instability near current interruption can cause current chopping. Although parameters for this model are best derived from short-line fault breaking tests, it is, however, feasible to provide some typical data. It has been observed through practical cases that the applicability range of Equation 7.24 is actually longer than the near current-zero region.

2. The Urbanek model can represent arc interruption and both thermal and dielectric failure [76]. Both current chopping and reignition are also represented. It is characterized by the following equation for the arc conductance:

$$\frac{1}{g}\frac{dg}{dt} = \frac{1}{\tau}\left(\frac{vi}{e^2 g} - 1 - \frac{P_0}{e^2 g}\left(1 - \left(\frac{v}{v_d}\right)^2 - 2\frac{\tau}{v_d^2}v\frac{dv}{dt}\right)\right) \qquad (7.27)$$

where
   e is the arc voltage for high currents
   P_0 is the minimum power to maintain the arc
   v_d is the dielectric breakdown voltage for cold arc channel

3. Another model for simulation of thermal breakdown is the Kopplin model, which is mostly suitable to represent generator circuit breakers [80]. It is characterized by a Mayr-type equation for the arc conductance:

$$\frac{1}{g}\frac{dg}{dt} = \frac{1}{\tau}\left(\frac{vi}{P_0} - 1\right) \qquad (7.28)$$

with

$$\tau = K_I \cdot (g + 0.0005)^{0.25}$$

$$P_0 = K_P \cdot (g + 0.0005)^{0.6}$$

(7.29)

where K_p and K_I are model parameters.

Testing and validation of this model was presented in Ref. [49].

From the modifications proposed to date, arc models can be divided into several groups. All models keep the basic idea of describing arc behavior using parameters τ and P, with different physical interpretation. Table 7.7 shows some of these equations as classified by the CIGRE Working Group [67]. A summary of the applicability of each group of models is given in the same reference.

The representation of the arc equation in a transients program, assuming that a circuit breaker is not available as a built-in model, depends on the capabilities of the program. Figure 7.37 depicts a diagram with three possible representations. As shown in the figure, for all cases arc equation is represented in the control section. The choice of one of these models should be made taking into account:

- The voltage source model is limited to those cases for which the circuit breaker is between a node and ground, unless the option of ungrounded source voltages is available
- The current source model can be numerically unstable if the simulation starts with a large arc current, as illustrated in Ref. [70]
- The controlled resistance model is the best one, if such a capability is available, since the implementation of the arc equation in the control section is straightforward

The following example illustrates arc modeling as described above.

Example 7.5

The arc model used in this example covers the thermal period of the interrupting process of a gas-filled circuit breaker and is based on a model proposed by Grütz and Hochrainer [74]:

$$\frac{1}{g}\frac{dg}{dt} = \frac{1}{\tau}\left(\frac{G}{g} - 1\right)$$

(7.30)

where G is the steady-state conductance, which is approached by the following expression:

$$G = \frac{i^2}{P_0 + v_0\,|i|}$$

(7.31)

where
P_0 is the steady-state heat loss
v_0 is the constant percentage of the steady-state v–i characteristic
i is the arc current

Figure 7.38 shows the representation used in this study to analyze the test system shown in Figure 7.33, see Ref. [104]. The circuit breaker is represented by means of a controlled resistance

TABLE 7.7

Arc Models and Related Parameters

Models	Dynamic Arc Equation	Arc Parameters	Reference
Two parameters	$\dfrac{1}{g}\dfrac{dg}{dt}=\dfrac{1}{\tau}\left(\left(\dfrac{v}{v_0}\right)^2-1\right)$	τ, v_0	[7]
	$\dfrac{1}{g}\dfrac{dg}{dt}=\dfrac{1}{\tau}\left(\dfrac{vi}{P_0}-1\right)$	τ, v_0	[8]
More than two parameters	$\dfrac{1}{g}\dfrac{dg}{dt}=\dfrac{1}{\tau}\left(\dfrac{vi}{P_0}-1\right)+\beta\dfrac{dv}{dt}$	τ, v_0, β	[73]
	$\dfrac{1}{g}\dfrac{dg}{dt}=\dfrac{1}{\tau}\left(\dfrac{vi}{e^2 g}-1-\dfrac{P_0}{e^2 g}\left(1-\left(\dfrac{v}{v_d}\right)^2-2\dfrac{\tau}{v_d^2}v\dfrac{dv}{dt}\right)\right)$	τ, v_0, e, v_d	[76]
	$\dfrac{1}{g}\dfrac{dg}{dt}=\dfrac{1}{T}\left(\dfrac{vi}{P_0+v_0\,|\,i\,|}-1\right)\quad T=\tau+\dfrac{\tau_1}{1+\left(\dfrac{g}{G_0}\right)^2}$	$\tau, v_0, P_0, G_0, \tau_1$	[80]
	$\dfrac{1}{g}\dfrac{dg}{dt}=\dfrac{1}{\tau}\left(\dfrac{vi}{a+b\,|\,i\,|+c\sqrt{|\,i\,|}}-1\right)$	τ, a, b, c	[86]
Functional models	$\dfrac{1}{g}\dfrac{dg}{dt}=\dfrac{1}{\tau(|\,i\,|)}\left(\dfrac{G_S(|\,i\,|)}{g}-1\right)$	$\tau(|\,i\,|), G_s(|\,i\,|)$	[74]
	$\dfrac{1}{g}\dfrac{dg}{dt}=\dfrac{1}{\tau(|\,g\,|)}\left(\dfrac{vi}{P(g)}-1\right)$	$\tau(g), P(g)$	[75]
	$\dfrac{1}{g}\dfrac{dg}{dt}=\dfrac{1}{\tau(g)}\left(\dfrac{G_S(|\,i\,|)}{g}-1\right)$	$\tau(g), G_s(|\,i\,|)$	[90]
Stochastic models	$\dfrac{1}{g}\dfrac{dg}{dt}=\dfrac{1}{\tau(g,t)}\left(\dfrac{vi}{P(g,t)}-1\right)$	$\tau(g,t), P(g,t)$	[78]
	$\dfrac{1}{g}\dfrac{dg}{dt}=\dfrac{1}{\tau}\left(\left|\dfrac{v\,|\,i\,|^a}{k}\right|^{2/(1+a)}-\chi(t)\right)$	$\alpha, k, \tau, \chi(t)$	[81]
	$\dfrac{1}{g}\dfrac{dg}{dt}=\dfrac{1}{\tau(g)}\left(\dfrac{vi}{g}-\gamma(t,g)\right)$	$\tau(g), P(g), \gamma(t,g)$	[82]

Source: CIGRE Working Group 13.01, Applications of black box modelling to circuit breakers, *Electra*, no. 149, pp. 40–71, 1993. With permission.

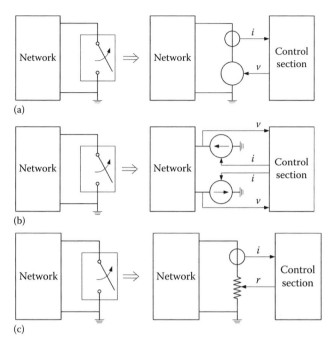

FIGURE 7.37
Modeling of a circuit breaker including arc dynamics. (a) Voltage source representation. (b) Current source representation. (c) Variable resistance representation.

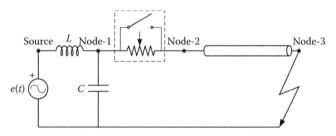

FIGURE 7.38
Example 7.5: Test system model.

in series and in parallel with a time-controlled switch, needed to obtain the steady-state initialization, as in the previous case.

Two cases using the same test circuit and the arc model have been simulated. Each case shows a different behavior of the circuit breaker: extinction and reignition of the arc. Parameters of the circuit breaker for each case are

1. Extinction: $P_0 = 360\,\text{kW}$, $v_0 = 8.0\,\text{kV}$, $\tau = 1.3\,\text{ms}$
2. Reignition: $P_0 = 80\,\text{kW}$, $v_0 = 4.5\,\text{kV}$, $\tau = 1.3\,\text{ms}$

Figure 7.39 show the TRV, the breaker current, and the arc conductance that result from the simulation of both cases.

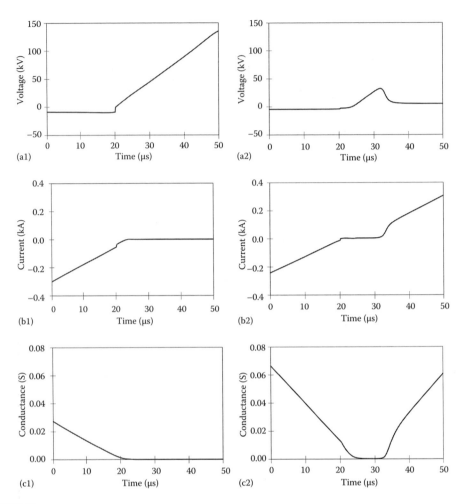

FIGURE 7.39
Example 7.5: Dynamic arc representation—Arc extinction (left side) and arc reignition (right side). (a) TRV across the circuit breaker. (b) Circuit breaker current. (c) Arc conductance.

7.6.2.4 Vacuum Circuit Breaker Models

7.6.2.4.1 Introduction

The vacuum arc can be generally divided into three regions: the cathode spot region, the interelectrode region, and the space charge sheath in front of the anode. The cathode spot region and the anode region have a constant thickness of several micrometers and are very small compared to the interelectrode space. Despite its small size, the cathode spot regions cover most of the arc voltage, which remains practically constant, independent from the value of the current [26,127,161,162].

The vacuum arc ceases when the cathode spots disappear. As the arc current approaches zero, the number of cathode spots reduces until only one is left. This spot continues to supply charge to the plasma, until finally the current reaches zero. At current zero, the interelectrode space still contains a certain amount of conductive charge. As the current reverses polarity, the old anode becomes the new cathode, but in the absence of

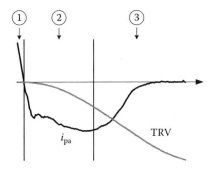

FIGURE 7.40
Phases of the post-arc current in a vacuum circuit breaker. (From van Lanen, E.P.A. et al., *IEEE Trans. Plasma Sci.*, 35, 925, August 2007. With permission.)

cathode spots, the overall breaker conductance drops, which allows a rise of the recovery voltage. The combination of the residual plasma conductance and this recovery voltage gives rise to a post-arc current, which contains a considerable scatter that disturbs the relationship between the arcing conditions and the post-arc current. This has to do with the final position of the last cathode spot: when its final position is near the edge of the cathode, a significant amount of charge is ejected away from the contacts, and less charge is returned to the external electric circuit, compared to the situation in which the final spot position is close to the center of the cathode. Since the cathode spots move randomly across the cathode surface, the final position is unknown. As a result, the post-arc plasma conditions are different for each measurement, which gives the post-arc current a random nature.

It is generally accepted that the post-arc current is divided into three phases (Figure 7.40):

1. Before current zero, ions are launched from the cathode towards the anode. At current zero, the ions that have just been produced continue to move towards the anode as a result of their inertia. Electrons match their velocity with the ion velocity to compensate for the ion current, so the total electric current is zero.

2. Immediately after current zero, the electrons reduce their velocity, and the flux of positive charge arrives at the post-arc cathode. This process continues until the electrons reverse their direction; until that moment, the net charge inside the gap is zero. With no charge, the voltage across the gap remains zero. As soon as the electrons reverse their direction, the post-arc current enters its second phase, in which the electrons move away from the cathode, leaving an ionic space charge sheath behind. Now, the gap between the electrodes is not neutral and the circuit forces a TRV across it. This potential difference stands almost completely across the sheath, which is not charge-neutral.

3. The sheath continues to expand until it reaches the new anode. At that moment, the post-arc current starts its third phase. The electric current drops, since all electrons have been removed from the gap. The electric field between the contacts moves the remaining ions towards the cathode, but the current that results from this process is negligible.

If ions and electrons move toward the post-arc cathode, they leave behind an empty gap between the plasma and the new anode. By definition, charge sources are absent after current zero; hence, no charge leaves the post-arc anode to fill up this gap. This implies breaking the electrical connection between the plasma and the anode, and hence, a voltage should rise across the gap during this phase. However, test measurements show a voltage-zero phase immediately following current zero. The voltage-zero period can be explained if the thermal velocity of the particles is taken into account [157]. In that case, the pressure of electrons and ions causes the plasma to remain neutral in the voltage-zero phase. The electrical behavior then resembles that of a Langmuir probe, for which the relation between current and voltage is known [163–165].

For the application of this theory, it is assumed that the electrodes are large, they are placed closely together, and the interelectrode properties only change in the direction perpendicular to them; this limits the problem to one dimension. It is further assumed that the plasma is stationary and equally distributed along the gap, and that all particles reaching an electrode are absorbed by it. The electron temperature is higher than the ion temperature, no collisions take place between particles in either the plasma or the sheaths, and no new plasma is generated. The process following current zero can be then described as follows:

- At $t = 0$, both the voltage across the gap and the current through the circuit are zero. Since their thermal energy is higher and their mass is lower, electrons have a higher velocity than ions. Because of this, the flux of electrons at an electrode should be larger than the flux of ions and result in an electrical current. However, since both electrodes experience the same flux in opposite direction, the current remains zero. An electric field in front of the electrodes repels the surplus of electrons to maintain a net charge flux of zero. This means that there is a potential difference between the plasma and the electrodes, which is mainly distributed across the sheath in front of the electrodes.

- When the TRV starts to rise, the electric field in front of the cathode increases and repels more electrons. However, due to their temperature, a large amount of electrons can cross the electric field to reach the cathode. The change of the flux of charged particles at the contacts may cause an electrical current, but because many electrons are still present in the cathodic sheath, the gap is quasineutral, and the voltage across the gap remains low.

- As the TRV continues to increase, the electric field in front of the cathode repels more electrons, until it becomes too strong for any electron to reach the cathode. At this point, the current is limited to the flux of ions reaching the cathode, and an increase in voltage would only lead to an increase of the cathodic sheath width, whereas the on saturation current remains constant.

7.6.2.4.2 Reignitions and Restrikes of Vacuum Circuit Breakers

When a breaker is unable to withstand a voltage after current interruption, a failure occurs and a new arc is formed. Present vacuum circuit breakers are capable of interrupting current reliably according to their current and voltage ratings. If nevertheless the breaker does not interrupt the current at the first current zero after contact separation, it most likely interrupts the current at the next current zero, since the conditions for a successful current interruption have generally improved. Since the breakdown voltage increases proportionally with the contact separation, and just before the first current zero the contacts have not reached their maximum separation, a breakdown at a voltage below the rated voltage can occur. As the contacts continue to separate, the breaker can interrupt the current successfully if the contacts reach its maximum separation at the following current zero.

As already mentioned, the arc current in vacuum is formed from electrons which are withdrawn from the metal surface by the electric field. The resulting current density increases locally the contact temperature, which may eventually lead to the formation of a cathode spot, and subsequently to a restrike. Since such a failure is initiated by an electric field, it is called dielectric restrike. For perfectly smooth contact surfaces, the electric field under normal operating conditions is generally too low to cause dielectric restrike. However, irregularities that increase locally the electric field are widely present on the contact surface. This concentrates the current density to smaller surface areas, increasing heating and creating conditions for a restrike. The restrike is not only caused by irregularities on the contact surface, but also by microscopic particles in the vacuum. Although manufacturers spend much effort on cleaning the interior of vacuum circuit breakers, the presence of these particles is inevitable. They may originate from protrusions on the contact surface, which are drawn under the influence of an electric field.

Thermal reignition occurs when a breaker fails in the period immediately following current zero. This term is originally used for gas-filled circuit breakers, where the probability of having a breakdown depends on the balance between forced cooling and heating of the residual charge in the hot gas. Charge and vapor densities in vacuum are much lower than in gas breakers. Immediately after current zero, the gap contains charge and vapor from the arc, the contacts are still hot, and there can be also pools of hot liquid metal on their surface. It takes several microseconds for the charge to remove, but it takes several milliseconds for the vapor to diffuse, and the pools to cool. A failure that occurs when vapor is still present, although charge has already decayed, is called dielectric reignition. With increased contact temperature, the conditions for a failure are improved. When an electron, accelerated in the electric field, hits a neutral vapor particle with sufficient momentum, it knocks out an electron from the neutral. This process reduces the kinetic energy of the first electron; both electrons accelerate in the electric field, hit other neutral particles and cause an avalanche of electrons in the gap, which eventually causes reignition. This process of charge multiplication enhances when the probability of an electron hitting a vapor particle increases. This can be prevented by either increasing the vapor pressure or the gap length. Both methods reduce the reignition voltage, but at some point electrons collide with particles before reaching the appropriate ionization energy. As a result, after reaching a minimum value, the reignition voltage eventually rises with increasing vapor pressure or gap length.

7.6.2.4.3 *Models of Vacuum Circuit Breakers*

The approaches used to represent arc extinction in gas-filled circuit breakers involve the balance between the internal energy of the arc and the cooling power of the surrounding gas; after current zero, the cooling power has to be stronger than the arc internal energy to successfully extinguish the arc. Vacuum circuit breakers lack an extinguishing medium and their arc internal energy is almost completely concentrated in the cathode spots. The reignition process is also quite different, since the source of the new arc (i.e., cathode spots) arises at a different location (the new cathode) before current zero. For this reason, post-arc vacuum modeling has been mainly focused on simulating the anode temperature during arcing, and the behavior of the cathodic space charge sheath after current zero, because both involve the conditions at the location of potentially new cathode spots.

The rest of this section has been divided into two parts dedicated to analyze the representation of vacuum circuit breakers during interruption of low currents and during the post-arc current period, respectively.

Low-current models. Vacuum circuit breakers show a high ability of interrupting currents, which can be the origin of switching overvoltages. The latest developments of vacuum switching technology reduce the chopping current level, and a constant current value of a few amperes can be assumed. When studying the interruption of small inductive currents, the model of the vacuum circuit breaker has to include (depending on the properties of the breaker and the surrounding network) the cold withstand voltage characteristic, the high-frequency quenching capability, and the chopping current level [166].

The cold withstand voltage characteristic of a vacuum circuit breaker is a function of the contact distance. A parameter that is of influence is the speed of contact separation [137,167]. It is known that the data vary with a statistical distribution. Smeets represented the withstand voltage characteristic with an exponential expression [168], while Glinkowski et al. showed that the failure can occur at short gaps (<1 mm) [137], and it is sufficient to use a straight line.

The high-frequency quenching capability is defined by the slope of the reignited current at current zero. Although this characteristic was initially assumed constant [123,135], it is now evident that it depends also on the reignited voltage [169] and exhibits a time-dependent behavior.

The chopping current depends mainly on the contact material, although the surge impedance of the load side may also have its effect [135].

The characteristics that describe whether or not reignition occurs are [137]

$$V_b = A\left(t - t_{open}\right) + B \tag{7.32}$$

$$\frac{di}{dt} = C\left(t - t_{open}\right) + D \tag{7.33}$$

where t_{open} is the moment of contact opening. The quantities V_b and di/dt represent the dielectric and arc quenching capability of the circuit breaker, respectively.

The performance of this model is tested in the following example [166].

Example 7.6

Figure 7.41 depicts the equivalent one-line circuit of a cable-fed load [170]. The circuit includes the source, the vacuum circuit breaker, the cable, and the load models. The data for the circuit parameters are

- Source $L_n = 5\,mH$, $C_n = 100\,nF$
- Circuit breaker $R_S = 100\,\Omega$, $L_S = 50\,nH$, $C_S = 100\,pF$
- Cable $R_\sigma = 2\,\Omega$, $L_\sigma = 40\,\mu H$
- Load $R_L = 1E5\,\Omega$, $L_L = 120\,mH$, $C_L = 10\,nF$.

In fact, C_L represents the capacitance of cable and load together.

The value of the constants in Equations 7.32 and 7.33 are $A = 1.7E7$ V/s, $B = 3400$ V, $C = -3.40E10$ A/s^2, and $D = 255E6$ A/s. These characteristics are randomly varied assuming a normal statistical distribution with a 10% standard deviation. The constant chopping current is 3 A.

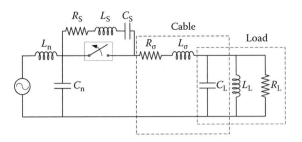

FIGURE 7.41

Example 7.6: Test circuit for reignition analysis. (From Popov, M. et al., *Eur. Trans. Electr. Power*, 11, 413, November/December 2001. With permission.)

After the circuit breaker contacts start opening, the current continues to flow until its absolute value is equal to the chopping current. Thereafter, the switch operation is controlled by Equations 7.32 and 7.33.

The results obtained are depicted in Figure 7.42, which shows the current and voltage of the vacuum circuit breaker, as well as the load side voltage, with two different timescales. The lower plots show the vacuum circuit breaker current and voltage, and load voltage of the first reignition.

After the current has been chopped the TRV occurs, resulting in two different oscillation frequencies. The oscillations depend on the parameters of the circuit. One frequency is in the range of a few megahertz (typically 2.5 MHz), and is approximately calculated by

$$f_1 \approx \left(2\pi\sqrt{L_\sigma C_S}\right)^{-1} \tag{7.34}$$

The other frequency is lower and is determined by the load parameters. It is the natural frequency of the load with values in the order of kilohertz (typically around 4.6 kHz) and can be estimated with

$$f_2 \approx \left(2\pi\sqrt{L_L C_L}\right)^{-1} \tag{7.35}$$

When the TRV surpasses the withstand voltage level of the vacuum gap, the vacuum circuit breaker reignites and a high-frequency current flows through the circuit. The reignited current has two components superimposed on the power-frequency current:

- One oscillation has a frequency of approximately 250 kHz and is approached by

$$f_3 \approx \left(2\pi\sqrt{L_\sigma C_L}\right)^{-1} \tag{7.36}$$

- The other oscillation has a very high frequency, finds its origin in the parasitic capacitance and inductance of the gap and can be approached with

$$f_4 \approx \left(2\pi\sqrt{L_S C_S}\right)^{-1} \tag{7.37}$$

This frequency is in the range of a few tens of megahertz (typically around 70 MHz). The associated high-frequency current normally causes voltage spikes as shown in the simulation results. However, measurements show that even higher frequency oscillations superimposed on the reignited current can be present [168].

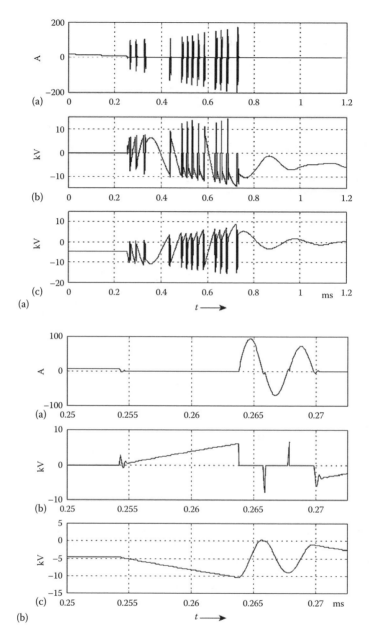

FIGURE 7.42

Example 7.6: Simulation results: (a) reignited current, (b) voltage across the vacuum circuit breaker, and (c) load voltage. (From Popov, M. et al., *Eur. Trans. Electr. Power*, 11, 413, November/December 2001. With permission.)

Post-arc Current Models

Black box modeling and arc resistance calculation for vacuum circuit breakers cannot rely on Cassie–Mayer-type equations. For many years, the post-arc current in vacuum circuit breakers has been modeled with the aid of Child's law [171], which was later modified by Andrews and Varey [172]. Their model is based on the kinetic motion of charged particles: the direction in which the movement of charged particles is controlled by the electric field, which implies that during the sheath growth, only ions arrive at the cathode, so the electrical current i can be written as

$$i = q_i n_i v_i A \tag{7.38}$$

where
   q_i is the average ion charge
   n_i is the ion density at the cathode surface
   v_i is the velocity of ions just before they arrive at the cathode
   A is the discharge area

The model assumes that the sheath is completely free of electrons, and the electric field at the plasma-sheath edge is zero, which makes this position the most convenient one to derive this equation. However, since the sheath grows into the plasma, stationary ions inside the plasma move through the sheath edge and an additional ion current flows through the sheath. The current at the sheath edge is derived as

$$i = q_i n_i A \left(v_i + \frac{ds}{dt} \right) \tag{7.39}$$

where
   v_i and n_i are now the ion velocity and density at the sheath edge
   ds/dt is the velocity of the sheath edge

According to the model proposed by Child, ions start at rest. However, as long as the sheath moves, Equation 7.39 implies that ions always have an initial velocity; therefore, that model cannot be used. Andrews and Varey found a way to include the initial velocity of ions [172]

$$s^2 = \frac{4\varepsilon_0 V_0}{9 q_i n_i} \left[\left(1 + \frac{v}{V_0} \right)^{3/2} + 3\frac{v}{V_0} - 1 \right] \tag{7.40}$$

where

$$V_0 = \frac{m_i}{2e} \left(v_i + \frac{ds}{dt} \right)^2 \tag{7.41}$$

where
   m_i is the ion mass
   v is the voltage
   V_0 is the ion potential at the sheath edge
   ε_0 is the dielectric constant of vacuum ($=8.854E{-}12\,F/m$)
   e is the electron charge ($=1.6E{-}19\,C$)

From Equation 7.39, the speed of the sheath length can be expressed as

$$\frac{ds}{dt} = \frac{i}{q_i n_i A} - v_i \tag{7.42}$$

The decay of the ion density n_i is assumed to be caused only by the post-arc current

$$\frac{dn_i}{dt} = -\frac{i}{q_i As}$$ (7.43)

where the product As is the volume of the discharge, with $A = \pi D^2/4$ and D is the discharge diameter.

Equation 7.40 is only valid in the second phase of the post-arc current. In the first phase, the voltage is assumed to be zero (perfect conduction). In the third phase, when the sheath thickness s is as long as the distance between contacts d, the current is assumed to have reached zero (perfect insulation). The transition from the first to the second phase is completed as soon as the current reaches the values obtained with Equation 7.38.

This model is based on Equations 7.40 through 7.43; although there are five unknown variables (s, v, i, n_i, V_0), the missing equation is the one that describes the external electric circuit. In addition, the model requires values for q_i, v_i, d, and A. q_i is usually taken as 1.8 times the elementary electron charge, which is the average charge value of copper ions that has been measured in vacuum arcs. The values for v_i, d, and A are generally assumed constant, and v_i is taken between 10^3 and 10^5 m/s, d is the contact distance, and A is a fraction of the anode area, which represents the possible constriction of the arc.

An exponential decay may be also assumed to model the ion density $n_i = n_i(0)e^{-t/\tau}$, which requires at least one additional parameter which represents the decay time constant τ. This value varies between 0.5 and 84 μs [173–177].

The following example uses a simple circuit to test the Andrews–Varey model for current interruption in a vacuum circuit breaker.

Example 7.7

Figure 7.43 shows the scheme of the circuit used to test the above arc model [178]. The main circuit consists of a capacitor C_L charged up to 25 kV, an adjustable air coil L, a trigger gap TG, and the test vacuum switch TS. C_E represents the capacitance of the test switch, which is permanently open. After the spark gap is fired, the voltage across C_E starts increasing. When the post-arc current density reaches a certain value the sheath starts growing and the current through C_E starts decreasing. The above model can be used from the moment the sheath starts growing from the cathode until the moment it reaches the anode, when the model loses its validity.

Circuit parameters are shown in the figure caption. Other parameters to be used in the model detailed above are $s = 0.1$ mm and $v_i = 1000$ m/s. The contacts are assumed to be circular and their diameter is 1 mm.

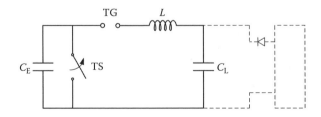

FIGURE 7.43
Example 7.7: Test circuit for post-arc current analysis in a vacuum circuit breaker. $C_L = 400$ pF, $C_E = 200$ pF, and $L = 40$ μH.

The equations of this circuit can be expressed as follows:

$$\frac{dv}{dt} = \frac{i_L - i}{C_E} \quad \frac{di_L}{dt} = \frac{V - v}{L} \tag{7.44}$$

where
i_L is the current through L
i and v are the current through C_E and the voltage across C_E, respectively
V is the voltage across C_L, which is assumed to be constant

An additional parameter, the electric field strength E_C at the cathode, can be calculated as follows [127]:

$$E_C = 2\sqrt{\frac{q_i n_i}{\varepsilon_0}} \sqrt{\sqrt{vV_0 + V_0^2} - V_0} \tag{7.45}$$

During arcing, ions move towards the anode, which becomes the new cathode after current zero. At $i = 0$, electrons move with the same speed of ions. As the post-arc current begins to rise, they become slower. In this case, the charge of electrons and ions is almost the same. Only when the post-arc current density becomes

$$j = q_i n_i v_i \tag{7.46}$$

the electrons reverse their direction. This is the instant of creation of the positive space charge sheath at the cathode and the starting point of the calculation according to the model previously described. Before this moment $v = 0$, and the capacitor current, which equals the inductor current, rises linearly and is calculated as follows:

$$i = \frac{V}{L} t \tag{7.47}$$

As already mentioned, the post-arc model is valid until the sheath reaches the new anode; thereafter, since the sheath cannot longer expand, the current drops to zero.

Calculated results are presented in Figure 7.44. Figure 7.44a shows the computed post-arc current and the TRV, while Figure 7.44b shows the ion density and the sheath expansion. As predicted in Ref. [127], the macroscopic electric field shows an early maximum during the post-arc flow. Figure 7.44c shows the calculated arc resistance; after approximately 0.3 µs, when the sheath reaches the new anode, the current drops to zero and the resistance automatically is set to a high value.

The above model simulates the growth of the cathodic space-charge sheath towards the post-arc anode; however, it lacks a proper theory for the voltage-zero phase, which can be explained, as detailed above when the thermal velocity of the particles is taken into account.

The electric field in front of a contact partly penetrates the quasineutral plasma. This field not only starts the deceleration of electrons, but it also accelerates ions in the direction of the sheath to velocities exceeding their thermal velocity. As a result, ions enter the sheath with Bohm velocity [163–165]:

$$v_B = \sqrt{\frac{kT_e}{m_i}} \tag{7.48}$$

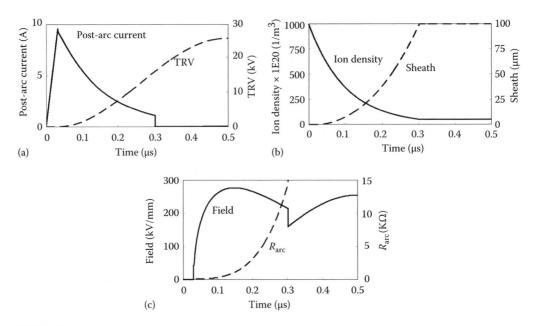

FIGURE 7.44
Example 7.7: Simulation results. (a) Post-arc current through C_E and TRV across the vacuum switch. (b) Ion density and sheath expansion. (c) Macroscopic field strength and arc resistance.

where
  k is the Boltzmann constant
  T_e is the electron temperature
  m_i is the ion mass

The ion saturation current is evaluated by

$$i_{i,sat} = q_i v_B n_i A \tag{7.49}$$

while the electron current i_e is found as

$$i_e = J_{e,sat} A \exp\left(\frac{q_i v_P}{k T_e}\right) \tag{7.50}$$

where
  $J_{e,sat}$ is the electron saturation current density
  v_P is the voltage drop across the sheath

Similar to the ion saturation current, the electron saturation current is reached when a contact potential becomes positive enough with respect to the plasma. In absolute values, it is much higher than the ion saturation current. The current is equal at both contacts, hence

$$i = i_e^a - i_{i,sat}^a = i_{i,sat}^c - i_e^c \tag{7.51}$$

where the superscripts a and c refer to the anode and cathode, respectively. Upon substitution of i_e from Equation 7.51 in Equation 7.50, and using for the voltage across the gap $v = v_P^a - v_P^c$, it results

$$\frac{i + i_{i,sat}^a}{i_{i,sat}^c - i} = \exp\left(\frac{q_i v}{kT_e}\right) \tag{7.52}$$

in which contacts are assumed to have equal size. This equation can be simplified to

$$i = i_{i,sat} \tanh\left(\frac{q_i v}{2kT_e}\right) \tag{7.53}$$

Equation 7.53 indicates that the total current is restricted to the ion saturation current at both contacts, and none will experience the much higher electron saturation current in this model.

The model needs to be upgraded for the post-arc period in order to estimate the variation of the postzero current. As the sheath grows in all directions, the plasma–sheath boundary grows, which effectively increases the cathode area. As a result, more ions are collected at the sheath boundary, and the ion-saturation current increases. For simplicity, it is assumed that the effective cathode area increases linearly with the voltage, thus

$$A_{ef}(t) = A + c_1 v \tag{7.54}$$

where c_1, in m²/V, is a parameter that needs to be determined.

The model assumes two components for the decay of the ion density after current zero. The first is a natural exponential decay, and the second is from the post-arc current. In this way, the ion density is defined as

$$\frac{dn_i}{dt} = -\left(\frac{n_i}{\tau} + \frac{i_{pa}}{q_i c_2}\right) \tag{7.55}$$

where
τ is the time constant for the natural decay
c_2, in m³, is a parameter

The post-arc current i_{pa} is considered to be positive only in Equation 7.55 to ensure an ion density decay.

The sheath capacitance for modeling the displacement current that results from the time change of the electric field inside the sheath is [157]

$$C_{sh} = \frac{K}{v^{3/4}} \tag{7.56}$$

where

$$K \approx 0.42 \left(q_i n_i \varepsilon_0\right)^{1/2} \left(\frac{kT_e}{q_i}\right)^{1/4} A_{ef} \tag{7.57}$$

By combining Equations 7.49, 7.53, 7.54, and 7.56 the post-arc current can be written as

$$i_{pa} = q_i \upsilon_B n_i A_{ef} \tanh\left(\frac{q_i \upsilon}{2kT_e}\right) + C_{sh}\frac{dv}{dt} \qquad (7.58)$$

This new model is described by Equations 7.54, 7.55, and 7.58. Since there are four unknown variables (A_{ef}, n_i, v, i_{pa}), the other equation is that which describes the external circuit.

If T_e, m_i, q_i, and A are considered constant, the model requires four parameters: $n_i(0)$, τ, c_1, and c_2. Although the above models are widely used to simulate the post-arc current in vacuum breakers, other models have been developed [127,179–181]. Most of them are based on the equations for particle motion between the contacts, similar to fluid dynamic equations with which the motion of particles in gases is modeled. When applied to plasma, those equations must be extended with the Poisson equation to account for the force that an electric field exhibits on the charged particles [161].

7.6.2.5 Examples

The following examples present the study of transients caused by the opening of circuit breakers in more complex systems.

Example 7.8

Arc instability in circuit breakers may occur only at high frequencies (from few tens up to hundreds of kilohertz) and near current zero (up to few tens of amperes). As the variable DC arc voltage excites all the circuit natural frequencies, arc instability happens at the frequencies where the amount of negative resistance it introduces cancels out the total frequency-dependent resistance of the circuit, thus creating a negatively damped current oscillation. It is of crucial importance to represent the surrounding network in details and through frequency-dependent models. This example demonstrates that complex problems can be studied by fine-tuning data and performing sensitivity analysis [160,182].

This example reports a repetitive failure of an SF_6 breaker (BR_9832 on phase a) during the opening of 100 MVAR shunt reactor in the 400 kV substation shown in Figure 7.45. Simulation results show that opposite-polarity high-frequency arc-instability-dependent oscillations, caused mainly by current transformers (CTs) on each side of the breaker, were responsible for its thermal failures and thus the noninterruption of the low 50 Hz reactor current by the 50 kA circuit breaker.

The whole substation is modeled in details using frequency-dependent models for bus bars and lines. The 150 and 500 VA CTs are modeled using a series inductance of 55 and 200 μH, respectively. Lead-to-ground capacitances of 0.25 nF and resistances are added in order to take into account their frequency dependence. The CT model is complex and represents a weak point in this study, due to uncertainties about the losses, mainly at those frequencies that govern arc instability. Parameters of the Cassie–Mayr model (Equation 7.23) were adjusted and checked for thermal capabilities under short-line faults using the test circuit of Figure 7.46. The fault is successfully cleared when the source inductance is 10 mH, but thermal failure occurs when 9 mH is used. The TRV shapes for both cases are shown in Figure 7.47, clearing at 28.9 A/μs and failing at 31 A/μs.

The study of the substation breaker must be started by performing a frequency scan for finding the poles and zeros of the network impedance seen by the arc. The initial simulations do not show a particular problem. Sensitivity analysis is performed by decreasing the cooling power of the arc,

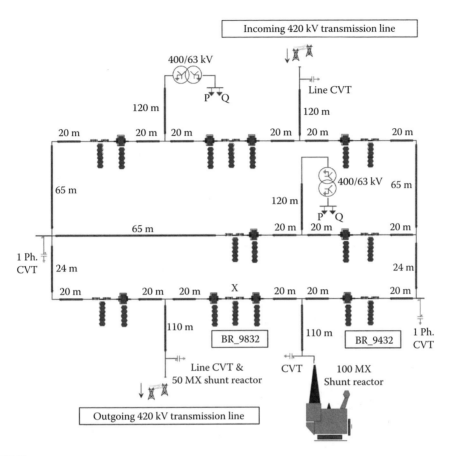

FIGURE 7.45
Example 7.8: Diagram of a 400/63 kV substation. (From Martinez, J.A. et al., *IEEE Trans. Power Deliv.*, 20, 2079, July 2005. With permission.)

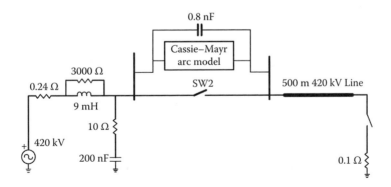

FIGURE 7.46
Example 7.8: Short-line fault test of a 420 kV circuit breaker. (From Martinez, J.A. et al., *IEEE Trans. Power Deliv.*, 20, 2079, July 2005. With permission.)

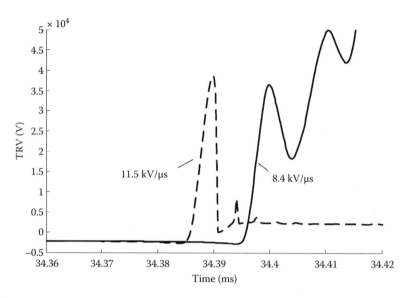

FIGURE 7.47
Example 7.8: Short-line fault strength of the modeled arc. (From J.A. Martinez, J.A. et al., *IEEE Trans. Power Deliv.*, 20, 2079, July 2005. With permission.)

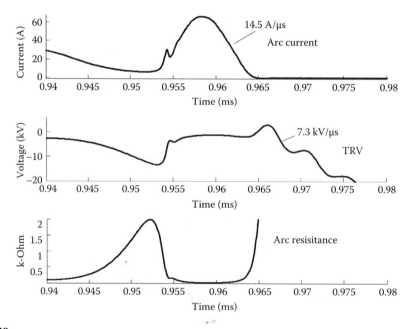

FIGURE 7.48
Example 7.8: Chopped current near current zero. (From Martinez, J.A. et al., *IEEE Trans. Power Deliv.*, 20, 2079, July 2005. With permission.)

which decreases the level of chopped current, without showing any special phenomenon. The simulations are based on increasing the damping of the CTs at high frequencies and those of the circuit elements (the two power transformers and the shunt reactor). These simulations exhibit a second arc instability at a higher frequency in the last current loop before current zero, see Figure 7.48.

The low-damped arc–circuit interaction at a frequency of 66 kHz excites a second arc instability. The current spike near 0.955 s corresponds to the instability created at 400 kHz due to the second low zero impedance of the circuit. This second high-frequency arc instability creates a high di/dt of 14.5 A/μs near current zero. A negative dynamic arc resistance of 500 Ω (corresponding to the circuit impedance at 400 kHz) is present for a short period of time [182].

The injected di/dt excites the 210 kHz parallel resonance (impedance main pole) seen across the breaker and creates a high dV/dt of 7.3 kV/μs. The fact that the arc produces a 0.4 A post-arc current after the current zero is another indication that the breaker is thermally stressed by this phenomena.

Although some of the parameters involved in the simulations were not too accurate, these results prove that the thermal failure of shunt-reactor circuit breakers can occur under certain conditions.

Example 7.9

When disconnecting an unloaded transformer by means of a vacuum circuit breaker, the transformer can be exposed to severe overvoltages due to virtual current chopping [183,184]. This happens when the induced high-frequency current in the uncleared pole is greater than the amplitude of the power-frequency load current, so chopping with a rather high current level (up to several tens of amperes) can occur. In general, the interruption of steady-state magnetizing currents is not a problem since the current amplitude is lower than the vacuum circuit breaker chopping current. This example analyzes the disconnection of a cable-fed unloaded transformer using detailed models of the system components [166]. Figure 7.49 shows the diagram of the test system. Details about each component are provided in the following paragraphs. The model and parameters of the vacuum breaker are the same that were used in Example 7.6.

Transformer Model
A transformer shows different behavior at different frequencies. At very high frequencies, the winding capacitances dominate and saturation can be neglected. At low frequencies, saturation and hysteresis cause losses and have a significant influence. The method used in this example to fit a transformer characteristic for a wide range of frequencies is presented in Ref. [185]. It is based on the two-port impedance, which can generally be expressed as

$$Z(s) = \frac{V(s)}{I(s)} = \frac{a_0 s^m + a_1 s^{m-1} + \cdots + a_{m-1}s + a_m}{b_0 s^n + b_1 s^{n-1} + \cdots + b_{n-1}s + b_n} \qquad (7.59)$$

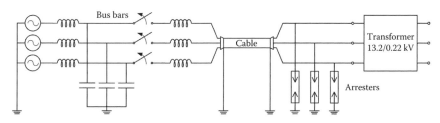

FIGURE 7.49
Example 7.9: Diagram of test system. (From Popov, M. et al., *Eur. Trans. Electr. Power,* 11, 413, November/December 2001. With permission.)

For transformers and motors, $m = n - 1$. The above expression can be also noted as follows:

$$Z(j\omega) = \sum_{i=1}^{n/2} \frac{\beta_i \delta_i + \alpha_i \gamma_i \omega}{\delta_i^2 + \gamma_i^2} + j \sum_{i=1}^{n/2} \frac{\alpha_i \delta_i \omega - \beta_i \gamma_i}{\delta_i^2 + \gamma_i^2} \qquad (7.60)$$

where

$$\delta_i = \sigma_i^2 + \omega_i^2 - \omega^2 \qquad (7.61a)$$

$$\gamma_i = 2\sigma_i \omega \qquad (7.61b)$$

Parameters α_i, β_i, σ_i, and ω_i describe a lumped-parameter Foster section and can be determined by the fitting procedure proposed in Ref. [185]. On the other hand, $n/2$ is the number of local maxima in the measured transformer characteristic.

The model used in this study is that depicted in Figure 7.50. A nonlinear inductance implies that the measurements of the two-port impedance at a different current level shift the impedance characteristic. Since measurements of the impedance characteristic were not available for the test transformer, the modified characteristic published in Ref. [186] is used. When saturation inductance is equal to zero (no Foster section for transformer core part), the other sections, which correspond to transformer windings, have their influence only for the higher frequencies and these values are closer to the values for different saturation inductances. Losses in this example are assumed to be constant although in reality they are frequency-dependent. The transformer core, represented as a nonlinear inductance and a parallel resistance in Figure 7.50, should therefore be modeled by means of parallel R–L elements, and the parameters should be calculated from the geometrical data and the permeability of the core [187]. The total no-load losses are eddy current and hysteresis losses, which must be accurately estimated because they are dampers to the system. For low-frequency switching, an estimation of the losses is done by a comparison of the used core model and a model with a dynamic hysteresis representation of the core [188].

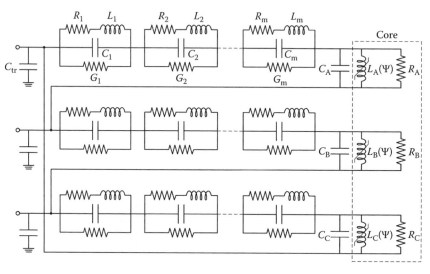

FIGURE 7.50
Example 7.9: Wideband transformer representation. (From Popov, M. et al., *Eur. Trans. Electr. Power*, 11, 413, November/December 2001. With permission.)

TABLE 7.8

Example 7.9: Transformer Magnetizing Curve

Current (A)	0	0.022	0.032	0.046	0.074	0.14	0.76
Flux (Vs)	0	34.66	39.62	44.56	49.52	54.4	59.4

Data for the three-phase, three-leg, Dyn, oil-filled transformer are as follows:

- Rated power   112.5 kVA
- Rated voltages   13.2/0.22 kV
- Rated frequency   60 Hz
- Iron core losses   350 W
- Winding losses   1.247 kW
- Primary/secondary leakage impedance ratio Z_0/Z_1 40%/60%
- Basic lightning insulation level   125 kV

The magnetizing curve is given in Table 7.8.

The magnetizing curve is approximated by a hyperbolic function of the form

$$\psi = A \cdot \tanh(Bi) + C \cdot \tanh(Di) + Ei \tag{7.62}$$

The constants are calculated by means of curve fitting. The values for which Equation 7.62 comes close to the measured curve are $A = 23.055$, $B = 13.948$, $C = 31.1566$, $D = 61.858$, and $E = 6.7973$.

Using the transformer model depicted in Figure 7.50, the resulting steady-state and inrush currents are those shown in Figures 7.51 and 7.52, respectively. Inrush currents have been obtained with two different initial phase angles. It is important to know the inrush currents for different initial phase angles if the instant at which the highest overvoltage occurs must be estimated.

Cable Model

The cross section of a conductor and insulation is shown in Figure 7.53. The sheaths of different phases are not interconnected. The losses of the mutual capacitance between the phases are ignored. Cable data are as follows:

- Inner and outer radius of the conductor r_1, r_2 0, 0.495 cm
- Inner and outer radius of the sheath r_3, r_4 1.2, 1.22 cm

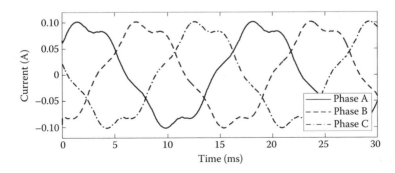

FIGURE 7.51

Example 7.9: Magnetizing currents. From Popov, M. et al., *Eur. Trans. Electr. Power*, 11, 413, November/December 2001. With permission.)

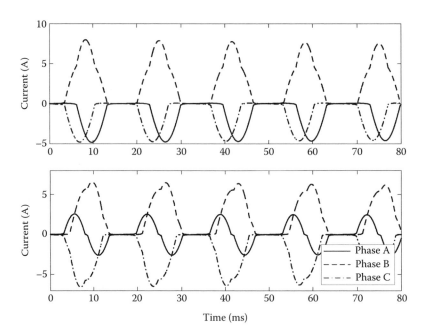

FIGURE 7.52
Example 7.9: Inrush currents: (a) initial phase = 0°; (b) initial phase = 30°. (From Popov, M. et al., *Eur. Trans. Electr. Power*, 11, 413, November/December 2001. With permission.)

- Outer radius of the outer insulator r_5 1.6 cm
- Resistivity of the conductor ρ_c 3.41E−8 Ω.m
- Permeability of the conductor ε_c 1
- Resistivity of the sheath ε_s 1.84E−8 Ω.m
- Permeability of the sheath μ_s 1
- Permeability of the inner and the outer insulator μ_{i1} and μ_{i2} 1
- Permittivity of the inner and the outer insulator ε_{i1} and ε_{i2} 3.02
- tan δ (at low frequencies)   0.01
- tan δ (at high frequencies)   <0.0006

The cable is modeled with pi sections, as shown in Figure 7.54. Each section has a length of 1 m. The dielectric losses of a cable are frequency-dependent. Since no information is available on how tan δ varies with the frequency, the dielectric losses are included by using its tan δ at power frequency and at a frequency of 1 MHz, making use of the relation tan δ = ωR′C. Measurements performed by Lindmayer and Helmer [189] confirm the validity of this model. On the other hand, its validity can be also checked by calculating the frequency response of the impedance and comparing it with the same frequency response derived from a more accurate frequency-dependent distributed-parameter model [166].

Surge Arrester Model
The effect of multiple reignitions and voltage escalation can be prevented by installing MO surge arresters [190]. Data for the arrester used in this example are as follows:

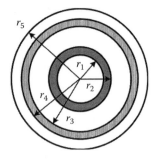

FIGURE 7.53
Example 7.9: Cross section of the cable. (From Popov, M. et al., *Eur. Trans. Electr. Power*, 11, 413, November/December 2001. With permission.)

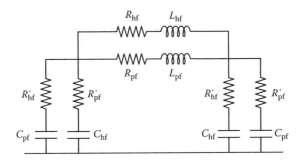

FIGURE 7.54
Example 7.9: Lumped-parameter representation of a single-phase cable section. (From Popov, M. et al., *Eur. Trans. Electr. Power*, 11, 413, November/December 2001. With permission.)

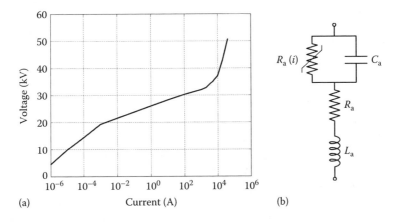

(a) (b)

FIGURE 7.55
Example 7.9: MO surge arrester model. (a) Measured voltage–current characteristic of the arresters. (b) Equivalent circuit. (From Popov, M. et al., *Eur. Trans. Electr. Power*, 11, 413, November/December 2001. With permission.)

- Rated voltage   12 kV
- Maximum continuous operating voltage   10 kV rms
- Discharge voltage for a 10 kA, 8/20 μs current   44 kV
- Discharge voltage for a 10 kA, 0.5 μs front time current   51.6 kV
- Discharge voltage for a 500 A, 30/60 μs current   32.3 kV
- Energy absorption   4.3 kJ/kV

The model used to represent surge arresters in this work is that proposed by Schmidt et al. [191]. Figure 7.55 shows the measured characteristic curve and the equivalent circuit, see also Chapter 6. The capacitance of the arrester block is nonlinear and depends on the rate of rise of the transient voltage wave; it is in the order of 1.5 nF·kV for distribution transformers and 5 nF·kV for station transformers. The parameters R_a and L_a represent the physical behavior of the ZnO grain. On the other hand, the arrester leads must also be included, and their inductances estimated from the distance of the arrester to the transformer terminals.

Arresters are attached to the transformer bushing by a copper cable of 10 cm, with the arrester grounding located on the transformer tank. This is equivalent to a winding loop with diameter of 20 cm. By making use of a special computer program, the inductance of this winding at 10 kHz is approximately 0.5617 μH.

The parameters used for the calculation with the arrester are $R_a = 0.6\,\Omega$, $L_a = 0.5\,\mu H$, and $C_a = 0.1\,nF$. These parameters were adjusted to match the arrester residual voltages for the 30/60 μs switching impulse, the 8/20 μs lightning impulse, and the 0.5 μs front-of-wave impulse.

Simulation Results

The switching overvoltages are simulated during steady-state and transient condition of the transformer. Because of the rather low magnetizing currents and the considerable core losses, the calculated overvoltages of this transformer do not reach a dangerous high value after switching steady-state currents and the vacuum circuit breaker can successfully withstand these TRVs. No reignition was noticed during the interruption of steady-state currents [166].

When the inrush currents are switched, a virtual current chopping occurs in almost all cases. This is because of the low value of the inrush currents compared with the high level of the reignited current in the circuit breaker. It is observed that the transformer capacitance plays an important role in the virtual current chopping process: the larger the transformer capacitance, the higher the reignited current. With the capacitance value used in computations, 500 pF, the reignited high-frequency currents are much higher than the power-frequency currents, which is the reason why virtual current chopping occurs. Figure 7.56 shows an example of overvoltages at transformer terminals obtained with a cable length of 30 m.

7.6.3 Circuit Breaker Modeling during Closing Operations

A circuit breaker is a mechanical device with a voltage-withstand capability between contacts that depends on the separation between them. As the contacts of a breaker close and the gap between them gets smaller, breakdown will occur if the voltage across the gap exceeds its dielectric strength. In other words, electrical closing can happen before mechanical closing. Figure 7.57 shows the prestrike phenomenon at the time the stress exceeds the strength [69].

An effect of prestrikes is that the probability distribution of closing instants will not be uniformly distributed, as shown in Figure 7.58 [69]. For a multiphase breaker, the

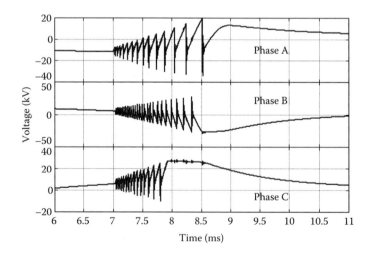

FIGURE 7.56

Example 7.9: Simulated voltages at transformer terminals in all phases after inrush current switching: (a) voltage in phase A, (b) voltage in phase B, and (c) voltage in phase C. (From Popov, M. et al., *Eur. Trans. Electr. Power*, 11, 413, November/December 2001. With permission.)

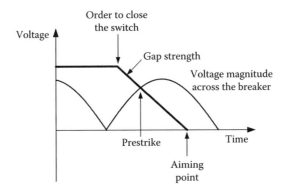

FIGURE 7.57
Prestrike phenomenon.

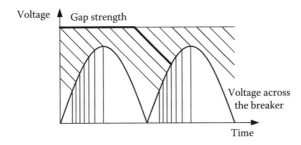

FIGURE 7.58
Distribution of the closing time during prestrike.

probability distribution of closing times is more complicated to determine since the order in which breaker poles close will be random in a cycle.

For some studies it may be of interest to represent also the prestrike arc conductance. Since an oscillatory arc current at a very high frequency may be caused during prestrike, such current may result in multiple quenchings and prestrikes prior to the ultimate contact.

A closing operation can produce transient overvoltages whose maximum peaks depend on several factors, for instance the network representation on the source side of the breaker or the charge trapped on transmission lines in a reclosing operation. One of the factors that have more influence on the maximum peak is the instant of closing, which can be different for every pole of a three-phase breaker. Most transients programs allow users to analyze the influence of this factor and obtain a statistical distribution of switching overvoltages, usually provided in the form of an accumulative frequency distribution. Therefore, several models can be used to represent a circuit breaker in closing operations [192]:

- The simplest model assumes that the breaker behaves as an ideal switch whose impedance passes instantaneously from an infinite value, when open, to a zero value at the closing time. This performance can be represented at any part of a power cycle. The effect of the operation (e.g., switching overvoltage) depends on the closing time of the breaker. In multiphase systems an additional factor is added since all poles do not close simultaneously. A further improvement of this approach assumes that the closing instant for either single- or multiphase breakers is randomly determined.

- A more advanced approach assumes that there is a closing time from the moment the contacts start to close to the moment that they finally make. The withstand voltage decreases as the separation distance between contacts decreases, an arc will strike before the contacts have completely closed if the voltage across them exceeds the withstand voltage of the dielectric medium. Modeling of the prestrike effect and its influence on the switching overvoltages produced during line energization has been analyzed in Ref. [193].

- As for the study of opening operations, a third approach includes the prestrike dynamic arc conductance.

The two first approaches are presented in Sections 7.6.3.1 and 7.6.3.2, in which some practical examples have been included to illustrate the application of circuit breaker models during closing operations.

7.6.3.1 Statistical Switch Models

The calculation of switching overvoltages is fundamental for the insulation design of many power system components. However, equipment insulation design is not usually based on the highest overvoltage because this particular event may have a low probability of occurrence, and the design would not be economical. Besides in many cases, it can be very difficult to determine the combination of parameters that will produce the highest overvoltage. The insulation design of power system equipment is based on the concepts stress and strength. The probability distribution of switching overvoltages (stress) is to be compared to equipment insulation (strength) which is also described statistically; the goal is to choose the insulation level so as to achieve a failure rate criterion.

Power system equipment strength can be described by a cumulative distribution, $P(V)$, which is the probability of failure of the insulation under a standard impulse of value V. Overvoltage stress is described by a probability distribution $f(V)$. The risk of failure is the probability that the stress exceeds the strength, and it can be obtained by means of the following expression [194]:

$$R = \int_0^\infty f(V)P(V)\mathrm{d}V \tag{7.63}$$

Figure 7.59 shows this calculation graphically, in which the overvoltage stress is described by a normal distribution.

The concept of statistical switch has been developed and implemented into transient programs to facilitate the calculation of the statistical distribution of switching overvoltages Two approaches can be considered [195]:

- The closing time is systematically varied from a minimum to a maximum instant in equal increments of time; this type of switch is known as *systematic* switch. Figure 7.60 shows the combinations that can correspond to a statistical calculation using a three-phase systematic switch. An accurate evaluation of the switching overvoltage probability distribution using these type of switches can be very laborious. If the closing operation is performed over the entire range of a cycle and a

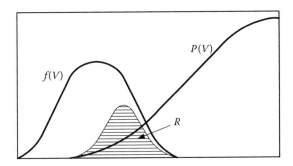

FIGURE 7.59
Stress–strength diagram.

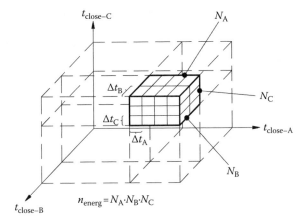

FIGURE 7.60
Three-dimension space for a systematic switch.

time increment of 0.1 ms is chosen, then the number of energizations per phase in a 50 Hz system is 200, and the total number is (200*200*200) = 8E+6.

- The closing time is randomly varied according to a given probability distribution; this type of switch is known as *statistics* switch. Data required to represent these switches are the mean closing time and the standard deviation. The closing time of a statistics switch is randomly selected according to a probability distribution, which is usually uniform or normal (Gaussian), see Figures 7.61 and 7.62. In general it is assumed that the closing of an independent breaker pole may occur at any point of the power-frequency cycle with equal likelihood.

A three-phase breaker can be represented as three single breakers each with an independent probability distribution. So the breaker is then represented as a master switch and two dependent switches whose mean closing times are dependent on the first pole to close

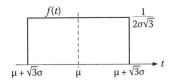

FIGURE 7.61
Uniform distribution.

$$T_{close(dependent)} = T_{close(master)} + T_{random} \qquad (7.64)$$

where T_{random} is the time delay, which is randomly deduced from an average time and a standard deviation. However, the three poles can be mechanically linked so that each pole attempts to close at the same instant, the aiming point. In reality, there will be a finite time or pole span between the first and the last pole to close. For instance, if the closing is described by a normal distribution, a random generation according to a normal distribution is performed to obtain the closing time from the aiming point, as shown in Figure 7.63. The normal distribution curve ranges from $-\infty$ to $+\infty$, but it can be truncated at -4σ and $+4\sigma$; in such case the pole span is 8 times the standard distribution.

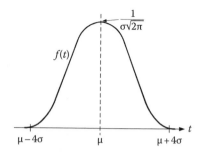

FIGURE 7.62
Normal (Gaussian) distribution.

If preinsertion resistors are used to mitigate switching overvoltages, the closing times of both main and auxiliary contacts are statistically determined, using a dependency similar to that presented above. Figure 7.64 shows a way to determine the closing times assuming that main contacts aim at the same closing time, both switches have a normal distribution, and the main contacts are those that close first. Another way to obtain the closing times is to consider that the main contacts are the latest to close [38].

If systematic switches are used to represent a circuit breaker with preinsertion resistors, the closing times of the auxiliary contacts are determined as follows:

$$T_{close(dependent)} = T_{close(master)} + T_{offset} \tag{7.65}$$

where T_{offset} is now a constant value.

The following example shows how to use statistical switches, what number of runs must be usually considered and what differences should be expected from different types of statistical distributions.

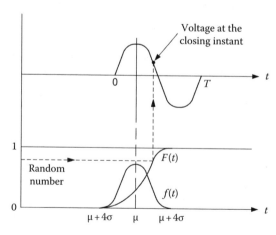

FIGURE 7.63
Selection of aiming point and closing time.

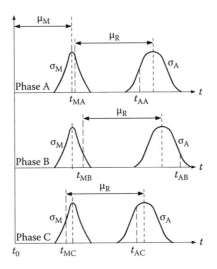

FIGURE 7.64
Closing times of main and auxiliary contacts.

Example 7.10

Figure 7.65 shows the schematic diagram of the test case. The parameters of the network equivalent and of the overhead line are given at 50 Hz.

Network equivalent:	Voltage	Peak value = 1 V
	Series impedance	$Z = 5.8 + j84\ \Omega$
Transmission line:	Length	$\ell = 250\,\text{km}$
	Sequential impedances	$Z_1 = (0.0371 + j0.3324)\ \Omega/\text{km}$
		$Z_0 = (0.3267 + j0.9398)\ \Omega/\text{km}$
	Charging capacitances	$C_1 = 11.10\,\text{nF/km}$
		$C_0 = 7.84\,\text{nF/km}$

The different parts of this example will be aimed at determining under what conditions the energization of the test line causes the maximum overvoltage, finding out the maximum overvoltage, and estimating the number of needed runs by using both systematic and statistical switching. A further study is included to compare the results derived from assuming different distributions for statistical switches.

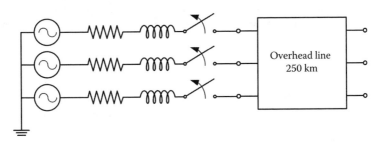

FIGURE 7.65
Example 7.10: One-line diagram of the test system.

The transmission line is represented in this study by means of a constant and lumped-parameter model, which is not the most accurate representation for an overhead line [68,196]. The main objective of the example is to present a methodology that could help to select the most adequate switch parameters for a statistical calculation of switching overvoltages.

Maximum Overvoltage during Energization
Usually unloaded systems are considered for calculation of switching overvoltages. However, under certain conditions, overvoltages caused by single-line-to-ground faults should be also considered in order to deduce the case for which the overvoltages are maxima. In this example it is assumed that the maximum overvoltage is produced when energizing the open line. Therefore, only this case is analyzed in this study.

Assuming that only phase-to-ground overvoltages are of concern, a three-phase closing operation produces three phase-to-ground overvoltages. Two methods can be used to obtain data from each operation [194]:

- Phase-case method, the highest peak value of the overvoltage of each phase-to-ground is taken into account.
- Case-peak method, only the highest peak value of the overvoltages is collected.

Using the first method, each operation contributes three values, while only one value is collected with the second approach. The results presented in the following example are based on the second method, without considering reclosing.

The phase-to-ground overvoltage magnitudes at various closing times are calculated over the duration of one 50 Hz cycle, assuming a cosine wave for phase A and the corresponding displaced voltage waves for the phases B and C. The resulting overvoltage distribution is presented in Table 7.9. One can note that the maximum overvoltage occurs when the breaker is closed before the voltage wave of phase C is at the peak. It can be also seen that the maximum overvoltages are produced when the poles are closed at 22 and 32 ms. Therefore, if the circuit breaker timing is chosen anywhere in between them the estimated overvoltages may not be realistic. Based on this result, the circuit breaker aiming time is chosen as 22 ms for further analysis.

Systematic Switching
In order to find out the maximum overvoltage by using a systematic switch, the energizations were performed over the entire range of a cycle, from 0° to 360°. With $\Delta t = 0.1$ ms, 20 steps in phase A, 10 steps in phase B, and 5 steps in phase C are chosen. The t_{mid} is chosen as 22 ms for all the phases. This requires a total energization of 1000 runs. The distribution of maximum overvoltages is presented in Figure 7.66.

In order to estimate the number of runs that are necessary to evaluate the maximum overvoltage, the results derived from 500 runs and 1000 runs are compared in Table 7.10. From this table, it can be seen that the selection of the closing time or the aiming point of the circuit breaker helps to obtain overvoltages above 2.45 pu. Therefore, if the circuit breaker closing time is chosen at a different location, the calculated overvoltages will be lower. On the other hand, the probability of capturing the maximum overvoltage during energization increases as the number of runs increases.

Statistics Switching
The energizations were again performed over the entire range of a cycle, which corresponds to 0–20 ms, and assuming that the three poles are independent. The parameters required to perform the simulations are the point of circuit breaker closing (aiming point of 22 ms) and the standard deviation, σ. To cover the entire range of a 50 Hz cycle, the standard deviation for a uniformly distributed switch must be $20/(2\sqrt{3}) = 5.77$ ms , see Figure 7.61. If a normal distribution is truncated at -4σ and $+4\sigma$, see Figure 7.62, more than 99.99% of the switching times will range within a cycle centered at the aiming instant; therefore, for a 50 Hz cycle, the selected standard deviation is 20/8 = 2.5 ms.

TABLE 7.9

Example 7.10: Magnitude of Overvoltages due to Simultaneous Pole Closing

Closing Time (ms)	Overvoltages (pu)			Maximum (pu)
	Phase A	Phase B	Phase C	
0.020	1.9881	1.7663	1.8060	1.9881
0.021	1.8353	1.5675	2.0669	2.0669
0.022	1.8716	1.5130	2.1632	2.1632
0.023	1.8160	1.7015	2.0644	2.0644
0.024	1.6362	1.9951	1.8356	1.9951
0.025	1.4902	2.1522	1.8519	2.1522
0.026	1.6016	2.1193	1.8515	2.1193
0.027	1.9068	1.8926	1.7049	1.9068
0.028	2.1198	1.8528	1.5125	2.1198
0.029	2.1523	1.8710	1.5479	2.1523
0.030	1.9881	1.7663	1.8060	1.9881
0.031	1.8353	1.5675	2.0669	2.0669
0.032	1.8716	1.5130	2.1632	2.1632
0.033	1.8160	1.7015	2.0644	2.0644
0.034	1.6362	1.9951	1.8356	1.9951
0.035	1.4902	2.1516	1.8518	2.1516
0.036	1.5988	2.1206	1.8523	2.1206
0.037	1.9039	1.8953	1.7069	1.9039
0.038	2.1185	1.8526	1.5138	2.1185
0.039	2.1530	1.8713	1.5469	2.1530

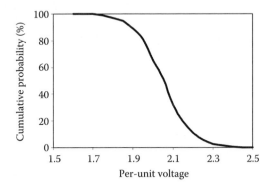

FIGURE 7.66
Example 7.10: Peak voltages with a systematic switch.

In order to estimate the number of runs necessary to evaluate the maximum overvoltage, the first study compares the overvoltages obtained with 200 runs and 1000 runs. The results are shown in Table 7.11. It can be seen that although there is a higher probability of capturing the highest possible overvoltage with a normal distribution and 1000 runs, the differences are not too large, and 200 runs provide an accurate enough estimation using either distribution. Figures 7.67 and 7.68 show the statistical distribution of overvoltages obtained with the two distributions when the total number of energizations is 1000 in both cases.

TABLE 7.10

Example 7.10: Overvoltage Distribution with
Systematic Switching

	Percentage of Surge Count Greater than the Level Indicated	
Overvoltage Limit (pu)	1000 Runs	500 Runs
1.60	100	100
1.65	99.9	100
1.70	99.8	99.6
1.75	99.0	98.8
1.80	97.1	94.6
1.85	94.7	91.4
1.90	89.4	82.4
1.95	81.6	68.8
2.00	64.7	55.0
2.05	51.3	40.6
2.10	31.8	21.0
2.15	19.4	9.6
2.20	11.0	3.8
2.25	5.9	1.0
2.30	2.7	0.4
2.35	1.4	0
2.40	0.7	0
2.45	0.2	0
2.50	0	0

A further study was aimed at analyzing the effect that the standard deviation value can have on a normally distributed switch. The values used in the study for σ ranged from 0.5 to 2.5 ms. The overvoltage distributions derived from each σ value are shown in Table 7.12.

From the results presented in this table is easy to deduce that a larger standard deviation helps to obtain, as expected, a broader spectrum of overvoltage values and therefore the highest overvoltages.

The results presented in this study can be summarized as follows. If the circuit breaker poles are assumed to be independently distributed, the maximum overvoltages are derived by selecting either an uniform or a normal distribution of closing times; however, in any case it is recommendable to choose a standard deviation that guarantees that the entire range of closing times within a cycle will be covered. Finally, differences will not be significant if the number of runs is above 200.

It is interesting to observe that both systematic and statistics switches, as used in this example, have captured the same maximum overvoltage when using the same number of runs.

7.6.3.2 Prestrike Models

For most purposes the prestrike phenomenon can be represented as an ideal switch for which the only information of concern is the closing time of each pole. This model can be represented using built-in capabilities in most transients programs. Figure 7.69 shows a possible representation in which the circuit breaker is modeled as an ideal switch whose operation is controlled by the voltage across terminals and taking into account the time variation of the gap strength [69].

An application of this approach is presented in the following example.

TABLE 7.11

Example 7.10: Overvoltage Distribution with Statistic Switching

Overvoltage Limit (pu)	Percentage of Surge Count Greater than the Level Indicated			
	Uniform (σ = 5.77 ms)		Normal (σ = 2.5 ms)	
	200 Runs	1000 Runs	200 Runs	1000 Runs
1.60	100	100	100	100
1.65	100	100	99.5	99.9
1.70	99.5	99.8	99.5	99.4
1.75	99.0	99.4	99.5	98.6
1.80	98.0	97.3	98.5	97.2
1.85	94.0	93.9	95.0	95.2
1.90	90.5	89.6	92.0	90.5
1.95	86.0	83.9	86.5	85.0
2.00	74.0	72.5	75.5	73.7
2.05	58.5	58.5	58.5	59.9
2.10	43.0	41.7	45.0	44.2
2.15	26.0	24.9	28.5	26.4
2.20	15.5	15.4	13.0	15.8
2.25	7.0	9.5	7.5	8.6
2.30	5.0	3.9	3.0	4.2
2.35	3.5	1.8	1.0	1.1
2.40	2.5	1.0	0	0.5
2.45	0.5	0.3	0	0.1
2.50	0	0	0	0

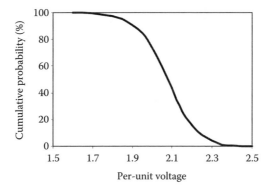

FIGURE 7.67
Example 7.10: Peak voltages with a normal distribution—1000 runs.

Example 7.11

Very fast front transients within a gas-insulated substation (GIS) are usually generated by disconnect switch operations [197,198]. A large number of pre- or restrikes can occur during a disconnector operation due to the relatively slow speed of the moving contact [199]. Figure 7.70 shows a very simple circuit that has been often used to explain the general switching behavior and the pattern of voltages

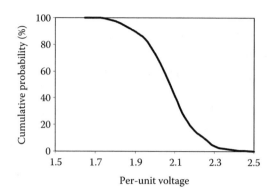

FIGURE 7.68
Example 7.10: Peak voltages with a uniform distribution—1000 runs.

TABLE 7.12

Example 7.10: Overvoltage Distribution: Normal Switch, 200 Runs

Surge Limit (pu)	Percentage of Surge Count Greater than the Level Indicated				
	σ = 2.5 ms	σ = 2.0 ms	σ = 1.5 ms	σ = 1.0 ms	σ = 0.5 ms
1.60	100	100	100	100	100
1.65	99.5	99.5	100	100	100
1.70	99.5	99.5	100	100	100
1.75	99.5	99.0	99.5	100	100
1.80	98.5	99.0	99.5	100	100
1.85	95.0	98.5	98.0	98.5	99.0
1.90	92.0	91.5	91.5	96.5	95.5
1.95	86.5	85.5	81.5	92.0	90.5
2.00	75.5	76.0	73.0	81.5	80.0
2.05	58.5	60.5	52.5	59.0	61.0
2.10	45.0	45.5	33.5	36.5	37.0
2.15	28.5	24.5	15.5	13.5	10.0
2.20	13.0	15.5	9.0	4.5	2.0
2.25	7.5	7.0	4.5	2.0	1.5
2.30	3.0	3.0	0	0.5	0
2.35	1.0	1.0	0	0	0
2.40	0	0	0	0	0

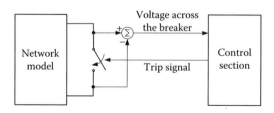

FIGURE 7.69
Modeling of a circuit breaker prestrike.

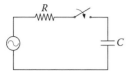

FIGURE 7.70
Example 7.11: Scheme of the test circuit.

on opening and closing of a disconnector [200]. During a closing operation, sparking occurs as soon as the voltage between the source and the load exceeds the dielectric strength across contacts. After a restrike, a high-frequency current will flow through the spark and equalize the capacitor voltage to the source voltage. The potential difference across the contacts will fall and the spark will extinguish. The subsequent restrike occurs when the voltage between contacts reaches the new dielectric strength level which is determined by the speed of the moving contact and other disconnector characteristics; e.g., polarity. The load side voltage will follow the supply voltage until the contact-make.

Figure 7.71 presents the simulation results obtained with two different values of the moving contact speed. The plots show the voltages at the source and the capacitor during the closing, as well as the dielectric strength and voltage across the switch.

Different results could be obtained by assuming that the connection of the capacitor is made with a trapped charge. On the other hand, a more rigorous model of the disconnector should consider an asymmetrical behavior since the dielectric strength is different for positive and negative voltages across the contacts. A detailed discussion of the physics involved in the restrikes and prestrikes of a disconnect switch operation is presented in Ref. [200].

A more sophisticated model may include the arc conductance and the possibility of multiple quenchings and restrikes prior to the ultimate metallic contact if during prestrike a very high-frequency oscillatory current is induced.

7.6.4 Discussion

This section has presented a summary of circuit breaker modeling techniques considering a black box approach. General features of these models can be summarized as follows:

- Arc models for gas-filled circuit breakers (air, SF_6) are only applicable for the thermal period of current interruption, and they link the arc conduction to the internal energy. The arc is described by a differential equation relating arc conductance with arc voltage and current. The equation contains some free parameters which are determined by tests. These models can be only used for simulating short-line fault interruption and current chopping when breaking small inductive currents. However, some models have also been proposed to represent arc conductivity during the dielectric period. Other mathematical formulations different from those presented in this chapter have been developed and applied, see for instance [87–89,98,99,105,106,109,111,112,116,121,122].

- Vacuum circuit breakers are of a different nature than high-pressure arcs such as those in SF_6 and other gas isolated circuit breakers; post-arc vacuum breaker models analyze the behavior of the cathodic space charge sheath after current zero.

The ideal switch approach can be used to obtain TRV, which can be postprocessed by comparing this result with the withstand voltage capability of the breaker and can be useful to predict whether the breaker will fail or not. This model cannot be used to analyze the effect that a breaker failure can have on the system. A model including a precomputed TRV characteristic can be used to reproduce the interaction, analyze the effect that a breaker failure can have on the system, and predict a dielectric failure without any further postprocessing. A more elaborate model can predict whether there will be arc extinction or arc reignition; such model will also reproduce the interaction between the breaker and the system.

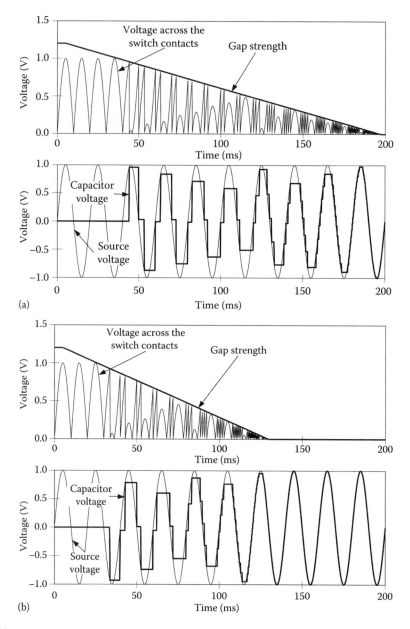

FIGURE 7.71

Example 7.11: Variation of load and source side voltages during disconnector closing. (a) Simulation results—base case. (b) Simulation results—the moving contact speed has been increased.

Although in some cases it is feasible to simulate arc model equations using control diagram blocks and sufficiently small time steps, the best solution is achieved using a hard-coded arc model with sophisticated iterative techniques, as explained in Ref. [108]. A highly nonlinear arc model requires a small integration time step to correctly account for its time constants and achieve nonlinear solution method convergence. Such a small time step (0.1 μs typically) requires a prohibitive computer time for statistical studies. Another difficulty is the availability of parameters from the breaker manufacturer. This is why a simplified arc model must be used in statistical studies [201,202]. Such a model combines a current–arc voltage characteristic $v_{arc}(i_{arc})$ per break and a time-dependent function $c_{arc}(t)$ to account for blast pressure, arc length, and arcing time. This simplified model cannot account for thermal reignition, but it can correctly compute the breaker's arcing time with a larger integration time step (typically 70 μs). The functions $v_{arc}(i_{arc})$ and $c_{arc}(t)$ must be provided by the manufacturer.

7.7 Parameter Determination

7.7.1 Introduction

The application of an arc modeling technique may consist of the following steps [67]: (1) choice of model equations; (2) tests with measurement of voltage and/or current during the period of interaction between arc and circuit; (3) evaluation of arc parameters; and (4) numerical simulations of the interruption processes in various circuits.

As for the practical application of a selected model, it may consist of two major parts [67]:

1. *Parameter determination*: Field tests provide traces of arc voltage and current. The analysis of these results is based on a specific arc equation and additional assumptions regarding particular parameters functions. The main objective of this analysis is the evaluation of arc parameters.

2. *Numerical experiments*: The interaction between circuit and arc is calculated for different system conditions by numerical simulation using the arc model.

A specific method of arc parameter evaluation has to be made in close connection with the selected arc model and the measurements to be made. An increasing number of parameters improve the model adaptability, but it also requires more information from measurements and does not necessarily provide a more accurate prediction.

The rest of this section is divided into two parts (Sections 7.7.2 and 7.7.3) dedicated to analyze the determination of arc parameters for gas-filled and vacuum circuit breakers, respectively.

7.7.2 Gas-Filled Circuit Breakers

7.7.2.1 Introduction

Black box models cover the thermal period of the switching process, the duration of which depends on the switching process itself and on the arc quenching medium. Typical values of the characteristic quantities for air blast and SF_6 circuit breakers are given in Table 7.13.

In general arc parameters are evaluated from voltage and current traces. Therefore, the accuracy of the modeling depends to a great extent on the resolution and accuracy of these measurements; moreover, the type of arc equation and the specific procedure of parameter evaluation are crucial, in particular if the quotient of voltage and current, or even the time derivative of arc conductance, are used. Finally, the measurement accuracy must be validated together with the whole modeling procedure. Achieving an adequate accuracy is mainly a resolution and sensitivity problem since the measurement devices have to withstand large current amplitudes before current zero and high voltage amplitudes after current zero.

TABLE 7.13

Quantities Characterizing the Thermal Period of a Switching Process

Switching Process	Characteristic Quantities	Air Blast	SF$_6$
Short-line fault	Thermal period duration	25 μs	5 μs
	Current range before zero	500 A	0 A
	Post-arc current peak value	20 A	2 A
	Voltage range after zero	40 kV	20 kV
Interruption of small inductive currents	Thermal period duration	2 μs	1 μs
	Current range before zero	100 A	50 A
	Voltage range after zero	10 kV	7 kV

Source: CIGRE Working Group 13.01, Applications of black box modelling to circuit breakers, *Electra*, no. 149, pp. 40–71, 1993. With permission.

The stochastic behavior of dynamic arcs deserves some special attention. Since stochastic arc models have been hardly applied to date, an average result must be deduced from the stochastic arc behavior. This can be made in several ways [67]:

- Averaging several current and voltage traces obtained under similar or identical conditions, provided that all tests lead to interruptions or reignitions, and evaluating arc parameters from averaged traces.

- Evaluating arc parameters from individual current and voltage traces, and averaging the calculated parameter functions.

- Calculating the interruption limits from parameter functions evaluated from individual tests. In this case, not only the average but also the distribution of the interruption limit may be calculated.

Therefore, the required number of experiments depends on both the number of degrees of freedom of the model and the method of treatment of the stochastic behavior.

7.7.2.2 Parameter Determination Methods

Under the condition that the parameter functions are evaluated in a mathematically unique way from arc current and recovery voltage measurements derived from a normal circuit breaker test, arc equations with two parameter functions will generally suffice. A larger number of free parameters require additional assumptions, the validity of which has to be separately proven. Any other possibility may lead to nonunique solutions.

A description of some of the methods developed to date for parameter estimation of gas-filled circuit breaker models is presented in the following paragraphs [67,203].

Amsinck's method [204]: It is suitable to evaluate parameters that are functions of the same arc quantity, either current or conductance. These parameters result from the solution of a set of two model equations at the same value of this arc quantity, but at different values of power input and time derivative of conductance.

Consider that the arc behavior is described by means of the Mayr–Schwarz equation, in which the time constant τ and the power loss P are functions of the arc conductance:

$$\frac{1}{g}\frac{dg}{dt} = \frac{1}{\tau(g)}\left(\frac{vi}{P(g)} - 1\right) \tag{7.66}$$

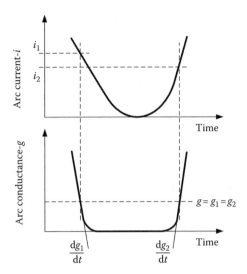

FIGURE 7.72
Application of the Amsinck's method during a reignition.

This method uses the information obtained during reignition. Figure 7.72 shows its principles. From the application of the arc equation to two points of the arc conductance with the same value, see figure, the following relationships are obtained:

$$\frac{dg_1}{dt} = \frac{1}{\tau(g_1)}\left(\frac{i_1^2}{P(g_1)} - g_1\right)$$ (7.67a)

$$\frac{dg_2}{dt} = \frac{1}{\tau(g_2)}\left(\frac{i_2^2}{P(g_2)} - g_2\right)$$ (7.67b)

where

$$\frac{dg_k}{dt} = \frac{dg}{dt}\bigg|_{t=t_k} \quad i_k = i(t_k)$$ (7.68)

Solving for the time constant and the power loss

$$\tau(g) = \frac{g\left(i_2^2 - i_1^2\right)}{i_1^2 \dfrac{dg_2}{dt} - i_2^2 \dfrac{dg_1}{dt}}$$ (7.69a)

$$P(g) = \frac{i_1^2 \dfrac{dg_2}{dt} - i_2^2 \dfrac{dg_1}{dt}}{g\left(\dfrac{dg_2}{dt} - \dfrac{dg_1}{dt}\right)}$$ (7.69b)

where $g = g_1 = g_2$.

This step is repeated several times along the conductance curve to obtain the functions $\tau(g)$ and $P(g)$.

To analyze successful interruptions, two different measurements of different current steepness are necessary.

Rijanto's method [205]: It is based on the fact that the parameter functions may be deduced from singular points of the dynamic arc traces. Consider again the Mayr–Schwarz equation:

- If $dg/dt = 0$, then

$$P(g) = vi \tag{7.70}$$

which allows the determination of P.

- If $i = 0$ or $v = 0$, then

$$\frac{1}{g}\frac{dg}{dt} = -\frac{1}{\tau(g)}\left(\tau(g) = -\frac{g}{dg/dt}\right) \tag{7.71}$$

which allows the determination of τ.

That is, to obtain dynamic arc parameter functions, several singular conditions in arc current/voltage (zero crossings) and arc conductance (minima/maxima) are required. Such conditions are found several times in a current chopping; for short-line fault interruption they have to be forced by using special test circuits.

Ruppe's method [206,207]: The Mayr equation can be rewritten as follows:

$$vi = P_0\left(\frac{1}{g}\frac{dg}{dt}\tau + 1\right) \tag{7.72}$$

which leads to a straight line if the product vi is plotted versus $(dg/dt)/g$. The intersection of this line with the two axes is P_0 on the vi-axis and $-1/\tau$ on the $(dg/dt)/g$-axis, see Figure 7.73.

Several conditions with the same arc conductance value, as for the Amsinck's method, can be deduced from a high number of traces. Using regression these points can be approached by a straight line with a constant arc conductance value, and then obtain the corresponding values of P_0 and τ. Functions $P_0(g)$ and $\tau(g)$ can be derived by repeating the procedure with other values of g.

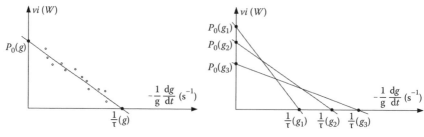

FIGURE 7.73
Application of the Ruppe's method.

An advantage of this method is that it does require data from only reignitions or from only the application of synthetic test circuits.

Application of curve-fitting techniques: An approach based on curve-fitting technique allows the formulation of parameters functions in a concise manner. Several methods using current and voltage records have been proposed for determination of arc parameters. Some of them are summarized in the following paragraphs.

St-Jean et al. developed a testing and measuring setup that was applied to one air-blast and two SF$_6$ breakers [92], whose behavior was assumed to be according to the modified arc resistance model:

$$\frac{dr}{dt} = \frac{r}{\tau(g)}\left(1 - \frac{vi}{P(g)}\right) \qquad (7.73)$$

where

$$\tau(g) = Ar^\alpha \quad P(g) = Br^\beta \qquad (7.74)$$

Therefore, the unknown parameters of this model are four: A, B, α, and β. Assume the error function is defined as follows [92]:

$$\int\left(\frac{dr}{dt} + \frac{vir}{\tau(g)P(g)} - \frac{r}{\tau(g)}\right)dt = \varepsilon \qquad (7.75)$$

Experimental data are then divided into n sections for integration purposes, and the total error is given by

$$\text{Total error} = \sum_{i=1}^{n} \varepsilon_i^2 \qquad (7.76)$$

where ε_i is the error of section i.

The best results are obtained when the total error is divided by the factor r^2, r being the arc resistance. This operation reduces this error in the region of high arc resistance and extremely low arc current, which can originate inaccurate calculations of the resistance value.

A similar approach was previously developed by Blahous [208], who established a procedure for determination of parameters of the Urbanek model defined by the following equation:

$$\frac{1}{g}\frac{dg}{dt} = \frac{1}{\tau}\left(\frac{vi}{e^2 g} - 1 - \frac{P_0}{e^2 g}\left(1 - \left(\frac{v}{v_d}\right)^2 - 2\frac{\tau}{v_d^2}v\frac{dv}{dt}\right)\right) \qquad (7.77)$$

The parameters that define this model are the minimum power input to maintain the arc, P_0, the voltage drop across the high current arc, e, the arc time constant, τ, and the breakdown voltage of the cold arc channel, v_d.

These parameters are derived from the minimization of the sum of squares:

$$f(P_0, \tau, e, v_d) = \sum_{i=1}^{n} \left[g_{theo}(P_0, \tau, e, v_d, t_i) - g_{meas}(t_i) \right]^2 \tag{7.78}$$

where

g_{theo} is the arc conductance calculated according to the Urbanek equation
g_{meas} is the arc conductance derived from measurements

A further simplification is introduced by assuming that the voltage across the breaker is much smaller than the parameter v_d after a thermal reignition, so the term $(v/v_d)^2$ may be neglected [208]. In such case, only three parameters have to be determined and the system of equations is

$$\frac{\partial f}{\partial P_0} = 0 \quad \frac{\partial f}{\partial e} = 0 \quad \frac{\partial f}{\partial \tau} = 0 \tag{7.79}$$

A more recent proposal uses also the derivative of the arc conductance, obtained from the measured values of current and voltage [209]

$$\frac{dg_m(t)}{dt} = \frac{g_m(t - \Delta t) - g_m(t + \Delta t)}{2\Delta t} \quad \left(g_m(t) = \frac{i(t)}{v(t)} \right) \tag{7.80}$$

This derivative is also calculated from the arc equation model. For the Mayr model it is

$$\frac{dg_c(t)}{dt} = \frac{g}{\tau} \left(\frac{vi}{P_0} - 1 \right) \tag{7.81}$$

The estimation of τ and P_0 may be made from the differences between these values by using a fitting technique based on the least square method.

Therefore, the evaluation of parameters can follow different principles. Table 7.14 shows the principle of each group and some examples.

7.7.2.3 Validation

Since arc parameters are normally evaluated from measured arc current and voltage traces, those traces should be recalculable provided that circuit data and mathematical procedure are correct. However, the agreement of measured and calculated curves is only an indication of the model validity. Such a procedure may be also helpful for sensitivity studies of an arc model with respect to the influence of errors in measuring and evaluating the arc parameters.

A real check of the model itself is possible if parameter evaluation and comparison of experimental and numerical results take place with different circuits. Subject of this comparison are current/voltage traces and the performance prediction with respect to the short-line fault interruption limit or the chopping level of small inductive current interruption.

A major goal of an arc model validation is the performance prediction under different stress levels or circuit conditions. It is important to keep in mind that a deterministic arc model shows a sharp boundary between successful interruption and reignition, while a real circuit breaker may show a continuous transition between these two behaviors due to its stochastic nature. Therefore, the predicted performance has to be corrected; for example, with the experimental 50% value.

TABLE 7.14

Principles for Arc Parameter Determination

Evaluation Principle	Examples
Parameter functions are calculated directly or step by step from measured current and voltage traces.	If one parameter is constant, the remaining parameter function may be calculated directly from the arc equation.
	The Amsinck's method is within this category [204]. It has been usually applied on reignitions. To analyze successful interruptions, two different measurements of different current steepness are necessary.
Special test circuits may be applied in order to produce singular conditions in arc current or arc conductance suitable for direct parameter evaluation.	The Rijanto's method is within this category [205]. It can be directly applied in current chopping studies; for short-line fault interruption, special circuits are required.
Parameter functions may be defined in general form; e.g., $P(g) = P + Bg^\beta$. The free parameters P, B, and β are determined by curve-fitting techniques.	Measurement quantities are the arc voltage before current zero and the surge impedance and parallel capacitance of the circuit only. The current before current zero is calculated from the voltage trace and the circuit parameters. Curve fitting is applied to the conductance curves calculating the mean square deviation [92].

Although black box arc models are used in both circuit breaker testing and application, it is important to consider that these models must be used within restricted validity limits: those established by validation.

The whole validation procedure of an arc model may consist of the steps mentioned at the beginning of this section: choice of an arc model, tests with measurements during the applicability period, arc parameter evaluation, and validation. Two practical examples of arc model validation related to small inductive current interruption and short-line fault interruption, respectively, are summarized in Table 7.15. The two examples were originally analyzed in a CIGRE publication [67].

In both cases, the arc behavior is described by means of the Mayr–Schwarz equation, and arc parameters are assumed to be power functions of arc conductance (or arc resistance). A quick verification of the arc model is possible, without recalculation of measured curves, by following the principles of the Ruppe's method. Assuming that arc parameters P and τ in the Mayr–Schwarz equation are power functions of g, $\tau(g) = Ag^\alpha$ and $P(g) = Bg^\beta$, Equation 7.66 can be rewritten as

$$\frac{1}{g^{1-\alpha}}\frac{dg}{dt} = \frac{1}{A}\left(\frac{vi}{Bg^\beta} - 1\right) \tag{7.82}$$

As for the Mayr equation, this equation leads to a straight line if $\frac{1}{g^{1-\alpha}}\frac{dg}{dt}$ is plotted versus $\frac{vi}{Bg^\beta}$, and the validity of an arc model in different parts of a switching process can be checked from the deviations of measured values from the straight line. As reasoned above, the values of the parameters A, B can be found from intersections of the line on the two axes. For the special case $\alpha = 0$, $\beta = 0$ (i.e., for the Mayr equation), the plot produces a straight line with intersections $-1/\tau$ and P_0.

TABLE 7.15

Estimation of Parameters of Black Box Arc Models

Application	Measurements	Parameter Evaluation	Validation
Prediction of current chopping [210]	Standardized laboratory tests providing current, arc voltage, and recovery voltage in the vicinity of arc interruption.	For a given value of arc resistance (r_0), the term $1/r \, dr/dt$ is a linear function of vi, so by plotting this function the coefficients $a \, (= 1/\tau(r_0))$ and $b \, (= 1/\tau(r_0)P(r_0))$ are obtained by a linear regression, see Figure 7.73. The process is repeated for several values of the arc resistance. Parameters functions are calculated by a curve-fitting technique.	Two different ways: • By comparing the chopped current, arc voltage, and overvoltage level of real experiments and numerical experiments. • By applying the model to a different circuit condition and verifying that the resulting chopped current is consistent with the experimental results.
Prediction of short-line fault interruption limit [92]	Post-arc current measurement using synthetic short-line fault testing of full-scale breakers with fast measurements of arc voltage and current. Capacitive current through the stray capacitances of test and auxiliary breakers is subtracted to assess the real arc current.	The four free parameters are determined by using a curve-fitting procedure aiming at a best fitting of measured and calculated arc resistance.	The calculated post-arc current is compared with the measured one. By varying the di/dt of the simulations at constant surge impedance, the interruption limit is found. Additional experiments beyond the limit stress are used to predict other experimental results.

Parameter determination and validation of gas-filled arc models have been subjects of several references, see for instance Refs. [95,100,211,212].

The following example illustrates the application of a method for evaluation of parameters and validation of an arc model.

Example 7.12

Consider the system shown in Figure 7.38 and assume that the dynamic arc of the circuit breaker behaves according to the Mayr model during the opening process

$$\frac{1}{g}\frac{dg}{dt} = \frac{1}{\tau}\left(\frac{vi}{P_0} - 1\right)$$

The goal of this example is to estimate the parameters of the arc model from the measured traces of current and voltage, although in this case they are obtained by simulation.

Figures 7.74 and 7.75 show the simulation results and the arc parameters used for each case. The goal is to apply a parameter estimation method to derive the values of P_0 and τ. The advantage in this case is that these values are known in advance.

The procedure used in this example is very similar to that proposed in Ref. [209], while the implementation is similar to that presented in Ref. [213].

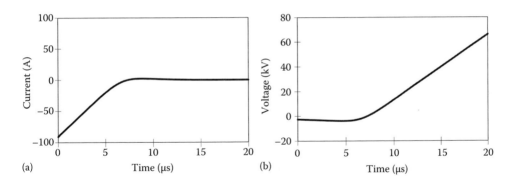

FIGURE 7.74
Example 7.12: Arc extinction (Mayr model: $P_0 = 360\,\text{kW}$, $\tau = 1.3\,\mu\text{s}$). (a) Arc current. (b) Arc voltage.

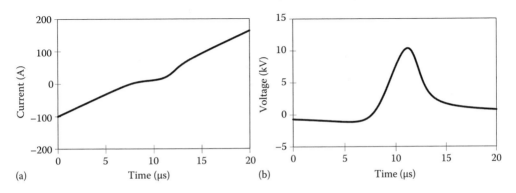

FIGURE 7.75
Example 7.12: Arc ignition (Mayr model: $P_0 = 100\,\text{kW}$, $\tau = 1.3\,\mu\text{s}$). (a) Arc current. (b) Arc voltage.

The first step is to obtain the arc conductance from the traces of current, $i(t)$ and voltage, $v(t)$

$$g(t) = \frac{i(t)}{v(t)} \tag{7.83}$$

Then the derivative of the arc conductance is obtained:

$$\frac{dg(t)}{dt} = \frac{g(t) - g(t - \Delta t)}{\Delta t} \tag{7.84}$$

The Mayr equation can be written as follows:

$$\frac{dg(t)}{dt} = \frac{g(t)}{\tau}\left(\frac{v(t)i(t)}{P_0} - 1\right) = \frac{1}{\tau}\left(\frac{i^2(t)}{P_0} - g(t)\right) \tag{7.85}$$

Upon using the following notations

$$F_m(t) = \frac{dg(t)}{dt} \qquad F_c(t) = Af_1(t) + Bf_2(t) \tag{7.86}$$

where

$$A = \frac{1}{\tau P_0} \quad f_1(t) = i^2(t) \quad B = \frac{1}{\tau} \quad f_2(t) = g(t) \tag{7.87}$$

the target function can be defined as follows:

$$F = \sum_k \left[F_m(t_k) - F_c(t_k) \right]^2 = \sum_k \left[F_m(t_k) - \left(A f_1(t_k) - B f_2(t_k) \right) \right]^2 \tag{7.88}$$

where t_k are the different time steps at which the values of $i(t)$ and $v(t)$ were measured.
The values of A and B that minimize F are calculated from the following formulas [213]:

$$A = \frac{\sum_k \left(F_m(t_k) \cdot f_2(t_k) \right) \cdot \sum_k \left(F_m(t_k) \cdot f_2(t_k) \right) - \sum_k \left(F_m(t_k) \cdot f_1(t_k) \right) \cdot \sum_k \left(f_2(t_k) \right)^2}{\left[\sum_k \left(f_1(t_k) \cdot f_2(t_k) \right) \right]^2 - \sum_k \left(f_1(t_k) \right)^2 \cdot \sum_k \left(f_2(t_k) \right)^2} \tag{7.89a}$$

$$B = \frac{\sum_k \left(F_m(t_k) \cdot f_2(t_k) \right) \cdot \sum_k \left(f_1(t_k) \right)^2 - \sum_k \left(F_m(t_k) \cdot f_1(t_k) \right) \cdot \sum_k \left(f_1(t_k) \cdot f_2(t_k) \right)}{\left[\sum_k \left(f_1(t_k) \cdot f_2(t_k) \right) \right]^2 - \sum_k \left(f_1(t_k) \right)^2 \cdot \sum_k \left(f_2(t_k) \right)^2} \tag{7.89b}$$

from where

$$\tau = \frac{1}{B} \quad P_0 = \frac{B}{A} \tag{7.90}$$

The resulting values of the two parameters for the two cases are shown in Figures 7.74 and 7.75, respectively

- Arc extinction—Figure 7.74 $P_0 = 360.24$ (kW) $\tau = 1.4\,\mu s$
- Arc ignition—Figure 7.75 $P_0 = 99.995$ (kW) $\tau = 1.4\,\mu s$

Note that the accuracy is very good for the parameter P_0 while the difference between the real value and the estimated value for τ are within a 7% error.

7.7.2.4 Conclusions

Although a large variety of black box arc models for gas-filled circuit breakers have been developed to date, there is not yet a general standard approach and they must be applied with caution. The progress in measuring techniques and in computer models has led to successful approaches and a common understanding of basic questions. The application of any arc modeling approach may consist of the same following steps: (1) choice of a model equation, (2) test(s) with measurements of voltage and/or current during the period under study, (3) evaluation of arc parameters, and (4) numerical simulations and validation of the arc model. Several calculations with variation of a circuit parameter (e.g., surge impedance) will yield an interruption limit. Several interruption limits at different values of a further circuit parameter (e.g., parallel capacitance) will yield a limiting curve.

The choice of a model, measurements and parameter evaluation are not independent from each other, but have to be seen as a whole. In general, an increasing number of free parameters improves the adaptability, but also requires more information from measurements and it is not necessarily linked with a higher predictive capacity. In practice, a broad range of modeling techniques has been developed reaching from very general models requiring sophisticated experimental techniques for parameter determination to simple models that require not much more than the standard measurements at the normal circuit breaker type tests. Most applications focus with 3–4 scalar parameters.

The validity of the procedure within a chosen application domain has to be checked. This validation has to establish the predictive power of the model and should contain at least the following steps:

- Identification of the model at an operating point (e.g., establishment of arc parameters)
- Prediction of performance in a changed circuit
- Experimental validation of the performance in the changed circuit

Successful applications have been reported mainly in two fields [67]:

- Prediction of short-line fault breaking capability for restricted modified circuit conditions (influence of parallel capacitance, line surge impedance, ITRV). Practical examples are prediction and comparison of breaking limits during circuit breaker development, comparison of the severity of actual circuit conditions in a substation with the severity of the corresponding standard test circuit or theoretical check for the validity of new results.
- Prediction of the influence of the circuit on chopping current levels at small inductive current switching. In cases where the circuit cannot be described by a simple parallel capacitance to the breaker, this is the only technique that uses laboratory test results for accurate prediction of the behavior in a substation.

7.7.3 Vacuum Circuit Breakers

Although the concept of black box model is also applied to the arc equations in vacuum, the models for this type of circuit breaker are based on a set of simplified equations aimed at describing the physical behavior of the vacuum arc.

Unlike black box models for gas-filled circuit breakers, not much work has been performed to date on the estimation of parameters which define a dynamic vacuum arc. Although fitting techniques similar to some applied to gas-filled breakers might be used, in general parameters for vacuum arcs are either measured or estimated taking into account typical values obtained from the experience.

Depending on whether a low- or a high-current switching is investigated, high-frequency restrike modeling or a postzero current model should be respectively applied.

When switching low inductive or capacitive circuits, the power-frequency current is interrupted as soon as the current amplitude through the circuit breaker reaches a low level so that it goes to current zero with a very steep di/dt. This level is called the chopping level and the amplitude of the current at that instant is the chopping current. The arc will be reignited when the TRV becomes greater than the withstand capability of the vacuum gap. The arc will be ceased when the reignited current passed through zero and the slope of this high-frequency current is lower than a predefined critical current slope. Therefore, for modeling high-frequency restrikes, the chopping current, the withstand voltage capability, and the critical current slope are needed. The chopping current in SF_6 circuit breakers is rather deterministic whilst the chopping current in vacuum circuit breakers has statistical nature, see for instance [214]. Chopping current behavior in vacuum circuit breakers was extensively studied by Smeets [215] and it can be determined by

$$I_{ch} \approx \left(\omega \hat{I} \alpha \beta \right)^q \tag{7.91}$$

where
  $\omega = 2\pi f$ (f is the power frequency)
  $\hat{I}$ is the load current amplitude, while α and β depend on the type of breaker contacts
  $q = (1-\beta)^{-1}$

For Cu/Cr contacts, $\alpha = 6.2E{-}16\,s$ and $\beta = 14.3$. The average chopping current values for different types of contact materials can be found in Ref. [26].

Because the arc voltage of the vacuum circuit breaker is low, the restrikes can be modeled efficiently by applying an ordinary switch which opens and closes according to specific conditions. The logic is fully described in Refs. [33,163]. The dielectric and arc quenching capabilities of vacuum circuit breakers have been investigated in Ref. [137]. The dielectric capability can be extracted from actual measurements: by recording the actual TRV during switching, the local maxima of the multiple restrikes define the voltage values where reignition takes place [216]. Critical current slope is more difficult to obtain and it requires more detailed investigation of the restrike behavior. The investigation of this phenomenon is reported in Ref. [217]. These characteristics were presented in Equations 7.32 and 7.33; Table 7.16 shows typical parameter values.

Post-arc current models were detailed in Section 7.6.2.4. Table 7.17 summarizes the equations that describe the different periods of a vacuum arc, the unknowns for each arc period, and the range of values for parameters to be specified in vacuum arc models [218–221].

It has to be pointed out that the parameter A represents the fraction of the anode area where the arc is constricted. The contact distance can be estimated from the measurements by multiplying the arcing time with the average speed of contact opening, which is normally considered constant and assumed to be $1\,m/s$.

A validation of vacuum arc models is presented in the following example where the parameters of the equations for the current-zero period and the post-arc period are taken from the literature.

TABLE 7.16

Typical Values for Vacuum Circuit Breakers

Quenching Capability	A (V/s)	B (V)	C (A/s²)	D (A/s)
High	1.7E7	3.4E3	−3.4E11	255E6
Medium	1.3E7	0.69E3	0.32E12	155E6
Low	0.47E6	0.69E3	1.00E12	190E6

TABLE 7.17

Equations and Parameters of a Vacuum Arc Model

Model	Equations	Unknowns	Typical Values of Parameters
Current zero model	$$\frac{ds}{dt} = \frac{i}{q_i n_i A} - v_i \quad (A = \pi D^2/4)(q_i = eZ)$$ $$\frac{dn_i}{dt} = -\frac{i}{q_i A s}$$ $$s^2 = \frac{4\varepsilon_0 V_0}{9 q_i n_i}\left[\left(1+\frac{v}{V_0}\right)^{3/2} + 3\frac{v}{V_0} - 1\right]$$ $$V_0 = \frac{m_i}{2e}\left(v_i + \frac{ds}{dt}\right)^2$$	s $v(t)$ $i(t)$ $n_i(t)$ V_0	$v_i \approx 1.0\text{E}+3 \div 1.0\text{E}+5(\text{m/s})$ $Z \approx 1.3 \div 1.5$ (for copper) $m_i \approx 1.06\text{E}-25$ (kg) $\tau \approx 0.5 \div 84$ (μs) $T_e \approx 2 \div 10$ (eV) $n_i(0) \approx 1.0\text{E}+17 \div 1.0\text{E}+19$ (m^{-3}) $c_1 \approx 0.5\text{E}-6 \div 2.0\text{E}-6$ (m^2/V) $c_2 \approx 30\text{E}-6 \div 60\text{E}-6$ (m^3)
Post-arc model	$$A_{ef}(t) = A + c_1 v$$ $$\frac{dn_i}{dt} = -\left(\frac{n_i}{\tau} + \frac{i_{pa}}{q_i c_2}\right)$$ $$i_{pa} = q_i v_B n_i A_{ef} \tanh\left(\frac{q_i v}{2kT_e}\right) + C_{sh}\frac{dv}{dt}$$	A_{ef} $n_i(t)$ $v(t)$ $i_{pa}(t)$	

Notes: e is the electron charge, $e = 1.602\text{E}-19$ C. ε_0 is the permittivity of vacuum, $\varepsilon_0 = 8.854\text{E}-12$ (F/m). k is the Boltzmann's constant, $k = 1.380\text{E}-23$ (J K^{-1}). The capacitance C_{sh} is obtained by means of Equations 7.56 and 7.57. Values to be considered for diameter D are respectively, 1 mm in the current zero model, and 4 cm in the post-arc model.

Example 7.13

The objective of this example is to compare measurements and simulation results obtained with a commercially available vacuum circuit breaker using the arc models detailed in Section 7.6.2.4. Figure 7.76 shows the simplified diagram of the test circuit. The circuit breaker has axial magnetic field contacts, a rated voltage of 24 kV, and a rated short-circuit current of 20 kA, with 35% direct current asymmetry [151,157].

A current and a metal vapor arc are flowing through the inductance and the breaker, respectively. After current zero, the arc disappears and the residual plasma between the breaker contacts is cleared by a TRV.

To validate the vacuum arc model taking into account the voltage-zero period and the post-arc current, some measurements and simulations were performed. First, the arc model for the period immediately after current zero is tested; next, the post-arc current model is tested. Finally, the validity of the model under different TRV conditions is checked.

The quantities T_e, m_i, and $n_i(0)$ (the initial ion density) are estimated or taken from the literature (see Table 7.17). T_e is normally in the order of 2–10 eV. It is assumed that only copper ions are present inside the gap, so $m_i = 1.06 \times 10^{-25}$ kg. The selected values for parameters T_e, q_i, and $n_i(0)$ are 3 eV, $1.6 \cdot 10^{-19}$ C, and $2.8 \cdot 10^{18}$ m^{-3}, respectively. The time constant τ for the ion decay is 20 μs. The contact diameter is 4 cm, and A is 12.5 cm^2.

Figure 7.77 presents the results of the applied model. The rise of the recovery voltage causes a fast change in the electric field inside the ionic space-charge sheath in front of the cathode. This

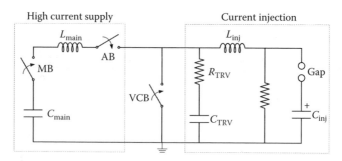

FIGURE 7.76
Example 7.13: Simplified diagram of the test circuit.

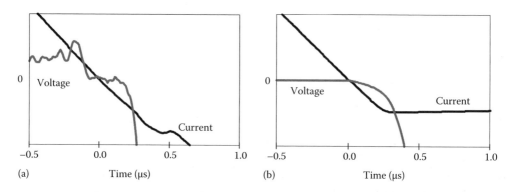

FIGURE 7.77
Example 7.13: Lower voltage period following current zero. (a) Measured data. (b) Simulation results. (From van Lanen, E.P.A. et al., *IEEE Trans. Plasma Sci.*, 35, 925, August 2007. With permission.)

yields a displacement current, which remains equal to the saturation current, whereas the measured current continues to increase for some time. One can observe that simulation results show a voltage-zero period that matches reasonably well the measurements.

To take into account the sheath growth in front of the cathode, the post-arc current model described by means of Equations 7.54, 7.55, and 7.58 is used. The values of the parameters are the same as in the voltage-zero period and c_1 and c_2 are 0.5×10^{-6} m²/V and 56×10^{-6} m³.

Figure 7.78 shows simulation results compared with the measurements. A good match between post-arc current is achieved, although the simulated TRV exhibits a less damped behavior than the measured TRV.

The final test involves three measurements with equal settings (same arcing time and same peak short-circuit current) but different TRVs. Figure 7.79a shows the corresponding measurements.

Since the three cases were performed with equal settings, it was assumed that plasma conditions would be equal at $t = 0$. The values of parameters used in simulations were $n_i(0) = 7 \cdot 10^{17}$, $\tau = 5\,\mu s$, and $c_2 = 33 \cdot 10^{-6}$ m³. However, the value of c_1 had to be adjusted for each case; it was $2 \cdot \times 10^{-6}$ m²/V for the case with the steepest TRV (case (1) in Figure 7.79), 1×10^{-6} m²/V for the case with the middle TRV (case (2) in Figure 7.79), $0.5 \cdot \times 10^{-6}$ m²/V for the case with the slowest TRV (case (3) in Figure 7.79). Figure 7.79b shows the simulation results, which match reasonably well the measurements depicted in Figure 7.79a. Therefore, the assumed relation between the effective cathode area and the voltage is not correct since the value of c_1 changes with the arc conditions.

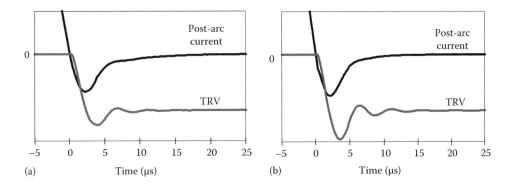

FIGURE 7.78

Example 7.13: Post-arc current. (a) Measured data. (b) Simulation results. (From van Lanen, E.P.A. et al., *IEEE Trans. Plasma Sci.*, 35, 925, August 2007. With permission.)

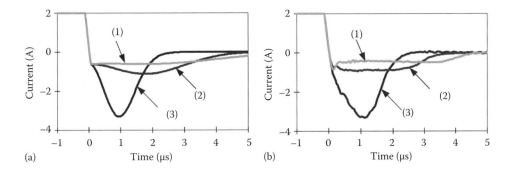

FIGURE 7.79

Example 7.13: Post-arc current with different TRV. (a) Measured data. (b) Simulation results. (From van Lanen, E.P.A. et al., *IEEE Trans. Plasma Sci.*, 35, 925, August 2007. With permission.)

References

1. L. van der Sluis, *Transients in Power Systems*, John Wiley & Sons, New York, 2001.
2. IEEE Std C37.04-1999, IEEE standard rating structure for AC high-voltage circuit breakers.
3. ANSI C37.06-2000, High-voltage circuit breakers rated on a symmetrical current basis-preferred ratings and related required capabilities.
4. IEEE Std C37.09-1999, IEEE standard test procedure for AC high-voltage circuit breakers rated on a symmetrical current basis.
5. IEC 60056, IEC standard for high-voltage alternating current circuit breakers, 1987.
6. IEC 60694, IEC standard with common specifications for high-voltage switchgear and controlgear standards, 2002.
7. A.M. Cassie, Arc rupture and circuit severity: A new theory, CIGRE Session, Report 102, Paris, 1939.
8. O. Mayr, Beitraege zur theorie des statischen und dynamischen lichtbogens, *Archiv für Elektrotechnik*, 37(12), 588–608, 1943.

9. R.D. Garzon, *High Voltage Circuit Breakers. Design and Applications*, 2nd edn., Marcel Dekker, New York, 2002.
10. K. Ragaller (Ed.), *Current Interruption in High-Voltage Networks*, Plenum Press, New York, 1978.
11. T.E. Browne (Ed.), *Circuit Interruption. Theory and Techniques*, Marcel Dekker, New York, 1984.
12. C.H. Flurscheim (Ed.), *Power Circuit Breaker Theory and Design*, Peter Peregrinus, London, U.K., 1985.
13. H.M. Ryan and G.R. Jones, SF_6 *Switchgear*, Peter Peregrinus, London, U.K., 1989.
14. K. Nakanishi (Ed.), *Switching Phenomena in High-Voltage Circuit Breakers*, Marcel Dekker, New York, 1991.
15. *Disjuntores e Chaves. Aplicaçao em Sistemas de Potência*, (in Portuguese) FURNAS/EDUFF, Rio de Janeiro, Brazil, 1995.
16. G.R. Jones, Gas filled interrupters—Fundamentals, Chapter 7 of *High Voltage Engineering and Testing*, H.M. Ryan (Ed.), 2nd edn., The Institution of Electrical Engineers, London, U.K., 2001.
17. S.M. Ghufran Ali, Switchgear design, development and service, Chapter 8 of *High Voltage Engineering and Testing*, H.M. Ryan (Ed.), 2nd edn., The Institution of Electrical Engineers, London, U.K., 2001.
18. B.M. Pryor, Distribution switchgear, Chapter 10 of *High Voltage Engineering and Testing*, H.M. Ryan (Ed.), 2nd edn., The Institution of Electrical Engineers, London, U.K., 2001.
19. S.M. Ghufran Ali, Differences in performance between SF_6 and vacuum circuit-breakers at distribution voltage levels, Chapter 11 of *High Voltage Engineering and* Testing, H.M. Ryan (Ed.), 2nd edn., The Institution of Electrical Engineers, London, U.K., 2001.
20. H.M. Ryan, Circuit breakers and interruption, Chapter 9 of *Advances in High Voltage Engineering*, A. Haddad and D. Warne (Eds.), The Institution of Electrical Engineers, London, U.K., 2004.
21. N.P. Allen, T. Harris, and C.J. Jones, Switchgear, Chapter 7 of *Electrical Power Engineer's Handbook*, D.F. Warne (Ed.), 2nd edn., Newnes, Burlington, MA, 2005.
22. IEEE Tutorial Course (R.D. Garzon, Coordinator), Application of power circuit breakers, IEEE Special Publication 93EH0 388-9-PWR, 1993.
23. S. Stewart, *Distribution Switchgear*, The Institution of Electrical Engineers, London, U.K., 2004.
24. S. Theoleyre, MV breaking techniques, Schneider Cahier Technique no. 193, June 1999.
25. A. Greenwood, *Vacuum Switchgear*, The Institution of Electrical Engineers, London, U.K., 1994.
26. P.G. Slade, *The Vacuum Interrupter. Theory, Design and Application*, CRC Press, Boca Raton, FL, 2008.
27. P. Picot, Vacuum switching, Schneider Cahier Technique no. 198, March 2000.
28. IEEE Tutorial Course (M.T. Glinkowski, Coordinator), IEEE tutorial on the vacuum switchgear, IEEE Special Publication TP-135-0, 1999.
29. IEEE Seminar/Report, Power systems transient recovery voltages, IEEE Special Publication 87TH0176-8-PWR, 1987.
30. CIGRE WG 13.02, Interruption of small inductive currents, S. Berneryd (Ed.), CIGRE Technical Brochure no. 50, December 1995.
31. CIGRE WG 13.03, Transient recovery voltages in medium voltage networks, CIGRE Technical Brochure no. 134, December 1998.
32. IEEE Std C37.011-2005, IEEE application guide for transient recovery voltage for AC high-voltage circuit breakers.
33. M. Popov, Switching three-phase distribution transformers with a vacuum circuit breaker—analysis of overvoltages and the protection of the equipment, PhD thesis, Delft University of Technology, the Netherlands, 2002.
34. E.W. Schmunk and R.P. O'Leary, Power system TRV characteristics, Chapter 2 of *Power Systems Transient Recovery Voltages*, IEEE Special Publication 87TH0176-8-PWR, 1987.
35. S.R. Lambert, Circuit breaker transient recovery voltage, Chapter 4 of *Application of Power Circuit Breakers*, (R.D. Garzon, Coordinator), IEEE Special Publication 93EH0 388-9-PWR, 1993.
36. C.L. Wagner, Transient recovery voltage standards, Chapter 3 of *Power Systems Transient Recovery Voltages*, IEEE Special Publication 87TH0176-8-PWR, 1987.

37. IEC 60909, Short-circuit currents in three-phase systems—Part 0: Calculation of currents, 2001.

38. EPRI Report EL-4202, Electromagnetic transients program (EMTP) primer, 1985.

39. C.L. Wagner, D. Dufournet, and G.F. Montillet, Revision of the application guide for transient recovery voltage for AC high-voltage circuit breakers of IEEE C37.011: A Working Group Paper of the High Voltage Circuit Breaker Subcommittee, *IEEE Transactions on Power Delivery*, 22(1), 161–166, January 2007.

40. IEEE Std C37.010-1999, IEEE application guide for AC high-voltage circuit breakers rated on a symmetrical current basis.

41. IEEE Std C37.012-2005, IEEE application guide for capacitance current switching for AC high-voltage circuit breakers.

42. IEEE Std C37.013-1997, IEEE standard for AC high-voltage generator circuit breakers rated on a symmetrical current basis.

43. IEC-62271-100, High-voltage switchgear and controlgear—Part 100: Alternating-current circuit-breakers, 2008.

44. ANSI/IEEE Std C37.081-1981, IEEE guide for synthetic fault testing of AC high-voltage circuit breakers rated on a symmetrical current basis.

45. IEEE Std C37.083-1999, IEEE guide for synthetic capacitive current switching tests of AC high-voltage circuit breakers.

46. G. St-Jean, Methods of generating and measuring TRV in laboratory, Chapter 4 of *Power Systems Transient Recovery Voltages*, IEEE Special Publication 87TH0176-8-PWR, 1987.

47. L. Blahous, The problem of defining a test circuit for circuit breakers in terms of prospective voltage, *IEE Proceedings*, 126(12), 1291–1294, December 1979.

48. W.A. van der Linden and L. van der Sluis, A new artificial line for testing high-voltage circuit breakers, *IEEE Transactions on Power Apparatus and Systems*, 102(4), 797–803, April 1983.

49. G. St-Jean and R.F. Wang, Equivalence between direct and synthetic short-circuit interruption tests on HV circuit breakers, *IEEE Transactions on Power Apparatus and Systems*, 102(7), 2216–2223, July 1983.

50. E. Thuries, P. Van Doan, J. Dayet, and B. Joyeux-Bouillon, Synthetic testing method for generator circuit breakers, *IEEE Transactions on Power Delivery*, 1(1), 179–184, January 1986.

51. S. Yamashita, N. Miyake, K. Suzuki, H. Ikeda, and S. Yanabu, Short-circuit testing method of 3-phase-in-one-tank-type SF_6 gas circuit breaker, *IEEE Transactions on Power Delivery*, 4(1), 349–354, January 1989.

52. W.P. Legros, A.M. Genon, R. Planche, and C. Guilloux, Computer aided design of synthetic test circuits for high voltage circuit-breakers, *IEEE Transactions on Power Delivery*, 4(2), 1049–1055, April 1989.

53. G. St-Jean and M. Landry, Comparison of waveshape quality of artificial lines used for short-line fault breaking tests on HV circuit breakers, *IEEE Transactions on Power Delivery*, 4(4), 2109–2113, October 1989.

54. R.P.P. Smeets and W.A. van der Linden, The testing of SF_6 generator circuit-breakers, *IEEE Transactions on Power Delivery*, 13(4), 1188–1193, October 1998.

55. G. Daigneault, G. St-Jean, and M. Landry, Comparing direct and synthetic test for interruption of line-charging capacitive current, *IEEE Transactions on Power Delivery*, 16(3), 409–414, July 2001.

56. A. Hochrainer, Synthetic testing for high power circuit-breakers, *Electra*, no. 6, pp. 51–70, 1968.

57. E. Slamecka, Synthetic make test requirements. I. Circuit-breaker not fitted with making resistors, *Electra*, no. 38, pp. 4–51, 1975.

58. W. Rieder, Aims and significance of comparison tests. Checking synthetic test schemes for high voltage circuit-breakers, *Electra*, no. 47, pp. 45–60, 1976.

59. CIGRE SC 13 WG 04, Synthetic testing of autoreclosing duties for circuit- breakers not equipped with parallel resistors, *Electra*, no. 54, pp. 21–26, 1977.

60. CIGRE SC 13 WG 04, Requirements for testing the making performance of high-voltage circuit-breakers at reduced applied voltage, *Electra*, no. 66, pp. 27–38, 1979.

61. CIGRE SC 13 WG 04, Synthetic make test requirements. II. Circuit-breaker fitted with making resistors, *Electra*, no. 66, pp. 40–57, 1979.

62. CIGRE SC 13 WG 04, Requirements for capacitive current switching tests employing synthetic test circuits for circuit-breakers without shunt resistors, *Electra*, no. 87, pp. 25–39, 1983.
63. CIGRE SC 13 WG 04, Requirements for switching tests of metal enclosed switchgear. First Part, *Electra*, no. 110, pp. 7–46, 1987.
64. CIGRE WG 13.04, Asymmetrical current breaking tests, *Electra*, no. 132, pp. 109–125, 1990.
65. CIGRE WG 13.04, Specified time constants for testing a symmetric current capability of switchgear, *Electra*, no. 173, pp. 18–31, 1997.
66. CIGRE Working Group 13.01, Practical application of arc physics in circuit breakers. Survey of calculation methods and application guide, *Electra*, no. 118, pp. 64–79, 1988.
67. CIGRE Working Group 13.01, Applications of black box modelling to circuit breakers, *Electra*, no. 149, pp. 40–71, 1993.
68. CIGRE Working Group 02 (SC 33), Guidelines for representation of network elements when calculating transients, CIGRE Technical Brochure no. 39, 1990.
69. EPRI Report EL-4651, *EMTP Workbook II*, 1987.
70. M. Kizilcay, Dynamic arc modelling in EMTP, *EMTP News*, 5(3), 15–26, July 1985.
71. L. van der Sluis, W.R. Rutgers, and C.G.A. Koreman, A physical arc model for the simulation of current zero behaviour of high-voltage circuit breakers, *IEEE Transactions on Power Delivery*, 7(2), 1016–1022, April 1992.
72. L. van der Sluis and W.R. Rutgers, Comparison of test circuits for high-voltage circuit breakers by numerical calculations with arc models, *IEEE Transactions on Power Delivery*, 7(4), 2037–2045, October 1992.
73. W. Rieder and J. Urbanek, New aspects of current-zero research on circuit-breaker reignition. A theory of thermal non-equilibrium arc conditions, *CIGRE Session*, Report 107, Paris, 1966.
74. A. Grütz and A. Hochrainer, Rechnerische untersuchung von leitungsschaltern mit hilfe einer verallgemeinerten lichtbogentheorie, *ETZ-A Elektrotechnik*, 92, 185–191, 1971.
75. J. Schwarz, Dynamisches verhalten eines gasbeblasenen, turbulenzbestimmten schaltlichtbogens, *ETZ-A Elektrotechnik*, 92, 389–391, 1971.
76. J. Urbanek, Zur berechnung des schaltverhaltens von leistungsschaltern, eine erweiterte Mayr-Gleichung, *ETZ-A Elektrotechnik*, 93, 381–385, 1972.
77. W. Hermann, U. Kogelschatz, L. Niemeyer, K. Ragaller, and E. Schade, Investigation on the physical phenomena around current zero in HV gas blast breakers, *IEEE Transactions on Power Apparatus and Systems*, 95(4), 1165–1176, July/August 1976.
78. B. Sporckmann, Ein stochastisches lichtbogenmodell zur berechnung von ausschaltvorgängen von hochspannungs-leitungsschaltern, *ETZ-A Elektrotechnik*, 99, 758–759, 1978.
79. T.E. Browne, Jr., Practical modelling of the circuit breaker arc as a short-line-fault interrupter, *IEEE Transactions on Power Apparatus and Systems*, 97, 838–847, 1978.
80. E. Schwalb, U. Habedank, H.H. Schramm, E. Slamecka, and K. Zückter, Investigation of unit testing and full-pole testing of a single pressure type double nozzle SF_6 circuit-breaker under short-line fault conditions, *CIGRE Session*, Paper 13-03, Paris, 1978.
81. W.P. Legros and A.M. Genon, Circuit breaker stochastic behaviour simulation, *Symposium on High Voltage Switching Equipment*, 38–43, Sydney, Australia, 1979.
82. E. von Bonin and K. Kriechbaum, Advanced methods of circuit-breaker engineering, *Symposium on High Voltage Switching Equipment*, 117–121, Sydney, Australia, 1979.
83. M.T.C. Fang and D. Brannen, A current-zero arc model based on forced convection, *IEEE Transactions on Plasma Science*, 7(4), 217–229, December 1979.
84. H. Kopplin, Mathematische modelle des schaltlichtbogens, *ETZ-A*, 2, 209–213, 1980.
85. A.V. Avdonin, S.V. Biryukov, A.L. Buynov, R.B. Dobrokhotov, V.G. Egorov, G.S. Puzyriysky, Yu. T. Sanivkov, and K. Seryakov, Some problems of EHV and UHV air-blast circuit breakers, *CIGRE Session*, Paper 13–04, Paris, France, 1980.
86. C. Portela, Study of problems related to switching of relatively small currents, *CIGRE Session*, Paper 13–05, Paris, France, 1980.

87. H. Kuwahara, T. Tanabe, K. Ibuki, M. Sakai, A. Sato, and S. Sakuma, New approach to analysis of arc interruption capability by simulation employed in the development of SF_6 GCB series with high capacity interrupter, *IEEE Transactions on Power Apparatus and Systems*, 102(7), 2262–2268, July 1983.

88. M. Ishikawa, H. Ikeda, S. Yanabu, and M. Yamamoto, Numerical study of delayed-zero-current interruption phenomena using transient analysis model for an arc in SF_6 flow, *IEEE Transactions on Power Apparatus and Systems*, 103(12), 3561–3568, December 1983.

89. T. Tanabe, K. Ibuki, S. Sakuma, and T. Yonezawa, Simulation of the SF_6 arc behavior by a cylindrical arc model, *IEEE Transactions on Power Apparatus and Systems*, 104(7), 1903–1909, July 1985.

90. T. Matsumura, W.T. Oppenlander, A. Schmidt-Harms, and A.D. Stokes, Prediction of circuit-breaker performance based on a refined cybernetic model, *IEEE Transactions on Plasma Science*, 14(4), 344–351, August 1986.

91. V. Phaniraj and A.G. Phadke, Modelling of circuit breakers in the electromagnetic transients program, *IEEE Transactions on Power Systems*, 3(2), 799–805, May 1988.

92. G. St-Jean, M. Landry, M. Leclerc, and A. Chenier, A new concept in post-arc analysis to power circuit breakers, *IEEE Transactions on Power Delivery*, 3(3), 1036–1044, July 1988.

93. K.P. Guruprasad, C.S. Sarma, M. Ramamoorty, and M.S. Naidu, Investigation of the characteristics of an SF_6 rotating arc by a mathematical model, *IEEE Transactions on Power Delivery*, 7(2), 727–733, April 1992.

94. C. Guilloux, Y. Therme, and P.G. Scarpa, Measurement of a post arc current of high voltage circuit breakers. Application to short circuit tests with ITRV, *IEEE Transactions on Power Delivery*, 8(3), 1148–1154, July 1993.

95. E. Zaima, S. Okabe, S. Nishiwaki, M. Ishikawa, K. Suzuki, and H. Toda, Application of dynamic arc equations to high-frequency arc extinctions in SF_6 gas circuit breakers, *IEEE Transactions on Power Delivery*, 8(3), 1199–1205, July 1993.

96. U. Habedank, Application of a new arc model for the evaluation of short-circuit breaking tests, *IEEE Transactions on Power Delivery*, 8(4), 1921–1925, October 1993.

97. L. van der Sluis and W.R. Rutgers, The comparison of test circuits with arc models, *IEEE Transactions on Power Delivery*, 10(1), 280–285, January 1995.

98. J.C. Vérité, T. Boucher, A. Comte, C. Delalondre, and O. Simonin, Arc modelling in circuit breakers: Coupling between electromagnetics and fluid mechanics, *IEEE Transactions on Magnetics*, 31(3), 1843–1848, May 1995.

99. J.C. Vérité, T. Boucher, A. Comte, C. Delalondre, P. Robin-Jouan, E. Serres, V. Texier, M. Barrault, P. Chevrier, and C. Fievet, Arc modelling in SF_6 circuit breakers, *IEE Proceedings. Science, Measurement and Technology*, 142(3), 189–196, May 1995.

100. G. Bizjak, P. Zunko, and D. Povh, Circuit breaker model for digital simulation based on Mayr's and Cassies's differential arc equations, *IEEE Transactions on Power Delivery*, 10(3), 1310–1315, July 1995.

101. P.H. Schavemaker and L. van der Sluis, The influence of the topology of test circuits on the interrupting performance of circuit breakers, *IEEE Transactions on Power Delivery*, 10(4), 1822–1828, October 1995.

102. D.W.P. Thomas, E.T. Pereira, C. Christopoulos, and A.F. Howe, The simulation of circuit breaker switching using a composite Cassie-modified Mayr model, *IEEE Transactions on Power Delivery*, 18(4), 1829–1835, October 1995.

103. J.A. Martinez-Velasco, Circuit breaker representation for TRV calculations, *EEUG Meeting*, Hannover, Germany, November 13–15, 1995.

104. J.A. Martinez-Velasco, Modelling of circuit breaker using the Type-94 component, *EEUG Meeting*, Hannover, Germany, November 13–15, 1995.

105. K.Y. Park and M.T.C. Fang, Mathematical modelling of SF_6 puffer circuit breakers. I: High current region, *IEEE Transactions on Plasma Science*, 24(2), 490–502, April 1996.

106. P. Chevrier, M. Barrault, and C. Fievet, Hydrodynamic model for electrical arc modeling, *IEEE Transactions on Power Delivery*, 11(4), 1824–1829, October 1996.

107. K.J. Tseng, Y. Wang, and D.M. Vilathgamuwa, An experimentally verified hybrid Cassie-Mayr electric arc model for power electronics simulations, *IEEE Transactions on Power Electronics*, 12(3), 429–436, May 1997.

108. J. Mahseredjian, M. Landry, and B. Khodabakhchian, The new EMTP breaker arc model, *IPST'97*, Seattle, WA, June 22–26, 1997.

109. X.B. Li, Q.P. Wang, Y.B. Li, and Y. Yang, Numerical analysis of flow field and the dynamic properties of arc in the interrupting chamber of an SF_6 puffer circuit breaker, *IEEE Transactions on Plasma Science*, 25(5), 982–985, October 1997.

110. Z. Ma, C.A. Bliss, A.R. Penfold, A.F.W. Harris, and S.B. Tennakoon, An investigation of transient overvoltage generation when switching high voltage shunt reactors by SF_6 circuit breaker, *IEEE Transactions on Power Delivery*, 13(2), 472–479, April 1998.

111. J.C. Vérité, T. Boucher, A. Comte, M. Barrault, P. Chivrier, and C. Fievet, Coupling between physical arc modelling and circuit modelling in a circuit-breaker, *IEEE Transactions on Magnetics*, 35(3), 1614–1617, May 1999.

112. L. Xiaoming, W. Erzhi, and C. Yundong, Dynamic arc modeling based on computation of couple electric field and flow field for high voltage SF_6 interrupter, *IEEE Transactions on Magnetics*, 40(2), 601–604, March 2004.

113. R.P.P. Smeets and V. Kertész, Evaluation of high-voltage circuit breaker performance with a validated arc model, *IEE Proceedings Generation, Transmission and Distribution*, 147(2), 121–125, March 2000.

114. P.H. Schavemaker and L. van der Sluis, An improved Mayr-type arc model based on current-zero measurements, *IEEE Transactions on Power Delivery*, 15(2), 580–584, April 2000.

115. P.H. Schavemaker, L. van der Sluis, and A. Pharmatrisanti, Post-arc current reconstruction based on actual circuit breaker measurements, *IEE Proceedings Science Measurement and Technology*, 149(1), 17–21, January 2002.

116. J.L. Zhang, J.D. Yan, A.B. Murphy, W. Hall, and M.T.C. Fang, Computational investigation of arc behavior in an auto-expansion circuit breaker contaminated by ablated nozzle vapor, *IEEE Transactions on Plasma Science*, 30(2), 706–719, April 2002.

117. P.H. Schavemaker and L. van der Sluis, Quantification of the interrupting performance of high-voltage circuit breakers, *IEE Proceedings Science Measurement and Technology*, 149(4), 153–157, July 2002.

118. G. Bizjak, P. Zunko, and D. Povh, Combined model of SF_6 circuit breaker for use in digital simulation programs, *IEEE Transactions on Power Delivery*, 19(1), 174–180, January 2004.

119. J.L. Guardado, S.G. Maximov, E. Melgoza, J.L. Naredo, and P. Moreno, An improved arc model before current zero based on the combined Mayr and Cassie arc models, *IEEE Transactions on Power Delivery*, 20(1), 138–142, January 2005.

120. H.A. Darwish and N.I. Elkalashy, Universal arc representation using EMTP, *IEEE Transactions on Power Delivery*, 20(2), 772–779, April 2005.

121. G.W. Chang, H.M. Huang, and J.H. Lai, Modeling SF_6 circuit breaker for characterizing shunt reactor switching transients, *IEEE Transactions on Power Delivery*, 22(3), 1533–1540, July 2007.

122. H. Takana, T. Uchii, H. Kawano, and H. Nishiyama, Real-time numerical analysis on insulation capability of compact gas circuit breaker, *IEEE Transactions on Power Delivery*, 22(3), 1541–1546, July 2007.

123. L. Czarnecki and M. Lindmayer, Measurement and statistical simulation of multiple reignitions in vacuum switches, *IEEE Transactions on Plasma Science*, 13(5), 304–310, October 1985.

124. E. Colombo, G. Costa, and L. Piccarreta, Results of an investigation on the overvoltages due to a vacuum circuit-breaker when switching an H.V. motor, *IEEE Transactions on Power Delivery*, 3(1), 205–213, January 1988.

125. B. Kulicke and H.H. Schramm, Application of vacuum circuit breakers to clear faults with delayed current zero, *IEEE Transactions on Power Delivery*, 3(4), 1714–1723, October 1988.

126. J.D. Gibbs, D. Koch, and P. Malkin, Investigations of prestriking and current chopping in medium voltage SF_6 rotating-arc and vacuum switchgear, *IEEE PES Winter Meeting*, New York, January 1989.

127. M.T. Glinkowski and A. Greenwood, Computer simulation of post-arc plasma behavior at short contact separation in vacuum, *IEEE Transactions on Plasma Science*, 17(1), 45–50, February 1989.

128. A.T. Roguski, Experimental investigation of the dielectric recovery strength between the separating contacts of vacuum circuit breakers, *IEEE Transactions on Power Delivery*, 4(2), 1063–1069, April 1989.

129. M.T. Glinkowski and A. Greenwood, Some interruption criteria for short high-frequency vacuum arcs, *IEEE Transactions on Plasma Science*, 17(5), 741–743, October 1989.

130. M. Lindmayer and E.D. Wilkening, The influence of circuit parameters and contact materials on the reignition of high-frequency vacuum arcs, *IEEE Transactions on Components, Hybrids, and Manufacturing Technology*, 13(1), 69–73, March 1990.

131. M. Glinkowski, A. Greenwood, J. Hill, R. Mauro, and V. Varneckes, Capacitance switching with vacuum circuit breakers - A comparative evaluation, *IEEE Transactions on Power Delivery*, 6(3), 1088–1095, July 1991.

132. E.D. Wilkening and M. Glinkowski, Spatial and time characteristics of short gap, high di/dt discharges, *IEEE Transactions on Plasma Science*, 21(5), 489–493, October 1993.

133. T. Kamikawaji, T. Shioiri, T. Funahashi, Y. Satoh, E. Kaneko, and I. Ohshima, An investigation into major factors in shunt capacitor switching performances by vacuum circuit breakers with copper-chromium contacts, *IEEE Transactions on Power Delivery*, 8(4), 1789–1795, October 1993.

134. R.K. Smith, Tests show ability of vacuum circuit breaker to interrupt fast transient recovery voltage rates of rise of transformer secondary faults, *IEEE Transactions on Power Delivery*, 10(1), 266–273, January 1995.

135. J. Kosmac and P. Zunko, A statistical vacuum circuit breaker model for simulation of transients overvoltages, *IEEE Transactions on Power Delivery*, 10(1), 294–300, January 1995.

136. M.T. Glinkowski and P. Stoving, Numerical modeling of vacuum arc interruption based on the simplified plasma equations, *IEEE Transactions on Magnetics*, 31(3), 1924–1927, May 1995.

137. M.T. Glinkowski, M.R. Gutierrez, and D. Braun, Voltage escalation and reignition behavior of vacuum circuit breakers during load shedding, *IEEE Transactions on Power Delivery*, 12(1), 219–226, January 1997.

138. T. Betz and D. Koenig, Fundamental studies on vacuum circuit breaker arc quenching limits with a synthetic test circuit, *IEEE Transactions on Dielectrics and Electrical Insulation*, 4(4), 365–369, August 1997.

139. J. Kaumanns, Measurements and modeling in the current zero region of vacuum circuit breakers for high current interruption, *IEEE Transactions on Plasma Science*, 25(4), 632–636, August 1997.

140. E. Huber, K. Fröhlich, and R. Grill, Dielectric recovery of copper chromium vacuum interrupter contacts after short-circuit interruption, *IEEE Transactions on Plasma Science*, 25(4), 642–646, August 1997.

141. E.F.J. Huber, K.D. Weltmann, and K. Fröhlich, Influence of interrupted current amplitude on the post-arc current and gap recovery after current zero-experiment and simulation, *IEEE Transactions on Plasma Science*, 27(4), 930–937, August 1999.

142. H. Weinert, Reignition phenomena after high-frequency current interruption with short vacuum gaps, *IEEE Transactions on Plasma Science*, 27(4), 944–948, August 1999.

143. M.B.J. Leusenkamp, Vacuum interrupter model based on breaking tests, *IEEE Transactions on Plasma Science*, 27(4), 969–976, August 1999.

144. S.W. Rowe, Post arc dielectric breakdown in vacuum circuit breakers, *High Voltage Engineering Symposium*, London, U.K., 22–27 August, 1999.

145. M. Popov and E. Acha, Overvoltages due to switching off an unloaded transformer with a vacuum circuit breaker, *IEEE Transactions on Power Delivery*, 14(4), 1317–1326, October 1999.

146. T. Fugel and D. Koenig, Switching performance of two 24 kV vacuum interrupters in series, *IEEE Transactions on Dielectrics and Electrical Insulation*, 9(2), 164–168, April 2002.

147. T. Fugel and D. Koenig, Influence of grading capacitors on the breaking performance of a 24-kV vacuum breaker series design, *IEEE Transactions on Dielectrics and Electrical Insulation*, 10(4), 569–575, August 2003.

148. J. Fontchastagner, O. Chadebec, H. Schellekens, G. Meunier, and V. Mazauric, Coupling of an electrical arc model with FEM for vacuum interrupter designs, *IEEE Transactions on Magnetics*, 41(5), 1600–1603, May 2005.

149. D. Pavelescu, G. Pavelescu, F. Gherendi, C. Nitu, G. Dumitrescu, S. Nitu, and P. Anghelita, Investigation of the rotating arc plasma generated in a vacuum circuit breaker, *IEEE Transactions on Plasma Science*, 33(5), 1504–1510, October 2005.

150. S.N. Kharin, H. Nouri, and D. Amft, Dynamics of arc phenomena at closure of electrical contacts in vacuum circuit breakers, *IEEE Transactions on Plasma Science*, 33(5), 1576–1581, October 2005.

151. E.P.A. van Lanen, M. Popov, L. van der Sluis, and R.P.P. Smeets, Vacuum circuit breaker current-zero phenomena, *IEEE Transactions on Plasma Science*, 33(5), 1589–1593, October 2005.

152. A. Horn and M. Lindmayer, Investigations on the series connection of two switching gaps in one tube in vacuum, *IEEE Transactions on Plasma Science*, 33(5), 1594–1599, October 2005.

153. Y. Niwa, T. Funahashi, K. Yokokura, J. Matsuzaki, M. Homma, and E. Kaneko, Basic investigation of a high-speed vacuum circuit breaker and its vacuum arc characteristics, *IEE Proceedings Generation, Transmission and Distribution*, 153(1), 11–15, January 2006.

154. B.K. Rao and G. Gajjar, Development and application of vacuum circuit breaker model in electromagnetic transient simulation, *IEEE Power India Conference*, New Delhi, India, April 10–12, 2006.

155. E.P.A. van Lanen, R.P.P. Smeets, M. Popov, and L. van der Sluis, Current-zero characteristics of a vacuum circuit breaker at short-circuit current interruption, *XXIInd International Symposium on Discharges and Electrical Insulation in Vacuum*, Matsue, Japan, 2006.

156. Y. Shiba, N. Ide, H. Ichikawa, Y. Matsui, M. Sakaki, and S. Yanabu, Withstand voltage characteristics of two series vacuum interrupters, *IEEE Transactions on Plasma Science*, 35(4), 879–884, August 2007.

157. E.P.A. van Lanen, R.P.P. Smeets, M. Popov, and L. van der Sluis, Vacuum circuit breaker post-arc current modelling based on the theory of Langmuir probes, *IEEE Transactions on Plasma Science*, 35(4), 925–932, August 2007.

158. L. Dubé and I. Bonfanti, MODELS: A new simulation tool in the EMTP, *European Transactions of Electrical Power (ETEP)*, 2(1), 45–50, January/February, 1992.

159. L. Dubé, M.T. Correia de Barros, I. Bonfanti, and V. Vanderstockt, Using the simulation language MODELS with EMTP, *11th Power Systems Computation Conference (PSCC)*, Avignon, France, August 30–September 3, 1993.

160. J.A. Martinez, J. Mahseredjian, and B. Khodabakhchian, Parameter determination for modeling systems transients. Part VI: Circuit breakers, *IEEE Transactions on Power Delivery*, 20(3), 2079–2085, July 2005.

161. E.P.A van Lanen, The Current Interruption Process in Vacuum: Analysis of the currents and voltages of current-zero measurements, PhD thesis, Delft University of Technology, the Netherlands, 2008.

162. S. Yanabu, M. Homma, E. Kaneko, and T. Tamagawa, Post arc current of vacuum interrupters, *IEEE Transactions on Power Apparatus and Systems*, 104(1), 166–172, January 1985.

163. Y.P. Raizer, *Gas Discharge Physics*, J.E. Allen (Ed.), Springer, Berlin, 1991.

164. R.L. Boxman, D.M. Sanders, and P.J. Martin, *Handbook of Vacuum Arc Science and Technology; Fundamentals and Applications*, Noyes, Park Ridge, NJ, 1995.

165. M.A. Liberman and A.J. Lichtenberg, *Principles of Plasma Discharges and Materials Processing*, John Wiley, Hoboken, NJ, 2005.

166. M. Popov, L. van der Sluis, and G.C. Paap, Investigation of the circuit breaker reignition overvoltages caused by no-load transformer switching surges, *European Transactions of Electrical Power (ETEP)*, 11(6), 413–422, November/December 2001.

167. T.A. Roguski, Experimental investigation of the dielectric recovery strength between the separating contacts of vacuum circuit breaker, *IEEE Transactions on Power Delivery*, 4(2), 1063–1069, April 1989.

168. R.P.P. Smeets, T. Funahashi, E. Kaneko, and I. Ohshima, Types of reignition following high-frequency current zero in vacuum interrupters with two types of contact material, *IEEE Transactions on Plasma Science*, 21(5), 478–483, April 1993.

169. J. Helmer and M. Lindmayer, Mathematical modelling of the high frequency behaviour of vacuum interrupters and comparison with measured transients in power systems, *XVIIth International Symposium on Discharges and Electrical Insulation in Vacuum*, Berkeley, CA, 1996.

170. A. Greenwood, *Electrical Transients in Power Systems*, John Wiley & Sons, New York, 1991.

171. C.D. Child, Discharge from hot CaO, *Physical Review*, 32(5), 492–511, May 1911.

172. J.G. Andrews and R.H. Varey, Sheath growth in a low pressure plasma, *Physics Fluids*, 14(2), 339–343, February 1971.

173. E. Düllni, E. Schade, and B. Gellert, Dielectric recovery of vacuum arcs after strong anode spot activity, *XIIth International Symposium on Discharges and Electrical Insulation in Vacuum*, Shoresh, Israel, 1986.

174. G. Lins, Influence of electrode separation on the ion density in vacuum arcs, *IEEE Transactions on Plasma Science*, 19(5), 718–724, October 1991.

175. G. Düning and M. Lindmayer, Energy and density of ions in vacuum arcs between axial and radial magnetic field contacts, *IEEE Transactions on Plasma Science*, 29(5), 726–733, October 2001.

176. A. Klajn, Experimental analysis of ion parameters during the forced vacuum arc interruption, *XXth International Symposium on Discharges and Electrical Insulation in Vacuum*, Tours, France, 2002.

177. H. Pursch and B. Jüttner, The behavior of the ion current at the extinction of high current vacuum arcs, *XVIIth International Symposium on Discharges and Electrical Insulation in Vacuum*, Berkeley, CA, 1996.

178. M. Lindmayer and E.D. Wilkening, Breakdown of short vacuum gaps after current zero of high-frequency arcs, *XIVth International Symposium on Discharges and Electrical Insulation in Vacuum*, Santa Fe, NM, 1990.

179. S. Baboolal, Boundary conditions and numerical fluid modelling of time evolutionary plasma sheaths, *Journal of Physics D: Applied Physics*, 35, 658–664, 2002.

180. S.W. Rowe and J.P. Boeuf, Simulation of plasma sheath dynamics in vacuum circuit breakers, *XXth International Symposium on Discharges and Electrical Insulation in Vacuum*, Tours, France, 2002.

181. S.-B. Wang and A.E. Wendt, Sheath thickness evaluation for collisionless or weakly collisional bounded plasmas, *IEEE Transactions on Plasma Science*, 27(5), 1358–1365, October 1999.

182. B. Khodabakhchian, J. Mahseredjian, M.R. Sehati, and M. Mir-Hosseini, Potential risk of failures in switching EHV shunt reactors in some one-and-a-half breaker scheme substations, *IPST'03*, New Orleans, LA, September 28–October 2, 2003.

183. G.C. Damstra, Virtual chopping phenomena switching three-phase inductive circuits, *Colloquium of CIGRE SC 13*, Helsinki, Finland, September 1981.

184. J. Panek and K.G. Fehrle, Overvoltage phenomena associated with virtual current chopping in three phase circuits, *IEEE Transactions on Power Apparatus and Systems*, 94(4), 1317–1325, July/August 1975.

185. A.O. Soysal, A method for wide frequency range modelling of power transformers and rotating machines, *IEEE Transactions on Power Delivery*, 8(4), 1802–1810, October 1993.

186. A.O. Soysal, Protection of arc furnace supply systems from switching surges, *IEEE PES Winter Meeting*, New York, January 31–February 4, 1999.

187. J. Avila-Rosales and F. Alvarado, Non-linear frequency dependent transformer model for electromagnetic transient studies in power systems, *IEEE Transactions on Power Apparatus and Systems*, 101(11), 4281–4288, November 1982.

188. M. Popov, L. van der Sluis, G.C. Paap, and P.H. Schavemaker, On a hysteresis model for transient analysis, *IEEE Power Engineering Review*, 20(5), 53–54, May 2000.

189. J. Helmer, Hochfrequente Vorgange zwischen Vakuum-Schaltstrecken und dreiphasigen Kreisen, PhD thesis, Technischen Universitat, Braunschweig, Germany, 1996.

190. G.P. Slade, Vacuum interrupters: The new technology for switching and protecting distribution circuits, *IEEE Transactions on Industry Applications*, 33(6), 1501–1511, November/December 1997.
191. W. Schmidt, J. Meppelink, B. Richter, K. Feser, L.E. Kehl, and D. Qui, Behaviour of MO-surge-arrester blocks to fast transients, *IEEE Transactions on Power Delivery*, 4(1), 292–300, January 1989.
192. J.A. Martinez-Velasco, Digital computation of electromagnetic transients in power systems. Current status, Chapter 1 of *Modeling and Analysis of System Transients using Digital Systems*, A. Gole, J.A. Martinez-Velasco, and A.J.F. Keri (Eds.), IEEE Special Publication, TP-133-0, 1998.
193. D.A. Woodford and L.M. Wedepohl, Impact of circuit breaker pre-strike on transmission line energization transients, *IPST'97*, Seattle, WA, June 22–26, 1997.
194. IEC 60071-2, Insulation co-ordination, Part 2: Application guide, 1996.
195. H.W. Dommel, *Electromagnetic Transients Program. Reference Manual (EMTP Theory Book)*, Bonneville Power Administration, Portland, 1986.
196. D.W. Durbak, A.M. Gole, E.H. Camm, M. Marz, R.C. Degeneff, R.P. O'Leary, R. Natarajan, J.A. Martinez-Velasco, K.C. Lee, A. Morched, R. Shanahan, E.R. Pratico, G.C. Thomann, B. Shperling, A.J.F. Keri, D.A. Woodford, L. Rugeles, V. Rashkes, and A. Sharshar, Modeling guidelines for switching transients, Chapter 4 of *Modeling and Analysis of System Transients using Digital Systems*, A. Gole, J.A. Martinez-Velasco and A.J.F. Keri (Eds), IEEE Special Publication, TP-133-0, 1998.
197. CIGRE WG 33/13-09, Very fast transient phenomena associated with gas insulated substations, *CIGRE Session*, Paper 33–13, Paris, France, 1988.
198. J.A. Martinez-Velasco, Very fast transients, Chapter 10 of *Power Systems*, L.L. Grigsby (Ed.), CRC Press, Boca Raton, FL, 2007.
199. A. Ecklin, D. Schlicht, and A. Plessl, Overvoltages in GIS caused by the operation of isolators, Chapter 6 of *Surges in High-Voltage Networks*, K. Ragaller (Ed.), Plenum Press, New York, 1980.
200. S.A. Boggs, F.Y. Chu, N. Fujimoto, A. Krenicky, A. Plessl, and D. Schlicht, Disconnect switch induced transients and trapped charge in gas-insulated substations, *IEEE Transactions on Power Apparatus and Systems*, 101(6), 3593–3602, October 1982.
201. B. Kulicke and H.H. Schramm, Clearance of short-circuits with delayed current zeros in the Itaipu 500 kV substation, *IEEE Transactions on Power Apparatus and Systems*, 99(4), July/August 1980.
202. Q. Bui-Van, B. Khodabakhchian, M. Landry, J. Mahseredjian, and J. Mainville, Performance of series-compensated line circuit breakers under delayed current-zero, *IEEE Transactions on Power Delivery*, 12(1), 227–233, January 1997.
203. A.C. Cavalcanti de Carvalho, C.M. Portela, M. Lacorte, and R. Colombo, A Teoria do Arco Elétrico nos Disjuntores de Alta-Tensão, (in Portuguese) Chapter 11 of *Disjuntores e Chaves. Aplicaçao em Sistemas de Potência*, FURNAS/EDUFF, Rio de Janeiro, Brazil, 1995.
204. R. Amsinck, Verfahren zur ermittlung der das ausschaltverhalten bestimmenden lichtbogen-kenngrössen, *ETZ-A Elektrotechn.*, 98, 566–567, 1977.
205. H. Rijanto, Experimentelle bestimmung der parameter der verallgemeinerten lichtbogengleichung zur berechnung von schaltvorgängen, *ETZ-A Elektrotechn.*, 95, 221–223, 1974.
206. R. Ruppe, Experimentelle und theoretische untersuchungen am axial beströmenden wechsel-stromlichtbögen vor dem stromnulldurchgang, PhD thesis, Ilmenau, Germany, 1980.
207. H. Drebnestedt, W. Rother, C.H. Weber, and P. Zahlmann, Ein verbessertes verfahren zur bestimmung der charakteristischen funktionen des zweipolmodells für schaltlichtbogen, *Int. Wissenschaftlichen Kolloquium*, Ilmenau, Germany, 1983.
208. L. Blahous, Derivation of circuit breaker parameters by means of Gaussian approximation, *IEEE Transactions on Power Apparatus and Systems*, 101(12), 4611–4616, December 1982.
209. W. Giménez and O. Hevia, Method to determine the parameters of the electric arc from test data, *IPST'99*, Budapest, Hungary, June 20–24, 1999.
210. D. Dufournet and M. Collet, Calculation of dynamic arc behaviour for small currents, *CIGRE Colloquium SC13*, Paper 13–89, Sarajevo, Bosnia and Herzegovina, 1989.

211. W. Widl, P. Kirchesch, and W. Egli, Use of integral arc models in circuit breaker testing and development, *IEEE Transactions on Power Delivery*, 3(4), 1685–1691, October 1988.
212. B. Blez and C. Guilloux, Post-arc current in high voltage SF_6 circuit breaker when breaking at up 63 kA, *IEEE Transactions on Power Delivery*, 4(2), 1056–1062, April 1989.
213. S. Nitu, C. Nitu, and P. Anghelita, Electric arc model for high power measurements, *EUROCON* 2005, Belgrade, Serbia, November 22–24, 2005.
214. J.D. Gibbs, D. Koch, P. Malkin, and K.J. Cornick, Comparison of performance of switching technologies on E Cigre motor simulation circuit, *IEEE Transactions on Power Delivery*, 4(3), 1745–1750, July 1989.
215. R.P.P. Smeets, The origin of current chopping in vacuum arcs, *IEEE Transactions on Plasma Science*, 17(2), 303–310, April 1989.
216. J. Lopez-Roldan, H. de Herdt, T. Sels, D. van Dommelen, M. Popov, L. van der Sluis, and J. Declercq, Analysis, simulation and testing of transformer insulation failures related to switching transients overvoltages, *CIGRE Session*, Paper 12–116, Paris, France, 2002.
217. A.N. Greenwood and M.T. Glinkowski, Voltage escalation in vacuum switching operations, *IEEE Transactions on Power Delivery*, 3(4), 1698–1706, October 1988.
218. M. Galonska, R. Hollinger, I.A. Krinberg, and P. Spaedtke, Influence of an axial magnetic field on the electron temperature in a vacuum arc plasma, *IEEE Transactions on Plasma Science*, 33(5), 1542–1547, October 2005.
219. S. Nam, B. Lee, S. Park, S. Kim, and Y. Han, Spectroscopic measurement of high current vacuum arc plasma in triggered vacuum switch, *Proceedings of PPPS*, 2, 1790–1793, June 2001.
220. K. Arai, S. Takahashi, O. Morimiya, and Y. Niwa, Probe measurements of residual plasma of magnetically confined high-current vacuum arc, *IEEE Transactions on Plasma Science*, 31(5), 929–933, October 2003.
221. A. Klajn, Langmuir probes in switching vacuum arc measurements, *IEEE Transactions on Plasma Science*, 33(5), 1611–1617, October 2005.

Appendix A: Techniques for the Identification of a Linear System from Its Frequency Response Data

Bjørn Gustavsen and Taku Noda

CONTENTS

A.1 Introduction 558
A.2 Problem Description..559
    A.2.1 Pole-Residue Modeling..559
    A.2.2 Pole-Residue Modeling with Inclusion of Delays 560
        A.2.2.1 Modal-Domain Modeling 561
        A.2.2.2 Phase-Domain Modeling...561
A.3 J. Marti's Fitting..561
A.4 Vector Fitting...562
    A.4.1 Classical Formulation 563
        A.4.1.1 Pole Relocation 563
        A.4.1.2 Initial Poles..564
        A.4.1.3 Column Fitting, Matrix Fitting 564
    A.4.2 Further Improvements and Extensions to Vector Fitting 564
        A.4.2.1 Relaxation 564
        A.4.2.2 Orthonormalization 565
        A.4.2.3 Modal Formulation 565
        A.4.2.4 Fast Implementation 566
        A.4.2.5 Multivariate Fitting 566
        A.4.2.6 Implementations in Time Domain and z-Domain..........566
A.5 Fitting by Frequency Response Partitioning...570
    A.5.1 Pole Identification by Rational Fitting...571
        A.5.1.1 Matrix Trace for Pole Identification 571
        A.5.1.2 Formulation of Overdetermined Linear Equations.........572
        A.5.1.3 Adaptive Weighting 573
        A.5.1.4 Column Scaling..574
        A.5.1.5 Solution of Equation A.38 by QR Decomposition 574
        A.5.1.6 Iteration Step Adjustment 575
        A.5.1.7 Calculation of Poles from Denominator 575
        A.5.1.8 Search for an Optimal Model Order..................................576
    A.5.2 Entire Identification Algorithm 576
        A.5.2.1 Partitioning of Frequency Response for
                Pole Identification...576
        A.5.2.2 Identification of Residue Matrices 577
    A.5.3 Other Applications 583

A.6   Passivity..584
    A.6.1   Passivity Assessment 584
        A.6.1.1   Unsymmetrical Models 584
        A.6.1.2   Symmetrical Models 584
    A.6.2   Passivity Enforcement..585
References...588

A.1   Introduction

Virtually every program for the simulation of electromagnetic transients (e.g., ATP, EMTP-RV, and PSCAD/EMTDC) have the capability of modeling overhead lines and cables by the traveling wave method, also known as the method of characteristics (MoC), while including their frequency-dependent effects due to skin effect in conductors and earth. This is achieved by fitting the functions of characteristic admittance and propagation [1–11] with rational functions, thereby enabling efficient time-domain simulations by recursive convolution [1]. A large number of papers have also shown how to use such techniques for the modeling of frequency-dependent network equivalents and transformers, see for instance [12–14]. Asymptotic fitting using real poles only is currently applied to the so-called J. Marti line model [2] but is limited to fundamentally nonoscillatory responses. Linear least-squares fitting via a ratio of two polynomials is a more general approach as it can also accommodate complex poles [15,16], but its application is limited to low-order cases due to poor numerical conditioning. The recent vector fitting (VF) approach [17] has overcome most of these limitations in robustness and is nowadays widely applied for the frequency-dependent modeling of transmission lines [10] and high-speed interconnects [18]. A time-domain counterpart is described in [19]. It is also possible to overcome the conditioning problem of the polynomial approach by subdividing the frequency response into a number of partitions that are each subjected to rational fitting [14,20]. The model must be passive in order to guarantee a stable time-domain simulation, which implies that the model cannot generate power whatever may be the terminal conditions. Several methods have been published which enforces passivity by perturbing the model parameters in a postprocessing step [21–29].

The first part of this appendix introduces the multiport pole-residue model, which is most amenable for inclusion in EMT-like programs. The inclusion of delays in the modeling is also explained.

The second part describes the classical VF approach. This method iteratively relocates a set of initial poles to better positions by solving a linear system where partial fractions (pole-residue terms) constitute the basis functions. The new poles are calculated in each iteration as the solution of an eigenvalue problem, and any unstable pole can be flipped into the left half plane to ensure a pole with stable poles only. We also describe some enhancements: relaxation of the nontriviality constraint, which can improve the convergence properties; orthonormalization, which can further improve the robustness; a modal formulation, which can retain the relative accuracy of the modes; and a fast implementation of the pole identification step, which can greatly reduce the computation time for cases with many ports. Mention is also made to a multivariate formulation, as well as to the application of VF to time-domain and z-domain data. The VF approach is demonstrated for the modeling of overhead lines by the traveling wave method. We also show the significance of least-squares weighting for an example where a transformer is to be modeled from measured frequency responses.

The third part presents an alternative algorithm for identifying a multiphase network equivalent for electromagnetic transient calculations based on partitioning of frequency response data [14]. Conventional rational fitting methods intrinsically get into ill-condition because the s^n terms in the rational function take a wide range of values which cannot be accurately treated within machine arithmetic accuracy. The proposed algorithm averts the ill-conditioning by partitioning the given frequency response into sections along the frequency axis. Rational fitting is applied to each section of the frequency response to identify the poles, and then the corresponding residue matrices are obtained by a standard least-squares procedure using the entire frequency response. Adaptive weighting, column scaling, and iteration step adjustment are utilized to facilitate the rational fitting. The proposed identification algorithm is applied to obtain a reduced-order equivalent of a 500 kV power network, and the equivalent is used in a switching transient simulation to demonstrate its performance.

The fourth part deals with the topic of model passivity. Passivity of the model is a requirement for guaranteeing a stable simulation. Methods are presented for assessing the model's passivity, based on frequency sweeping and on algebraic tests. A half-size test matrix is described for use with the symmetrical pole-residue model. Passivity enforcement methods are discussed that are based on perturbing the model's residues, with emphasis on an approach that perturbs the residue matrix eigenvalues [24]. Computational results are shown for the wideband modeling of a power transformer.

A.2 Problem Description

A.2.1 Pole-Residue Modeling

The frequency response data of an M-input M-output linear time invariant multi-input multi-output (LTI MIMO) system are given in the form of numerical transfer function matrices:

$$\mathbf{H}_k = \begin{bmatrix} h_{k,11} & h_{k,12} & \cdots & h_{k,1M} \\ h_{k,21} & h_{k,22} & \cdots & h_{k,2M} \\ \vdots & \vdots & & \vdots \\ h_{k,M1} & h_{k,M2} & \cdots & h_{k,MM} \end{bmatrix} \in \mathbf{C}^{M \times M} \tag{A.1}$$

defined at discrete angular frequency points $\omega = \omega_k (k = 1, 2, \ldots, K)$. At the kth frequency point, the frequency-domain input vector $\mathbf{u}_k \in \mathbf{C}^M$ is related to the output vector $\mathbf{y}_k \in \mathbf{C}^M$ by

$$\mathbf{y}_k = \mathbf{H}_k \mathbf{u}_k \tag{A.2}$$

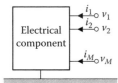

FIGURE A.1
Multiterminal electrical component.

Figure A.1 shows an electrical component with M terminals. This is a subnetwork to be reduced or a network component like a transformer whose equivalent will be identified. When the voltages (to ground) and the currents (injected into the component) of the terminals are considered to be the input and the output respectively, $\mathbf{H}_k$ becomes the admittance matrix of the component.

If the transfer function matrices $\mathbf{H}_k$ in Equation A.1 are formally denoted by $\mathbf{H}(s)$ (function of the complex frequency $s = j\omega$), the discrete input–output relation (Equation A.2) is expressed by

$$\mathbf{y}(s) = \mathbf{H}(s)\,\mathbf{u}(s) \tag{A.3}$$

where $\mathbf{u}(s)$ and $\mathbf{y}(s)$ are the frequency-domain input and output vectors of size M. It can be approximated by the matrix partial fraction expansion model (referred to as MPFE hereafter):

$$\mathbf{H}(s) \cong \sum_{n=1}^{N} \frac{\mathbf{R}_n}{s - p_n} + \mathbf{D} + s\mathbf{E} \tag{A.4}$$

where

$\mathbf{D}$ and $\mathbf{E}$ are possibly zero

$\mathbf{R}_n$ is the residue matrix for the nth pole p_n ($\mathbf{R}_n \in \mathbf{R}^{M \times M}$ for real poles and $\mathbf{R}_n \in \mathbf{C}^{M \times M}$ for complex poles)

Thus, our objective is to identify the poles p_n, the corresponding residue matrices $\mathbf{R}_n$, and the constant term $\mathbf{D}$ from the given frequency response matrix $\mathbf{H}(s)$.

The terms $\mathbf{D} + s\mathbf{E}$ correspond to the asymptotic value of $\mathbf{H}(s)$ for $f \to \in$. If this value is *a priori* known, the final form of our objective is to identify the poles p_n and the residue matrices $\mathbf{R}_n$ of the strictly proper MPFE model

$$\mathbf{F}(s) \cong \sum_{n=1}^{N} \frac{\mathbf{R}_n}{s - p_n} \tag{A.5}$$

from the transfer function matrix $\mathbf{F}(s) = \mathbf{H}(s) - \mathbf{D} - s\mathbf{E}$. If the response at $f \to \in$ is unknown, we can use $\mathbf{D} = 0$ assuming no response at $f \to \in$.

It is remarked that one can also do the modeling using a private pole set for each matrix column (columnwise fitting). This often gives a lower order model, for a given accuracy. On the other hand, the columnwise fitting leads to a slightly unsymmetrical model.

A.2.2   Pole-Residue Modeling with Inclusion of Delays

In transmission line modeling, it is necessary to model the transfer of waves from one line end to the other end. This propagation is defined by the propagation function,

$$\mathbf{A}(s) = e^{-\sqrt{\mathbf{ZY}}\,l} \tag{A.6}$$

where

$\mathbf{Z}$ and $\mathbf{Y}$ are the per-unit-length values of series impedance and shunt admittance

l is the line length

A.2.2.1 Modal-Domain Modeling

Direct pole-residue modeling of $\mathbf{A}(s)$ will in most cases lead to very high-order fittings (and thus slow simulations), due to the delay effect of the line. It is therefore a common practice to factor out the delay effect before doing the fitting, thus permitting low-order modeling. For a single line model this leads to the fitting problem:

$$a(s) \cong \left(\sum_{m=1}^{N} \frac{r_m}{s - a_m} \right) e^{-s\tau} \tag{A.7}$$

The same fitting problem also arises when modeling multiconductor lines via modal decomposition since the multiconductor problem is effectively decoupled into a set of single-phase lines.

A.2.2.2 Phase-Domain Modeling

Some transmission line models formulate the transmission line equations directly in the phase domain, without modal decomposition. Some models attempt at using a common delay for all matrix elements

$$\mathbf{H} \cong \sum_{m=1}^{M} \left(\frac{\mathbf{R}_m}{s - a_m} \right) e^{-s\tau} \tag{A.8}$$

while the universal line model (ULM) [10] makes use of several delays in the following approximation:

$$\mathbf{H} \cong \sum_{g=1}^{G} \left(\sum_{m=1}^{N_g} \frac{\mathbf{R}_{m,g}}{s - a_{m,g}} \right) e^{-s\tau_g} \tag{A.9}$$

A.3 J. Marti's Fitting

J. Marti's frequency-dependent transmission line model [2], which has been implemented in the EMTP, is widely used for practical electromagnetic transient simulations. It is a modal-domain transmission line model, and thus, the characteristic impedance and the propagation function of each mode are, respectively, replaced by a reduced-order linear equivalent as in Equation A.5 (scalar version) and Equation A.7 of Section A.2. A reduced-order equivalent is identified so that the frequency response of the equivalent closely reproduces that of the modal-domain characteristic impedance or propagation function. Note that this identification process is a scalar problem since the line model is a modal-domain one.

This fitting (identification) process basically uses Bode's method [30]. When J. Marti implemented the fitting algorithm in the EMTP, he fully automated the process as a computer program. The basic idea of Bode's method is simple:

1. If a −20 dB/decade slope is found in the given frequency response, extrapolate the slope to obtain the angular frequency ω where the slope intersects the 0 dB line. Then, place a pole on the s-plane (complex frequency plane) at $p = -\omega$.

2. If a +20 dB/decade slope is found in the given frequency response, obtain the angular frequency ω in the same way as mentioned earlier, and then, place a zero on the s-plane at $z = -\omega$.

When the given frequency response is plotted as a Bode diagram (angular frequency versus response in a log–log scale), the response curve can be divided into parts each of which can be approximated by a straight line. Then, if one of the aforementioned rules is applied to each part, a reduced-order linear equivalent is identified in the following form:

$$20\log|H(s)| \cong 20\log k + 20\log|s + z_1| + \cdots + 20\log|s + z_n|$$

$$- 20\log|s + p_1| - \cdots - 20\log|s + p_n| \tag{A.10}$$

where k is the overall gain. This leads to the partial fraction form

$$H(s) \cong k\frac{(s+z_1)\cdots(s+z_n)}{(s+p_1)\cdots(s+p_n)} = k_0 + \frac{k_1}{s+p_1} + \cdots + \frac{k_n}{s+p_n} \tag{A.11}$$

How the frequency response is divided into many straight lines is the key to obtain good approximation result, and this has been realized in an automatic way as a computer program in the EMTP [2,31].

The following points should be noted when this fitting method is used. The algorithm uses only the magnitude information of the given frequency response and assumes that the response is of the minimum-phase shift type. Thus, the method is applicable only to the fitting of minimum-phase shift type responses. In addition, since the method identifies real poles only, fitting an oscillatory response is difficult. If these limitations can be a problem, vector fitting (Section A.4) or fitting by frequency response partitioning (Section A.5) should be used instead.

A.4 Vector Fitting

A widely applied technique for calculating rational approximations is the pole relocating VF method [17]. This approach relocates a set of initial poles to better positions by repeatedly solving a well-conditioned set of equations.

A.4.1 Classical Formulation

A.4.1.1 *Pole Relocation*

The VF method approximates the given frequency response $h(s)$ with a pole-residue model as follows:

$$h(s) \cong \sum_{m=1}^{N} \frac{r_m}{s - a_m} + d + se \tag{A.12}$$

where terms d and e are optional. The poles $\{a_m\}$ are given some initial values and the following linear problem is solved as an overdetermined linear least-squares problem:

$$\sigma(s)h(s) \cong \sum_{m=1}^{N} \frac{r_m}{s - a_m} + d + se \tag{A.13a}$$

$$\sigma(s) = \sum_{m=1}^{N} \frac{\tilde{r}_m}{s - a_m} + 1 \tag{A.13b}$$

It can be shown [17] that the zeros of $\sigma(s)$ represent an improved estimate for the poles of $h(s)$. The zeros are calculated by solving the eigenvalue problem (Equation A.13) [4],

$$\{a_m\} = \text{eig}(\mathbf{A} - \mathbf{b} \cdot \mathbf{c}^{\mathrm{T}}) \tag{A.14}$$

where
   $\mathbf{A}$ is a diagonal matrix holding the poles $\{a_m\}$
   $\mathbf{b}$ is a column of ones
   $\mathbf{c}$ holds the residues $\{\tilde{r}_m\}$

Repeated application of Equations A.13 and A.14 relocates the initial pole set into better positions. Convergence implies $\{\tilde{r}_m = 0\}$. In practice, one will usually terminate the iterations before the convergence is complete. The final residues are calculated by solving Equation A.12 with known poles.

During the iterations, unstable poles may occur. Any unstable pole is flipped into the left half plane by inverting the sign of the real part, thereby guaranteeing a fitting with stable poles only.

The pole identification and subsequent residue identification both lead to solving an overdetermined problem in the form

$$\mathbf{A}\mathbf{x} \cong \mathbf{b} \tag{A.15}$$

A transformation of variables is used in [17] to ensure that complex poles and residues come in conjugate pairs. The conditioning of $\mathbf{A}$ is improved by introducing column scaling,

$$\mathbf{A}_{\text{col}} := \mathbf{A}_{\text{col}} / \|\mathbf{A}_{\text{col}}\|_2 \tag{A.16}$$

After solving Equation A.16, the corresponding elements of the solution vector are recovered:

$$\mathbf{x}_{col} := \mathbf{x}_{col} / \|\mathbf{A}_{col}\|_2 \tag{A.17}$$

The obtained result can be sensitive to the applied solver used in the pole identification and residue identification steps. A solver based on QR decomposition with rank-revealing column pivoting is usually the preferred choice. This solver is used by MATLAB® when specifying the backslash operator, when **A** has more rows than columns (overdetermined).

A.4.1.2 Initial Poles

The initial pole set to be specified for the first iteration should, in general, be taken as complex-conjugate pairs with weak attenuation that are distributed over the frequency band,

$$a_n = -\alpha + j\beta, \quad a_{n+1} = -\alpha - j\beta \tag{A.18}$$

where

$$\alpha = k \cdot \beta \tag{A.19}$$

The parameter k should be chosen as 0.01 or smaller. This choice of initial poles results in that the system to be solved is well conditioned, leading to fast convergence. Also, the distribution of the poles over the entire frequency band reduces the probability that poles must be relocated over long distances, thus reducing the required number of iterations. In the case of intrinsically smooth responses such as those encountered in transmission line modeling by the MoC, logarithmically spaced real poles can also be used.

A.4.1.3 Column Fitting, Matrix Fitting

VF can also be applied to a vector of elements (hence the designation VF). As a consequence, all elements in the vector become fitted with a common pole set; see appendix in [17] for details. This can be used for columnwise fitting (Equation A.6). By stacking the elements of the upper (or lower) triangle of the matrix into a single vector and subjecting it to VF, a model is obtained which has a common pole set for all elements. Rearranging coefficients leads to the pole-residue model defined in Equation A.4.

A.4.2 Further Improvements and Extensions to Vector Fitting

A.4.2.1 Relaxation

Practical experience with VF has shown that the convergence can be slow when the fitting errors are nonnegligible. This happens when the response $h(s)$ is contaminated with noise, and when trying to fit with a too low order. The underlying cause was in [32] found to be the normalization used in Equation A.13 where the factor $\sigma(s)$ is enforced to approach unity at infinite frequency. It was shown that by freeing up the asymptotic value, faster convergence and a more accurate end result is achieved.

In this relaxed VF (RVF) [32], the asymptotic value $\tilde{d}$ is a free variable (Equation A.20a) and a relaxed normalization (Equation A.20b) is included as an additional row in the system matrix.

$$\left(\sum_{m=1}^{N} \frac{\tilde{r}_m}{s - a_m} + \tilde{d} \right) h(s) = \sum_{m=1}^{N} \frac{r_m}{s - a_m} + d + se \qquad \text{(A.20a)}$$

$$\text{Re} \left\{ \sum_{k=1}^{N_s} \left(\sum_{m=1}^{N} \frac{\tilde{r}_m}{s_k - a_m} + \tilde{d} \right) \right\} = N_s \qquad \text{(A.20b)}$$

Equation A.20b is in the LS problem when weight is given as:

$$\text{weight} = \|w(s) \cdot f(s)\|_2 / N_s \qquad \text{(A.21)}$$

where w is the user-specified weight for the fitting of $h(s)$.

The new pole set is obtained by solving the following equation:

$$\{a_m\} = \text{eig}(\mathbf{A} - \mathbf{b} \cdot \tilde{d}^{-1} \cdot \mathbf{c}^{\mathsf{T}}) \qquad \text{(A.22)}$$

A.4.2.2 Orthonormalization

If the initial poles are specified with too large real parts, the resulting basis functions (partial fractions) will result in a poorly conditioned system matrix. This problem can be alleviated by replacing the original basis functions with an orthonormal set [33]. The computation is done analytically via Gram–Schmidt orthonormalization. The result by this orthonormal VF (OVF) usually converges to the same result as VF, since the conditioning with VF and a poor initial pole set improves with each iteration. Another advantage of the orthonormalization is that it permits to fit frequency responses that have multiple poles. Combining OVF and RVF leads to a very powerful method (ROVF) [34].

A.4.2.3 Modal Formulation

One difficulty in the modeling of multiport devices via the nodal admittance matrix $\mathbf{Y}$ is that the small eigenvalues of $\mathbf{Y}$ may be weakly observable in its elements. At the same time, the eigenvector matrix of $\mathbf{Y}$ is in general frequency dependent, making diagonalization via a constant transformation matrix inapplicable. In order to deal with this situation, modal VF (MVF) was introduced [35]. This approach explicitly introduces eigenvalue–eigenvector pairs in the VF formulation, thereby allowing relative error control of the modes by inverse weighting of the eigenvalue magnitude. That way, the small eigenvalues can be accurately represented along with the large eigenvalues. The relaxed nontriviality constraint can also be introduced in MVF (RMVF), for the purpose of faster convergence. The disadvantage of (R) MVF is a less-sparse matrix structure of the least-squares problems associated with pole identification and residue identification, which leads to slower computations.

A.4.2.4   Fast Implementation

The sparse structure of VF was utilized in [36] to achieve a fast solving for the pole identification step, leading to fast VF (FVF). This leads to large savings in computation time when fitting vectors having many elements. Usage of FVF is particularly useful when fitting a symmetrical pole-residue model to a device/system having many ports. The same procedure can also be used in the pole identification step of MVF [37].

A.4.2.5   Multivariate Fitting

OVF has been extended to multivariate cases, where one or more geometrical or material parameters are used as free variables (in addition to frequency) [38]. That way, a single model can represent many variations of the same structure/component.

A.4.2.6   Implementations in Time Domain and z-Domain

VF was in [39] reformulated for use with time-domain data, so-called time-domain VF (TD-VF). Usage of TD-VF has been used for rational modeling of components and systems from time-domain responses, mainly in the fields of electronic packaging. VF can also be used for handling z-domain data [40].

Example A.1:   Transmission Line Modeling

1. Phase-domain fitting
Frequency-dependent transmission line modeling by the MoC requires fitting the functions of characteristic admittance $\mathbf{Y}_c$ and propagation function $\mathbf{H}$ with rational functions. One option is to fit the elements of $\mathbf{H}$ directly in the phase domain with extraction of a common delay factor [7,8].

   In the following, the column of $\mathbf{H}$ that corresponds to the fourth conductor (Figure A.2) is fitted. The column elements are multiplied with a factor $\exp(s\tau)$ before doing the fitting, with τ being the lossless delay of the line. These elements are seen to oscillate quite strongly (Figure A.3), due to the high ground resistivity (10,000 Ω.m) which causes the delay of the ground mode to be significantly higher than that of the aerial modes. Figure A.3 also shows the resulting rational fitting as calculated by the RVF using a 40th order common pole set. The initial poles were taken as complex conjugate as in Equation A.18 with $k = 0.01$, logarithmically distributed over the frequency

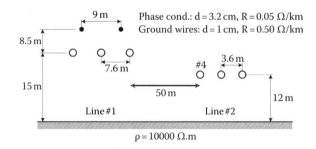

FIGURE A.2
Example A.1: Two parallel overhead lines ($l = 50$ km).

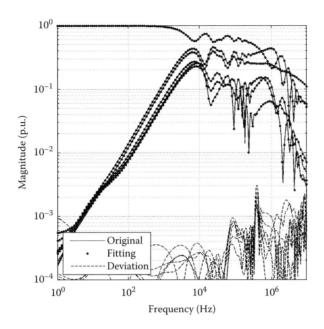

FIGURE A.3
Example A.1: Propagation function (4th column)—40th order fitting.

band. It is seen that the magnitude of the complex deviation is smaller than 2*E*–3, implying a quite accurate result.

Figure A.4 shows the RMS error as a function of the iteration count with the different VF versions, when the order is either 30 or 40. It is seen that usage of relaxation in VF greatly improves

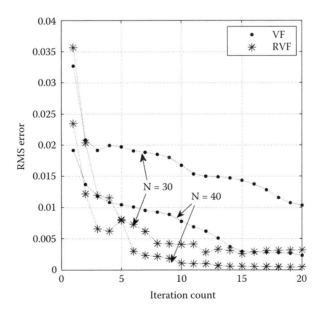

FIGURE A.4
Example A.1: RMS error versus iteration count.

the convergence speed. The main reason for this improvement is that the standard VF algorithm is biased to relocating poles from high to low frequencies, whereas in this example the initial poles need to be relocated toward higher frequencies.

2. Modal-domain fitting with delay optimization
One category of transmission line models [1,2] is based on fitting the modes of the propagation function by a rational function plus a single time delay (Equation A.23). The fitted modes can then be directly utilized in the model together with a real transformation matrix.

$$a(s) \cong \sum_m \frac{r_m}{s-a}\, e^{-s\tau} \tag{A.23}$$

The accuracy of the fitting can be greatly improved by refining the time delays during the fitting process. This is easily achieved by combining VF with a single variable search method for optimizing the delay. One implementation that uses Brent's method is detailed in [41].

Figure A.5 compares the fitting result for the ground mode of the configuration in Figure A.2 when the delay is taken as the lossless delay, and when obtained via optimization. It is seen that a great improvement in accuracy is achieved by delay optimization, which in this example increased the delay from 167 to 192 μs.

3. *The universal line model*
In the ULM [10,11], the poles and delays obtained from mode fitting are used as known quantities for a final fitting of residues **R** in the phase domain.

$$\mathbf{A}(s) \cong \sum_{g=1}^{G} \left(\sum_{m=1}^{N_g} \frac{\mathbf{R}_{m,g}}{s-a_m} \right) e^{-s\tau_g} \tag{A.24}$$

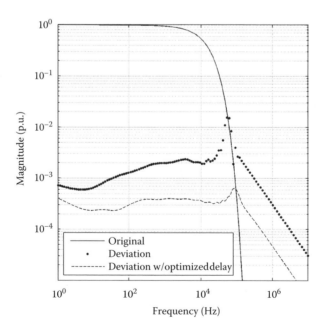

FIGURE A.5
Example A.1: Tenth order fitting of ground mode.

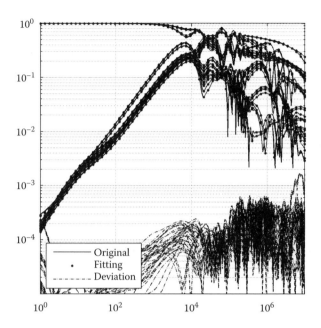

FIGURE A.6
Example A.1: Rational fitting (all columns).

where
   G denotes the number of (lumped) modes
   N_g is the number of poles used for fitting the gth mode

Figure A.6 shows the resulting fitting when the modes are each fitted using 10 poles. The accuracy is better than in Figure A.3 (direct phase-domain fitting), but the effective order is now 50 (five delay groups) compared to 40 in Figure A.3. The advantage of using ULM over direct phase-domain fitting is mainly when modeling cable systems. Here, the delays are often widely different, thereby requiring an excessive order with a direct fitting approach.

Example A.2:   Transformer Modeling

Frequency responses are often characterized by a large variation in magnitude, as a function of frequency. In such situations, the resulting fitting accuracy can be greatly improved by introducing weighting in the least-squares problems for pole identification and subsequent residue identification.

Figure A.7 shows an example of 3 × 3 block of a power transformer. A 40th order pole-residue model (Equation A.12) is calculated by simultaneously fitting the nine responses by the RVF. All responses are given weight unity, at all samples. Since the responses contain about 0.1% noise, the poles tend to be shifted toward low frequencies in an attempt to fit the noise associated with the large elements. This leads to large fitting errors at higher frequencies.

This problem is effectively alleviated by appropriate weighting. Figure A.8 shows the same result when each element $h(s)$ is weighted by the inverse of the element magnitude

$$w(s) = \frac{1}{h(s)} \tag{A.25}$$

Clearly, a much better result is achieved.

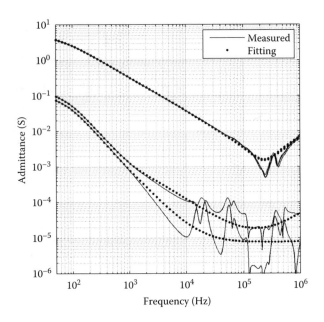

FIGURE A.7
Example A.2: No weighting.

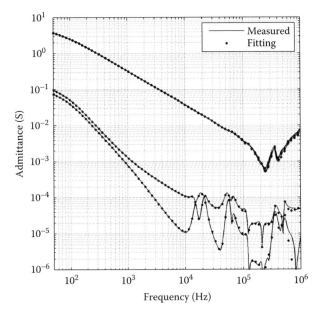

FIGURE A.8
Example A.2: Inverse magnitude weighting.

A.5   Fitting by Frequency Response Partitioning

The limitations of the polynomial fitting techniques can be overcome by partitioning the frequency response into a number of segments, thus being an alternative to VF. This section describes a complete procedure for achieving this goal.

A.5.1   Pole Identification by Rational Fitting

The flowchart of the pole identification procedure used in the present fitting method is shown in Figure A.9. It basically uses rational fitting. Each process of the flowchart is described below.

A.5.1.1   Matrix Trace for Pole Identification

According to linear system theory, any element of the matrix $\mathbf{F}(s)$ contains sufficient information to identify all poles, see Equation A.5; this theoretically means that all poles can

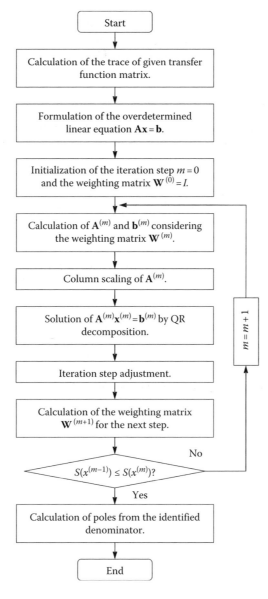

FIGURE A.9
Pole identification procedure by rational fitting. (From Noda, T., *IEEE Trans. Power Deliv.*, 20, 1134, April 2005. With permission.)

be identified using the frequency response of any element of $\mathbf{F}(s)$. In practice, however, the identification of poles is carried out by a numerical procedure, and if, for instance, an element in which particular poles are dominant is used, we cannot expect good accuracy for nondominant poles. With this in view, we use the trace of $\mathbf{F}(s)$ for the pole identification. It is defined as the sum of the diagonal elements:

$$\mathrm{tr}\{\mathbf{F}(s)\} \cong \sum_{i=1}^{M} f_{ii}(s) \tag{A.26}$$

where $f_{ij}(s)$ denotes the (i, j) element of $\mathbf{F}(s)$. It is expected that the trace of a transfer function matrix evenly contains information of all poles. Hereafter, the trace of a matrix is denoted by the corresponding lowercase letter, i.e., $f(s) = \mathrm{tr}\{\mathbf{F}(s)\}$.

A.5.1.2 Formulation of Overdetermined Linear Equations

To identify the poles p_n, we first fit the frequency response of the trace $f(s)$ with the rational function (the strictly proper scalar case is shown)

$$f(s) \cong \frac{b_0 + b_1 s + \cdots + b_{N-1} s^{N-1}}{1 + a_1 s + \cdots + a_{N-1} s^{N-1} + a_N s^N} \tag{A.27}$$

where N is the model order.

Multiplying both sides of Equation A.27 by the denominator of the right-hand side gives

$$(1 + a_1 s + \cdots + a_{N-1} s^{N-1} + a_N s^N) f(s) \cong b_0 + b_1 s + \cdots + b_{N-1} s^{N-1} \tag{A.28}$$

Since $f(s)$ is defined at discrete frequency points, all equations of Equation A.28 for $k = 1, 2, \ldots, K$ can be brought into the overdetermined linear equations

$$\mathbf{A}\mathbf{x} \cong \mathbf{b} \tag{A.29}$$

where
   $\mathbf{A} \in \mathbf{R}^{2K \times 2N}$
   $\mathbf{x} \in \mathbf{R}^{2N}$
   $\mathbf{b} \in \mathbf{R}^{2K}$

The unknown vector $\mathbf{x}$ contains the parameters of the rational function in the form

$$\mathbf{x} = \begin{bmatrix} b_0 & a_1 & b_1 & a_2 & \cdots & b_{N-1} & a_N \end{bmatrix}^{\mathrm{T}} \tag{A.30}$$

The matrix $\mathbf{A}$ in Equation A.29 is given by

$$
\mathbf{A} = \begin{bmatrix}
\beta_{r0,1} & \alpha_{r1,1} & \beta_{r1,1} & \alpha_{r2,1} & \cdots & \beta_{r(N-1),1} & \alpha_{rN,1} \\
\beta_{i0,1} & \alpha_{i1,1} & \beta_{i1,1} & \alpha_{i2,1} & \cdots & \beta_{i(N-1),1} & \alpha_{iN,1} \\
\vdots & \vdots & \vdots & \vdots & & \vdots & \vdots \\
\beta_{r0,k} & \alpha_{r1,k} & \beta_{r1,k} & \alpha_{r2,k} & \cdots & \beta_{r(N-1),k} & \alpha_{rN,k} \\
\beta_{i0,k} & \alpha_{i1,k} & \beta_{i1,k} & \alpha_{i2,k} & \cdots & \beta_{i(N-1),k} & \alpha_{iN,k} \\
\vdots & \vdots & \vdots & \vdots & & \vdots & \vdots \\
\beta_{r0,K} & \alpha_{r1,K} & \beta_{r1,K} & \alpha_{r2,K} & \cdots & \beta_{r(N-1),K} & \alpha_{rN,K} \\
\beta_{i0,K} & \alpha_{i1,K} & \beta_{i1,K} & \alpha_{i2,K} & \cdots & \beta_{i(N-1),K} & \alpha_{iN,K}
\end{bmatrix} \tag{A.31}
$$

where

$$
\alpha_{rn,k} = \mathrm{Re}\{-s_k^n f(s_k)\}, \quad \alpha_{in,k} = \mathrm{Im}\{-s_k^n f(s_k)\},
$$

$$
\beta_{rn,k} = \mathrm{Re}\{s_k^n\}, \quad \beta_{in,k} = \mathrm{Im}\{s_k^n\}, \tag{A.32}
$$

and the vector $\mathbf{b}$ is given by

$$
\mathbf{b} = \begin{bmatrix} \mathrm{Re}\{f(s_1)\} & \mathrm{Im}\{f(s_1)\} & \cdots & \mathrm{Re}\{f(s_k)\} & \mathrm{Im}\{f(s_k)\} & \cdots & \mathrm{Re}\{f(s_K)\} & \mathrm{Im}\{f(s_K)\} \end{bmatrix}^{\mathrm{T}}. \tag{A.33}
$$

If the response of the system at $f = 0$ is known, we can eliminate b_0 from $\mathbf{x}$ using the relation $b_0 = f(0)$. In this case, the first column of $\mathbf{A}$ multiplied by $f(0)$ is subtracted from the right-hand side vector b.

A.5.1.3 Adaptive Weighting

Since both sides in Equation A.27 have been multiplied by the denominator of the right-hand side, the least-squares problem (Equation A.29) is biased by the denominator. To cancel this bias, an iterative process is introduced in [42]. The process sequentially multiplies an estimate of the denominator to both sides of Equation A.28, until the solution converges.

In this appendix, however, we introduce a further improved iterative process. Let us consider the error function for the kth frequency sample

$$
e_k(\mathbf{x}) = f(s_k) - \hat{f}(s_k, \mathbf{x}) \tag{A.34}
$$

or, if relative-error evaluation is desired,

$$
e_k(\mathbf{x}) = \frac{f(s_k) - \hat{f}(s_k, \mathbf{x})}{|f(s_k)|} \tag{A.35}
$$

where $\hat{f}(s_k, \mathbf{x})$ is the frequency response of identified rational function. It should be noted that our ultimate objective is, rather than to cancel the bias, to find a solution $\mathbf{x}$ which minimizes

$$S(\mathbf{x}) = \sum_{k=1}^{K} |e_k(\mathbf{x})|^2 \tag{A.36}$$

Thus, we iteratively solve Equation A.29 so that at an iteration step we apply heavier weightings for frequency samples which have exhibited larger errors at the previous iteration step. In other words, the value of the weighting function is changed adaptively according to the solution obtained at the previous step.

This can be achieved by the weighting function defined by

$$w_k^{(m)} = w_k^{(m-1)} |e_k(\mathbf{x}^{(m-1)})| \tag{A.37}$$

where $m = 0, 1, 2, \ldots$ denotes iteration steps and at the initial step $w_k^{(0)} = 1$. With this weighting function, we can expect that $\mathbf{x}$ reaches the desired solution faster than by multiplying an estimate of the denominator. If the weighting values for $k = 1, 2, \ldots, K$ are arranged in the diagonal matrix $\mathbf{W}^{(m)}$, the multiplication by the weighting function can be written in the matrix notation

$$\mathbf{A}^{(m)}\mathbf{x}^{(m)} = \mathbf{b}^{(m)} \tag{A.38}$$

by defining $\mathbf{A}^{(m)}$ and $\mathbf{b}^{(m)}$ as

$$\mathbf{A}^{(m)} = \mathbf{W}^{(m)}\mathbf{A}, \quad \mathbf{b}^{(m)} = \mathbf{W}^{(m)}\mathbf{b} \tag{A.39}$$

In short, Equation A.38 is sequentially solved for $m = 0, 1, 2, \ldots$ with the weighting matrix obtained by the solution at the previous iteration step.

This iterative process is not a smoothly converging process. Therefore, if the weighting function reaches the value that minimizes $S(\mathbf{x})$, the solution at the next step increases $S(\mathbf{x})$. Thus, when $S(\mathbf{x}^{(m-1)}) \leq S(\mathbf{x}^{(m)})$ is satisfied, $\mathbf{x}^{(m-1)}$ is regarded as the final solution.

A.5.1.4   Column Scaling

Since $\mathbf{A}^{(m)}$ in Equation A.38 is badly scaled due to widespread values of the s^n terms, the condition number of $\mathbf{A}^{(m)}$ can be improved by column scaling [43–46], which balances the Euclidean norms of the columns of $\mathbf{A}$ to unity. Thus, column scaling is applied prior to the solution of Equation A.38 at each iteration step. Section IV-B of [42] describes the detailed procedure of column scaling used in the present method.

A.5.1.5   Solution of Equation A.38 by QR Decomposition

The present method solves Equation A.38 by QR decomposition with Householder transformations. This is one of the best numerical algorithms currently available to solve this kind of least-squares problems in terms of accuracy and efficiency [45,46] (MATLAB's operator "\" is used for implementation [47]).

A.5.1.6 Iteration Step Adjustment

In a sequential iteration process, we often observe unwanted overshoots of the solution. This can be minimized by the following iteration step adjustment [42]. The solution vector **x** of Equation A.38 is linearly interpolated with respect to a parameter θ using the preceding $(m - 1)$ and present (m) iteration steps:

$$\mathbf{x} = (1-\theta)\mathbf{x}^{(m-1)} + \theta\mathbf{x}^{(m)} \tag{A.40}$$

where $\theta = 0$ gives the preceding solution $\mathbf{x}^{(m-1)}$ and $\theta = 1$ the present solution $\mathbf{x}^{(m)}$.

We assume that the residual vector **r** can also be approximated by a linear interpolation with the same parameter:

$$\mathbf{r} = (1-\theta)\mathbf{r}^{(m-1)} + \theta\mathbf{r}^{(m)} \tag{A.41}$$

The value of θ that minimizes S is obtained by setting $\partial S/\partial\theta = 0$:

$$\theta_{best} = \frac{\mathbf{r}^{(m-1)\mathrm{T}}\mathbf{r}^{(m-1)} - \mathbf{r}^{(m-1)\mathrm{T}}\mathbf{r}^{(m)}}{\mathbf{r}^{(m-1)\mathrm{T}}\mathbf{r}^{(m-1)} - 2\mathbf{r}^{(m-1)\mathrm{T}}\mathbf{r}^{(m)} + \mathbf{r}^{(m)\mathrm{T}}\mathbf{r}^{(m)}} \tag{A.42}$$

After calculating the solution vector $\mathbf{x}^{(m)}$ by solving Equation A.38, the value of θ_{best} is calculated by Equation A.42 and substituted into Equation A.40 to obtain an improved solution $\mathbf{x}_{best}$. Finally, the improved solution $\mathbf{x}_{best}$ replaces the present solution $\mathbf{x}^{(m)}$. In a simplified two-dimensional case where **x** has only two elements, contours of S in the solution plane are shown in Figure A.10. θ_{best} gives the solution point on the straight line of θ where the gradient of S is perpendicular to the line. Thus, the best step avoids unnecessary overshooting during the iterations.

A.5.1.7 Calculation of Poles from Denominator

The poles p_n, that is, the roots of the denominator polynomial

$$1 + a_1 s + \cdots + a_{N-1}s^{N-1} + a_N s^N \tag{A.43}$$

can be obtained as the eigenvalues of the companion matrix of the polynomial. The eigenvalues are calculated by the QR algorithm via upper Hessenberg matrix form with prebalancing (MATLAB's function "roots()" is used for implementation [47]).

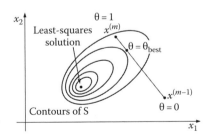

FIGURE A.10
Contours of S in the solution plane in a simplified two-dimensional case. (From Noda, T. et al., *IEEE Trans. Power Deliv.*, 18, 1478, October 2003. With permission.)

If the given frequency response is of a stable system and the fitting process achieves a high accuracy, all identified poles should be on the left-hand side of the s-plane. However, it is possible to get a few poles on the right-hand side, even for a stable frequency response, when a given model order is too small to represent the frequency response. In this case, the real parts of the unstable poles are inverted for stabilization in the present method. This is reasonable due to the following reason. When the real part of a pole is inaccurately identified to be positive, the contribution from the pole to the $j\omega$ axis should be small compared with other adjacent poles. Therefore, even if the real part is inverted, still the contribution from the pole to the $j\omega$ axis is similar with same magnitude and inverted phase angle.

A.5.1.8　Search for an Optimal Model Order

In the present method, an optimal model order is searched for by increasing the model order N one by one in the user-defined range; from the initial order N_{min} to the final order N_{max}. For each order between N_{min} and N_{max}, the pole identification process shown in Figure A.9 is performed, and the standard deviation

$$\bar{e}(\mathbf{x}) = \sqrt{\frac{1}{K} S(\mathbf{x})}$$

(A.44)

is calculated for the final solution of each order. Assuming that the error tolerance ε is user-defined, if $\bar{e}(\mathbf{x}) \leq \varepsilon$ is satisfied at order N, this order is identified to be the optimal order and the order search is terminated. If $\bar{e}(\mathbf{x}) \leq \varepsilon$ is not satisfied for all orders between N_{min} and N_{max}, an order which minimizes $\bar{e}(\mathbf{x})$ is considered to be the optimal order.

A.5.2　Entire Identification Algorithm

A.5.2.1　Partitioning of Frequency Response for Pole Identification

The rational fitting method described in Section A.5.1 includes adaptive weighting, column scaling, and iteration step adjustment. However, if a response with a wide frequency range is fitted, the s^n terms take on widespread values, and thus Equation A.38 cannot be solved with a satisfactory condition number. To overcome this problem, we partition the given frequency response into N_p sections along the frequency axis and apply the rational fitting method to each frequency section for identifying the poles that belong to the section. If each section is narrow enough to limit the largest ratio of the s^n terms within a value which keeps sufficient number of digits to solve Equation A.38, then it can be solved with a satisfactory condition number. It should be noted, however, that ill-conditioning comes not only from the bandwidth but also from the shape of the frequency response in the section. Thus, explicitly calculating an appropriate bandwidth would be difficult, and a binary partitioning algorithm based on a trial-and-error process is used [48].

First, the rational fitting for the pole identification is applied to the entire frequency region without partitioning. If the rational fitting achieves the specified accuracy, the pole identification is completed. Otherwise, the frequency region is divided into two sections and the same procedure is recursively applied to both sections. The subdivision is repeated until all subsections achieve specified accuracy. The algorithmic description of the proposed binary partitioning is shown in Table A.1. In this way, it is ensured that all poles are identified within specified accuracy.

TABLE A.1

Binary Partitioning Algorithm

Step 1	Set the current section to the entire frequency region.
Step 2	Apply the rational fitting to the current section.
Step 3	If the specified accuracy is achieved in the current section, this section is completed. Otherwise, go to Step 4.
Step 4	Divide the current section into two subsections, and go to Step 2 for both subsections.

Source: Noda, T., *IEEE Trans. Power Deliv.*, 22, 1257, April 2007. With permission.
Note: If all subsections achieve specified accuracy in Step 3, the algorithm is completed.

In the algorithm described earlier, how a section is divided into two subsections is not mentioned. Since two neighboring subsections should not identify (share) same poles, the boundary of two subsections should be placed at a local minimum of the magnitude of $\mathbf{H}_k$. Therefore, the following procedure is used. First, the boundary is placed at the midpoint of the frequency section of interest ("current section" in Table A.1). Starting from the midpoint, the closest local minimum is searched for. This search can be coded by a simple loop with comparison of values. If a local minimum is found within a distance from the midpoint closer than half the bandwidth of the section, then the boundary is moved to the local minimum. Otherwise (if a local minimum is not found nearby), the boundary remains at the midpoint.

A.5.2.2 Identification of Residue Matrices

Let us assume that N_r real poles and N_c complex-conjugate pole pairs have been identified. Here, we rewrite the real poles to be $p_n(n = 1, 2, \ldots, N_r)$ and the complex poles with positive imaginary parts to be $\hat{p}_n(n = 1, 2, \ldots, N_c)$. Using this notation, the MPFE model (Equation A.5) can be expressed by (the symbol "*" denotes complex conjugate but not transposed)

$$\mathbf{F}(s) \cong \sum_{n=1}^{N_r} \frac{\mathbf{R}_n}{s - p_n} + \sum_{n=1}^{N_c} \left(\frac{\hat{\mathbf{R}}_n}{s - \hat{p}_n} + \frac{\hat{\mathbf{R}}_n^*}{s - \hat{p}_n^*} \right) \tag{A.45}$$

where $\mathbf{R}_n$ and $\hat{\mathbf{R}}_n$ are the residue matrices of p_n and $\hat{p}_n$, respectively. By defining $r_{n,ij}$ and $\hat{r}_{n,ij}$ as the (i, j) elements of $\mathbf{R}_n$ and $\hat{\mathbf{R}}_n$, the (i, j) element of Equation A.45 is expressed by

$$f_{ij}(s) \cong \sum_{n=1}^{N_r} \frac{r_{n,ij}}{s - p_n} + \sum_{n=1}^{N_c} \left(\frac{\hat{r}_{n,ij}}{s - \hat{p}_n} + \frac{\hat{r}_{n,ij}^*}{s - \hat{p}_n^*} \right) \tag{A.46}$$

Since $f_{ij}(s)$ is defined at the discrete frequency points $k = 1, 2, \ldots, K$, we have K equations of Equation A.46 which can be brought into overdetermined linear equations of the form

$$\mathbf{A}\mathbf{x} \cong \mathbf{b} \tag{A.47}$$

where
$\mathbf{A} \in \mathbf{R}^{2K \times (N_r + 2N_c)}$
$\mathbf{x} \in \mathbf{R}^{N_r + 2N_c}$
$\mathbf{b} \in \mathbf{R}^{2K}$

The unknown vector $\mathbf{x}$ includes, as its elements, $r_{n,ij}$ for $n = 1,2,\ldots, N_r$, and the real and imaginary parts of $\hat{r}_{n,ij}$ for $n = 1,2,\ldots, N_c$. Equation A.47 can be solved by QR decomposition with column scaling as described before, and we obtain the residue matrices $\mathbf{R}_n$ and $\hat{\mathbf{R}}_n$ by repeating the solutions of Equation A.47 for all matrix elements, i.e., all i and j. If relative-error evaluation is desired, the weighting $1/|f_{ij}(s_k)|$ is applied to the equations for the kth frequency sample.

The matrix $\mathbf{A}$ in Equation A.47 is given by

$$\mathbf{A} = \begin{bmatrix} \gamma_{r1,1} & \cdots & \gamma_{rN_r,1} & \alpha_{r1,1} & \beta_{r1,1} & \cdots & \alpha_{rN_c,1} & \beta_{rN_c,1} \\ \gamma_{i1,1} & \cdots & \gamma_{iN_r,1} & \alpha_{i1,1} & \beta_{i1,1} & \cdots & \alpha_{iN_c,1} & \beta_{iN_c,1} \\ \vdots & & \vdots & \vdots & \vdots & & \vdots & \vdots \\ \gamma_{r1,k} & \cdots & \gamma_{rN_r,k} & \alpha_{r1,k} & \beta_{r1,k} & \cdots & \alpha_{rN_c,k} & \beta_{rN_c,k} \\ \gamma_{i1,k} & \cdots & \gamma_{iN_r,k} & \alpha_{i1,k} & \beta_{i1,k} & \cdots & \alpha_{iN_c,k} & \beta_{iN_c,k} \\ \vdots & & \vdots & \vdots & \vdots & & \vdots & \vdots \\ \gamma_{r1,K} & \cdots & \gamma_{rN_r,K} & \alpha_{r1,K} & \beta_{r1,K} & \cdots & \alpha_{rN_c,K} & \beta_{rN_c,K} \\ \gamma_{i1,K} & \cdots & \gamma_{iN_r,K} & \alpha_{i1,K} & \beta_{i1,K} & \cdots & \alpha_{iN_c,K} & \beta_{iN_c,K} \end{bmatrix} \tag{A.48}$$

where

$$\gamma_{rn,k} + j\gamma_{in,k} = \frac{1}{s_k - p_n}$$

$$\alpha_{rn,k} + j\alpha_{in,k} = \frac{2(s_k - \mathrm{Re}\{\hat{p}_n\})}{(s_k - \hat{p}_n)(s_k - \hat{p}_n^*)}$$

$$\beta_{rn,k} + j\beta_{in,k} = \frac{2\,\mathrm{Im}\{\hat{p}_n\}}{(s_k - \hat{p}_n)(s_k - \hat{p}_n^*)} \tag{A.49}$$

The vectors $\mathbf{b}$ and $\mathbf{x}$ are given by

$$\mathbf{b} = \begin{bmatrix} \mathrm{Re}\{f(s_1)\} & \mathrm{Im}\{f(s_1)\} & \cdots & \mathrm{Re}\{f(s_k)\} & \mathrm{Im}\{f(s_k)\} & \cdots & \mathrm{Re}\{f(s_K)\} & \mathrm{Im}\{f(s_K)\} \end{bmatrix}^{\mathrm{T}}$$

$$\tag{A.50}$$

$$\mathbf{x} = \begin{bmatrix} r_1 & \cdots & r_{N_r} & \mathrm{Re}\{\hat{r}_1\} & \mathrm{Im}\{\hat{r}_1\} & \cdots & \mathrm{Re}\{\hat{r}_{N_c}\} & \mathrm{Im}\{\hat{r}_{N_c}\} \end{bmatrix}^{\mathrm{T}} \tag{A.51}$$

The suffix ij in Equation A.46 is omitted here.

Example A.3

1. Test network
The 500 kV transmission network shown in Figure A.11, hereafter referred to as the test network, is used in this example for the numerical validation. Although the network diagram is shown on a one-line diagram, the network is modeled by full three-phase representations.

Figure A.12a shows the equivalent circuit for the generators G1–G4. The electromotive forces are represented by the three-phase sinusoidal voltage source E, and the subtransient impedance

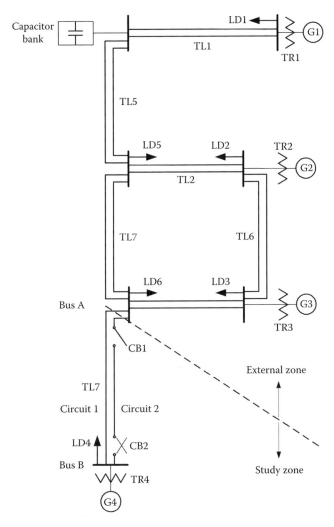

FIGURE A.11
Example A.3: Test network (500 kV transmission network). (From Noda, T., *IEEE Trans. Power Deliv.*, 20, 1134, April 2005. With permission.)

by the circuit block consisting of L_1, R_2, and L_2. The circuit block closely reproduces the frequency dependence of the subtransient impedances (the time constant $\tau = L/R$ of the subtransient impedance with respect to frequency is given for typical generators in Figure 3.5 of [49], and the L_1–R_2–L_2 block matches the impedance calculated from the time constant curve in the frequency range 60 Hz – 1 kHz). Figure A.12b shows the equivalent circuit used to represent the transformers TR1–TR4, consisting of the Δ–Y ideal transformer and the L_1–R_2–L_2 circuit block. This circuit block represents the frequency dependence of the leakage impedances of the transformers (the time constant versus frequency curve of the leakage impedance for typical transformers is given in Figure 3.4 of [49], and the L_1–R_2–L_2 block matches the impedance calculated from the curve in the frequency range 60 Hz – 1 kHz). The loads LD1–LD6 are modeled by the simple R–L circuit in Figure A.12c. As shown in Figure A.12d, the capacitor bank consists of the delta-connected capacitors and the step-down transformer. The step-down transformer is represented in the same way as TR1–TR4, and the stray capacitances to the ground are assumed to be 1 nF.

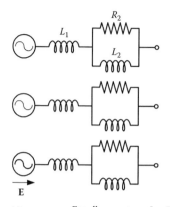

	E (15 kV base)
G1	1.14 ∠ 37.6°
G2	1.10 ∠ 12.4°
G3	1.14 ∠ 10.4°
G4	1.45 ∠ 24.5°

(a)   For all generators, $L_1 = 7.96$ μH, $R_2 = 0.0262$ Ω, $L_2 = 18.1$ μH

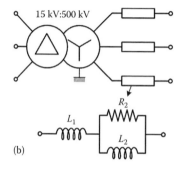

	L_1 (mH)	R_2 (Ω)	L_2 (mH)
TR1	10.0	2.75	0.587
TR2	14.5	3.97	0.847
TR3	14.5	3.97	0.847
TR4	4.89	1.34	0.286

(b)

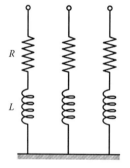

	R (Ω)	L (mH)
LD1	58.0	57.4
LD2	86.6	86.6
LD3	85.1	89.4
LD4	14.8	6.71
LD5	130	3.45
LD6	129	3.41

(c)

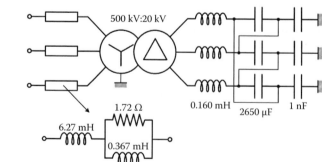

(d)

FIGURE A.12

Example A.3: Representations of the network components: (a) generator representation, (b) transformer representation, (c) load representation, and (d) capacitor bank representation. (From Noda, T., *IEEE Trans. Power Deliv.*, 20, 1134, April 2005. With permission.)

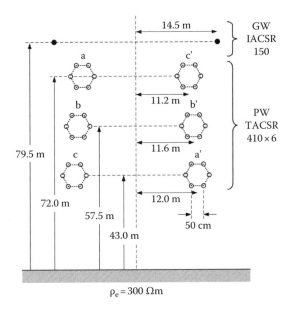

FIGURE A.13
Example A.3: Conductor configuration of the transmission lines TL 1–7. The phases a and a', b and b', and c and c' are respectively connected at both ends. (From Noda, T., *IEEE Trans. Power Deliv.*, 20, 1134, April 2005. With permission.)

Figure A.13 shows the conductor configuration of the transmission lines TL 1–7 (all the lines are untransposed double-circuits). The lengths of TL 2 and TL 3 are 50 km, and those of the remaining lines are 100 km. The skin effects of the earth [50] and the conductors [51] are rigorously taken into account to calculate the frequency response.

2. Simulation scenario
In this study, the switching transient on TL 4 when closing CB 1 is calculated. The lower part seen from Bus A is the study zone that is represented in detail. The remaining upper part is considered to be the external zone, which will be replaced by an equivalent.

3. Simulation results
The admittance matrix $\mathbf{Y}(s)$ of the test network's external zone is calculated at equidistant 2000 frequency points between 0 Hz and 10 kHz. The trace of $\mathbf{Y}(s)$ is then calculated and partitioned into nine frequency sections using the binary partitioning algorithm, as shown in Figure A.14.

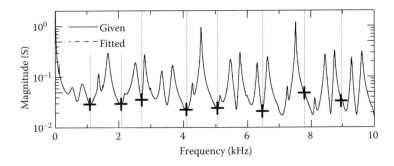

FIGURE A.14
Example A.3: Result of the pole identification process with the result of frequency-region partitioning by the binary algorithm. (From Noda, T., *IEEE Trans. Power Deliv.*, 20, 1134, April 2005. With permission.)

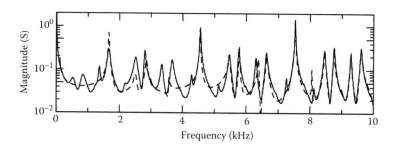

FIGURE A.15

Example A.3: Result of the pole identification process without frequency-region partitioning. (From Noda, T., *IEEE Trans. Power Deliv.*, 20, 1134, April 2005. With permission.)

The boundaries, shown by the + symbols, are placed on a local minimum if it exists nearby. Each partition of the trace is fitted by the rational fitting method described earlier with the following conditions: relative-error evaluation is used with Equation A.35, and an optimal model order is searched between $N_{min} = 10$ and $N_{max} = 20$, with an error tolerance $\varepsilon = 0.01$. Figure A.14 is the fitted result where good agreement is observed. In all partitions, $\bar{e}(\mathbf{x}) \leq \varepsilon$ is satisfied.

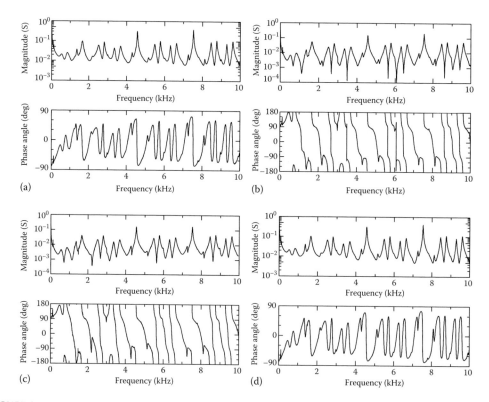

FIGURE A.16

Example A.3: Fitted result for some elements of the admittance matrix: (a) (1,1) element, (b) (1,2) element, (c) (1,3) element, and (d) (2,2) element. (The given responses are shown by the solid lines. The fitted responses are superimposed with the dashed lines and the difference from the solid lines cannot be observed). (From Noda, T., *IEEE Trans. Power Deliv.*, 20, 1134, April 2005. With permission.)

Figure A.15 shows the result obtained without frequency partitioning. The model order was increased one by one from 10, but at orders $N \geq 35$ the overdetermined linear equations to be solved exhibited rank deficiency (ill-conditioning) and we were unable to obtain solutions for $N = 35$ and higher orders. Thus, $N = 34$, which gives the smallest value of $\bar{e}(\mathbf{x})$, is considered to be the optimal order in this case, although $\bar{e}(\mathbf{x}) \leq \varepsilon$ is not satisfied ($\varepsilon = 0.01$). The poor result in Figure A.15 is due to the widespread values of the s^n terms in the rational function (Equation A.5), and thus the effectiveness of frequency partitioning is clear.

Then, the residue matrices are identified for the above identified poles with the following condition: relative-error evaluation is used, and the terms $\mathbf{D} + s\mathbf{E}$ in Equation A.4 are assumed to be zero. Figure A.16 shows the fitted result, where only some elements of the admittance matrix are shown; the other elements are also fitted with the same degree of accuracy. The responses have many resonance peaks mainly due to the transmission lines, and all those peaks are fitted accurately.

The switching transient scenario, CB 1 is closed at 7.09 [ms], is simulated. Figure A.17 compares the calculated voltages at CB 2 obtained by the full system representation and by the identified equivalent, where these results are practically the same. Computational efficiency can roughly be evaluated by the number of poles used in the computation. The equivalent uses 115 poles (7 real poles and 54 complex-conjugate pole pairs), while the full system representation uses 4064 poles (counted by the number of L, C elements and poles to represent the transmission lines by the J. Marti model [2]). The speed-up factor achieved by this equivalent is roughly estimated to be 35 as per the numbers of poles.

A.5.3   Other Applications

The fitting technique based on frequency response partitioning described above has also been used for the following applications:

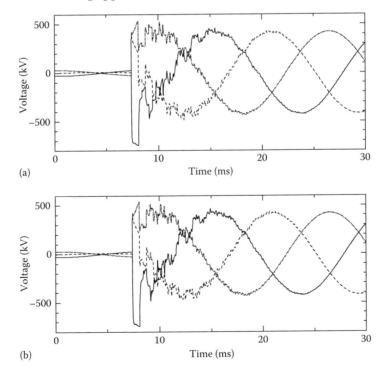

(a)

(b)

FIGURE A.17
Example A.3: Switching transient simulation results by (a) the full system representation and (b) the identified equivalent. (From Noda, T., *IEEE Trans. Power Deliv.*, 20, 1134, April 2005. With permission.)

- Identification of a reduced-order equivalent of the external system seen from a synchronous generator for the design of the generator's power system stabilizer (PSS) [52].
- Identification of a reduced-order equivalent seen from an AC terminal of an HVDC link for the simulation of electromagnetic transients [53,54].

A.6 Passivity

A.6.1 Passivity Assessment

A.6.1.1 Unsymmetrical Models

Although VF gives a model with guaranteed stable poles, the simulation may still result unstable due to passivity violations. A model is passive if it cannot generate power under any terminal conditions. This implies that the model's terminal admittance matrix $\mathbf{Y}(j\omega)$ satisfies the following condition [21]:

$$\mathrm{eig}(\mathbf{Y}(j\omega) + \mathbf{Y}^H(j\omega)) = \mathrm{eig}(\mathbf{Y}_H(j\omega)) > 0 \quad \forall \omega \tag{A.52}$$

where the superscript "H" denotes the Hermitian (transpose and conjugate).

The passivity of a model can be checked by sweeping Equation A.52 over a grid of frequencies. Application of sweeping is sometimes not easy since passivity violations may occur between sample points. It has therefore become an established practice to check passivity via the eigenvalues of the following Hamiltonian matrix $\mathbf{M}$ [22–24]:

$$\mathbf{M} = \begin{bmatrix} (\mathbf{A} - \mathbf{B}(\mathbf{D}+\mathbf{D}^T)^{-1}\mathbf{C}) & \mathbf{B}(\mathbf{D}+\mathbf{D}^T)^{-1}\mathbf{B}^T \\ -\mathbf{C}^T(\mathbf{D}+\mathbf{D}^T)^{-1}\mathbf{C} & (-\mathbf{A}^T+\mathbf{C}^T(\mathbf{D}+\mathbf{D}^T)^{-1}\mathbf{B}^T) \end{bmatrix} \tag{A.53}$$

Any purely imaginary eigenvalue of $\mathbf{M}$ defines a crossover frequency where an eigenvalue of $\mathbf{Y}_H$ in Equation A.52 changes sign. Thus, the imaginary eigenvalues of $\mathbf{M}$ can be used for establishing intervals of passivity violations. For a rigorous treatment of passivity conditions we refer to [25].

A.6.1.2 Symmetrical Models

In the case of the symmetrical pole-residue model, the passivity criterion (Equation A.52) simplifies to

$$\mathrm{eig}(\mathrm{Re}\{\mathbf{Y}(j\omega)\} = \mathrm{eig}(\mathbf{G}(j\omega)) > 0 \quad \forall \omega \tag{A.54}$$

A half-size test matrix $\mathbf{S}$

$$\mathbf{S} = \mathbf{A}(\mathbf{B}\mathbf{D}^{-1}\mathbf{C} - \mathbf{A}) \tag{A.55}$$

can now be used as a replacement for $\mathbf{M}$ [26]. The square-root of any positive-real eigenvalue of $\mathbf{S}$ defines a crossover frequency where an eigenvalue of $\mathbf{G}$ in Equation A.54

changes sign. Thus, the imaginary eigenvalues of $\mathbf{S}$ can be used for establishing intervals of passivity violations. The advantage of using Equation A.55 over Equation A.53 is computational efficiency: A reduction in computation time by a factor of about eight is often achieved due to the cubic complexity of eigenvalue computations.

A.6.2 Passivity Enforcement

Once the passivity characteristics of the model have been established, the model may be subjected to passivity enforcement where the model's parameters are subjected to perturbation in an iterative procedure. The main challenges are the computational efficiency and robustness of the procedure, and the accuracy of the perturbed model.

Most methods are based on perturbing the model's residues (or elements of $\mathbf{C}$ of the state-space model) while minimizing the change to the model's behavior [22,24,27]. It has also been proposed to perturb the residues indirectly via the Hamiltonian matrix eigenvalues [23]. The latter approach is substantially faster but requires more iterations [28], and it is more restricted regarding the use of least-squares weighting for controlling the change to the model's behavior. Perturbation of poles has been proposed [29], but comes at the price of larger perturbations.

In what follows we outline an approach to perturb the residue matrix eigenvalues [24] for use with the symmetrical pole-residue model. This method minimizes the change to the admittance matrix elements (Equation A.56a) while enforcing (Equation A.56b) that the perturbed model obeys the passivity criterion (Equation A.54). In addition, asymptotic passivity is enforced (Equation A.56c) as well as positive definiteness of $\mathbf{E}$ (Equation A.56d).

$$\Delta\mathbf{Y} = \sum_{m=1}^{N} \frac{\Delta\mathbf{R}_m}{s-a_m} + \Delta\mathbf{D} + s\Delta\mathbf{E} \cong \mathbf{0} \tag{A.56a}$$

$$\text{eig}\left(\text{Re}\left\{\mathbf{Y} + \sum_{m=1}^{N} \frac{\Delta\mathbf{R}_m}{s-a_m} + \Delta\mathbf{D} + s\Delta\mathbf{E}\right\}\right) > 0 \tag{A.56b}$$

$$\text{eig}(\mathbf{D}+\Delta\mathbf{D}) > 0 \tag{A.56c}$$

$$\text{eig}(\mathbf{E}+\Delta\mathbf{E}) > 0 \tag{A.56d}$$

First-order perturbation is used in Equation A.56 for relating the perturbation of a (symmetrical) matrix $\mathbf{F}$ to its eigenvalues. With $\mathbf{v}_i$ denoting a right eigenvector of (the unperturbed) $\mathbf{F}$, we get for the eigenvalue perturbation

$$\Delta\lambda_i = \frac{\mathbf{v}_i^T\Delta\mathbf{F}\mathbf{v}_i}{\mathbf{v}_i^T\mathbf{v}} \tag{A.57}$$

The number of free (perturbed) variables is reduced by separately diagonalizing the residue matrices $\{\mathbf{R}_m\}$, (and $\mathbf{D}$ and $\mathbf{E}$), and perturbing only their eigenvalues. Consider the eigenvalue decomposition of a symmetrical real matrix, $\mathbf{F} = \mathbf{S}_F\Gamma_F\mathbf{S}_F^T$. Using first-order approximation we get for the perturbation, $\mathbf{F}+\Delta\mathbf{F} \approx \mathbf{S}_F(\Gamma_F+\Delta\Gamma_F)\mathbf{S}_F^T$. We can therefore introduce the following relations:

$$\Delta \mathbf{R}_m = \mathbf{S}_m \Delta \mathbf{\Gamma}_{Rm} \mathbf{S}_m^{\mathrm{T}}$$

(A.58a)

$$\Delta \mathbf{D} = \mathbf{S}_{\mathrm{D}} \Delta \mathbf{\Gamma}_{\mathrm{D}} \mathbf{S}_{\mathrm{D}}^{\mathrm{T}}$$

(A.58b)

$$\Delta \mathbf{E} = \mathbf{S}_{\mathrm{E}} \Delta \mathbf{\Gamma}_{\mathrm{E}} \mathbf{S}_{\mathrm{E}}^{\mathrm{T}}$$

(A.58c)

Implementation of Equations A.56 and A.57 leads to

$$\min_{\Delta \mathbf{x}} \tfrac{1}{2} (\Delta \mathbf{x}^{\mathrm{T}} \mathbf{A}_{\mathrm{sys}}^{\mathrm{T}} \mathbf{A}_{\mathrm{sys}} \Delta \mathbf{x})$$

(A.59a)

$$\mathbf{B}_{\mathrm{sys}} \Delta \mathbf{x} < \mathbf{c}$$

(A.59b)

where $\Delta \mathbf{x}$ is a vector holding the perturbed parameters. Equation A.59 is solved using quadratic programming. The formulation (Equation A.59) allows implementing advanced weighting schemes for improving the accuracy of the perturbed model, including relative error control and weighting of the individual modes. The passivity enforcement by Equation A.59 is combined with passivity assessment in an iterative procedure as described in [24].

Example A.4

In the following, we consider an example from [24] for the passivity enforcement of a wideband two-winding transformer model. Figure A.18 shows the result based on rational modeling from measurements and using the RVF method. Figure A.19 shows the eigenvalues of **G**, before and after passivity enforcement. It is observed that large out-of-band violations have been corrected without corrupting the in-band behavior (Figure A.20).

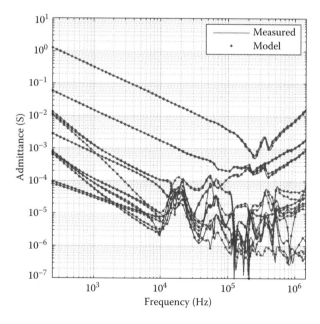

FIGURE A.18
Example A.4: Rational fitting of transformer admittance matrix by relaxed VF. (From Gustavsen, B., *IEEE Trans. Power Deliv.*, 23, 2278, October 2008. With permission.)

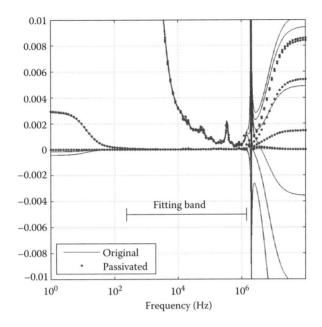

FIGURE A.19
Example A.4: Passivity enforcement of wideband transformer model. (From Gustavsen, B., *IEEE Trans. Power Deliv.*, 23, 2278, October 2008. With permission.)

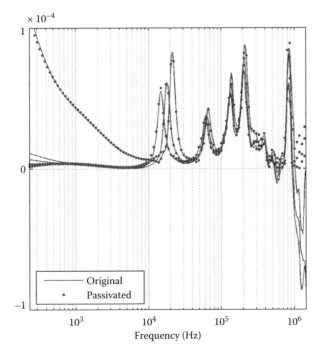

FIGURE A.20
Example A.4: Close-up of Figure A.19. (From Gustavsen, B., *IEEE Trans. Power Deliv.*, 23, 2278, October 2008. With permission.)

References

1. A. Semlyen and A. Dabuleanu, Fast and accurate switching transient calculations on transmission lines with ground return using recursive convolutions, *IEEE Transactions on Power Apparatus and Systems*, 94(2), 561–575, March/April 1975.
2. J.R. Marti, Accurate modelling of frequency-dependent transmission lines in electromagnetic transient simulations, *IEEE Transactions on Power Apparatus and Systems*, 101(1), 147–157, January 1982.
3. L. Marti, Simulation of transients in underground cables with frequency-dependent modal transformation matrices, *IEEE Transactions on Power Delivery*, 3(3), 1099–1110, July 1988.
4. B. Gustavsen and A. Semlyen, Simulation of transmission line transients using vector fitting and modal decomposition, *IEEE Transactions on Power Delivery*, 13(2), 605–614, April 1998.
5. G. Angelidis and A. Semlyen, Direct phase-domain calculation of transmission line transients using two-sided recursions, *IEEE Transactions on Power Delivery*, 10(2), 941–949, April 1995.
6. T. Noda, N. Nagaoka, and A. Ametani, Phase domain modeling of frequency-dependent transmission lines by means of an ARMA model, *IEEE Transactions on Power Delivery*, 11(1), 401–411, January 1996.
7. H.V. Nguyen, H.W. Dommel, and J.R. Marti, Direct phase-domain modelling of frequency-dependent overhead transmission lines, *IEEE Transactions on Power Delivery*, 12(3), 1335–1342, July 1997.
8. B. Gustavsen and A. Semlyen, Combined phase and modal domain calculation of transmission line transients based on vector fitting, *IEEE Transactions on Power Delivery*, 13(2), 596–604, April 1998.
9. B. Gustavsen and A. Semlyen, Calculation of transmission line transients using polar decomposition, *IEEE Transactions on Power Delivery*, 13(3), 855–862, July 1998.
10. A. Morched, B. Gustavsen, and M. Tartibi, A universal model for accurate calculation of electromagnetic transients on overhead lines and underground cables, *IEEE Transactions on Power Delivery*, 14(3), 1032–1038, July 1999.
11. B. Gustavsen, G. Irwin, R. Mangelrød, D. Brandt, and K. Kent, Transmission line models for the simulation of interaction phenomena between parallel AC and DC overhead lines, *International Conference on Power System Transients (IPST)*, Budapest, Hungary, June 20–24, 1999, pp. 61–67.
12. A. Morched, J. Ottevangers, and L. Marti, Multi-port frequency dependent network equivalents for the EMTP, *IEEE Transactions on Power Delivery*, 8(3), 1402–1412, July 1993.
13. B. Gustavsen, Wide band modeling of power transformers, *IEEE Transactions on Power Delivery*, 19(1), 414–422, January 2004.
14. T. Noda, Identification of a multiphase network equivalent for electromagnetic transient calculations using partitioned frequency response, *IEEE Transactions on Power Delivery*, 20(2), 1134–1142, April 2005.
15. C.K. Sanathanan and J. Koerner, Transfer function synthesis as a ratio of two complex polynomials, *IEEE Transactions on Automatic Control*, 8, 56–58, 1963.
16. R. Pintelon, P. Guillaume, Y. Rolain, J. Schoukens, and H.V. Hamme, Parametric identification of transfer functions in the frequency domain: A survey, *IEEE Transactions on Automatic Control*, 39(11), 2245–2260, November 1994.
17. B. Gustavsen and A. Semlyen, Rational approximation of frequency domain responses by vector fitting, *IEEE Transactions on Power Delivery*, 14(3), 1052–1061, July 1999.
18. S. Grivet-Talocia, On the generation of large passive macromodels for complex interconnect structures, *IEEE Transactions on Advanced Packaging*, 29(1), 39–54, February 2006.
19. S. Grivet-Talocia, Package macromodeling via time-domain vector fitting, *IEEE Microwave and Wireless Components Letters*, 13(11), 472–474, November 2003.

20. S.H. Min and M. Swaminathan, Construction of broadband passive macromodels from frequency data for simulation of distributed interconnect networks, *IEEE Transactions on Electromagnetic Compatibility*, 46(4), 544–558, November 2004.

21. B. Gustavsen and A. Semlyen, On passivity tests for unsymmetrical models, *IEEE Transactions on Power Delivery*, 24(3), 1739–1741, July 2009.

22. D. Saraswat, R. Achar, and M.S. Nakhla, A fast algorithm and practical considerations for passive macromodeling of measured/simulated data, *IEEE Transactions on Advanced Packaging*, 27(1), 57–70, February 2004.

23. S. Grivet-Talocia, Passivity enforcement via perturbation of Hamiltonian matrices, *IEEE Transactions on Circuits and Systems I*, 51(9), 1755–1769, September 2004.

24. B. Gustavsen, Fast passivity enforcement for pole-residue models by perturbation of residue matrix eigenvalues, *IEEE Transactions on Power Delivery*, 23(4), 2278–2285, October 2008.

25. P. Triverio, S. Grivet-Talocia, M.S. Nakhla, F.G. Canavero, and R. Achar, Stability, causality, and passivity in electrical interconnect models, *IEEE Transactions on Advanced Packaging*, 30(4), 795–808, November 2007.

26. A. Semlyen and B. Gustavsen, A half-size singularity test matrix for fast and reliable passivity assessment of rational models, *IEEE Transactions on Power Delivery*, 24(1), 345–351, January 2009.

27. B. Gustavsen and A. Semlyen, Enforcing passivity for admittance matrices approximated by rational functions, *IEEE Transactions on Power Systems*, 16, 97–104, February 2001.

28. S. Grivet-Talocia and A. Ubolli, A comparative study of passivity enforcement schemes for linear lumped macromodels, *IEEE Transactions on Advanced Packaging*, 31(4), 673–683, November 2008.

29. A. Lamecki and M. Mrozowski, Equivalent SPICE circuits with guaranteed passivity from nonpassive models, *IEEE Transactions on Microwave Theory and Techniques*, 55(3), 526–32, March 2007.

30. H.W. Bode, *Network Analysis and Feedback Amplifier Design*, Van Nostrand, New York, 1945.

31. H.W. Dommel, *Electro-Magnetic Transients Program Reference Manual (EMTP Theory Book)*, Bonneville Power Administration, Portland, OR, 1986.

32. B. Gustavsen, Improving the pole relocating properties of vector fitting, *IEEE Transactions on Power Delivery*, 21(3), 1587–1592, July 2006.

33. D. Deschrijver, B. Haegeman, and T. Dhaene, Orthonormal vector fitting: A robust macromodeling tool for rational approximation of frequency domain responses, *IEEE Transactions on Advanced Packaging*, 30(2), 216–225, May 2007.

34. D. Deschrijver, B. Gustavsen, and T. Dhaene, Advancements in iterative methods for rational approximation in the frequency domain, *IEEE Transactions on Power Delivery*, 22(3), 1633–1642, July 2007.

35. B. Gustavsen and C. Heitz, Modal vector fitting: A tool for generating rational models of high accuracy with arbitrary terminal conditions, *IEEE Transactions on Advanced Packaging*, 31(4), 664–672, November 2008.

36. D. Deschrijver, M. Mrozowski, T. Dhaene, and D. de Zutter, Macromodeling of multiport systems using a fast implementation of the vector fitting method, *IEEE Microwave and Wireless Components Letters*, 18(6), 383–385, June 2008.

37. B. Gustavsen and C. Heitz, Fast realization of the modal vector fitting method for rational modeling with accurate representation of small eigenvalues, *IEEE Transactions on Power Delivery*, 24(3), 1396–1405, July 2009.

38. D. Deschrijver, T. Dhaene, and D. de Zutter, Robust parameteric macromodeling using the multivariate orthonormal vector fitting, *IEEE Transactions on Microwave Theory and Techniques*, 56(7), 1661–1667, July 2008.

39. S. Grivet-Talocia, Package macromodeling via time-domain vector fitting, *IEEE Microwave and Wireless Components Letters*, 13(11), 472–474, November 2003.

40. N. Wong and C.U. Lei, IIR approximation of FIR filters via discrete-time vector fitting, *IEEE Transactions on Signal Processing*, 56(3), 1296–1302, March 2008.

41. B. Gustavsen, Time delay identification for transmission line modeling, in *Eighth IEEE Workshop on Signal Propagation on Interconnects*, Heidelberg, Germany, May 9–12, 2004, pp. 103–106.
42. T. Noda, A. Semlyen, and R. Iravani, Reduced-order realization of a nonlinear power network using companion-form state equations with periodic coefficients, *IEEE Transactions on Power Delivery*, 18(4), 1478–1488, October 2003.
43. A.O. Soysal and A. Semlyen, Practical transfer function estimation and its application to wide frequency range representation of transformers, *IEEE Transactions on Power Delivery*, 8(3), 1627–1637, July 1993.
44. A. Van der Sluis, Condition numbers and equilibration of matrices, *Numerical Mathematics*, 14, 14–23, 1969.
45. C.L. Lawson and R.J. Hanson, *Solving Least Squares Problems, Classics in Applied Mathematics*, 15, SIAM, Prentice Hall, Englewood Cliffs, NJ, 1995.
46. A. Björck, *Numerical Methods for Least Squares Problems*, SIAM, Philadelphia, PA, 1996.
47. *MATLAB Function Reference Volume 1–3*, Version 6, The MathWorks, Inc., Natick, MA, 2002.
48. T. Noda, A binary frequency-region partitioning algorithm for the identification of a multi-phase network equivalent for EMT studies, *IEEE Transactions on Power Delivery*, 22(2), 1257–1258, April 2007.
49. CIGRE WG 13.05, The calculation of switching surges, *Electra*, no. 32, 17–42, 1974.
50. J.R. Carson, Wave propagation in overhead wires with ground return, *Bell System Technical Journal*, 5, 539–554, 1926.
51. S.A. Schelkunoff, The electromagnetic theory of coaxial transmission line and cylindrical shields, *Bell System Technical Journal*, 13, 532–579, 1934.
52. T. Noda, M. Takasaki, H. Okamoto, K. Fujibayashi, H. Nishigaito, and T. Okada, Low-order linear model identification method of power system by frequency-domain least-squares approximation, *Transactions of IEE Japan*, 121-B(1), 52–59, July 2001.
53. T. Noda and M. Takasaki, Electromagnetic transient analysis of a power system with a sub-network reduction method, *Proceedings of the 2005 Annual Meeting Record*, IEE Japan, Paper # 6–115, 2005.
54. T. Noda and M. Takasaki, A sub-network reduction method for the calculation of electromagnetic transients in a power system (Part 1), CRIEPI Report #T03013, 2004.

Appendix B: Simulation Tools for Electromagnetic Transients in Power Systems

Jean Mahseredjian, Venkata Dinavahi, Juan A. Martinez-Velasco, and Luis D. Bellomo

CONTENTS

B.1 Introduction 591
B.2 Applications 593
    B.2.1 Range of Applications 593
    B.2.2 Modeling Guidelines 594
B.3 Off-Line Simulation Tools 594
    B.3.1 Generalities 594
    B.3.2 Solution Methods 594
    B.3.3 Graphical User Interface 596
    B.3.4 Initialization 598
    B.3.5 Statistical and Parametric Methods 600
    B.3.6 External Interface 600
    B.3.7 Time-Domain Module 600
    B.3.8 Control Systems 601
    B.3.9 Available Software Packages 602
    B.3.10 General-Purpose Modeling Environments 603
    B.3.11 Hybrid Tools 604
    B.3.12 Frequency-Domain Methods 605
    B.3.13 Case Studies 605
        B.3.13.1 Switching Overvoltages 605
        B.3.13.2 Lightning Overvoltages 606
        B.3.13.3 Electromechanical Transients 607
        B.3.13.4 Wind Generation 608
B.4 Real-Time Simulation Tools 609
    B.4.1 Transient Network Analyzers 610
    B.4.2 Real-Time Digital Simulators 610
    B.4.3 Real-Time Playback Simulators 611
    B.4.4 Case Study 613
References 616

B.1 Introduction

The design and operation of modern power systems is more and more based on simulation and analysis tools. The increasing speed of modern computer CPUs, multicore systems, parallel processing, real-time simulation, more memory and better models, and numerical methods are allowing to study modern power systems with higher precision and for multiple scenarios.

The most common power system study tools are based on load-flow and steady-state computations. Transient stability analysis is performed to study electromechanical transients. It is common to assume balanced conditions and to apply positive sequence networks at fundamental frequency. Approximations are necessary and acceptable in many cases, but the availability of data and faster computations allow using more generic methods and study the power system at the detailed "circuit level" without making any assumptions on frequency content and network topology. Such methods are called electromagnetic transients programs or EMT-type (also denoted as EMTP-type) tools. The complete system is studied using time-domain waveforms for the state variables of concern. It is comparable to the connection of an oscilloscope at any point in the power system.

Simulation and analysis of electromagnetic transients in modern power systems is widely used for the determination of component ratings such as insulation levels and energy absorption capabilities, in the design and optimization process; for testing control and protection systems; and for analyzing power systems under various operating scenarios. With the increasing speed of computers and improvements in computational methods, the computation of electromagnetic transients is capable of including efficiently the analysis of electromechanical transients. In fact, such transients can be studied with higher precision, as compared to traditional positive-sequence-based methods, under unbalanced network conditions and by including power electronics components and nonlinear models. In the time-domain approach, there are no inherent limitations in studying harmonics, nonlinear effects, and balanced or unbalanced networks. As for control systems, they are usually represented using block diagrams. It is also possible to model control systems at the circuit level, although such requirements are less common.

Since EMT-type programs are able to represent the actual phase-domain circuits of a network, they are much more general than traditional power system analysis tools. It is important to emphasize that some traditional power system analysis tools may encounter important limitations for studying practical network problems through sequence networks. In the case of short circuit programs, the presence of an arrester in parallel with a series compensation capacitor may cause coupling between sequence networks for a fault near the capacitor bank. Such a condition is not honored by a traditional short circuit package, but it is not a problem when studied with an EMT-type application.

EMT-type tools can be designed to work on standard computers in off-line mode. This means that the simulation is performed for the given time interval without any constraints on the speed of computations. Although an off-line tool must be designed to be highly efficient using powerful numerical methods and programming techniques, it does not have any time constraints and can be made as precise as possible within the available data, models, and related mathematics.

Another family of tools is based on the real-time paradigm. Real-time simulation programs are available in specific hardware environments and designed to be capable of generating results in synchronism with a real-time clock. Such tools have the advantage of being capable of interfacing with physical devices and maintaining data exchanges within the real-time clock. The capability to compute and interface within real time imposes important restrictions on the design of such tools. The real-time simulation programs have specific advantages for testing external physical hardware, such as control systems, within the simulated network. This is called hardware-in-the-loop (HIL) simulation. It is ultimately feasible to study power hardware-in-the-loop (PHIL).

The objective of this appendix is to provide an overview of simulation tools and methods for the computation and analysis of electromagnetic transients [1–3]. Since the number of variants in available methods and programs can become very high, this appendix concentrates only on the most widely recognized and available tools. Generic and proven methods are considered. Examples are used to demonstrate the concepts. In some cases, this appendix also discusses limitations and research topics for practical simulation needs.

B.2 Applications

B.2.1 Range of Applications

The traditional application of EMT-type tools is the computation of overvoltages in power systems. There are four categories of overvoltages [4]: very fast front, fast front, slow front, and temporary. The very fast front category includes restrikes in gas-insulated substations. The frequencies range from 100 kHz to 50 MHz. The lightning overvoltages fall into the fast front category with the typical frequency content from 10 kHz to 3 MHz. The switching overvoltages fall into the slow front category with the frequencies ranging from fundamental frequency to 20 kHz. Switching events can be of controlled or uncontrolled types. Typical controlled events are line switching actions. Faults on buses or in transmission lines fall into the list of uncontrolled events. Transformer energization transients and ferroresonance contain frequencies from 0 Hz to 1 kHz. For the temporary overvoltages, the typical causes are single-line-to-ground faults causing overvoltages on live phases, open line energization, and load-shedding. It is possible to observe temporary overvoltages with ferroresonance. The frequency content for temporary overvoltages is typically from 0.1 to 1 kHz.

Frequencies above the fundamental frequency usually involve electromagnetic phenomena. Frequencies below the fundamental frequency may also include electromechanical modes (synchronous or asynchronous machines).

The above categories can be expanded to list specific important study topics in power systems:

- Switchgear, transient recovery voltage (TRV), shunt compensation, current chopping, delayed-current zero conditions
- Insulation coordination
- Saturation and surge arrester influences
- Harmonic propagation, power quality
- Interaction between compensation and control
- Wind generation, distributed generation
- Precise determination of short circuit currents
- Detailed behavior of synchronous machines and related controls, auto-excitation, subsynchronous resonance, power oscillations
- Protection systems
- High voltage direct current (HVDC) systems, power electronics, FACTS, and Custom Power controllers

These applications are in a wideband range of frequencies, from dc to 50 MHz. This range is different from the classical studies of electromechanical transients performed using transient stability (stability-type) programs. Although separate and more widely used packages are available for studying electromechanical transients (from 0 to 10 Hz), it is feasible to apply EMT-type programs to study transient stability or even small signal stability problems. EMT-type programs can produce more precise simulation results for such studies due to inherent modeling capabilities to account for network nonlinearities and unbalanced conditions. Frequency-dependent and voltage-dependent load models can be also incorporated. The main disadvantages, especially in off-line tools, remain the computational speed and requirements for data. In EMT-type programs, the network equations are solved in time domain and not with phasors as in transient stability solution methods, which is the main cause of reduced computational speed.

B.2.2   Modeling Guidelines

As it became apparent in the previous section, in EMT-type programs it is possible and sometimes necessary to model network components for the entire range of frequencies. In many cases, it is neither simple nor practical to develop and maintain unique models for the entire range of frequencies. The main reason is available data and computer timings. It is thus necessary to select models adapted to the simulation type and frequency content of the studied phenomenon. Studies are performed in a layered approach. It is emphasized however that the greater availability of wideband models and data has contributed to the reduction in the number of layers. But even if all data layers are conveniently arranged in a graphical user interface (GUI), the engineering approach may still select the required layer for the given study.

Several publications ([5–7], for example) are available to help users of EMT-type programs on the correct representation of power system components according to the studied phenomenon. Other publications, such as [8] (see also its references), are available for providing guidelines on needed and typical data.

B.3   Off-Line Simulation Tools

Off-line simulation tools are available on generic computer systems on which they can be easily installed and integrated within the working environment and operating system of the user computer.

B.3.1   Generalities

The scope of this section is to provide a high-level overview on the most important computational modules currently available in EMT-type tools. The building blocks that constitute an EMT-type program are shown in Figure B.1. This figure is labeled as "ultimate" since some of the presented modules or internal features are still at the research level or not available in industrial grade applications.

B.3.2   Solution Methods

The purpose of this appendix is not to present details on solution methods, but to provide generic information on existing and applied methodologies.

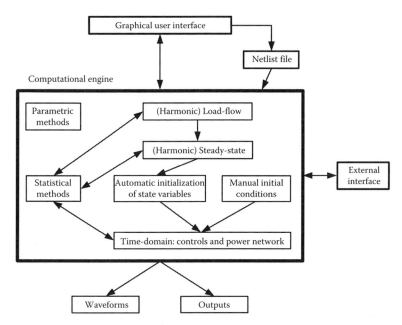

FIGURE B.1
Ultimate building blocks of an EMT-type simulation tool. (From Mahseredjian, J. et al., *IEEE Trans. Power Deliv.*, 24, 1657, July 2009. With permission.)

Most EMT-type programs are based on nodal analysis

$$\mathbf{Yv} = \mathbf{i} \qquad (B.1)$$

where bold characters are used to denote the admittance symmetrical matrix $\mathbf{Y}$ and the voltage and current vectors $\mathbf{v}$ and $\mathbf{i}$, respectively. Basically the left-hand side represents the sum of currents exiting an electrical node and the right-hand side represents the sum of currents entering the node. There are as many equations as nodes.

As explained in many papers (e.g., [9]), such a representation is restrictive and causes several formulation problems for devices (network components) that are modeled with branch-type or branch-dependency-type equations. Such devices can be connected between arbitrary nodes and use coefficients for node voltages in addition to internal unknowns. The ideal voltage source and the ideal transformer are typical examples of devices with branch equations. The inclusion of such equations in Equation B.1 creates the modified-augmented-nodal analysis formulation used extensively in [9] and which can be written symbolically as

$$\mathbf{Ax} = \mathbf{b} \qquad (B.2)$$

where the vector $\mathbf{x}$ includes in addition to node voltages unknown variables related to various device models and the vector $\mathbf{b}$ contains independent current sources and various other fixed variables for device models. The matrix $\mathbf{Y}$ is now only a submatrix of the more general matrix $\mathbf{A}$. The programming and solution of Equation B.2 is based on sparse-matrix techniques which are essential to solve large networks.

The linear steady-state solution vector **x** is found directly from Equation B.2 using complex values for all variables and the steady-state models for all network devices. It is also possible to find the harmonic steady-state solution using superposition. Each member x_i of **x** is found in the form of a Fourier series which corresponds to the time-domain steady-state solution and can be used to initialize all state-variables.

In many cases, the studied network may contain nonlinearities. Nonlinear functions are encountered when modeling various devices, such as arresters and saturable transformers. A nonlinear function can be linearized at its solution (or operation) point to provide a linear equation entered into the system of equations (Equation B.2). It is noticed that matrix **A** is actually the Jacobian matrix and can be used directly through an iterative Newton solution process.

B.3.3 Graphical User Interface

The GUI plays an important role in modern computer programs. It is the main access from the user level to the study case setup, to simulation procedures, and to simulation options. The GUI (also called schematic capture) must be designed to be sufficiently simple to use to allow assembling simple and complex networks, running simulation and visualizing results for analysis purposes. Normally the GUI is used to generate the complete network model data in the form of a Netlist file which is sent to the computational engine. Some programs can also communicate directly with the computational engine memory through data objects.

An example of GUI-based design is shown in Figure B.2. Modern GUIs are based on the hierarchical design approach with subnetworks and masking. Subnetworks allow simplifying the drawing and hiding details while masking provides data encapsulation. The design shown in Figure B.2 uses several subnetworks. The 230 kV network is interconnected with a 500 kV network evacuated with all its details into the subnetwork shown in the figure. In a hierarchical design, subnetworks can also contain other subnetworks. Subnetworks can be also used to develop complete models. In Figure B.2 the synchronous machine symbols are subnetworks containing the load-flow constraints, machine data, and also voltage regulator and governor controls in separate subnetworks.

In addition to selecting and assembling existing models in a given software library, GUIs allow building user-defined models and can provide complex customization features.

Several advanced GUIs are currently available in EMT-type programs. A complicated problem from the user's point of view remains the absence of interoperability standards between various software applications. Networks or user-defined models assembled in one EMT-type package are not portable to another EMT-type package which is sometimes concurrently used in another organization or even within the same organization. There are no applicable standards for transient data fields which complicates even manual copying of models between GUIs. In some cases, the standardization problem is directly linked to the complexity of models and solution methods for electromagnetic transients.

Some applications provide external access functions and might be called directly from other applications for performing simulations on assembled networks. The programming aspects of such applications are not complex, but interfacing networks solved in different computational engines may become error prone or create numerical divergence due to inherent lack of simultaneous solution capability.

A possible solution to data portability between applications is the utilization of the Common Information Model (CIM [10]) format in EMT-type programs. The CIM format is an open standard for representing power system components. It could be used for electromagnetic transients if augmented with many data fields related to such models.

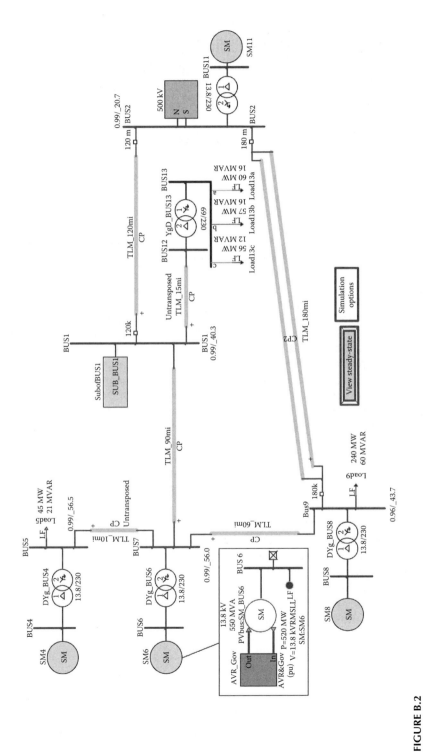

FIGURE B.2
A sample 230 kV network study case. (From Mahseredjian, J. et al., *IEEE Trans. Power Deliv.*, 24, 1657, July 2009. With permission.)

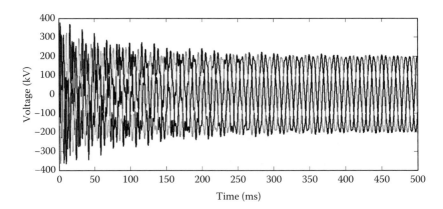

FIGURE B.3
Transmission line phase voltages at the receiving end without initialization.

Standardization can also help solving the data exchange problems [11] with conventional power system applications and result in significant benefits to the power industry.

B.3.4 Initialization

The importance of initialization can be illustrated through the simple example of Figure B.3. The presented waveforms are the voltages at the receiving end of an arrester-protected transmission line. Since the simulation starts with zero initial conditions the transmission line transients do not reach steady state even after 0.5 s of simulation time. This will have dramatic computing time consequences on large systems. Automatic initialization procedures allow starting immediately with steady-state waveforms. In most cases, the study of transients is conducted from an existing steady-state condition in the network.

The problem becomes more complex in the presence of synchronous or asynchronous machines within multiple generator networks. Machine phasors can be made available from an external load-flow program, but since the actual network may be unbalanced or use models specific to the simulation of transients, the best approach is the implementation of a load-flow solution directly before the steady-state solution and within the same simulation tool [9]. This is also depicted in Figure B.1.

As demonstrated in Figure B.4 (test case of Figure B.2, machines SM6 and SM3 in SubofBUS1), without automatic initialization and even after 5 seconds of simulation the machines do not reach steady state, whereas the automatic initialization starts from the load-flow solution where the machines are given PV constraints. In some cases, without initialization, the simulated network may reach abnormal operating conditions.

The Load-flow module shown in Figure B.1 is used to compute the operating conditions of the power system. It must employ a multiphase solution since the objective is to use the same network data and initialize the corresponding time-domain network. The load-flow solution replaces all load-flow constraints (PQ, PV, and slack buses) by specific equations in a Newton solution method. It can be achieved by simply augmenting Equation B.1 with the constraint equations [9].

Upon convergence of the load-flow solution, all steady-state phasors become available. The synchronous machine phasors are used to calculate internal state variables. For asynchronous machines, it is required to calculate the slip parameter for a given mechanical power or torque.

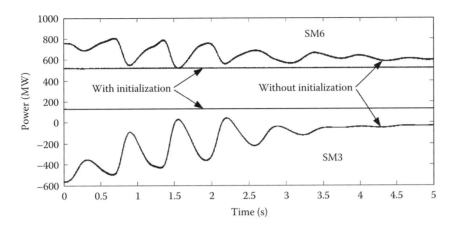

FIGURE B.4
Two synchronous machine 3-phase powers, with and without initialization. (From Mahseredjian, J. et al., *IEEE Trans. Power Deliv.*, 24, 1657, July 2009. With permission.)

In some cases, the network may contain harmonic sources or nonlinearities and it is necessary to perform a harmonic load flow. It is feasible to program such a method [12] for specific applications.

The steady-state module follows the load-flow solution and replaces all devices by lumped equivalents to proceed with a phasor solution. Phasors are used for initializing all state variables at the time-point $t = 0$. The solution at $t = 0$ is only from the steady state and all history terms for all devices are initialized for the first solution time-point in the time-domain module.

When the solved network is linear or is running under linear operating conditions, the initialization procedure will allow mapping directly and almost perfectly the steady-state Fourier series into the waveforms computed in time domain. In some special conditions, such as different rotor frequencies, initialization is still possible by solving the rotor networks independently. A more significant programming effort is needed to account for nonlinearities using an iterative Newton method. It can have a significant impact on computing time under some particular conditions [13] or when analyzing multiple harmonic sources (see also [14]).

If there is no calculated steady-state solution there could be manual initial conditions, such as trapped charge, or all variables can be at zero state. Manual initial conditions are also useful for reproducing complex conditions such as ferroresonance.

A complex subject in automatic initialization is the initialization of systems with power electronics switching devices. It is not obvious to automatically predict commutation patterns in a given operating mode and initialize state variables for harmonic waveforms. A programmed initialization method should find steady-state conditions in significantly less computing time than the brute force approach. In some cases, such as wind generation installations with power electronics devices connected on the rotor side, the best approach is to start with mean-value models or tricked equivalents and to switch onto actual commutating functions after establishing steady-state operation.

To complete the picture, it is important to mention that initialization concerns also the control system diagrams. It is usually a more complex, but essential feature, since, for example, initialization of synchronous machine variables without initialization of its controls can become worthless. Fully automatic methods do not yet exist, but backward propagation of variables in control blocks from specified initial condition variables is a practical option.

B.3.5   Statistical and Parametric Methods

Statistical methods are used for simulating with random data and evaluating worst case overvoltages or other probabilities for any system variable. A new trend in power system applications is to provide Parametric study options. These options can incorporate arbitrary solution search rules through statistical and/or systematic data functions (data variation rules).

Parametric and statistical studies are particularly useful for estimating failure risks due to lightning and switching events or for evaluating performance limits for controllers.

B.3.6   External Interface

Modern applications have some means of interfacing with external packages or codes. The interfacing methods are either object oriented or capable of calling Dynamic Link Library (DLLs), or both. Such interfaces provide a simple interoperability and expandability path. An important user-defined modeling application is the connection of advanced controllers or relay models available in actual programming language codes. It is also possible to develop complete and complex models and interface them directly with the computational engine of the hosting application.

B.3.7   Time-Domain Module

The time-domain module is the heart of an EMT-type program. It starts from zero state (all devices are initially de-energized) or from given automatic or manual initial conditions and computes all variables as a function of time.

Since component models may have differential equations, it is needed to select and apply a numerical integration technique. Since many electrical circuits result in a stiff system of equations, the chosen numerical integration method must be stiffly stable. Such a need excludes explicit methods. In the list of implicit numerical integration methods, the most popular method in industrial applications remains the trapezoidal integration method. It is a polynomial method that can be programmed very efficiently. If an ordinary differential equation is written as

$$\frac{\mathrm{d}x}{\mathrm{d}t} = f(x,t) \quad (x(0) = x_0) \tag{B.3}$$

then the trapezoidal integration solution is given by

$$x_t = x_{t-\Delta t} + \frac{\Delta t}{2}\left(f_t + f_{t-\Delta t}\right) \tag{B.4}$$

The terms found at $t - \Delta t$ constitute history terms and all quantities at time-point t are also related through network equations. For the case of an inductance, for example, the basic equation $v = L\dfrac{\mathrm{d}i}{\mathrm{d}t}$ is transformed into

$$i_t = i_{t-\Delta t} + \frac{\Delta t}{2L}\left(v_t + v_{t-\Delta t}\right) \tag{B.5}$$

A similar approach is used for the capacitor. Using such a transformation for all network models allows writing the real version of Equation B.2. This process is called discretization with a given time-step. The vector **b** holds the independent variables and the history terms are updated after each solution time-point [9].

The integration time-step Δt can be fixed or variable. The fixed (set by the user) approach has several advantages in power systems. It avoids the time-consuming reformulation of system equations and programming issues related to the models. In the case of transmission line models, for example, it is necessary to maintain history buffers for interpolating for propagation delays. The time-step variability will affect the buffer sizes continuously thus slowing down the computations. Fixing the size for the smallest time-step will create memory problems for large cases.

The automatic computation of time-steps can be based on the local truncation error [3]. The variable time-step approach provides an important advantage for riding through various system time constants with the required precision. Another advantage is for the solution of nonlinear functions. Reducing the time-step may help convergence. It also provides more precise function fitting in time domain. Changing the time-step can, however, become time consuming as explained above.

Using a variable time-step does not fix the numerical oscillation problems, but it can minimize them. It will also minimize errors related to interpolation issues (see [15] and its references), but may become extremely time consuming for such problems. A complete solution for numerical oscillations due to discontinuities and interpolation for events occurring within the fixed time-step must be able to correctly account for nonlinear functions and distributed parameter models. The efficient and precise treatment of discontinuities remains an ongoing research topic.

There are other numerical integration methods, such as multistep methods and the backward-differentiation formula [16]. Some of these methods can be more precise or provide other advantages over the trapezoidal method for a given integration time-step. The backward-differentiation formula, for example, has the advantage of providing an extremely simple equation for evaluating the local truncation error. The polynomial Gear method can be used in a variable order setup to increase the integration time-step. It must be, however, restarted at each breakpoint and requires the maintenance of more history terms. The difficulty is with the added computational burden due to added number of coefficients, history terms, and restarting procedures. The theoretical advantages become overshadowed by the computational overhead specially since lowering the integration time-step in the trapezoidal method allows attaining similar precision while still remaining more efficient in most cases.

As explained earlier, in addition to linear functions, it is necessary to account for nonlinear functions. Such functions are solved through linearization at the last solution time-point for updating the Jacobian matrix **A**. An iterative solution method is required to maintain precision.

B.3.8 Control Systems

The simulation of control system dynamics is fundamental for studying power system transients. The development of control system solution algorithms based on the block-diagram approach has been initially triggered by the modeling of synchronous machine exciter systems. It was then extensively used in HVDC applications. Control elements can be transfer functions, limiters, gains, summers, integrators, and many other mathematical functions. In many applications, the block-diagram approach is also used to build and interface user-defined models with the built-in power system components.

A typical control diagram taken from the AVR_Gov block shown in Figure B.2 is shown in Figure B.5. Such diagrams are drawn in the GUI and solved directly. The GUI must allow drawing arbitrary control systems. Several commonly required functions may be available through GUI libraries.

A complicated problem in oriented-graph systems is the capability to solve the complete system simultaneously without inserting artificial (one time-step) delays in feedback loops. A solution to this problem is available in some applications [17,18].

In most applications, the control system diagram equations are solved separately from network equations. The control system uses its own set of equations similar to Equation B.1. Although this is not a significant source of errors in most cases, it can become an important drawback for user-defined network models and in the simulation of power electronic systems. The combination of both systems into a unique system of iteratively solved equations is complex. A fixed-point approach where both systems are solved sequentially is more efficient and acceptable in many cases [17].

B.3.9   Available Software Packages

The development of the first fully working EMT-type program was reported in [19]. The programs evolved with various sophistications in GUIs and solution methods. There are many tools currently available for the simulation of power system transients; here are some well known packages: ATP, EMTDC [20], and EMTP-RV [9]. These tools are all based on the fixed time-step trapezoidal integration method using Equation B.1. EMTP-RV has introduced the nonsymmetric and modified-augmented version given by Equation B.2.

The electromechanical modeling aspect is covered in most EMT-type packages through multimass machine models. More complexity might be added by interfacing with external packages specific to the simulation of mechanical motion or torque computation problems, such as in wind generator modeling [21].

The main advantage in EMT-type tools is the availability of a large number of validated models specific to power system studies. The most complex models are machine models, frequency-dependent transmission line models, and transformer models. Some models are designed for a wide range of frequencies. Built-in models can be used as building blocks for elaborate modeling of complex installations.

In addition to the above, there are also many simulation tools used for simulating electronic or power electronic circuits. It is difficult to enumerate all such tools, but the most powerful and popular tools are based on the original algorithms of SPICE [22]. SPICE uses the modified version of Equation B.1, which is called modified-nodal analysis. It also uses

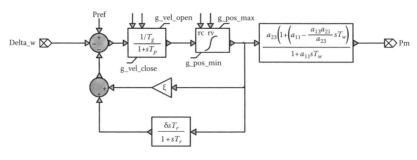

FIGURE B.5
Typical control system diagram.

the trapezoidal integration method, but with a variable time-step algorithm for controlling truncation error [3] and for controlling the convergence of nonlinear functions. Some versions may provide extra integration techniques, but the trapezoidal method remains the most popular choice.

SPICE-type (used hereinafter to regroup such tools) packages are not designed for power system applications but for elaborate electronic switching device models and electronic circuits. Such models must account much more precisely for the stresses and losses in semiconductor devices. In EMTP-type solution methods, switching devices, such as thyristors or transistors, are modeled as ideal switches with extra components included externally to account for losses. Although it is also possible to include nonlinear behavior, the level of model sophistication is limited since the target is the study of surrounding circuit system behavior. SPICE-type applications are targeting the detailed analysis of the semiconductor device behavior in the simulated circuits. In some versions of SPICE-type programs, it is possible to access directly semiconductor device libraries from various manufacturers providing data for all model parameters including even temperature effects.

In SPICE-type programs, it is usually possible to use a variable integration time-step which can have important advantages for solving nonlinearities. The inconvenience however is that changing the time-step requires reformulation and may become extremely demanding in computer time. It is possible to fix the time-step by fixing its limits, but this may affect the behavior of the nonlinear solver.

In EMTP-type applications, the built-in nonlinearities are monotonically increasing and crossing zero. In SPICE-type applications it is possible to use nonmonotonically increasing characteristics and search for multiple solutions.

Although it is feasible to use SPICE (or SPICE-type) for the computation of power system transients, it is not designed for this field of applications. The readily available models for transmission lines and rotating machines are usually much less sophisticated. Many specialized fields, such as lightning transients and switching transients benefit from advanced modeling capabilities available only in EMT-type applications.

Contrary to power systems in the case of electronic circuits and microchips, it is easier to obtain data and maintain advanced databases of models from various manufacturers. Advanced packages such as SABER [23] are used for analog, digital, mixed-signal, and mixed-technology simulations. A specialized language named MAST is used to model complex electrical circuits. It is capable of interfacing with Fortran and C++ codes and reuse existing models. MAST is a hierarchical language.

The electronic industry uses Very High Speed Integrated Circuit Hardware Description Language (VHDL) [24] for the purpose of synthesizing and simulating digital circuit designs. VHDL designs can be simulated and translated into a form suitable for hardware implementation. There are several IEEE standard extensions to VHDL for analog, mixed signals, and mathematics. VHDL borrows heavily from Ada (programming language) in both concepts and syntax. VHDL has constructs to handle the parallelism inherent in hardware designs.

B.3.10 General-Purpose Modeling Environments

The most popular general-purpose modeling environment is MATLAB®/Simulink. There are no built-in stand-alone programs in MATLAB for simulating transients, but its programming language has advanced functions for solving large-scale linear systems, which allows programming complete solvers [25]. There are many advantages in such codes since they provide a completely open and high-level architecture which can be used for rapid testing of new solution methods and prototyping of new models. The programming environment of MATLAB can be used as a laboratory for programming compiled code applications using standard computer languages. It also offers many advantages for programming and compiling visualization and analysis tools [26].

Simulink [18] is a block-diagram-based package available in MATLAB. It is a general-purpose application, widely used for simulating control systems in time domain. Simulink offers many advantages with a large library of control blocks and various design functions. Both fixed time-step and variable time-step integration methods are available. The state-space block can be used for entering electrical network equations in the state-space format:

$$\frac{dx}{dt} = Ax + Bu \tag{B.6a}$$

$$y = Cx + Du \tag{B.6b}$$

where
x is the vector of states
u is the vector of inputs

The C and D matrices are used to obtain the vector of outputs y. This is the main concept behind the development of a specialized tool named SimPowerSystems [27] for the simulation of power systems transients. This tool offers advanced flexibility for customization and definition of user-defined models. It is also a powerful tool for designing and testing control systems. The drawbacks are in the usage of the state-space formulation for network equations. The computer time required for the formulation of Equation B.6a can become unaffordable for larger systems. This is not the case with nodal analysis where the assembly of Y or A matrices (for Equation B.1 or B.2) is a straightforward process and requires minimum computer time for very large networks.

Other implementation methods are also available [28].

B.3.11 Hybrid Tools

The term hybrid is used for designating simulation tools or methods based on the following combinations: different types of solution methods and different simulation environments.

A typical example for different types of solution methods is when a frequency-domain solution for the network equations is combined with a time-domain solution. Such an approach offers many advantages in modeling and computational speed. In some cases, hybrid methods are used for initialization purposes. A frequency-domain solution of the network in the steady-state module is set to call the time-domain solution of a nonlinear component. The time-domain solution generates harmonics which are sent back to the network solution in the form of a Fourier series [13].

For "different simulation environments" the meaning is the simulation of physical problems in different engineering domains. Packages such as [18] and [23] fall into this category. In [23], for example, it is possible to simulate hydraulic, electronic, and thermal effects. The ultimate objective is to reduce the need for physical prototypes.

As explained before, hybrid methods can be also established by connecting and interfacing specialized applications from different domains. In [29], an EMT-type program is linked with an external package based on the finite element method (FEM) for detailed transformer energization studies. In the FEM-based software, it is possible to use a highly precise model to account for the material nonlinearity, winding connections, and anisotropy. Such software, however, does not offer advanced power system models which become available on the EMT-type application side.

Another application example is shown in [30], where the CIGRE HVDC benchmark is modeled using an interface between EMTDC and MATLAB/Simulink. Such an approach also allows creating model portability between applications [31].

Interfacing has also been used to incorporate optimization when multiple simulations are involved in design applications in power electronics or simulation of transients. The references [32,33] provide examples of interfacing the SABER and the PSCAD/EMTDC respectively, with optimization routines.

As explained in a previous section, the time-domain approach for solving network equations is more precise, but offers a significantly reduced performance. To provide significant acceleration in the solution of large networks or to combine with solvers for electromechanical transients (lower frequency oscillations), a given network can be separated into fast (precise) and slow areas. Relaxation methods or stability time-frame methods can be applied in the slower regions. The main difficulty is related to the interfacing of methods between regions (see [34,35] and related references). In some applications, it is possible to solve the same system separately in both frequency- and time domain [36].

B.3.12 Frequency-Domain Methods

Although time-domain techniques have been usually preferred for calculation of electromagnetic transients, frequency-domain (FD) techniques have some advantages that can be considered in some applications; e.g., modeling and calculations can be more rigorous with frequency-dependent distributed-parameter elements, and numerical errors can be better determined and controlled.

There are many examples of cases where FD analysis has been instrumental for advancing time-domain techniques; some of these are the development of full frequency-dependent line and cable models, convolution techniques, rational fitting, and network equivalent synthesis.

Several FD techniques have been developed over the years; they are based on the Fourier Transform, the Numerical Laplace Transform, or the z-Transform. A comprehensive coverage of this topic is out of the scope of this appendix. The reader can consult the partial list of references [37–47].

B.3.13 Case Studies

B.3.13.1 *Switching Overvoltages*

A typical study type is the computation of overvoltages on transmission lines for switching transients. Such studies can result in the selection of line arresters or usage of pre-insertion resistors during the line energization, and they require statistical analysis, see Sections 2.6.4 and 7.6.3 of this book.

The network of Figure B.2 contains complete data with synchronous machines and related controls. In the case of line energization, it is sufficient to model the network with simple equivalents: ideal sources with Thevenin impedances followed by transmission lines up to the substation where energization is performed. The amount of details will improve the precision in the computation of overvoltage waveforms. Inclusion of load and transformer models will improve the precision on damping. It is also more precise to select frequency-dependent models for transmission lines. The waveforms shown in Figure B.6 are from the energization study of the transmission line TLM_120mi shown in Figure B.2. The line is protected with surge arresters at both ends. Trapped charge conditions capable of causing worst overvoltages are imitated by opening the line from steady-state conditions at 2 ms and reclosing at 13 ms. Since there is coupling in the line, its phase voltages continue changing until all phases are isolated. The overvoltages are effectively limited by using arresters.

B.3.13.2 Lightning Overvoltages

This case demonstrates that EMT-type programs allow reaching high levels of sophistication for modeling direct lightning strokes on transmission lines. The studied line shown in Figure B.7 is a double-circuit line with phase 1 subjected to a direct lightning stroke. The line model is frequency dependent. The lightning current is the CIGRE concave lightning current source as defined in [48]. The tower at each span is modeled using 6 insulator chains connected from phase wires to the constant-parameter (CP) transmission line model of the tower, which has a characteristic impedance of 90 Ω with a length of 15.5 m. The insulator chains are represented with the leader propagation model [48]. See also Chapter 2 of

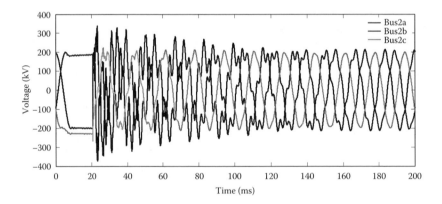

FIGURE B.6
Transmission line overvoltages at receiving end phases.

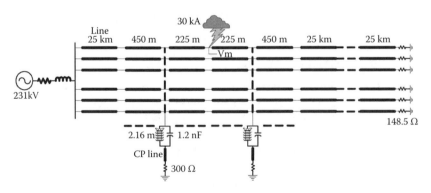

FIGURE B.7
Double circuit transmission line subjected to a lightning stroke.

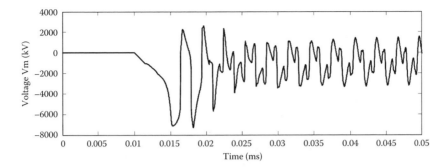

FIGURE B.8
Overvoltage on a double-circuit transmission line submitted to a lightning stroke.

this book. The resulting overvoltage at the location Vm is shown in Figure B.8. Insulator flashovers are observed. In this case, the integration time-step is 0.02 μs, which is required to account for the short line lengths and waveform steepness.

B.3.13.3 Electromechanical Transients

In the following study, the objective is to evaluate the transient stability of the network shown in Figure B.2. The full network is simulated with synchronous machine models with field voltage and mechanical power controls. In this case, the controls also included a stabilizer. The event is a single-phase-to-ground fault occurring on the transmission line TLM_120mi at 200 ms. The fault is cleared at 0.3 s and the line is reclosed at about 0.6 s. It appears that the system is able to regain stability (see 3-phase powers in Figure B.9) after reclosing into the line. The first step in this study is a load-flow solution that establishes the machine phasors and thus all power transfers. The second step is the steady-state solution where loads and load-flow constraints are replaced by actual equivalents. This step is used to initialize all network variables including internal machine variables and related controls.

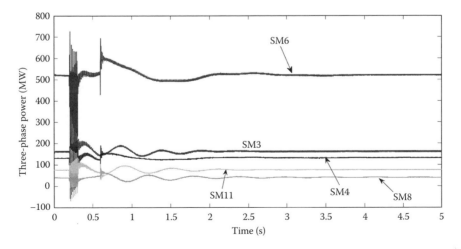

FIGURE B.9
Synchronous machines, 3-phase powers.

The integration time-step for the case of Figure B.9 was chosen as 50 μs. Although it was possible to select larger time-steps for the machine models, the limiting factor was the propagation delay on the short transmission line TLM_120mi. In classical stability studies, propagation delays are not modeled since the line is a simple pi-section.

B.3.13.4 Wind Generation

In this case, the objective is to study wind generation. The setup is shown in Figure B.10. There are three groups of wind generators. Aggregation is used to model 10 generators of 1.5 MW. The wind generators are of doubly fed induction generator (DFIG) type and modeled with details including power electronics converters and protection systems, such as crowbar and voltage protection. A simplified model is used for the connected network at the left-hand side.

The simulated event is a single-phase fault on BUS12. The waveform shown in Figure B.11 is measured on the right-hand side of the main breaker MAIN_SW, which receives a trip signal at 1.15 s. The resulting overvoltages include harmonics due to transformer saturation. The voltage measured in the wind turbine WTG1 is shown in Figure B.12 where the overvoltage trip signal opens the wind generator breaker and the crowbar protection signal opens the rotor side inverter and connects short circuit resistances on the rotor.

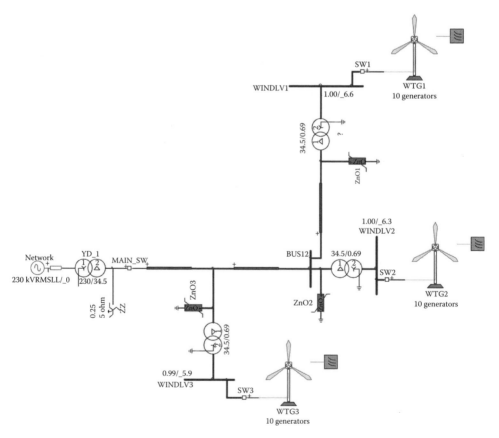

FIGURE B.10
Wind generator study.

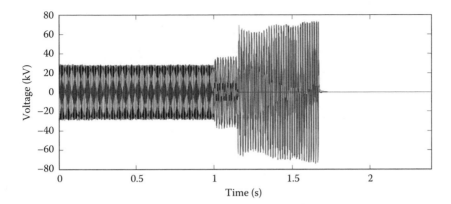

FIGURE B.11
Wind generator study: Voltages at the interconnection point.

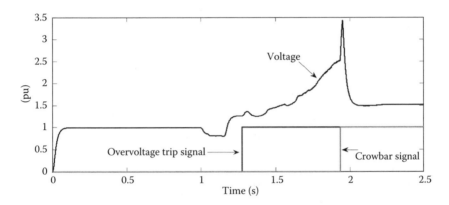

FIGURE B.12
Wind generator study: Wind generator WTG1 positive sequence voltage, overvoltage trip signal, and crowbar protection signal.

Due to the presence of power electronics converters, it is required to use an integration time-step of 10–20 μs for achieving high-precision results. Initialization of wind generators is a complicated problem, since significant and sometimes unaffordable computing time can be required if the simulation starts without proper initialization. In this example, a load-flow solution is combined with specific initializing scripts. As illustrated by the voltage waveform of Figure B.12, the simulation reaches steady state very quickly.

In addition to validating wind generator behavior and related protection systems, this case study is also used to verify the energy absorption capabilities for the wind park surge arresters.

B.4   Real-Time Simulation Tools

In contrast to off-line transient simulation tools, real-time simulators are useful for testing hardware equipment by interfacing them to the simulator. Real-time simulators can be made up of analog components or digital computers. For over 70 years, real-time analog simulators have been used for various applications, but over the last 10 years significant advances have been made in real-time digital simulators. The three main types of real-time simulators include the Transient Network Analyzers (TNA), the Real-Time Digital Simulators, and the Real-Time Playback Systems.

B.4.1 Transient Network Analyzers

The TNA is a collection of scaled down models of analog power system equipment. The main strength of a TNA, as with any analog set-up, lies in its real-time capability thus allowing a comprehensive HIL testing of control and protection equipment. However, TNAs suffer from several drawbacks which have limited their application. First, they require significant resources and time to build and maintain which is why they are owned and operated by large utilities. Once a transient study has been completed and the setup disassembled, it can take several days to prepare the TNA for another test scenario by reconfiguring various components and rewiring the connections between them. Third, TNAs generally lack the scalability for an accurate system representation. Traveling wave effects cannot be reproduced faithfully using only a few cascaded pi segments to model a long line. Some of these drawbacks can be overcome by using a *hybrid* simulator which is a combination of analog components and digital computer models.

B.4.2 Real-Time Digital Simulators

The best alternative to an analog TNA is a digital simulator that can solve the system equations in real time. Due to rapid advances in digital processor and parallel processing technology in the last two decades, real-time digital transient simulators are increasingly popular. Real-time digital simulators are required to solve the system differential equations within the time-step selected for simulation. For example, if a transient event happens in 50 μs in the actual system, the real-time simulator should be able to perform the necessary computations for the transient and output the results within 50 μs. It is not sufficient for the end of the simulation run to coincide with the real-time clock. Instead, the computation of every time-step must be executed within the same corresponding interval of real time. The reason for this stringent requirement is that the simulator must be able to interface and synchronize with actual external control or protection hardware.

There are several industrial grade real-time digital simulators such as HYPERSIM [49], RT-LAB [50], and RTDS [51]. These simulators are based on various types of digital processors such as DSPs, RISC processors, and general-purpose processor-based PC-Clusters. Currently the main applications of real-time digital simulators are three-fold: closed-loop testing of digital controllers for power electronic-based FACTS and HVDC systems, closed-loop testing of protective relays, and simulation of transients specifically for analyzing a large number of operating scenarios and fault conditions. Other applications are harmonics and power quality evaluation [52,53].

Although earlier efforts at real-time simulation were more or less an extension of the off-line simulators such as EMTP [54], the latest developments in real-time simulation have a distinct flavor of their own in terms of newer models and algorithms. This is especially true when performing hardware-in-the loop (HIL) simulations. There are several important issues that need to be addressed regarding the interfacing of real-time digital simulators and external hardware, such as digital controllers, for power electronic apparatus [55]. A real-time digital simulator simulating power electronic systems takes discrete switching signals as external inputs from the digital controller. Digital simulation being itself discrete in nature is unable to cope effectively with switching signals that arrive between two calculation steps of the simulator. The conventional off-line approach of using small step-sizes for simulation to overcome the problem is not a favorable option under real-time conditions. Several algorithms have been proposed for correcting firing errors and extra delays for power electronics in real-time digital simulators [56,57]. There are also several commercially available packages such as ARTEMIS [58] that address this issue. The fixing of interpolated signals within fixed time-step simulations is an ongoing research topic (see also references in [15]). Wideband real-time electromagnetic transient simulation of large-scale systems using accurate and efficient frequency-dependent network equivalents is currently feasible [59].

A combination of rapid advances in PC technology and the development of accurate power system models in mathematical modeling packages such as MATLAB/Simulink are driving the current trend of using PC-clusters for real-time and HIL simulation which previously could only be implemented by expensive high-end technologies [60]. The PC-cluster-based real-time simulator is built entirely from high performance commodity-off-the-shelf (COTS) components to sustain performance at a reasonable cost. Real-time simulation and off-line model preparation are divided between two groups of computers comprising the *target cluster* and *hosts*. Gigabit Ethernet is used for fast data transfer between the target and host computers, while InfiniBand technology is used for inter-node communication in the targets. Dedicated A/D and D/A signal conditioning modules formed the physical connection to the real-world hardware. The real-time model construction and validation is carried out on the hosts under MATLAB/Simulink with customized toolkits. The C-code generation is based on MATLAB's Real-Time Workshop (RTW) and the compiled executable is downloaded to the *targets* for real-time simulation.

This target cluster and hosts configuration is truly flexible and scalable. When more computation power is needed for real-time simulation, additional cluster nodes can be added by directly connecting through the InfiniBand switching hub. For more users to share the real-time computation power, properly equipped PCs are simply connected to the gigabit network from the Ethernet switch.

An upcoming hardware trend in real-time simulator design is the use of field programmable gate arrays (FPGAs) as the core computational engines. The parallel processing hardwired architecture and large resource count of these devices is enabling this development. Time steps of the order of a few nanoseconds are now possible for highly accurate real-time simulation of power electronics and variable-speed motor drives [61]. FPGAs can also enable highly accurate real-time electromagnetic transient simulation of conventional power systems with frequency-dependent transmission lines [62].

B.4.3 Real-Time Playback Simulators

In this type of simulators, the transient waveform data are first generated by an off-line EMTP-type program. The stored data are then played back and synchronized in real-time to the device under test. A playback simulator can test the device under open-loop conditions only [63]. This is the main drawback of real-time playback simulators in contrast to an analog TNA or a fully digital real-time simulator.

Real-time playback systems have been used for testing protective relays by subjecting the relay to simulated fault currents and voltages. In addition to simulated waveforms, it is also possible to record field results and play them back in real-time. The main advantage of real-time playback systems lies in the fact that they can utilize the full capabilities of off-line EMTP-type programs. Since the transient data is not collected in the real-time mode, the complexity or the size of the model is not an issue. Multiple test runs can be

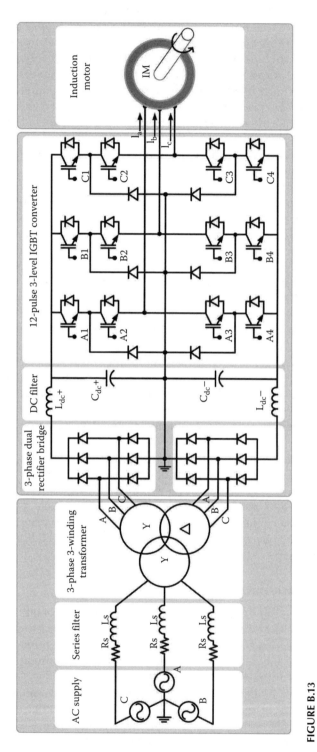

FIGURE B.13

Vector-controlled variable-speed ac drive. (From Pak, L.-F. et al., *IEEE Trans. Power Syst.*, 21, 455, 2006. With permission.)

scheduled in the real-time playback equipment enabling an automated evaluation of the test equipment under a large number of fault scenarios.

B.4.4 Case Study

Figure B.13 shows the 3-level 12-pulse vector-controlled variable-speed ac drive system which has been used as a case study for the real-time simulation on the PC-Cluster described in the previous section. The complete system consists of a squirrel-cage induction motor, the drive, and the digital controller. Decoupled vector-control technique has been employed to realize the motor torque/speed control. The electrical system consists of two 3-phase diode rectifier bridges which are connected to the 3-phase ac supply through a 3-winding transformer. After LC filters, the dc sides of the rectifiers are connected to a 3-level 12-pulse IGBT (insulated-gate bipolar transistors) converter which produces the controlled voltages for the induction machine. All parts of the electrical system were first modeled using built-in models in the SimPowerSystems blockset [27]. The system model was then customized for real-time simulation on the PC-Cluster. The power system is modeled in two parts: a state-space (SS) model for the linear circuit and a current injection feedback model for the nonlinear elements. In this case, the electrical model was divided into three parts (Figure B.14). All the linear circuit elements, such as the ac supply, the series filter, the 3-winding transformer, the 3-phase dual rectifier bridge, and the dc filter were modeled by SS equations. The IGBT converter was modeled using a first-level feedback interfacing with the main SS model through the input voltages and output currents. With the output voltages of the converter as inputs, the custom machine model would then generate the stator line currents and feed them back to the converter model.

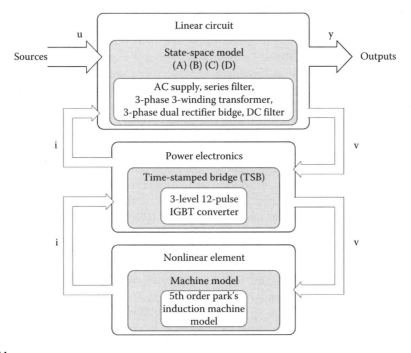

FIGURE B.14
System model separation for real-time simulation. (From Pak, L.-F. et al., *IEEE Trans. Power Syst.*, 21, 455, 2006. With permission.)

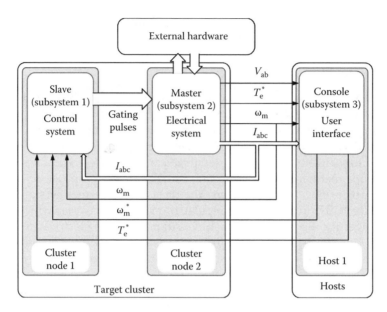

FIGURE B.15
Implementation subsystems for real-time simulation. (From Pak, L.-F. et al., *IEEE Trans. Power Syst.*, 21, 455, 2006. With permission.)

The objective of the controller is to maintain a constant machine speed ω_m while the mechanical torque T_m is changing. The independent control of ω_m is achieved in the synchronously rotating dq frame with the d-axis oriented along the rotating magnetic-flux vector.

The 3-level 12-pulse ac drive and its control system was implemented in real-time using multiple subsystems with multirate sampling as shown in Figure B.15; a *Master* subsystem for the linear part of the electrical system, a *Slave* subsystem for the machine, 12-pulse converter and the controller, and a *Console* subsystem for command and monitoring of the various inputs and outputs. Out of these three subsystems, only the Master and the Slave are allowed to run in real-time using two target nodes in the PC-Cluster. The Console runs on the host computer and communicates with the targets for controlling the inputs and outputs. The real-time simulation is fully interactive and all variables can be either through the console or through external devices such as an oscilloscope connected to the FPGA-based I/Os. The real-time simulation utilizes multinode multirate sampling.

A time-step of 10 μs was assigned for the real-time simulation, and a carrier frequency of 2.5 kHz was used for the pulse width modulation (PWM). The simulation results have been recorded for the two operating conditions: steady-state and transient situations. Figure B.16 shows the real-time scopes of the steady-state waveforms (up to 40 ms). These results were corroborated using off-line simulations.

For the transient shown in Figure B.17 the reference torque is maintained at 140 Nm, while the reference speed is changed from 120 to 160 rad/s at 10 s.

The model has been sufficiently optimized so that the model step-size of real-time simulation can be brought down as small as possible. A time-step dissection revealed that using a single target node, real-time simulation has been achieved with a step-size of 10 μs, without causing any overruns, of which the maximum computation time is only 5.35 μs. A detailed breakdown of the execution time also revealed a processor idle time of 2.49 μs

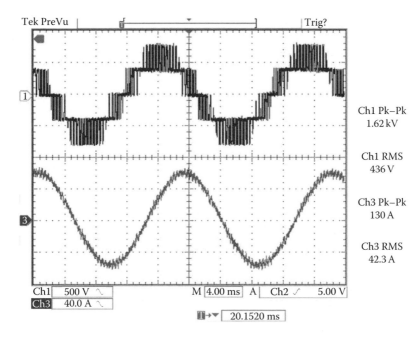

FIGURE B.16
Traces for steady-state line-to-line voltage and line current. (From Pak, L.-F. et al., *IEEE Trans. Power Syst.*, 21, 455, 2006. With permission.)

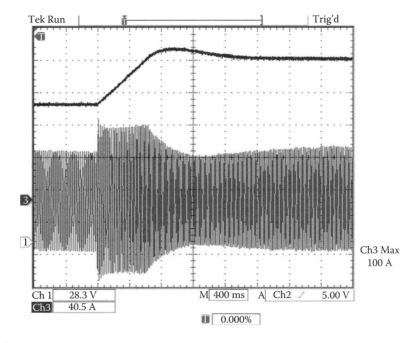

FIGURE B.17
Traces of the transient response: Machine speed and line current. (From Pak, L.-F. et al., *IEEE Trans. Power Syst.*, 21, 455, 2006. With permission.)

which indicated the possibility of further reduction in the step-size. The experience gained during this case study indicates that practical ac drive controllers can be simulated and tested with great accuracy and efficiency.

References

1. J. Mahseredjian, V. Dinavahi, and J.A. Martinez, Simulation tools for electromagnetic transients in power systems: Overview and challenges, *IEEE Transactions on Power Delivery* 24(3), 1657–1669, July 2009.
2. J. Mahseredjian, V. Dinavahi, and J.A. Martinez, An overview of simulation tools for electromagnetic transients in power systems, *Proceedings of IEEE Power Engineering Society General Meeting*, Tampa, FL, June 24–28, 2007.
3. J. Mahseredjian, Computation of power system transients: Overview and challenges, *Proceedings of IEEE Power Engineering Society General Meeting*, Tampa, FL, June 24–28, 2007.
4. IEC 60071-1, Insulation co-ordination—Part 1: Definitions, principles and rules, January 2006.
5. CIGRE WG 33.02, Guidelines for Representation of Network Elements when Calculating Transients, CIGRE Technical Brochure no. 39, 1990.
6. IEC 60071-4, Insulation co-ordination—Part 4: Computational guide to insulation co-ordination and modelling of electrical networks, June 2004.
7. Modeling and analysis of system transients using digital programs, A.M. Gole, J.A. Martinez-Velasco, and A.J.F. Keri (Eds.), IEEE PES Special Publication, TP-133-0, 1999.
8. J.A. Martinez, J. Mahseredjian, and R.A. Walling, Parameter determination for modeling system transients, *IEEE Power and Energy Magazine*, 3(5), 16–28, September/October 2005.
9. J. Mahseredjian, S. Dennetière, L. Dubé, B. Khodabakhchian, and L. Gérin-Lajoie, On a new approach for the simulation of transients in power systems, *Electric Power Systems Research*, 77(11), 1514–1520, September 2007.
10. Common Information Model (CIM): CIM 10 version, EPRI, Palo Alto, CA, 2001.
11. G. Irwin and D. Woodford, E-TRAN: Translation of load-flow/stability data into electromagnetic transients programs, *Proceedings of International Conference on Power Systems Transients*, IPST 2003, New Orleans, LA, September 28–October 2, 2003.
12. W. Xu, J. Marti, and H. W. Dommel, A multiphase harmonic load flow solution technique, *IEEE Transactions on Power Systems*, 6(1), 174–182, February 1991.
13. X. Lombard, J. Mahseredjian, S. Lefebvre, and C. Kieny, Implementation of a new harmonic initialization method in the EMTP, *IEEE Transactions on Power Delivery*, 10(3), 1343–1352, July 1995.
14. J.A. Martinez-Velasco, Computational methods for EMTP steady-state initialization, *Proceedings of International Conference on Power Systems Transients*, IPST 1999, Budapest, Hungary, June 20–24, 1999.
15. M. Zou, J. Mahseredjian, G. Joos, B. Delourme, and L. Gérin-Lajoie, Interpolation and reinitialization in time-domain simulation of power electronic circuits, *Electric Power Systems Research*, 76(8), 688–694, May 2006.
16. R.K. Brayton, F.G. Gustavson, and G.D. Hachtel, A new efficient algorithm for solving differential-algebraic systems using implicit backward-differentiation formulas, *Proceedings of the IEEE*, 60(1), 98–108, January 1972.
17. J. Mahseredjian, L. Dubé, M. Zou, S. Dennetière, and G. Joos, Simultaneous solution of control system equations in EMTP, *IEEE Transactions on Power Systems*, 21(1), 117–124, February 2006.
18. Simulink, 1994–2007. The MathWorks, Inc.
19. H.W. Dommel, Digital computer solution of electromagnetic transients in single- and multiphase networks, *IEEE Transactions on Power Apparatus and Systems*, 88(4), 734–741, April 1969.

20. D.A. Woodford, A.M. Gole, and R.Z. Menzies, Digital simulation of dc links and ac machines, *IEEE Transactions on Power Apparatus and Systems*, 102(6), 1616–1623, June 1983.
21. www.garradhassan.com.
22. L.W. Nagel, SPICE2 A computer program to simulate semiconductor circuits, Memorandum No. UCB/ERL M520, May 9, 1975.
23. www.synopsys.com.
24. www.vhdl.org.
25. J. Mahseredjian and F. Alvarado, Creating an electromagnetic transients program in MATLAB: MatEMTP, *IEEE Transactions on Power Delivery*, 12(1), 380–388, January 1997.
26. J. Mahseredjian, F. Alvarado, G. Rogers, and B. Long, MATLAB's power for power systems, Invited paper, *IEEE Journal on Computer Applications in Power*, 14(1), 13–19, January 2001.
27. SimPowerSystems User's Guide, Version 4, The MathWorks, Inc., 2006.
28. PLECS, Electrical Systems and Power Electronics in Simulink, ETH Zurich, 2006.
29. S. Dennetière, Y. Guillot, J. Mahseredjian, and M. Rioual, A link between EMTP-RV and FLUX3D for transformer energization studies, *Proceedings of the International Conference on Power Systems Transients*, IPST 2007, Lyon, France, June 4–7, 2007.
30. M.O. Faruque, Y. Zhang, and V. Dinavahi, Detailed modeling of CIGRE HVDC benchmark system using PSCAD/EMTDC and PSB/SIMULINK, *IEEE Transactions on Power Delivery*, 21(1), 378–387, January 2006.
31. S. Casoria, J. Mahseredjian, R. Roussel, J. Beaudry, and G. Sybille, A portable and unified approach to control system simulation, *Proceedings of the International Conference on Power Systems Transients*, IPST 2001, Rio de Janeiro, Brazil, June 24–28, 2001.
32. H. Kragh, F. Blaabjerg, and J.K. Pedersen, An advanced tool for optimized design of power electronic circuits, *Proceedings of IEEE Industry Applications Conference*, St. Louis, MO, 991–998, 1998.
33. A.M. Gole, S. Filizadeh, R.W. Menzies, and P.L. Wilson, Optimization-enabled electromagnetic transient simulation, *IEEE Transactions on Power Delivery*, 20(1), 512–518, January 2005.
34. H.T. Su, L.A. Snider, K.W. Chan, and B.R. Zhou, A new approach for integration of two distinct types of numerical simulators, *Proceedings of the International Conference on Power Systems Transients*, IPST 2003, New Orleans, LA, September 28–October 2, 2003.
35. R. Shintaku and K. Strunz, Branch companion modeling for diverse simulation of electromagnetic and electromechanical transients, *Proceedings of the International Conference on Power Systems Transients*, IPST 2005, Montréal, Quebec, Canada, June 19–23, 2005.
36. B. Kulicke, E. Lerch, O. Ruhle, and W. Winter, NETOMAC—Calculating, analyzing and optimizing the dynamic of electrical systems in time and frequency domain, *Proceedings of the International Conference on Power Systems Transients*, IPST 1999, Budapest, Hungary, June 20–24, 1999.
37. S.J. Day, N. Mullineux, and J.R. Reed, Developments in obtaining transient response using Fourier transforms. Part I: Gibbs phenomena and Fourier integrals, *International Journal of Electrical Engineering Education*, 3, 501–506, 1965.
38. S.J. Day, N. Mullineux, and J.R. Reed, Developments in obtaining transient response using Fourier transforms. Part II: Use of the modified Fourier transform, *International Journal of Electrical Engineering Education*, 4, 31–40, 1966.
39. L.M. Wedepohl and S.E.T. Mohamed, Multiconductor transmission lines. Theory of natural modes and Fourier integral applied to transient analysis, *Proceedings of the IEE*, 116(9), 1553–1563, September 1969.
40. A. Ametani, The application of the fast Fourier transform to electrical transient phenomena, *International Journal of Electrical Engineering Education*, 10, 277–287, 1973.
41. J.P. Bickford, N. Mullineux, and J.R. Reed, *Computation of Power System Transients*, Peter Peregrinus Ltd., London, 1976.
42. N. Nagaoka and A. Ametani, A development of a generalized frequency domain transient program – FTP, *IEEE Transactions on Power Delivery*, 3(4), 1996–2004, October 1988.
43. A.M. Cohen, *Numerical Methods for Laplace Transform Inversion*, Springer, New York, 2007.

44. P. Moreno and A. Ramirez, Implementation of the numerical Laplace transform: A review, *IEEE Transactions on Power Delivery*, 23(4), 2599–2609, October 2008.

45. W. Derek Humpage and K. Wong, Electromagnetic transient analysis in EHV power networks, *Proceedings of the IEEE*, 70(4), 379–402, April 1982.

46. W.D. Humpage, *z-Transform Electromagnetic Transient Analysis in High-Voltage Networks*, Peter Peregrinus Ltd., London, 1982.

47. T. Noda and A. Ramirez, z-Transform-based methods for electromagnetic transient simulations, *IEEE Transactions on Power Delivery*, 22(3), 1799–1805, July 2007.

48. CIGRE WG 33-01 Guide to procedures for estimating the lightning performance of transmission lines, CIGRE Brochure no. 63, 1991.

49. D. Paré, G. Turmel, J.-C. Soumagne, V.A. Do, S. Casoria, M. Bissonnette, B. Marcoux, and D. McNabb, Validation tests of the Hypersim digital real time simulator with a large ac-dc network, *Proceedings of the International Conference on Power Systems Transients*, IPST 2003, New Orleans, LA, September 28–October 2, 2003.

50. S. Abourida, C. Dufour, J. Belanger, G. Murere, N. Lechevin, and B. Yu, Real-time PC-based simulator of electric systems and drives, *Proceedings of 17th IEEE Annual Applied Power Electronics Conference Exp.*, New Orleans, LA, APEC, 1, 433–438, March 10–14, 2002.

51. R. Kuffel, J. Giesbrecht, T. Maguire, R.P. Wierckx, and P.G. McLaren, RTDS-A fully digital power system simulator operating in real-time, *Proceedings of EMPD'95*, 2, 498–503, 1995.

52. Y. Liu, M. Steurer, S. Woodruff, and P. Ribeiro, A novel power quality assessment method using real time hardware-in-the-loop simulation, *Proceedings of International Conference on Harmonics and Quality of Power*, ICHQP'04, New York, September 2004.

53. L.-F. Pak, V. Dinavahi, G. Chang, M. Steurer, and P.F. Ribeiro, Real-time digital time-varying harmonic modeling and simulation techniques, *IEEE Transactions on Power Delivery*, 22(2), 1218–1227, April 2007.

54. J.R. Marti and L.R. Linares, Real-time EMTP-based transients simulation, *IEEE Transactions on Power Systems*, 9(3), 1309–1317, August 1994.

55. V.R. Dinavahi, M.R. Iravani, and R. Bonert, Real-time digital simulation of power electronic apparatus interfaced with digital controllers, *IEEE Transactions on Power Delivery*, 16(4), 775–781, October 2001.

56. M.O. Faruque, V. Dinavahi, and W. Xu, Algorithms for the accounting of multiple switching events in the digital simulation of power electronic apparatus, *IEEE Transactions on Power Delivery*, 20(2), 1157–1167, April 2005.

57. R. Champagne, L.-A. Dessaint, H. Fortin-Blanchette, and G. Sybille, Analysis and validation of a real-time ac drive simulator, *IEEE Transactions on Power Electronics*, 19(2), 336–345, March 2004.

58. C. Dufour and J. Belanger, Discrete time compensation of switching events for accurate real-time simulation of power systems, *Proceedings of 27th IEEE Industrial Electronics Society Conference*, Denver, CO, *IECON'01*, 2, pp. 1533–1538, November–December 2001.

59. X. Nie, Y. Chen, and V. Dinavahi, Real-time transient simulation based on a robust two-layer network equivalent, *IEEE Transactions on Power Systems*, 22(4), 1771–1781, November 2007.

60. L.-F. Pak, M.O. Faruque, X. Nie, and V. Dinavahi, A versatile cluster-based real-time digital simulator for power engineering research, *IEEE Transactions on Power Systems*, 21(2), 455–465, May 2006.

61. G.G. Parma and V. Dinavahi, Real-time digital hardware simulation of power electronics and drives, *IEEE Transactions on Power Delivery*, 22(2), 1235–1246, April 2007.

62. Y. Chen and V. Dinavahi, FPGA-based real-time EMTP, *IEEE Transactions on Power Delivery*, 24(2), 892–902, April 2009.

63. P. McLaren, J. Song, B. Xiao, E. Dirks, L. Arendt, R. Wachal, and P. Wilson, Incorporating PSCAD/EMTDC into a real-time playback test set, *Proceedings of International Conference on Power Systems Transients*, IPST 1997, Seattle, WA, June 22–26, 1997.

Index

A

Air–Air circuit elements, 190
Air-blast circuit breakers, 471, 473; *see also*
    Circuit breaker
Air-core reactance, 193
Air-gap flux linking, 255
Air gap, function, 353
Air-insulated station (AIS), 416
Air insulation, behavior, 93
Amsinck's method, 532–534; *see also* Gas-filled
    circuit breaker models
Analog models, of MO surge arresters, 378–379
Arc control devices, development, 472
Arc current in vacuum, formation, 501
Arc extinction in gas-filled circuit breakers,
    representation, 501
Arc models
  applications, 486
  approach application, steps, 540
  comparison, 493
  and related parameters, 496
  technique, steps for, 531
  validation, 536
Arc parameter determination, principles, 537
Armature current
  ac components, 275, 276
  produced by application of three-phase
    voltages, 276
  time periods, 276
Armature-to-field transfer function, 285
Arrester; *see also* Metal oxide (MO) surge
    arrestors
  applications, 403–404
    distribution systems protection, 422–438
    transmission systems protection, 404–422
  classification, 357–358
  class selection
    coordinating currents, 391–392
    location, 392–393
  current, time-to-crest effect, 361
  model
    CIGRE model, 367–368
    construction methods, 364
    hysteretic model, 370–371
    IEEE model, 368–370
    simplified model, 371–374
  protective level, 387
  ratings, determination, 357–358
  selection procedures, 387–393
  thermal performance, simulation, 384
Auto-expansion interruption technique, 475
Auto-Regressive Moving Average (ARMA)
    model, 32

B

Backfeed overvoltages, occurrence, 403
Basic impulse level, 477
Basic lightning impulse insulation
    level (BIL), 91
Basic switching impulse insulation
    level (BSL), 91
Binary Partitioning Algorithm, 577
Black box arc models, parameters
    estimation, 538
Black box models, 486, 488, 531
Bode's method, 562
Breakdown voltages, 126

C

Cable-connected stations, protection, 417
Cable Constants (CC) routines, 138; *see also*
    Insulated cables
  core data, 161
  differentiation of cable designs, 142–143
  semiconducting screens, 161–162
Cable designs; *see also* Insulated cables
  pipe-type, 143
  single-core self-contained, 142
  three-phase self-contained, 142–143
Cable geometry, role, 138
Cable installation, in tunnel, 155–156
Cable model, for vacuum circuit breakers,
    515–516; *see also* Circuit breaker
Cable propagation characteristics, sensitivity,
    164
Cables parameters, calculation; *see also*
    Insulated cables
  coaxial cables, 144–153
  pipe-type cables, 153–155
  tunnel and trench cables, 155–156
  wire sheath, spiral effect, 156

Canay's characteristic inductance, 292
Capacitance, calculation and initial voltage
    distribution, 238–240
Capacitive currents, interruption, 461–463
Capacitor banks, protection, 424
Capacitor switching, 180
Carson approximation, in Z_g calculation, 172
Carson integral method, 152
Carson's ground impedance, 23
Cassie–Mayr model, 493
Cauer network, 329
CFO, *see* Critical flashover voltage
Charge–voltage curves, 43
Chopped wave withstand, 478
Chopping current behavior, in vacuum circuit
    breakers, 542
CIGRE model, of MO surge arrester,
    367–368
CIM, *see* Common Information Model
Circuit breaker, 448–449
    breaking technologies, 469–472
        air-blast circuit breakers, 473
        oil circuit breakers, 472–473
        SF$_6$ circuit breakers, 473–475
        vacuum circuit breakers, 476
    current interruption, principles, 449–450
        electric arc, 450–451
        fault conditions in, 463–469
        normal operating conditions, 454–463
        thermal and dielectric characteristics, 452
    modeling, 486–487, 529–531
        current interruption, 487–488
        gas-filled circuit breaker models, 489–498
        objective, 488–489
        prestrike models, 526–529
        statistical switch models, 520–526
        vacuum circuit breaker models, 498–518
    parameter determination
        gas-filled circuit breakers, 531–541
        vacuum circuit breakers, 541–545
    protection, 422
    standards and ratings
        capacitance switching and mechanical
            requirements, 482
        current-related ratings, 479–481
        voltage-related ratings, 477–479
    testing, 482–483
        direct test, 483
        synthetic testing methods, 483–485
Coaxial cables, calculation, 144–153; *see also*
    Insulated cables
    approximate parameter calculations,
        149–150

ground-return impedance evaluation
    method accuracy, 150–153
series impedance matrix, 144–148
shunt admittance matrix, 148
Coaxial wave, sensitivity, 157
"Cobra" flux–current curves, 194
Coefficient of grounding (COG), 389
Cole–Cole model, 140
Commodity-off-the-shelf, 611
Common Information Model, 596
Complex permittivity, of paper–oil
    insulation, 141
Computerized real-time simulations, 1
Computer simulation tools, for power system
    transient analysis, 1
Concave current source, simulations, 373
Conductive materials, 139–140; *see also*
    Insulated cables
Conductor systems, different configurations, 56
Constant-parameter (CP) line, 5
Constant-parameter (CP) transmission line
    model, 606
Contamination flashovers, on transmission
    systems, 123
Continuous operating voltage, 357
Conventional rational fitting methods, 559
Copper and core impedances, frequency
    dependence, 240
Corona capacitance, 46
Corona effect, 43–45
    models for
        dynamic models, 46–47
        parabolic models, 45
        parameter, 46
        piecewise models, 45
        polynomial models, 45–46
    wave propagation on, 43
Corona inception voltage, 19, 43, 44, 96
Corona streamers, 93
COTS, *see* Commodity-off-the-shelf
Counterpoises, to limit lightning-caused
    overvoltages, 88
COV, *see* Continuous operating voltage
Critical flashover voltage, 92
    lightning impulses for, 113
    practical gap configurations, with and
        without insulators, 114
Critical time-to-crest, 99
Critical wave front, 99
CTs, *see* Current transformers
Current chopping, importance, 455
Current interruption, principles, 449–450
    arc–power system interaction, 453

fault conditions in, 463–469
   normal operating conditions in, 454–463
electric arc, 450–451
thermal and dielectric characteristics, 452
Current interruption process, 450
Current transformers, 510
Curve-fitting techniques, application, 535–536;
   see also Gas-filled circuit breaker
   models
CWF, *see* Critical wave front

D

Damper winding current, 278
Deep-earth electrodes, 63
Delta-connected transformers, 196
DFIG, *see* Doubly fed induction generator
Dielectric breakdown
   gas, 93
   insulation, 95
Dielectric failure, cause, 452
Dielectric recovery phase, characteristics, 452
Dielectric reignition, definition, 501
Dielectric strength
   under lightning overvoltages, 111
   power-frequency voltage for, 123–125
   switching surges for, 98–101
Direct test circuits, 483; *see also* Circuit breaker
Discharge current waveform, 66
Discharge mechanisms, fundamentals of
   description of phenomena, 93
   physical–mathematical models, 93–94
      lightning impulse strength models, 94–98
      switching impulse strength models, 98
Discharge–voltage characteristic tests, aim, 359
Distribution arresters
   protection, 422–438
   selection, 401–403
Distribution lines, protection, 423
DLL, *see* Dynamic Link Library
Doubly fed induction generator, 608
Dry-type transformers, 182
   tests for, 184
Durability and capability tests, for MO surge
   arresters, 359
Dynamic Link Library, 600

E

Earth-fault factor, 389
Eddy-current
   Cauer circuit for representation, 216
   core in, 214–218

effect due to flux penetration in core, 237
losses, 182, 210
synchronous machines in, 272
windings in, 214–215
EFF, *see* Earth-fault factor
EHV, *see* Extra high voltage
Electrical analog model, for MO surge arrestors,
   378–379
Electrical breakdown, in soil, 68
Electrical endurance, definition, 481
Electrical network, synthesization, 230
Electric arc, 450–451; *see also* Current
   interruption, principles
   in circuit breaker, 448
   discharge, 488
Electrogeometric model, 406
Electromagnetic transient simulations, J.
   Marti's fitting, 561
Electromagnetic Transients Program, 1, 138,
   194, 289, 489
   applications, 593–594
   off-line simulation tools
      case studies, 605–609
      control systems, 601–602
      external interface, 600
      FD techniques, 605
      GUI, 596–598
      hybrid tools, 604–605
      initialization importance, 598–599
      MATLAB®/Simulink, 603–604
      software packages, 602–603
      solution methods, 594–596
      statistical and parametric methods, 600
      time-domain module, 600–601
   real-time simulation tools, 609–610
      case study, 613–616
      real-time digital simulators, 610–611
      real-time playback simulators, 611–613
      TNA, 610
   simulation and analysis, 592
Electromagnetic transients, simulation, 138
Electromagnetic transient tools, 3
Electrothermal models, for MO
   surge arrestors, 375
   analog models, 378–379
   electric power input, 383–385
   solar radiation in power input, 383
   thermal parameters determination, 379–383
EMTP, *see* Electromagnetic Transients Program
EMTP-RV software, 602
Energy absorption limit, determination, 375
Energy discharge by surge arrester,
   estimation, 362

Energy withstand capability, 360–362; *see also*
    Metal oxide (MO) surge arrestors
Extra high voltage, 123

F

Fast front arrester model, parameter
    determination, 369–370
Fast front model parameters, determination,
    369–370
Fast front surges
    IEEE MO surge arrester model, 369
    MO surge arrester model, 368, 370
Fast transients, models for, 366–374
Fast VF (FVF), 566
FEM, *see* Finite element method
Ferroresonance, 179
Ferroresonant overvoltages, 402–403
Field programmable gate arrays, 611
Finite element method, 604
    contour of electric potential obtained by, 239
    to evaluate elements of winding capacitance
        matrix, 233
Five-legged transformers, 202
Flashover voltage, 114
    atmospheric effects on, 125–126
        air density, 126
        humidity, 126–127
        other atmospheric effects, 127–128
        external insulation, 125–126
Flux linking damper circuit, 272
Forced convection in heat transfer, thermal
    resistances, 382
Foster network, 329
Foundation electrodes, 63
FOW, *see* Front-of-wave protective level
FPGAs, *see* Field programmable gate arrays
Free charges, formation, 93
Free electrons, 93
Frequency-dependent (FD) line, 5
Frequency-dependent transmission line
    modeling, 566
Frequency-domain (FD) techniques, 605
Frequency ranges, classified groups, 3
Frequency response partitioning, 570, 583–584
    identification algorithm
        pole identification, 576–577
        residue matrices identification, 577–583
    pole identification
        adaptive weighting, 573–574
        column scaling, 574
        iteration step adjustment, 575
        matrix trace, 571–572

optimal model order, 576
overdetermined linear equations, 572–573
poles calculation, 575–576
QR decomposition, 574
Frequency–time transformation, 224
Front-of-wave protective level, 360

G

Gap factor, definition, 102
Gapless metal oxide arresters, 354–355
Gapped metal oxide arresters, 355–357
Gapped silicon carbide arresters, 353–354
Gas, dielectric breakdown, 93
Gas-filled circuit breaker models; *see also*
    Circuit breaker
    dynamic arc representation, 492–498
    ideal switch models, 489–492
    parameter determination, 531–532
        methods, 532–536
        validation, 536–540
Gas insulated substations (GIS), 220, 416, 527
Gaussian distribution, 91
Generator synchronization, 253
Gram–Schmidt orthonormalization, 565
Graphical user interface, 594, 596–598
Ground electrodes
    characteristic dimensions, 85, 86
    harmonic impedance, 76
    impulse impedance, 87
    for overhead lines, 63
    power-frequency ground resistance, 85
    types and sizes, 87
Grounding impedance, 87
Grounding system
    high-frequency models
        discussion, 78
        distributed-parameter grounding model,
            74–77
        lumped-parameter grounding model,
            77–78
    low-frequency models
        compact grounding systems, 68–71
        extended grounding systems, 71–72
        grounding resistance in
            nonhomogeneous soils, 72–74
    of power line, 68
Ground resistance
    compact ground electrodes, 85
    elementary electrodes, 73
Ground-return impedance; *see also*
    Insulated cables
    accuracy, 150–153

estimation, 145
sensitivity analysis, 158–160
terms, 149
Ground-return resistance, horizontal distance
variations, 159
Ground rods, 88
Ground wires, 63
GUI, *see* Graphical user interface

H

Hamiltonian matrix, 584
Hardware-in-the-loop (HIL) simulation,
592, 610
Harmonic impedance, 76
Heat loss-input diagram, usage, 377
Heat transfer by conduction, thermal
resistances, 381
Heidler model, 373
High-frequency quenching capability,
definition, 502
High-frequency transformer models, 221–223
admittance model from terminal
measurements, 228–230
characteristics, 231
distributed-parameter model based on
combined STL and MTL theories, 226
multiconductor transmission line theory,
225–226
single-phase transmission line theory, 225
lumped-parameter model
based on state-space equations, 223–224
from network analysis, 224
modeling for transference analysis, 226–228
representation of two discs, 232
High voltage arresters, IEC standards, 357
High-voltage cables
insulation, 141
SC cables designing, 142
High voltage direct current, 593
High-voltage (HV) applications, power
transformers for, 182
High-voltage systems, current interruption,
449–450
HVDC, *see* High voltage direct current
Hysteresis cycle, 7
Hysteretic model, of MO surge arrester, 370–371

I

Ideal switch models, for circuit breaker, 489–492
IEC standard, for transformers, 186–187
IEEE model, of MO surge arrester, 368–370

Impedance calculation, challenge, 138
Impedance transmission line, 50
Impulse bias test, 478
Impulse efficiency, definition, 76
Inductive current interruption with premature
current interruption, case studies,
457–461
Inductive switching, 454
Infinite Earth Model, 150, 172
Institute of Electrical and Electronics Engineers
(IEEE), 181, 353
circuit breaker standards, 449
classification for arresters, 357
Insulated cables, 138–139
cable parameters and design, transients
sensitivity to
ground-return impedance sensitivity,
158–160
transient response sensitivity, 157–158
cases studies, 162–172
designs, 142–143
input data preparation, 160–162
material properties
conductive materials, 139–140
insulating materials, 140–141
semiconducting materials, 141
parameters, calculation
coaxial cables, 144–153
pipe-type cables, 153–155
tunnel and trench cables, 155–156
wire sheath, spiral effect, 156
sheaths and armors grounding, 156–157
Insulating materials, 140–141; *see also* Insulated
cables
Insulator contamination, in transmission-line
operation, 124
Internal impedance, of round wire, 26
International Electrotechnical Commission
(IEC), 181, 353
circuit breaker standards, 449
classification for arresters, 357
Inverse Fourier transforms, 76
Inverse Laplace transforms, 76
Iron-core lamination, 216

J

J. Marti's fitting line model, 558, 561–562
Joule losses, in transformer windings, 189

K

Korsuncev curve, 86

L

Laplace ratio
  d-axis armature flux linkages, 282
  d-axis rotor voltages, 282
Leader current, 97
Leader progression model (LPM), 98, 124
Leakage fluxes, 179
Leakage inductances, 189
  first and second damper windings, 273
Lightning impulses
  CFO for, 113
  insulation strength, 112
  protection level, 360
  standard deviation for, 111
  strength, 99
    nonstandard impulses, 115
    standard lightning impulses, 112–115
  tests, 478
  waveshape, 90–91
  withstand, 477–478
Lightning overvoltages
  calculation, 19
  caused by
    strokes to a tower, 119
    strokes to phase conductors, 120
  dielectric strength under, 111–112
    lightning impulse strength, 112–115
  distribution, 123
  recommendations to limit, 88
  tall structures in, 88
Lightning return stroke current, 119
Lightning stroke
  current, 118
  flashovers caused by, 122
  and grounding impedance, 66
  influence of grounding resistance
    on, 119
  overvoltages caused by, 116, 119
  peak current magnitude, 406
  protection
    capacitor banks, 424
    distribution lines, 423
    equipment on underground systems,
      424–425
    regulators, 424
    switches, reclosers, and sectionalizers,
      424
    transformers, 423–424
  statistical variation, 121
  tower response, 50
  traveling surges in overhead lines, 112
  waveform, 121

Lightning surge
  energy discharge, 362
  voltages, 386–387
Linear coupled circuits, 254, 298
Linear time invariant multi-input
    multi-output, 559
Line conductors, 21
Line constants routine, application to obtain
    overhead line models, 9
Line models
  for different ranges of frequency, 20
  implementation, 138
  used in lightning overvoltage
    calculations, 62
Line parameters, sensitivity analysis,
    34–36
Line surge arresters, 404; *see also* Metal oxide
    (MO) surge arrestors
  energy analysis, 410–413
  lightning flashover rate, 413–414
  model, 410
  selection, 409–410
Line-to-line voltage, 314
Liquid-immersed transformers, 182
  tests for, 183
Load loss, for transformers, 182, 184
Load rejection test, for synchronous
    machines, 326
Loss factor, of oil-impregnated cellulose *vs.*
    frequency, 238
Lossless distributed-parameter transmission
    line, 51
Low-current models, for vacuum circuit
    breakers, 502–513; *see also* Circuit
    breaker
Lower-voltage XLPE cables, in
    metallic sheath, 156
Low-frequency and slow-front transients,
    models, 365–366
Low-frequency transformer models, 182
LTI MIMO, *see* Linear time invariant multi-
    input multi-output
Lumped multimass model, of generator driven
    by steam turbine, 331–334
Lumped multimass shaft system,
    structure, 332

M

Magnetic air-blast type, 473
Magnetic core saturation, 179
Magnetic flux density, 220
Magnetic hysteresis, 254

Magnetic saturation, in synchronous
machines
open circuit and short circuit characteristics
in, 298–300
representation, 300–303
Magnetizing current chopping, 192
Magnetizing inductance, 192
Magnetomotive forces, 197, 254
Mass–spring–damper model, 334
Mass–spring–damper system, 332
MATLAB® software, 603–604
Matrix trace, for pole identification, 571–572;
see also Vector fitting (VF) technique
Maximum continuous operating voltage,
357, 388
Maximum operating voltage, of
circuit breaker, 477
Mayr–Schwarz equation, 532, 537
MCOV-COV, *see* Maximum continuous
operating voltage
Mechanical damping, decrements tests for
determining, 337
Metallic sheaths
and armors, grounding, 156–157
lower-voltage XLPE cables in, 156
Metal oxide arrester disk, characteristics, 363
Metal oxide blocks performance, improvement,
355–356
Metal oxide (MO) surge arrestors, 352
applications, 403–404
distribution systems protection, 422–438
transmission systems protection, 404–422
electrothermal models, 375
analog models, 378–379
electric power input, 383–385
solar radiation in power input, 383
thermal parameters determination,
379–383
gapless, 354–355
gapped, 355–357
heat sinks, usage, 385
models for, 362–364
fast transients, 366–374
low-frequency and slow-front transients,
365–366
models mathematics, 364–365
very fast front transients, 374
requirements
arrester classification, 357–358
durability and capability tests, 359
energy capabilities, 360–362
protective characteristic tests,
359–360

selection, 385–386
distribution arresters, 401–403
overvoltages, 386–387
phase-to-phase transformer protection,
400–401
protective levels and arrester selection
procedure, 387–400
thermal stability, 376–378
Metal oxide (MO) varistor material, 352
Metal oxide surge arrester, 10
Method of characteristics (MoC), 558
Minimum oil circuit breakers, 472
MMFs, *see* Magnetomotive forces
Modal-domain modeling, 561; *see also* Pole-
residue modeling
Modal VF (MVF), 565
Model leakage inductances, computation, 212
Monte Carlo method, 408
Multiconductor lines
calculation of electrical parameters, 25
model with frequency-dependence of
winding parameters, 329
Multiconductor transmission line (MTL),
222, 225
Multiterminal electrical, component, 559
Mutual ground-return impedances, 145
Mutual ground-return resistance
cable depth variations, 159
ground resistivity variations, 160
MVA transformer, modeling, 217

N

Natural convection in heat transfer, thermal
resistances, 381–382
Nodal analysis and EMT-type programs, 595;
see also Electromagnetic Transients
Program
Non-self-restoring insulations, 91
Normal-duty and heavy-duty surge
arresters, 402
Numerical Laplace transform (NLT)
technique, 172

O

Off-line simulation tools, in EMT programs;
see also Electromagnetic Transients
Program
case studies, 605–609
control systems, 601–602
external interface, 600
FD techniques, 605

GUI, 596–598
hybrid tools, 604–605
initialization importance, 598–599
MATLAB®/Simulink, 603–604
software packages, 602–603
solution methods, 594–596
statistical and parametric methods, 600
time-domain module, 600–601
Ohmic discharge channels, 81–82
Oil circuit breakers, 472–473; *see also* Circuit
  breaker
Online frequency response test, for
  synchronous machines, 326
Online test, for synchronous machines,
  325–326
Open circuit characteristic (OCC),
  of machine, 298
Orthonormalization, advantage, 565; *see also*
  Vector fitting (VF) technique
Orthonormal VF (OVF), 565
Overhead line
  application of line constants routine for
    modeling, 9
  calculation of parameters for, 20
  data input and output for, 33–34
  differential section of three-phase, 22
  ground electrodes for, 63
  influence of voltage stress in design, 18
  line constants for, 33
  line equations for, 21–24
  modeling guidelines for, 21
  modeling in transient simulations, 20
  single-conductor, 22
  switching-impulse strength, 99
  techniques for solving equations of
    multiconductor frequency-dependent
    alternate techniques, 33
    modal-domain techniques, 29–31
    phase-domain techniques, 31–33
  time-domain models developed for, 19
Overvoltages; *see also* Lightning overvoltages
  armor permeability effect, 171
  insulation losses effect, 170
  types, 386–387

P

Paper–oil insulation, 169
Passivity assessment, models
  symmetrical, 584–585
  unsymmetrical, 584
Passivity enforcement, 585–587
Peak closing current, definition, 481

PFRV, *see* Power-frequency recovery voltage
Phase conductors
  calculation of line parameters for
    series impedance matrix, 24–28
    shunt capacitance matrix, 23–24
  and shield wires, 20–21
  voltages, 19
Phase currents, dc components, 313
Phase-domain modeling, 561; *see also* Pole-
  residue modeling
Phase-to-ground fault, surge arrester, 389
Phase-to-ground insulations, 104
Phase-to-ground surges, 106
Phase-to-ground switching surges, 104
Phase-to-ground voltages, 104
  statistical distribution, 110
Phase-to-phase insulations, 104, 108
Phase-to-phase surges, 106
Phase-to-phase transformer protection,
  400–401
Phase-to-phase voltage, 106
PHIL, *see* Power hardware-in-the-loop
Piece-wise linear inductance, 195
Pi-models, 19
Pipe-type cables, 143; *see also* Insulated cables
  series impedance matrix, 153–155
  shunt admittance matrix, 155
Pole identification, frequency response,
  576–577; *see also* Frequency response
  partitioning
Pole-mounted shunt capacitor banks,
  protection, 424
Pole relocation, 563–564; *see also* Vector fitting
  (VF) technique
Pole-residue modeling, 559–560
  inclusion of delays, 560
    modal-domain modeling, 561
    phase-domain modeling, 561
Pollaczek integral, 150
Polynomial fitting techniques, limitation, 570
Polynomial Gear method, advantage, 601
Post-arc current
  models, 504–506
  phases, 499–500
Power circuit breakers, categorization, 448
Power component
  modeling for transient simulations, 4
  modeling guidelines for, 2
  parameter determination, 6–10, 14
  procedure to obtain complete
    representation, 2
Power-frequency dielectric withstand, 477
Power-frequency recovery voltage, 488

Power-frequency voltage
  dielectric strength for, 123–125
  stresses, 89
Power hardware-in-the-loop, 592
Power line
  grounding design, 61
  insulation, 89, 111
  lightning impulse strength, 112
Power system
  classification of methods for parameter
    determination, 8
  international standards for, 12–13
  origin and frequency ranges of
    transients in, 3
  transient analysis, 1
    modeling guidelines for, 2–6
Power systems, EMT-type programs
  applications, 593–594
  off-line simulation tools
    case studies, 605–609
    control systems, 601–602
    external interface, 600
    FD techniques, 605
    GUI, 596–598
    hybrid tools, 604–605
    initialization importance, 598–599
    MATLAB®/Simulink, 603–604
    software packages, 602–603
    solution methods, 594–596
    statistical and parametric methods, 600
    time-domain module, 600–601
  real-time simulation tools, 609–610
    case study, 613–616
    real-time digital simulators, 610–611
    real-time playback simulators,
      611–613
    TNA, 610
  simulation and analysis, 592
Power transformers, types, 182
Prestrike models, for circuit breaker,
    526–529
Prestrikes, effect, 518
Propagation function, goal, 560
Protective characteristic tests, for MO
    surge arrestors, 359–360; *see also*
    Metal oxide (MO) surge arrestors
Puffer interrupter principle, application, 474
Pulse width modulation (PWM), 614

Q

QR decomposition, 574; *see also* Vector fitting
  (VF) technique

R

Radiation in heat transfer, thermal resistances,
    380–381
Rated continuous current, of circuit breaker,
    479–480
Rated dielectric strength, of circuit breaker,
    477–478
Rated operating duty cycle, 481
Rated short-circuit current, of circuit breaker,
    480–481
Rated transient recovery voltage, of circuit
    breaker, 479
Rate of rise of recovery voltage, 452
Rational fitting, in pole identification; *see also*
    Vector fitting (VF) technique
  adaptive weighting, 573–574
  column scaling, 574
  iteration step adjustment, 575
  matrix trace, 571–572
  optimal model order, 576
  overdetermined linear equations, 572–573
  poles calculation, 575–576
  QR decomposition, 574
Rational fitting, of transformer admittance
    matrix, 586
Reactance drop, 186
Real system testing, advantage, 482–483
Real-time simulation programs, advantage, 592
Real-time simulation tools, in EMT programs,
    609–610; *see also* Electromagnetic
    Transients Program
  case study, 613–616
  real-time digital simulators, 610–611
  real-time playback simulators, 611–613
  TNA, 610
Real-Time Workshop, 611
Recovery voltage, components, 488
Reignited current, components, 503
Reignition, Amsinck's method, 533
Relaxed VF, 565
Residue matrices identification, 577–583; *see also*
    Frequency response partitioning
Resistive switching, 454
Resistivity, of conductive materials, 139
Rijanto's method, 534; *see also* Gas-filled circuit
    breaker models
RMVF, disadvantage, 565; *see also* Vector fitting
    (VF) technique
Rotating machine coil model, 327
Rotor angular velocity, 6
Rotor circuits, 278, 280
Rotor–rotor inductances, 259–260

Rotor windings, 259
Round–rotor machines, 325
Round wire steel armors permeability,
    calculation, 139–140
RRRV, *see* Rate of rise of recovery voltage
RTW, *see* Real-Time Workshop
Ruppe's method, 534–535; *see also* Gas-filled
    circuit breaker models
RVF, *see* Relaxed VF

S

Saad, Gaba, and Giroux, 149
Salient-pole machines, 280, 325
Schelkunoff's surface impedance, 23
Schematic capture, *see* Graphical user
    interface
Self-ground-return resistance
    cable depth variations, 158
    ground resistivity variations, 159
Self-restoring insulation, characteristic, 91
Semiconducting layers, parameters, 141
Semiconducting materials, 141; *see also*
    Insulated cables
Semiconducting screens
    effect, 161
    improved modeling, 165–167
Semiconductor thickness sensitivity, 168–169
Series capacitor banks, protection, 415
Series-gapped MO arrester, characteristic, 356
Series impedance, calculation, 330
Series impedance matrix
    coaxial cable, 144–148
    pipe-type cables, 153–155
Series voltage-injection synthetic
    test circuit, 485
Service capability, definition, 481
SF$_6$ circuit breakers, 471–475; *see also*
    Circuit breaker
SFFR, *see* Shielding failure flashover rate
SGG, *see* Saad, Gaba, and Giroux
SGG formula, in Z_g calculation, 172
Sheath resistance sensitivity, 167–168
Sheaths and armors, grounding, 156–157
Shielding failure flashover rate, 405
Short circuit characteristic (SCC), of
    synchronous machines, 299
Short-circuit current interruption, 463–467
Short circuit currents, 181
Short-circuiting breaker, 314
Short circuit phase currents, 281
Short circuit ratio (SCR), definition, 299

Short circuit tests, for parameter determination
    of synchronous machines
    analysis of ac components, 309
    machine operating open-circuited at
        rated speed
        sudden short circuit of armature and field
            with, 313–314
        sudden short circuit with, 307–313
    sudden short circuit with machine operating
        on load at low voltage, 314–315
Short-line fault current interruption, 467–469
Short time current, 481
Shunt admittance matrix
    coaxial cable, 148
    pipe-type cables, 155
Shunt capacitance, 20
Shunt capacitor banks, protection, 415
Shunt-gapped MO surge arrester,
    characteristic, 356
Silicon carbide (SiC) valve arrestors, 352–354
Simplified model, of MO surge arrester, 371–374
Simulink software, 603–604
Single-conductor line, time-domain
    equations, 21
Single-core (SC) coaxial cable systems, 139
Single-core self-contained cables, 142; *see also*
    Insulated cables
Single-phase corona model, 47
Single-phase transformers
    assembling of models for, 218–219
    calculation of parameter for, 210
    dual circuit model for, 219
    duality-derived equivalent circuit, 190
    polarity, 186
    π-shaped model for, 189
Single-phase transmission line (STL), 225
Single-pressure puffer-type interrupter,
    principle, 474
Slow-front and low-frequency transients,
    models, 365–366
Small inductive currents, interruption, 454–
    461; *see also* Current interruption,
    principles
Soil breakdown, 87
Soil conductivity, 87
Soil ionization, 67
    hemisphere electrode before and after, 83
    treatment, 81–87
Soil resistivity, within ionization zone, 82
Solid rotor machines, 300
Space equation realization (SER), 229
Spark gaps, role, 352

Speed voltages, 265
SPICE software, 602–603
SPL, *see* Switching impulse protective level
Stacked-core transformers, 197
Standstill frequency response (SSFR), 9,
        304, 321
Standstill frequency response tests, for
        synchronous machines,
        320–321
    parameter identification from SSFR test
        results, 322–325
    SSFR test measurements, 321–322
Statistical switch models, for circuit breaker,
        520–526
Statistics switch, 521
Stator leakage impedance, 287
Stator–rotor inductances, 259
Stator–stator inductances, 255–258
Stator-to-rotor mutual reactance, 306
Stator-to-rotor turns ratio, 322
Steady-state tests, for parameter determination
        of synchronous machines
    determination of d-axis synchronous
        impedance from OCC and SCC,
        305–306
    open circuit characteristic curve, 304–305
    short circuit characteristic curve, 305
    slip test and measuring q-axis synchronous
        impedances, 306–307
Steep current impulse discharge voltage, 360
Steep-fronted transient voltages, on multiturn
        coils of rotating machines, 326
Stochastic behavior of dynamic arcs, 532
Stranded conductors, modeling, 139
Stray loss
    definition, 182
    for transformers, 182
Streamers, 93
Submarine cables, designing, 139
Subsynchronous resonance, 253, 331
Subtransient inductance, definition, 278
Superposition principle, 453
Surface arcing, for limiting lightning-caused
        overvoltages, 88–89
Surface-earth electrodes, 63
Surge arresters; *see also* Circuit breaker
    electrothermal models, 375, 383
        analog models, 378–379
        electric power input, 383–385
        solar radiation in power input, 383
        thermal parameters determination,
            379–383

heat sinks, usage, 385
model, for vacuum circuit breakers,
        516–518
model parameters, adjustment, 372
thermal stability of metal oxide arresters,
        376–378
types, 352
Surge current, 88
Surge impedance, 43
Surge propagation velocity, 51
Switching impulse protective level, 387
Switching impulse
    strength, 102
    strength models, 98
    waveshape, 90, 91
    withstand, 478
Switching overvoltages, 19
    simulation, 518, 520
Switching surges, 18
    design procedure for various tower and line
        configurations, 102
    dielectric strength for, 98–101
    encountered on electric power systems, 99
    flashover strength, 99
    phase-to-ground, 104
    on transmission lines, 104
Synchronous generator, field winding current
        in, 6
Synchronous impedance, 299
Synchronous inductance, 285
Synchronous machines
    data conversion procedures for, 286
        determination of fundamental
            parameters, 287, 297–298
        procedure for complete second-order
            d-axis model, 287–288, 292–295
        procedure for second-order *q*-axis model,
            295–297
        procedure for simplified second-order
            d-axis model, 289–291
    determination of electrical parameters, 10
    electrical performance, 261
    equations
        circuit equations, 253–254
        electrical torque equation, 261
        flux linkage equations, 254–260
        voltage equations, 260–261
    equivalent circuits, 271–274
    magnetic saturation effects in
        open circuit and short circuit
            characteristics, 298–300
        representation, 300–303

mechanical system equations for
   determination of damping factors of
      mass–spring–damper model,
      339–340
   determination of mechanical parameters,
      338–339
   lumped multimass model, 331–334
   modal parameters, 334–338
modeling guidelines for, 253
models for high-frequency transient
   simulations for
   calculation of electrical parameters,
      329–331
   winding models, 326–329
models for transient studies, 274
parameters for, 274–275
   armature time constant, 281–282
   operational parameters, 282–286
   subtransient and transient
      inductances and time constants,
      276–281
   three-phase short circuit at terminals,
      275–276
park transformation for, 261–262
   choice of k_d, k_q, and k_0 constants in,
      267–268
   electrical torque equation in $dq0$
      components, 266–267
   flux linkage equations in $dq0$ components,
      261–262
   voltage equations in $dq0$ components,
      264–265
per unit representation for
   rotor quantities, 269–270
   stator quantities, 268–269
with rotating armature windings, 264
stator and rotor circuits, 254
test procedures for parameter determination
   for, 303–304
   decrement test, 315–320
   online tests, 325–326
   short circuit tests, 307–315
   standstill frequency response tests,
      320–321
   steady-state tests, 304–307
   three-phase short circuit at terminals,
      275–276
   type of data sets in, 286
Synthetic test circuits, 483–485; *see also*
      Circuit breaker
System voltage, circuit breaker
      selection, 448

T

TD-VF, *see* Time-domain VF
Temporary overvoltages, 355, 388–390
T-equivalent circuit, for single-phase
      transformer, 188
Test network, 578–581; *see also* Frequency
      response partitioning
Thermal breakdown, definition, 452
Thermal parameters determination, 379–383
Thermal recovery phase, characteristics, 452
Thermal reignition, occurrence, 501
Thermal stability
   improvement, effects, 385
   of MO arrester
      achievement, 377
      definition, 376
Thermal steady-state analysis, electrical analog
      circuit, 380
Three-legged stacked-core delta-wye
      transformers, 199
Three-legged stacked-core transformer, 197
Three-legged stacked-core wye-wye
      transformers, 199
Three-phase self-contained cables, 142–143;
      see also Insulated cables
Time-domain large disturbance test, for
      synchronous machines, 326
Time-domain module, in EMT programs,
      600–601; *see also* Electromagnetic
      Transients Program
Time-domain small disturbance test, for
      synchronous machines, 326
Time-domain VF, 566
TNA, *see* Transient Network Analyzers
Torsional mechanics, typical mode shapes, 335
Torsional mode damping, 336
Torsional oscillations, 331
TOVs, *see* Temporary overvoltages
Tower grounding impedance, 65
Tower model, performance against lightning
      current, 65
Transformer admittance matrix, rational fitting,
      586
Transformers
   eddy current losses in, 179
   electromagnetic field simulations for
      designing, 188
   electromagnetic transients in, 221
   energization, 188
   impedance voltage, 185
   load losses, 184

losses, classification, 182
modeling, 569–570
modeling guidelines for, 179–181
nonlinear behavior, 178
parameter determination for
    fast and very fast transients, 220–230
    high-frequency models for, 230–238
    low-frequency and switching transients,
        187–220
parameters calculated from on-site tests
    data, 191
protection, 415, 423–424
saturation curve, 193
standard tests to obtain parameters for
    IEC standard, 186–187
    IEEE standards, 182–186
    introduction, 181–182
three-phase core designs for, 180
three-phase transformer, 184
    assembling of models for, 219–220
    matrix representation, 195–196
    topology-based models for, 197–204
Transformers, three-winding
    duality models, 207–208
        duality model matching terminal
            measurements, 208–211
    introduction, 204–205
    leakage inductance tests for, 205
    saturable transformer component for, 207
    traditional equivalent circuit of
        model derived from design information,
            206–207
        model derived from terminal
            measurements, 205–206
Transformer windings
    capacitances for, 230–233
    conductor losses for, 236–237
    core losses for, 237
    distributed-parameter model for, 222
    equivalent circuit per unit length, 221
    inductances for, 233–236
    series and shunt elements, 327
    unit-step response, 244
    waveshape of applied voltage on, 326
Transient Network Analyzers, 1, 610
Transient processes, characteristics, 461–462
Transient recovery voltages, 415, 451, 453, 593
    fault conditions in, 463–469
    normal operating conditions in,
        454–463
Transient response, sensitivity analysis,
    157–158; *see also* Insulated cables

Transmission line grounding, 61–66
    grounding design, 87–89
    high-frequency models of
        discussion, 78
        distributed-parameter grounding model,
            74–77
        lumped-parameter grounding model,
            77–78
    impedance
        discussion, 68
        high-frequency models, 67–68
        low-frequency models, 66–67
    low-frequency models of
        compact grounding systems, 68–71
        extended grounding systems, 71–72
        grounding resistance in
            nonhomogeneous soils, 72–74
    treatment of soil ionization, 81–87
Transmission line insulation, 89–90
Transmission line insulators, protection, 404
Transmission line modeling
    modal-domain fitting, 568
    phase-domain fitting, 566–568
    role, 568
    universal line model, 568–569
Transmission lines, ultra high voltage, 57
Transmission line tower, 50–51
    multiconductor vertical line models for,
        53–57
    multistory model for, 57–59
    single vertical lossless line models for,
        51–53
Transmission systems
    contamination flashovers on, 123
    lumped-parameter line models for, 19
Transverse electromagnetic mode
    (TEM) wave, 50
TRVs, *see* Transient recovery voltages
TSR, *see* Two-sided recursions
Tunnel and trench cables, 155–156
Turbine–generator rotor, 331, 336
Two-pole machines, 261
Two-sided recursions, 31
Two-winding transformers, 188–190
    nonlinear core model, 192–195

U

ULM, *see* Universal Line model
Ultra high voltage (UHV) transmission lines, 57
Underground cables, protection, 415–416
Universal line model, 32, 165, 561, 568–569

V

Vacuum arc; *see also* Arc models; Electric arc
    model, equations and parameters, 543
    regions, 498
Vacuum circuit breakers, 472, 476; *see also*
        Circuit breaker
    model, 498–502
        cable model, 515–516
        low-current models, 502–513
        reignitions and restrikes, 500–501
        surge arrester model, 516–518
        transformer model, 513–515
    parameter determination, 541–545
    values, 542
Valve arresters, types, 352
Valve element operating voltage,
        expression, 377
Valve type arresters, 352; *see also* Metal oxide
        (MO) surge arrestors
    gapped metal oxide, 355–357
    metal oxide gapless, 354–355
    silicon carbide gapped, 353–354
Vector fitting (VF) technique, 558, 562
    classical formulation
        initial poles, 564
        matrix fitting, 564
        pole relocation, 563–564
    extension and improvement
        modal formulation, 565
        multivariate fitting and fast
            implementation, 566
        orthonormalization, 565
        relaxation, 564–565
        time domain and z-domain, 566–570
Very fast front transients, models, 374
Very High Speed Integrated Circuit Hardware
        Description Language (VHDL), 603
V-I characteristic
    linear resistor and SiC block, 354
    MO arresters, regions, 355, 363
    nonlinear resistors for, 369
    SiC valve arrestors, 353
    surge arrester, 364
    ZnO valve arrestors, 353

Virtual chopping, arc in, 456
Voltage and current waveforms, in current
        interruption, 463
Voltage escalation, in capacitive current
        interruption, 461
Voltage-injection method, 485
Voltage recovery test
    analysis, 316
    EMTP-RV simulation, 318–319
Voltage-zero period, 500
Volt–time curve, 89, 94–96, 114

W

Waveform distortion, 183
Weibull cumulative distribution
        function, 361
Weibull distribution, 111
Wideband transformer model, passivity
        enforcement, 587
Wind generation, case study, 608–609;
        see also Electromagnetic
        Transients Program
Winding designs, for transformers, 181
Winding losses, frequency-dependent, 7
Winding resonance, 179
Winding stray loss, for transformer, 182
Wire sheath, spiral effect, 156
WW formula, in Z_g calculation, 172

Y

Yoke–Yoke circuit elements, 190

Z

z-Domain data, VF, 566
Zero-phase-sequence impedance test, for
        modeling transformers, 182
Zero-sequence energization, of untransposed
        overhead line, 38
Zero-sequence fluxes, 197
Z_g calculation, methods, 172
Zinc oxide (ZnO) valve arrestors, 352–353